Continuous System Modeling

François E. Cellier

Continuous System Modeling

With 288 Figures

Springer-Verlag
New York Berlin Heidelberg London
Paris Tokyo Hong Kong Barcelona

François E. Cellier, Ph.D.
Department of Electrical and Computer Engineering
  and Applied Mathematics Program
University of Arizona
Tucson, AZ 85721
USA

Library of Congress Cataloging-in-Publication Data
Cellier, François E.
    Continuous system modeling / François E. Cellier.
          p.   cm.
    Includes bibliographical references (p.) and index.
    Contents: Modeling
    ISBN 0-387-97502-0
    1. Simulation methods.   2. Computer simulation.   3. Mathematical
models.   I. Title.
T57.62.C26   1991
      620'.001'13—dc20   90-25286

Printed on acid-free paper.

Camera-ready copy prepared by the author.
Printed and bound by Edwards Brothers, Inc., Ann Arbor, MI.
Printed in the United States of America.

9 8 7 6 5 4 3 2 1

ISBN 0-387-97502-0 Springer-Verlag New York Berlin Heidelberg
ISBN 3-540-97502-0 Springer-Verlag Berlin Heidelberg New York

to Ursula
for all these years of friendship

Parasitism:

In the beginning, there was Chaos.
Chaos nurtures Progress.
Progress enhances Order.
Order tries to defy Chaos at all cost.
... But the day Order wins the final battle
against Chaos, there will be mourning.
'Cause Progress is dead.

François E. Cellier
September, 1990

Parasitism:

In the beginning, there was Chaos.
Chaos nurtures Progress.
Progress enhances Order.
Order tries to defy Chaos at all cost.
... But the day Order wins the final battle
against Chaos, there will be mourning.
'Cause Progress is dead.

François E. Cellier
September, 1990

Preface

Modeling and simulation have become endeavors central to all disciplines of engineering and science. They are used in the analysis of physical systems where they help us gain a better understanding of the functioning of our physical world. They are also important to the design of new engineering systems where they enable us to predict the behavior of a system before it is actually built. Modeling and simulation are the only techniques available that allow us to analyze arbitrarily nonlinear systems accurately and under varying experimental conditions.

The two books, *Continuous System Modeling* and *Continuous System Simulation*, introduce the student to an important subclass of these techniques. They deal with the analysis of systems described through a set of ordinary or partial differential equations or through a set of difference equations. These books do not describe the techniques of discrete–event modeling and simulation since those are quite distinct from the techniques that are at the heart of our discussion, and since excellent texts can be found on the book market that deal with discrete–event modeling and simulation explicitly and exclusively. However, this does not mean that the systems which can be tackled by either of the two classes of modeling and simulation techniques are necessarily different. One and the same system can often be modeled either through differential equations or through discrete events, depending on the granularity level of the desired model. Yet, as the reader may have already noticed, my two books are quite bulky as they are, and adding the concepts of discrete–event modeling and simulation to the text would have meant adding several hundred pages more. I felt that this was not justified.

This book introduces the concepts of *modeling*, i.e., it describes the transition from the physical system itself down to an abstract description of that system in the form of a set of differential and/or difference equations. This text has a flavor of the mathematical discipline of dynamical systems, and is strongly oriented towards the Newtonian physical sciences. The concepts of mass and energy conservation, and the laws of thermodynamics are at the heart of the discussion. Various modeling techniques such as bond graphs and System Dynamics are introduced, and various specialized software tools such as DYMOLA and STELLA are presented, techniques and tools which support the process of modeling. While some chapters introduce new modeling techniques, others exercise previously introduced concepts by means of virtually hundreds of practical examples. I believe strongly in the idea of "learning through practice." While the basic modeling concepts are introduced in methodology chapters, many of the tricks —the blows and whistles of modeling— are presented in the flow of discussing examples.

The companion book, *Continuous System Simulation*, introduces the concepts of *simulation*, i.e., it describes the transition from the mathematical model to the trajectory behavior. This text has a flavor of the mathematical discipline of applied numerical analysis. It introduces some of the techniques currently available for the numerical integration of differential equations, it talks about the generation of random numbers and the design of statistical experiments, and it introduces parameter estimation techniques. Finally, the basic principles behind the design of simulation software and simulation hardware are presented from a computer engineering point of view.

Here at the University of Arizona, the material is taught in two consecutive courses which are open to both graduating seniors and graduate students, and this is the ideal setting for use of these books in a classroom environment. The texts are organized in such a manner that they contain only a few cross references between them. Consequently, it is feasible to study the material described in the companion book prior to the material contained in this book, although I recommend starting with this book. The material is perfectly understandable to juniors as well (no junior level prerequisites for these courses exist), but the teacher will have to keep a fairly rapid pace in order to make it through an entire volume in one semester, and he or she must rely on home reading. This pace may be too stiff for most juniors. That is, when teaching these topics at junior level, the teacher will probably have to restrict himself or herself to covering the first few chapters of each text only, and for that purpose, better textbooks are on the market, books which cover these introductory topics in broader depth.

The two texts have been written by an engineer for engineers. They teach the essence of engineering design. Engineering intuition is often preferred over mathematical rigor in the derivation of formulae and algorithms. My home department is the Department of Electrical and Computer Engineering. However, the two courses are cross–listed in the Computer Science Department, and they are regularly attended by graduate students from the Aeronautical and Mechanical Engineering Department, the Systems and Industrial Engineering Department, and the Applied Mathematics Program. These students perform equally well in the classes as our own students, which reflects the fact that the applications covered in these courses stem from a wide variety of different topic areas.

Due to the interdisciplinary nature of the subject matter, each chapter has been organized as a separate entity with a "front matter" and a "back matter" of its own. Thus, each chapter is in fact a minibook. The front matter consists of a *Preview* section which tells the student what she or he can expect from the chapter. The bulk of the chapter contains a number of subsections. It often starts with an *Introduction*, and it always ends with a *Summary*. The back matter starts with the *Reference* section, often followed by a *Bibliography* section which is not meant to be exhaustive, but which can provide the student with a selected set of pointers for further reading. Following the bibliography, all chapters offer a *Homework Problems* section. More difficult homeworks are marked with an asterisk. Most homeworks can be solved within a few hours. Most chapters also have a *Projects* section which is meant to provide the student with ideas for senior projects and/or masters' theses. This section shows to the student that the topics mastered can be directly applied to larger and more serious problems. Finally, many chapters end with a *Research* section in which open research problems are presented which might be tackled by M.S. or Ph.D. students in their theses and/or dissertations. The projects and research sections are meant to enhance the student's motivation since they show that the topics learned can be used in serious research. These sections are not meant to be discussed in class, but are recommended for home reading.

Studying science is *fun*, and understanding science by means of mathematical models is even more fun. I experience the development of these textbooks as an extremely creative and rewarding endeavor which provides me with much satisfaction. I hope that my own fascination with the subject matter is reflected in my books and will prove to be highly contagious.

François E. Cellier
Tucson, Arizona, October 1990

About This Book

This text introduces concepts of modeling physical systems through a set of differential and/or difference equations. The purpose of this endeavor is twofold: it enhances the *scientific understanding* of our physical world by codifying (organizing) knowledge about this world, and it supports *engineering design* by allowing us to assess the consequences of a particular design alternative before it is physically built.

The text contains 15 chapters which only partially depend on each other. A flowchart of the chapter dependencies is shown below.

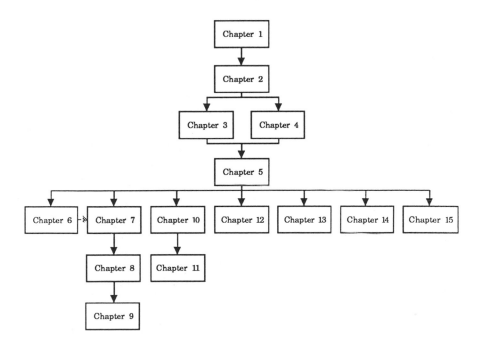

Chapter 1 introduces the basic terminology and provides the student with a motivation for the class.

Chapter 2 introduces the primary concepts behind the user interface of state–of–the–art simulation software. At the end of Chapter 2, the student should be able to use continuous–system simulation languages such as DARE–P, DESIRE, and ACSL, and she or he should be able to formulate simple differential equation models in terms of these software systems.

Chapters 3 and 4 introduce basic modeling concepts used in electrical engineering (Kirchhoff's laws applied to RLC circuits) and in mechanical engineering (Newton's law applied to spring–damper–mass systems). These concepts are very basic, and every engineer can be expected to master these concepts irrespective of his or her home department. I usually defer the last few pages of Chapter 4 (the Euler equation, the Lagrangian, and the Hamiltonian) to home reading and do not request this material in any tests.

Chapter 5 deals with the concepts of modular and hierarchical modeling. It is shown how hierarchical models can be implemented in current simulation languages using macros, and the shortcomings of this approach are demonstrated. The modeling language DYMOLA is introduced as an alternative, which provides for true modular modeling.

Chapter 6 is the first of a series of more specialized chapters. It deals with advanced concepts of electronic circuit modeling. It introduces SPICE, the most common among today's circuit simulators, and it shows how DYMOLA might be used as an alternative. DYMOLA provides much more insight into the internal structure of submodels than SPICE does. When I teach this course, I skip over the details of the bipolar transistor model and refer this to home reading not requested on any test. This chapter can be easily skipped except for the section "How DYMOLA Works," which should be read in preparation for the next chapter.

Chapter 7 is an important methodological chapter. It introduces the concept of bond graph modeling which provides the student with a tool essential to the understanding of the functioning of physical systems in general. The chapter contains a short survey of bond graph modeling software and discusses in greater detail how DYMOLA can be used for bond graph modeling.

Chapter 8 applies the freshly mastered methodological tool of the bond graph to phenomena of thermodynamics, namely, conduction, convection, and radiation. This chapter sheds light on the mechanisms of power flow and energy conservation in general. Mastering the knowledge provided in this chapter can be expected of all engineers, physicists, and applied mathematicians irrespective of their home department. This chapter presents a project (modeling of a solar–heated house) which may be worthwhile discussing in class since it introduces a new concept: hierarchical modeling with bond graphs.

Chapter 9 is also a more specialized chapter. It is aimed at chemical engineers. Here, bond graphs are applied to the thermodynamic modeling of chemical reaction systems. When I teach this course, I treat this chapter in a somewhat skimpy fashion, and I leave the sections dealing with chemical reactors, photochemistry, and electrochemistry away. However, this chapter contains a number of very important lessons. For the first time, mixed energy systems are treated in a unified and uniform manner, and the *laws of thermodynamics* as taught in this chapter are fundamental to the understanding of physics in general. From this perspective, Chapter 9 may easily be the most important chapter of the entire text.

Chapter 10 deals with population dynamics. This chapter addresses a number of important methodological issues such as the differences between *deductive modeling* and *inductive modeling*. It also introduces the concept of chaotic motion, and it finally deals with the exotic issue of modeling the evolution.

Chapter 11 introduces another methodological tool, *System Dynamics,* and presents another modeling software, STELLA, which is based on this modeling methodology. This chapter also discusses basic problems in structure characterization and causality determination. The applications include Forrester's world model.

Chapters 12 to 15 present the student with an introduction to various interfaces between modeling, simulation, and artificial intelligence. The techniques presented include the naïve physics approach to modeling, inductive reasoning, a tool for optimal forecasting of poorly understood systems, and artificial neural networks and genetic algorithms, two techniques that attempt to mimic the behavior of adaptive biological systems. Chapter 15 introduces a methodology for management of models of complex systems. These four chapters are fairly independent of each other and can be dealt with in an arbitrary sequence. They can be skipped or included depending on the teacher's emphasis and interests. All of these chapters are strongly research–oriented and open up more questions than they provide answers for.

According to my judgment, this text presents the most comprehensive account of modeling methodologies for continuous systems currently available. While many individual texts can be found that deal with particular aspects of any one of these techniques, they are often difficult to read because they all use a different terminology, a natural consequence of the interdisciplinary nature of modeling and simulation methodology. In this text, major emphasis has been placed on unifying the terminologies and the presentation style of the various techniques discussed.

I hope that this text, besides providing students with an easily digestable overview of continuous systems modeling techniques, may stimulate further work in this fascinating research area and will also prove of use to the accomplished practitioner.

After working on this text intensively for over two years, I feel like I am giving birth to a long–expected baby. The book is a source of pride and joy ... and finishing it leaves an empty spot in my stomach.

François E. Cellier
Tucson, Arizona, October 1990

Acknowledgments

Many individuals have contributed to this text. The book contains hundreds of practical examples, many of which I have not developed from scratch. The following individuals have contributed the original versions of some of the examples:

- Michael Amrhein: [H9.1], [H9.2], [H9.3], [H9.4]
- Spyros Andreou: [P8.1]
- Milan Bier: [R9.2]
- George Cootellumi [P13.1]
- Lee Chic Cheung: [H10.7], [H10.8]
- Andy Cutler: [P9.1]
- Hilding Elmqvist: [H5.6], [H6.7]
- Andreas Fischlin: The Larch Bud Moth Model
- Jay Forrester: [H11.2], The World Model
- Dean Frederick: [H6.1]
- King–Sun Fu: Fig.P4.2a–b
- Michael Gilpin: The Gilpin Model
- Gene Hostetter: [H1.2], [H3.6], [H4.3], [H7.3]
- Mac Hyman: [R11.1]
- Walter Karplus: Fig.1.5
- George Klir: [H13.2], [H13.3]
- Granino Korn: [H2.2], [H14.6], Backpropagation
- Donghui Li: [H13.1]
- Mohammed Mansour: [H4.4], [H7.4], The Crane Crab Model, The Inverse Pendulum Model, The Ward–Leonard Group Model
- Ed Mitchell: [H2.5], [H5.2]
- Norman Nise: [H4.5], Fig.4.2, Fig.4.5, Fig.7.18
- Ya–Dung Pan: [P4.2]
- Steve Peterson: [H11.1], [H11.5], The Influenza Model

- Alan Pritsker: [H10.2]
- Nicolas Roddier: [R13.4]
- Stu Shacter: [H6.4], [H6.5]
- Greg Smith: The Opamp Model
- Hugo Uyttenhove: The Heart Monitor Model
- Jan van Dixhoorn: Table 7.1
- Pentti Vesanterä: [P4.1], [P13.2]
- Jakob Vogel: [H2.1], [H3.7], [H3.8], [H3.9], [H5.3], [H5.4], [H5.5], [H7.8], [H10.1], The Cable Reel Model, The Lunar Lander Model
- Hubert Wong: [H10.7], [H10.8]

I wish to thank the following individuals for their critical review of parts of my manuscript:

> Heinz Borchardt
> Peter Breedveld
> Heriberto Cabezas
> Ursula Cellier
> Walt Fahey
> Paul Fishwick
> Jerry Graeme
> Herbert Green
> Dean Karnopp
> George Klir
> Granino Korn
> Ed Mitchell
> Tony Morgan
> Mark Smith
> Jean Thoma
> Terry Triffet
> Tom Vincent
> Bernard Zeigler

Finally, I wish to thank many of my seniors and graduate students who have contributed significantly to the success of this text. They are:

> Michael Amrhein
> Spyros Andreou
> Chris Beamis

George Castellum
Lee Chic Cheung
Sung–Do Chi
Tim Dennison
Glenn Farrenkopf
Mark Ferris
Steve Kosier
Pam Kury
Donghui Li
Ya–Dung Pan
Nicolas Roddier
Hessam Sarjoughian
Paul Sundberg
Pentti Vesanterä
Qingsu Wang
Hubert Wong
David Yandell

In particular, I wish to thank Hessam Sarjoughian for hundreds of hours of work, which he spent drawing figures for this text with SuperPaint.

The poem at the end of this book has been printed with the explicit permission of the poet, Ann Droid, and her creator, Ray Moses. My warmly felt thanks go to both of them.

Some of the research results reported in Chapter 6 were made possible by partial financial support from Burr Brown Corp. Some of the results reported in Chapters 9 and 15 were made possible by partial financial support from NASA through our Space Engineering Research Center for Utilization of Local Planetary Resources, SERC/CULPR. The development of the robot model reported in Chapter 4 was sponsored by NASA/USRA/RIACS through their Telescience Testbed Pilot Program (Subcontract 800–62). Many results reported in Chapters 7, 12, 13, and 15 were made possible by generous financial support of NASA's Ames Research Center through Cooperative Agreement No. NCC 2-525, "A Simulation Environment for Laboratory Management by Robot Organizations". My students and I are grateful for the generous support received.

Last but not least, I wish to thank the crew of Springer–Verlag, New York, for their support of this project. In particular, I wish to express my gratitude to my scientific editor for his trust in me and this project.

Contents

Preface ix

About This Book xiii

Acknowledgments xvii

1   **Introduction, Scope, and Definitions** 1

         Preview 1
     1.1   What is a System? 1
     1.2   What is an Experiment? 4
     1.3   What is a Model? 5
     1.4   What is a Simulation? 6
     1.5   Why is Modeling Important? 7
     1.6   Why is Simulation Important? 8
     1.7   The Dangers of Simulation 9
     1.8   Good Reasons to Use Simulation 10
     1.9   The Types of Mathematical Models 11
     1.10 Direct Versus Inverse Problems 18
     1.11 Summary 19
         References 19
         Bibliography 20
         Homework Problems 21
         Projects 22
         Research 22

2   **Basic Principles of Continuous System Modeling**   23

         Preview 23
     2.1   Introduction 23
     2.2   The Algebraic Loop Problem 25
     2.3   Memory Functions 26

2.4   Explicit Versus Implicit Integration   28
2.5   Implicit Loop Solvers   29
2.6   Procedural Sections   30
2.7   The Basic Syntax of Current CSSLs...............   32
2.8   Discontinuity Handling   41
2.9   Model Validation   45
2.10   Summary...   46
       References.......................................   46
       Bibliography   47
       Homework Problems..............................   48
       Projects   50
       Research   50

3   Principles of Passive Electrical Circuit Modeling .   51

       Preview   51
3.1   Introduction......................................   51
3.2   Mesh Equations   52
3.3   Node Equations   58
3.4   Disadvantages of Mesh and Node Equations.......   62
3.5   State–Space Models...............................   64
3.6   Algebraic Loops   66
3.7   Structural Singularities   69
3.8   Disadvantages of State–Space Models   70
3.9   Summary...   71
       References.......................................   72
       Homework Problems.............................   72

4   Principles of Planar Mechanical System Modeling   79

       Preview   79
4.1   Introduction......................................   79
4.2   Newton's Law for Translational Motions   81
4.3   Newton's Law for Rotational Motions   90
4.4   The Crane Crab Example.........................   96
4.5   Modeling Pulleys   99
4.6   The Inverse Pendulum Problem...................   100
4.7   Modeling Electromechanical Systems..............   102
4.8   Summary...   105
       References   109

Homework Problems........................... 110
Projects 116
Research 132

**5   Hierarchical Modular Modeling of
Continuous Systems** 133

Preview.. 133
5.1   Modeling Transfer Functions..................... 133
5.2   Modeling Static Characteristics 137
5.3   Dynamic Table Load............................. 139
5.4   Modular and Hierarchical Modeling 139
5.5   The Macro Facility.............................. 140
5.6   Modular State–Space Models..................... 156
5.7   The Equation Solver 164
5.8   Code Optimization.............................. 167
5.9   Linear Algebraic Loops.......................... 167
5.10 Nonlinear Algebraic Loops 169
5.11 Structural Singularities.......................... 170
5.12 Large–Scale System Modeling 179
5.13 Graphical Modeling............................. 180
5.14 Summary....................................... 182
      References..................................... 182
      Homework Problems............................. 184
      Projects 195
      Research 199

6   Principles of Active Electrical Circuit Modeling.. 201

Preview.. 201
6.1   Topological Modeling 201
6.2   Models of Active Devices in SPICE 203
6.3   Hierarchical Modeling........................... 210
6.4   Transient Analysis in SPICE..................... 212
6.5   Graphical Modeling............................. 215
6.6   Circuit Design Using DYMOLA................... 219
6.7   How DYMOLA Works........................... 229
6.8   Summary....................................... 238
      References..................................... 239
      Bibliography 239

Homework Problems 240
Projects 248
Research 249

7   **Bond Graph Modeling** 251

Preview 251
7.1   Block Diagrams 251
7.2   Signal Flow Graphs 255
7.3   Power Bonds 258
7.4   Bond Graphs for Electrical Circuits 260
7.5   Bond Graphs for Mechanical Systems 265
7.6   Generalization to Other Types of Systems 268
7.7   Energy Transducers 270
7.8   Bond Graph Modeling in DYMOLA 274
7.9   The Dual Bond Graph 282
7.10   Summary 287
References 287
Bibliography 288
Homework Problems 289
Projects 296
Research 296

8   **Modeling in Nonequilibrium Thermodynamics** ... 297

Preview 297
8.1   Power Flow 298
8.2   Thermal Conduction 303
8.3   Thermal Convection 316
8.4   Thermal Radiation 317
8.5   Thermal Inertance: The Missing Link 320
8.6   Irreversible Thermodynamics 321
8.7   Summary 330
References 332
Bibliography 333
Homework Problems 333
Projects 334
Research 344

9   Modeling Chemical Reaction Kinetics 347

        Preview 347
9.1   Introduction 347
9.2   Chemical Reaction Kinetics..................... 349
9.3   Chemical Thermodynamics 359
9.4   The Equation of State 366
9.5   Chemical Reaction Bond Graphs 371
9.6   Energies of Formation 386
9.7   Continuous Reactors 392
9.8   Photochemistry and Electrochemistry........... 398
9.9   Summary 403
        References 405
        Bibliography 406
        Homework Problems 407
        Projects 412
        Research 415

10   Population Dynamics Modeling 417

        Preview 417
10.1   Growth and Decay 417
10.2   Predator–Prey Models 422
10.3   Competition and Cooperation 428
10.4   Chaos 430
10.5   The Forces of Creation 443
10.6   Summary 446
        References 447
        Bibliography 448
        Homework Problems 448
        Projects 453
        Research 454

11   System Dynamics 455

        Preview 455
11.1   Introduction 455
11.2   The Laundry List 457
11.3   The Influence Diagram 458
11.4   The Structure Diagram........................ 461

11.5   Structure Characterization 464
11.6   Causality 472
11.7   Differential Versus Difference Equations 476
11.8   The Larch Bud Moth Model 480
11.9   The Influenza Model 487
11.10 Forrester's World Model 491
11.11 Model Validation 498
11.12 Summary 500
       References 500
       Bibliography 501
       Homework Problems 501
       Research 504

12   Naïve Physics 507

       Preview 507
12.1   Introduction 507
12.2   Definitions 511
12.3   State Discretization and Landmarks 514
12.4   Operations on Qualitative Variables 516
12.5   Functions of Qualitative Variables 519
12.6   Qualitative Simulation 520
12.7   Qualitative Discrete–Time Simulation 544
12.8   Pros and Cons 546
12.9   Summary 548
       References 548
       Bibliography 549
       Homework Problems 550
       Projects 551
       Research 553

13   Inductive Reasoning 555

       Preview 555
13.1   Introduction 555
13.2   The Process of Recoding 558
13.3   Input/Output Behavior and Masking 569
13.4   Inductive Modeling and Optimal Masks 575
13.5   Forecasting Behavior 583
13.6   A Linear System — An Example 586

13.7 Gambling the Stock Market 593
13.8 Structure Characterization 601
13.9 Causality 610
13.10 Summary 612
   References 612
   Bibliography 614
   Homework Problems 614
   Projects 619
   Research 621

**14 Artificial Neural Networks and
      Genetic Algorithms** 623

   Preview 623
14.1 Introduction 623
14.2 Artificial Neurons 625
14.3 Artificial Neural Engineering Networks 629
14.4 The Pattern Memorizing Power of Highly
      Rank–Deficient Matrices 634
14.5 Supervised and Unsupervised Learning 637
14.6 Neural Network Software 647
14.7 Neural Networks for Dynamical Systems 659
14.8 Global Feedback Through Inverse Networks 670
14.9 Chaos and Dreams 671
14.10 Internalization Processes and Control Mechanisms 672
14.11 Genetic Learning 679
14.12 Neurobiological Learning 690
14.13 Summary 694
   References 694
   Bibliography 697
   Homework Problems 698
   Projects 700
   Research 701

15 Automated Model Synthesis 703

   Preview 703
15.1 Introduction 704
15.2 Level One: Classical Simulation Models 707
15.3 Level Two: Object–Oriented Modeling 709

15.4   Level Three: The System Entity Structure 711
15.5   Level Four: The Generalized SES 715
15.6   Level Five: Goal–Driven Pruning 716
15.7   The Cable Reel Problem 718
15.8   Summary... 732
         References 735
         Bibliography 736
         Homework Problems.............................. 737
         Projects 739
         Research 739

Index 743

1

Introduction, Scope, and Definitions

Preview

This chapter attempts to motivate the student for the course. Why should he or she study modeling and simulation? What can these techniques do for him or her that other techniques might not? We shall start out with some basic definitions of terms that will be used in this text over and over again, such as the terms "system," "experiment," "model," and "simulation." We shall then discuss good and bad reasons for using modeling and simulation as problem–solving tools. Finally, we shall list areas of science and engineering to which modeling and simulation have been successfully applied, and we shall explain what makes these various application areas different from each other and why the modeling and simulation approaches taken in these application areas vary so drastically from each other.

1.1 What is a System?

What is it that we focus on when we talk about a "system"? Brian Gaines gave the following interesting (and verbose) definition of what a system is [1.2]:

> " 'A system is what is distinguished as a system.' At first sight this looks to be a nonstatement. Systems are whatever we like to distinguish as systems. Has anything been said? Is there any possible foundation here for a systems science? I want to answer both these questions affirmatively and show that this definition is full of content and rich in its interpretation.

"Let me first answer one obvious objection to the definition above and turn it to my advantage. You may ask, 'What is peculiarly systemic about this definition?' 'Could I not equally well apply it to all other objects I might wish to define?' i.e.,

"A rabbit is what is distinguished as a rabbit. 'Ah, but,' I shall reply, 'my definition is adequate to define a system but yours is not adequate to define a rabbit.' In this lies the essence of systems theory: that to distinguish some entity as being a system is a necessary and sufficient criterion for its being a system, and this is uniquely true for systems. Whereas to distinguish some entity as being anything else is a necessary criterion to its being that something but not a sufficient one.

"More poetically, we may say that the concept of a system stands at the supremum of the hierarchy of being. That sounds like a very important place to be. Perhaps it is. But when we realize that getting there is achieved through the rather negative virtue of not having any further distinguishing characteristics, then it is not so impressive a qualification. I believe this definition of a system as being that which uniquely is defined by making a distinction explains many of the virtues, and the vices, of systems theory. The power of the concept is its sheer generality; and we emphasize this naked lack of qualification in the term general systems theory, rather than attempt to obfuscate the matter by giving it some respectable covering term such as mathematical systems theory. The weakness, and paradoxically the prime strength of the concept is in its failure to require further distinctions. It is a weakness when we fail to recognize the significance of those further distinctions to the subject matter in hand. It is a strength when those further distinctions are themselves unnecessary to the argument and only serve to obscure a general truth through a covering of specialist jargon. No wonder, general systems theory is subject to extremes of vilification and praise."

Brian Gaines expresses here in very nice words simply the following: The largest possible system of all is the universe. Whenever we decide to cut out a piece of the universe such that we can clearly say what is *inside* that piece (belongs to that piece), and what is *outside* that piece (does not belong to that piece), we define a new "system."

A system is characterized by the fact that we can say what belongs to it and what does not, and by the fact that we can specify how it interacts with its environment. System definitions can furthermore be hierarchical. We can take the piece from before, cut out a yet smaller part of it, and we have a new "system."

Let me quote another famous definition which is due to Ross Ashby [1.1]:

"At this point, we must be clear about how a 'system' is to be defined. Our first impulse is to point at the pendulum and to say 'the system is that thing there.' This method, however, has a fundamental disadvantage: every material object contains no less than an infinity of variables, and therefore, of possible systems. The real pendulum, for instance, has not only length and position; it has also mass, temperature, electric conductivity, crystalline structure, chemical impurities, some radioactivity, velocity, reflecting power, tensile strength, a surface film of moisture, bacterial contamination, an optical absorption, elasticity, shape, specific gravity, and so on and on. Any suggestion that we should study 'all' the facts is unrealistic, and actually the attempt is never made. What is necessary is that we should pick out and study the facts that are relevant to some main interest that is already given ...

"... The system now means, not a thing, but a list of variables."

Clearly, the two definitions are in contradiction with each other. According to the former definition (by Brian Gaines), the pendulum certainly qualifies for a system, and I would agree with him on that. However, taking the pendulum, we can now "cut out" a smaller piece by declaring that we are only interested in certain properties of the pendulum, say, its mass and its length, and thereby define another "system." The cutting does not necessarily denote a separation in the physical world, it can also take place at the level of a mathematical abstraction ... and in the context of modeling, this is actually most commonly the case.

Another property of a "system" is the fact that it can be "controlled" and "observed." Its interactions with the environment naturally fall into two categories:

(1) There are variables that are generated by the environment and that influence the behavior of the system. These are called the *"inputs"* of the system.

(2) There are other variables that are determined by the system and that in turn influence the behavior of its environment. These are called the *"outputs"* of the system.

In general, we ought to be able to assign values to at least some of the "inputs" of the system, and observe the behavior of the system by recording the resulting "outputs."

This leads to yet another definition for the term "system":

"A system is a potential source of data."

I personally like this definition best, because it is very short and concise. I am not entirely sure who gave this definition first, but I believe it was Bernard Zeigler [1.10].

1.2 What is an Experiment?

The last definition for "system" immediately leads to a definition for the term "experiment":

"An experiment is the process of extracting data from a system by exerting it through its inputs."

Experimenting with a system thus means to make use of its property of being "controllable" and "observable" (please notice that these terms are used here in a plausible sense rather than in the more stringent sense of linear system theory [1.4]). To perform an experiment on the system means to apply a set of external conditions to the accessible inputs and to observe the reaction of the system to these inputs by recording the trajectory behavior of the accessible outputs.

One of the major disadvantages of experimenting with real systems is the fact that these systems usually are under the influence of a large number of additional inaccessible inputs (so–called disturbances), and that a number of really useful outputs are not accessible through measurements either (they are internal states of the system).

One of the major motivations for simulation is the fact, that, in the simulation world, *all* inputs and outputs are accessible. This allows us to execute simulation runs that lie outside the range of experiments that are applicable to the real system.

1.3 What is a Model?

Given the preceding definitions for systems and experiments, we can now attempt to define what we mean by the term "model." I shall give the definition that was first coined by Marvin Minsky [1.8]:

> "A model (*M*) for a system (*S*) and an experiment (*E*) is anything to which *E* can be applied in order to answer questions about *S*."

Notice that this definition does not imply that a model is a computer program. It could as well be a piece of hardware or simply an understanding of how a particular system works (a so–called mental model). However, in this text, we shall concentrate on the subclass of models that are codable as computer programs (the so–called mathematical models).

Notice that this definition clearly qualifies any model to be called a system. This automatically implies that models are hierarchical in nature, i.e., we can cut a smaller portion out, and thereby generate a new model which is valid for a subset of the experiments for which the original model was valid. It is thus common to create models of models. Jack Kleijnen calls such models "meta–models" [1.6]. Bernard Zeigler talks about "pruning" particular features out of a model to create a simplified version of the previous model [1.12].

Notice finally that the definition does not describe "models for systems" *per se*. A model is always related to the tuple *system and experiment*. If people say that "a model of a system is invalid" (as can be frequently read), they don't know what they are talking about. A model of a system may be valid for one experiment and invalid for another, that is: the term "model validation" always relates to an experiment or class of experiments to be performed on a system, rather than to the system alone. Clearly, *any* model is valid for the "null experiment" applied to *any* system (if we don't want to get *any* answers out of a given simulation, we can use any model for that purpose). On the other hand, no model of a system is valid for *all* possible experiments except the system itself or an identical copy thereof.

1.4 What is a Simulation?

Again, many definitions exist for the term "simulation," but I shall quote the one that I like best. It has been coined by Granino Korn and John Wait [1.7]:

> "A simulation is an experiment performed on a model."

As before, this definition does not imply that the simulation is coded in a computer program. However, in this text, we shall concentrate on the subset of simulations which are codable as computer programs (the so–called mathematical simulations).

A *mathematical simulation* is a coded description of an experiment with a reference (pointer) to the model to which this experiment is to be applied.

It was Bernard Zeigler who first pointed out the importance of the physical separation between the *model description* on the one hand and the *experiment description* on the other [1.10]. We want to be able to experiment with models as easily and conveniently as with real systems. We want to be able to use our simulation tool in exactly the same way as we would use an oscilloscope in the lab.

However, a certain danger lies in this separation. It makes it all too easy to apply an experiment to a model for which the model is not valid. In the lab environment, this can never happen since the real system is valid for *all* experiments, whereas the model is not. Bernard Zeigler realized this problem and therefore demanded that the model description contain, as an intrinsic and unseparable part, an *experimental frame* definition [1.10]. The "experimental frame" establishes the set of experiments for which the model is valid. When a simulation refers to that model, the actual experiment is then compared with the experimental frame of the model, and the execution of the simulation will only be allowed if the simulation experiment to be performed is established as belonging to the set of applicable experiments.

Unfortunately, today's realities don't reflect these conceptual demands very well. Most commercially available simulation software systems are *monolithic*. They do not support the concept of separating the model description from the experiment description. While the model description mechanisms have seen quite a bit of progress over the years, most software systems do not allow the user to specify models in a truly hierarchical manner. Mechanisms for describing simulation experiments are meager, and mechanisms for describing experimental frames barely exist. In fact, we are still lacking an appropriate "language" to express experimental frames in general terms. All this still belongs to the area of open research.

1.5 Why is Modeling Important?

Let me quote yet another definition of the term "modeling" which is also attributed to Bernard Zeigler [1.11]:

"Modeling means the process of organizing knowledge about a given system."

By performing experiments, we gather knowledge about a system. However, in the beginning, this knowledge is completely unstructured. By understanding what are *causes* and what are *effects*, by placing observations both in a *temporal* as well as a *spatial* order, we organize the knowledge that we gathered during the experiment. According to the preceding (very general) definition, we are thereby engaged in a process of *modeling*. No wonder that every single discipline of science and engineering is interested in modeling, and uses modeling as a problem–solving tool.

It can thus be said that modeling is the single most central activity that unites *all* scientific and engineering endeavors. While the scientist is happy to simply *observe* and *understand* the world, i.e., create a model of the world, the engineer wants to *modify* it to his or her advantage. While science is all *analysis*, the essence of engineering is *design*. As this text will demonstrate, *simulation* can be used not only for analysis (the so–called *direct problems*), but also for design (the so–called *inverse problems*).

1.6 Why is Simulation Important?

Except by experimentation with the real system, simulation is the only technique available for the analysis of arbitrary system behavior. Analytical techniques are great, but they usually require a set of simplifying assumptions to be made before they become applicable, assumptions that cannot always be justified, and even if they might be, whose justification cannot be verified except by experimentation or simulation. In other words, simulation is often not used alone but in an interplay with other analytical or semianalytical techniques.

The typical scenario of a scientific discovery is as follows:

(1) The scientist performs experiments on the real system to extract data (to gather knowledge).

(2) She or he then looks at the data, and postulates a number of hypotheses relating to the data.

(3) Simplifying assumptions help to make the data tractable by analytical techniques to test these hypotheses.

(4) A number of simulation runs with different experimental parameters are then performed to verify that the simplifying assumptions were justified.

(5) He or she performs the analysis of his or her system, verifies (or modifies) the hypotheses, and finally draws some conclusions.

(6) Finally, a number of simulation runs are executed to verify the conclusions.

Most of today's simulation software systems live in isolation. They do not allow us to easily combine simulation studies with other techniques applied to the same set of data. Even the data gathered in the experiment must often be retyped (or at least edited) to fit into the framework of the simulation software system. This text, however, presents the reader with tools (CTRL–C and MATLAB) that are much more flexible than the old–fashioned CSSL–type simulation languages (such as ACSL or DARE–P), and that strongly support a mixed environment of different analysis techniques including simulation as one of them.

Simulation is applicable where analytical techniques are not. Since such situations are very common, simulation is often the only game in town. No wonder simulation is the single most frequently used problem–solving tool throughout all disciplines of science and engineering.

1.7 The Dangers of Simulation

The most important strengths of simulation, but also ironically its most serious drawbacks, are the generality and ease of its applicability. It does not require much of a genius to be able to utilize a simulation program. However, in order to use simulation intelligently, we must understand what we are doing.

All too often, simulation is a love story with an unhappy ending. We create a model of a system, and then fall in love with it. Since love is usually blind, we immediately forget all about the experimental frame, we forget that this is *not* the real world, but that it represents the world only under a very limited set of experimental conditions (we become "model addicts"). We find a control strategy that "shapes" our model world the way we want it to be, and then apply that control strategy back to the real world, convinced that we now have the handle to make the real world behave the way we want it to ... and here comes the unhappy ending — it probably doesn't.

In this way, the Australians introduced the rabbit to the Continent ... and found that these rabbits were soon all over the place and ate all the food that was necessary for the local species to survive ... since the rabbits did not have a natural enemy. Thereafter, the Australians introduced the fox to the Continent (foxes feed on rabbits — the model) ... to see that the foxes were soon all over the place ... but left the rabbits alone ... since they found the local marsupialia much easier to hunt.

Simulations are rarely enlightening. In fact, running simulations is very similar to performing experiments in the lab. We usually need many experiments, before we can draw legitimate conclusions. Correspondingly, we need many simulations before we understand how our model behaves. While analytical techniques (where they are applicable) often provide an understanding as to how a model behaves under *arbitrary* experimental conditions, one simulation run tells us only how the model behaves under the one set of experimental conditions applied during the simulation run.

Therefore, while analytical techniques are generally more restricted (they have a much smaller domain of applicability), they are more powerful where they apply. So, whenever we have a valid alternative to simulation, we should, by all means, make use of it. Only an idiot uses simulation *in place* of analytical techniques.

1.8 Good Reasons to Use Simulation

Let me state a number of good reasons for using simulation as a problem–solving tool.

(1) The physical system is not available. Often, simulations are used to determine whether a projected system should ever be built. So obviously, experimentation is out of the question. This is common practice for *engineering* systems (for example, an electrical circuit) with well–established and widely applicable meta–knowledge. It is very dangerous to rely on such a decision in the case of systems from soft sciences (the so–called *ill–defined systems*) since the meta–knowledge available for these types of systems is usually not validated for an extension into unknown territory.

(2) The experiment may be dangerous. Often, simulations are performed in order to find out whether the real experiment might "blow up," placing the experimenter and/or the equipment under danger of injury/damage or death/destruction (for example, an atomic reactor or an aircraft flown by an inexperienced person for training purposes).

(3) The cost of experimentation is too high. Often, simulations are used where real experiments are too expensive. The necessary measurement tools may not be available or are expensive to buy. It is possible that the system is used all the time and taking it "off–line" would involve unacceptable cost (for example, a power plant or a commercial airliner).

(4) The time constants (eigenvalues) of the system are not compatible with those of the experimenter. Often, simulations are performed because the real experiment executes so quickly that it can hardly be observed (for example, an explosion) or because the real experiment executes so slowly that the experimenter is long dead before the experiment is completed (for example, a transgression of two galaxies). Simulations allow us to speed up or slow down experiments at will.

(5) Control variables (disturbances), state variables, and/or system parameters may be inaccessible. Often, simulations are performed because they allow us to access *all* inputs and *all* state variables, whereas, in the real system, some inputs (disturbances) may not be accessible to manipulation (for example, the time of sunrise) and some state variables may not be accessible to measurement. Simulation allows us to manipulate the model

outside the feasible range of the physical system. For example, we can decide to change the mass of a body at will from 50 kg to 400 kg and repeat the simulation at the stroke of a key. In the physical system, such a modification is either not feasible at all or it involves a costly and lengthy alteration to the system.

(6) Suppression of disturbances. Often, simulations are performed because they allow us to suppress disturbances that are unavoidable in the real system. This allows us to isolate particular effects, and may lead to a better insight (intuition) into the generic system behavior than would be possible through obscured measurements taken from the real process.

(7) Suppression of second–order effects. Often, simulations are performed because they allow us to suppress second–order effects (such as nonlinearities of system components). Again, this can help with the understanding of the primary underlying functionality of the system.

1.9 The Types of Mathematical Models

What types of mathematical models do exist? A first category is the set of *continuous–time models*. Figure 1.1 shows how a state variable x changes over time in a continuous–time model.

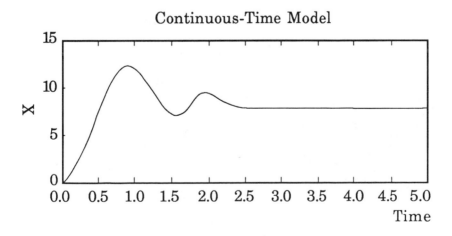

Figure 1.1. Trajectory behavior of a continuous–time model.

We can give the following definitions for continuous–time models:

> "Continuous–time models are characterized by the fact that, within a finite time span, the state variables change their values infinitely often."

No other mathematical model shares this property.

Continuous–time models are represented through sets of differential equations. Among the continuous–time models, two separate classes can be distinguished: the *lumped parameter models,* which are described by *ordinary differential equations* (ODEs), in general:

$$\dot{\mathbf{x}} = \mathbf{f}(\mathbf{x}, \mathbf{u}, t) \tag{1.1}$$

and for the special case of *linear systems*:

$$\dot{\mathbf{x}} = \mathbf{A}\mathbf{x} + \mathbf{B}\mathbf{u} \tag{1.2}$$

and the *distributed parameter models,* which are described by *partial differential equations* (PDEs) such as the diffusion equation:

$$\frac{\partial u}{\partial t} = \sigma \cdot \frac{\partial^2 u}{\partial x^2} \tag{1.3}$$

Both types will be encountered in this text and it is indeed the set of the continuous–time models that is at the center of our interest.

The second class of mathematical models to be mentioned is the set of *discrete–time models.* Figure 1.2 depicts the trajectory behavior exhibited by discrete–time models.

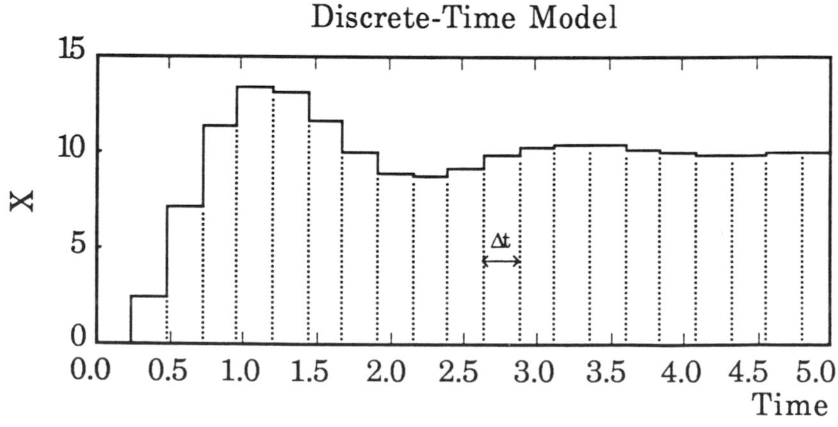

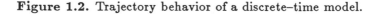

Figure 1.2. Trajectory behavior of a discrete–time model.

In these type of models, the time axis is discretized. Discrete–time models are commonly represented through sets of *difference equations*, at least if the discretization is equidistantly spaced. Such models can be represented as:

$$\mathbf{x}_{k+1} = \mathbf{f}(\mathbf{x}_k, \mathbf{u}_k, t_k) \tag{1.4}$$

In the case of nonequidistantly spaced discrete–time models, a *discrete–event representation* is generally preferred (cf. later).

Discrete–time models can occur naturally. For example, the population dynamics of an insect population is commonly represented through a set of difference equations, since the insects breed only during a short period of the year, i.e., a discretization interval of one year is natural.

Discrete–time models occur frequently in engineering systems, most commonly in *computer-controlled systems*. If a digital computer is used in a control system to compute one or several control signals, it cannot do so on a continuous basis since the algorithm to compute the next set of values of the control signals requires time. It is therefore most natural to apply "time–slicing," i.e., to cut the time axis into short and equidistant intervals where each interval is usually chosen to be just long enough to allow the digital computer to compute one new set of values. If the system to be controlled is itself a continuous–time system (as is often the case), we call this a *sampled–data control system*.

Alternatively, discrete–time models can also be discretized versions of continuous–time models. This is in fact very common. For instance, if we discretize the time axis of the continuous–time state–space model:

$$\dot{\mathbf{x}} = \mathbf{f}(\mathbf{x}, \mathbf{u}, t)$$

with a discretization interval Δt, the state derivative becomes

$$\frac{\mathbf{x}_{k+1} - \mathbf{x}_k}{\Delta t} \approx \mathbf{f}(\mathbf{x}_k, \mathbf{u}_k, t_k) \tag{1.5}$$

or

$$\mathbf{x}_{k+1} \approx \mathbf{x}_k + \Delta t \cdot \mathbf{f}(\mathbf{x}_k, \mathbf{u}_k, t_k) \tag{1.6}$$

which immediately leads us to a discrete–time model. In fact, whenever we use a digital computer to simulate a continuous–time model, we actually *must* discretize the time axis in some way in order to avoid the problem with the infinitely many state changes in a finite time span. However, any garden variety simulation language (such as ACSL or DARE–P) will do this discretization for us, and hide the fact from us ... except when something goes wrong. Then, we must understand how the program works in order to be able to help it get back on track. This will be one of the major topics to be discussed in the companion book on continuous system simulation.

The third class of models is the set of *qualitative models* which are by nature discrete–time models (although not necessarily with equidistant time–slicing). Figure 1.3 shows the trajectory behavior of a qualitative model

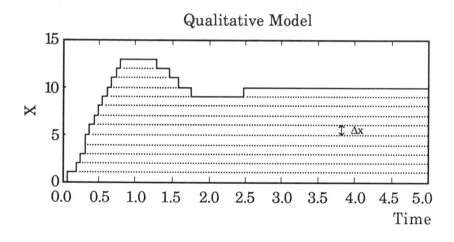

Figure 1.3. Trajectory behavior of a qualitative model.

In a qualitative model, the dependent variables are discretized. Qualitative models are usually coded using a *finite state representation*, and this type of model will be covered in the text as well.

The fourth and final class of models is the set of *discrete–event models*. Paradoxically, both the time axis and the state axis of discrete–event models are usually "continuous" (i.e., *real* rather than *integer*), but discrete–event models differ from the continuous–time models by the fact that, in a finite time span, only a finite number of state changes may occur. Figure 1.4 depicts the typical trajectory behavior of a state variable in a discrete–event simulation.

Discrete–event models have been the main topic of a series of previous simulation texts [1.9,1.10,1.11] and will not be covered here.

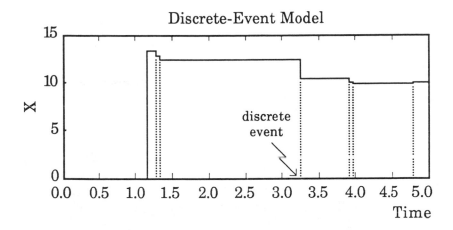

Figure 1.4. Trajectory behavior of a discrete–event model.

When should we use what type of model? Walter Karplus generated a "rainbow" (the way children draw it) that answers this question in a systematic way [1.5]. Figure 1.5 represents a slightly modified version of that "rainbow."

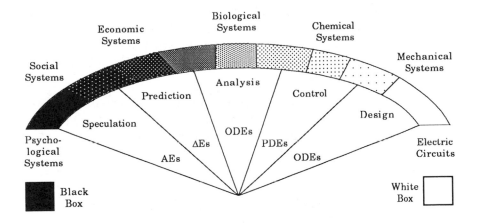

Figure 1.5. Spectrum of modeling and simulation.

Above the rainbow, various application areas of modeling and simulation are shown. They range from electrical circuits to psychological systems. The application areas shown are exemplary. Areas that are not shown include the thermal, hydraulic, and pneumatic systems which should be located somewhere between the mechanical and the chemical systems. In this text, we shall proceed along the rainbow from the right to the left, i.e., from well–defined ("white box") systems to ill–defined ("black box") systems.

Immediately below the rainbow, common purposes for modeling and simulation are specified. Remember that modeling and simulation are always goal–driven, i.e., we should know the purpose of our potential model before we sit down to create it.

Electrical circuits are so well understood that it is possible to use a model to design an overall circuit, i.e., once the performance of the model is satisfactory, we can build the real system, and, in all likelihood, it will work just as predicted by the model. This is also true for some of the mechanical systems (except where nonlinearities and friction effects become dominant factors).

This is, however, no longer true for chemical systems. Many factors influence a chemical reaction, factors which are all of approximately equal importance. Therefore, models that are valid for a large set of experiments cannot be specified. Thus, a theoretically derived model of a chemical process may predict one thing while the real system that is built after the model may react quite differently. Yet, if we build the system first and match the model to the system, the model contains sufficient internal validity to allow us to build a model of a controller for the system that, when applied to the real system, may still work nicely. This is due to the fact that feedback controllers have a tendency to reduce the system's sensitivity to parameter variations.

When we proceed further to the left, we find that the internal validity of our models decays further and further. Eventually, we come to a point where the available models no longer contain sufficient internal validity to allow us to use the model for any design purposes. Yet, we can still use the model for analyzing the system behavior, i.e., the internal structure of the model is still sufficiently valid to allow us to reason competently about cause–effect relationships among the variables that are captured by the model.

Advancing further to the left, we come to a point where even this statement is no longer true. Such models are constructed in a mostly inductive manner and a decent match between model and system behavior no longer guarantees that the internal structure of the model represents the internal structure of the real system in any meaningful way. Yet, we may still be able to predict the future of the real system from simulating the model beyond the current time.

Finally, systems exist where even this is no longer true. All we can achieve is to speculate about possible futures, maybe with probability tags attached to the various possible outcomes. This is true in particular for social and psychological systems since they are retroactive. These systems include humans who, due to their knowledge of the model predictions, will adjust their behavior to modify that same outcome. In some cases, we end up with self–fulfilling prophecy. If I have a "good" model of the stock market that predicts the growth of a particular stock and if many people have access to that model and believe in its value, then all these people will wish to buy that particular stock, and sure enough, the stock will gain value (at least for a while). The opposite can also occur. If my model predicts a major disaster and if a sufficiently large number of influential people know about that prediction and believe in the accuracy of my model, they will do their best to modify the system behavior to prevent that very disaster from happening. Good examples are George Orwell's book *1984* and Jay Forrester's world model, which predicted clearly undesirable futures. Consequently, legislative actions were taken that hopefully will prevent those very predictions from ever becoming a reality. Walter Karplus wrote rightly that the major purpose of such models is to "arouse public opinion" [1.5].

Below the purpose spectrum, a methodology spectrum is presented. Electrical circuits can be accurately described by ordinary differential equations, since the influence of geometry is usually negligible. This is true except for very high frequencies (microwaves) or very small dimensions (integrated circuits).

When geometry becomes important, we must introduce the space dimensions as additional independent variables and we end up with distributed parameter models that are described by partial differential equations. This is true for mechanical systems with finite stiffness, for thermodynamics, fluid dynamics, optics, and diffusion processes in chemistry.

Advancing further to the left, the available data and the limited knowledge of the meta–laws of these systems no longer warrant the specification of distributed parameter models and we use again ODEs, not because that is how these systems really behave, but because we cannot validate any more complex models with our limited understanding of the processes and with the limited experimental data available.

When even less information is present, the accuracy that ODEs provide (and that we must pay for in terms of computing time) is no longer warranted. It makes sense to use very high–order integration algorithms only for the best–defined systems, such as those in celestial mechanics. When we simulate a celestial mechanics problem, we like to use an eighth–order Runge–Kutta algorithm, since it allows us to select a large integration step size and yet integrates the model equations with high accuracy. Fourth–order algorithms are optimal for most engineering tasks. As a rule of thumb, we use a kth–order algorithm if we wish to obtain results with an accuracy of k decimals. For systems with an inherent accuracy of several percent (such as in biology), it does not make sense to use any integration algorithm higher than first order, i.e., the forward Euler algorithm shown in Eq.(1.6) is appropriate. Such models are therefore often represented in the form of *difference equations* (ΔEs).

Finally, in the "darkest" of all worlds, i.e., in social and psychological modeling, the models used are mostly static. They are described by *algebraic equations* (AEs). They are usually entirely inductive and depend on "gut feeling" or the position of the stars in the sky.

1.10 Direct Versus Inverse Problems

Envisage a system as depicted in Fig.1.6.

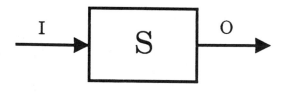

Figure 1.6. Block diagram of a system.

The system is characterized by a set of *inputs* (I) (including both control inputs and disturbances), by a set of *outputs* (O), and by a set of *internal variables* (S) including both the state variables and any auxiliary algebraic variables.

The "normal" situation for a simulation is given, when all inputs are known as functions over time, and when the system structure and the initial conditions of all state variables are specified. The task of the simulation is to determine the trajectory behavior of all outputs, i.e.,

$$I, S = \text{known} ; O = \text{unknown}$$

This problem is called the *direct problem.*

However, two types of inverse problems exist as well. For instance, it could be that the system under study is a "black box." While all inputs and outputs are known, the internal structure of the system and/or the initial values of the state variables are unknown, i.e.,

$$I, O = \text{known} ; S = \text{unknown}$$

These problems are referred to as the *structure identification problem* and the *state estimation problem*, respectively. We shall demonstrate in the companion book how simulation can be used to solve identification problems.

A third type of problem is given if:

$$S, O = \text{known} ; I = \text{unknown}$$

This is referred to as the *control problem* and is the major subject of the area of automatic control [1.3,1.4]. In the companion book, we shall also demonstrate how simulation can be used to solve control problems.

1.11 Summary

In this chapter, we have given some basic definitions, outlined the scope of our undertaking, and tried to answer the question why students might be interested in this subject and why they might want to continue with this course.

References

[1.1] W. Ross Ashby (1956), *An Introduction to Cybernetics*, John Wiley, New York.

[1.2] Brian Gaines (1979), "General Systems Research: Quo Vadis," *General Systems Yearbook*, **24**, pp. 1–9.

[1.3] Gene H. Hostetter, Clement J. Savant, Jr., and Raymond T. Stefani (1982), *Design of Feedback Control Systems*, Holt, Rinehart and Winston, New York.

[1.4] Thomas Kailath (1980), *Linear Systems*, Information and System Sciences Series, Prentice–Hall, Englewood Cliffs, N.J.

[1.5] Walter J. Karplus (1976), "The Spectrum of Mathematical Modeling and Systems Simulation," Proceedings Eighth AICA Congress on Simulation of Systems (L. Dekker, ed.), North–Holland, Amsterdam, pp. 5–13.

[1.6] Jack P. C. Kleijnen (1982), "Experimentation with Models: Statistical Design and Analysis Techniques," in: *Progress in Modelling and Simulation* (F.E. Cellier, ed.), Academic Press, London, pp. 173–185.

[1.7] Granino A. Korn and John V. Wait (1978), *Digital Continuous–System Simulation*, Prentice–Hall, Englewood Cliffs, N.J.

[1.8] Marvin Minsky (1965), "Models, Minds, Machines," Proceedings IFIP Congress, pp. 45–49.

[1.9] A. Alan B. Pritsker (1985), *Introduction to Simulation and SLAM–II*, third edition, Halsted Press, New York.

[1.10] Bernard P. Zeigler (1976), *Theory of Modeling and Simulation*, John Wiley, New York.

[1.11] Bernard P. Zeigler (1984), *Multifaceted Modeling and Discrete Event Simulation*, Academic Press, London.

[1.12] Bernard P. Zeigler (1990), *Object–Oriented Simulation with Hierarchical, Modular Models: Intelligent Agents and Endomorphic Systems*, Academic Press, Boston, Mass.

Bibliography

[B1.1] Rutherford Aris (1978), *Mathematical Modelling Techniques*, Pitman, London and San Francisco.

[B1.2] Edward Beltrami (1987), *Mathematics for Dynamic Modeling*, Academic Press, Boston, Mass.

[B1.3] Edward A. Bender (1978), *An Introduction to Mathematical Modeling*, John Wiley, New York.

[B1.4] François E. Cellier, Ed. (1982), *Progress in Modelling and Simulation*, Academic Press, London.

[B1.5] Charles M. Close and Dean K. Frederick (1978), *Modeling and Analysis of Dynamic Systems*, Houghton Mifflin, Boston, Mass.

[B1.6] Naim A. Kheir, Ed. (1988), *Systems Modeling and Computer Simulation*, Marcel Dekker, New York.

[B1.7] William J. Palm III (1983), *Modeling, Analysis, and Control of Dynamic Systems*, John Wiley, New York.

[B1.8] Jon M. Smith (1987), *Mathematical Modeling and Digital Simulation for Engineers and Scientists*, John Wiley, New York.

[B1.9] Jan A. Spriet and Ghislain C. Vansteenkiste (1982), *Computer–Aided Modeling and Simulation*, Academic Press, London.

Homework Problems

[H1.1] Sampled–Data System

Given the following linear differential equation system:

$$\dot{x}_1 = x_2 \qquad (H1.1a)$$
$$\dot{x}_2 = x_3 \qquad (H1.1b)$$
$$\dot{x}_3 = -2x_1 - 3x_2 - 4x_3 \qquad (H1.1c)$$
$$y = 7x_1 - 5x_2 \qquad (H1.1d)$$

Use the forward Euler integration algorithm to convert this set of differential equations to a set of difference equations. Use a step size of $\Delta t = 0.1$ sec.

[H1.2] Signal Types [1.3][†]

For the following systems, try to identify the inputs, outputs, and disturbances (where applicable):

(a) The water level in a reservoir.

(b) The power supply of a city.

(c) A car being driven along a mountain road.

(d) A toaster.

[H1.3]* Meta–Models

Given the following nonlinear second–order model:

$$\dot{x}_1 = -3x_1 + 1.5x_1 x_2 \qquad (H1.3a)$$
$$\dot{x}_2 = 4.5x_2 - 1.5x_1 x_2 \qquad (H1.3b)$$

[†] Adapted excerpt from DESIGN OF FEEDBACK CONTROL SYSTEMS, First Edition by Gene Hostetter *et al.*, ©1982 by Saunders College Publishing, a division of Holt, Rinehart and Winston, Inc., reprinted by permission of the publisher.

Analyze the behavior of this system (a so–called Lotka–Volterra model) in the vicinity of all of its steady–state points (*hint:* steady–state points are those points in which all derivatives are equal to zero). This task is accomplished through the use of linear meta–models. The problem can be decomposed into the following subtasks:

(a) Determine all steady–state points of this system.

(b) For any steady–state point that is not the origin ($[x_1, x_2] = [0, 0]$) itself, apply a linear variable transformation that moves the steady-state point to the origin, i.e., if $[x_1, x_2]_{ss} = [a, b]$, we introduce the new set of state variables: $\xi_1 = x_1 - a$ and $\xi_2 = x_2 - b$, and rewrite our set of differential equations in terms of these new variables. In the new coordinate system $[\xi_1, \xi_2]$, the steady–state point will be the origin.

(c) We can now linearize the models around their origins by simply throwing out all nonlinear terms.

(d) The resulting linear meta–models are so simple that they are amenable to an analytical treatment (i.e., they have closed–form solutions). Find these solutions and determine qualitatively the behavior of the nonlinear system in the vicinity of these steady–state points. Sketch a graph of what you expect the trajectories to look like in the $[x_1, x_2]$ plane (the so–called phase plane) in the vicinity of the steady–state points.

Projects

[P1.1] Definitions

Get a number of simulation and/or system theory textbooks from your library and compile a list of definitions of "What is a System"? Write a term paper in which these definitions are critically reviewed and classified. (Such a compilation has actually been published once.)

Research

[R1.1] Experimental Frames

Study the separation of the model description from the experiment description. Analyze under what conditions such a separation is meaningful and/or feasible. Develop mechanisms to ensure the compatibility of a proposed experiment with a given model (mechanisms to code the experimental frame) and develop a generic language to code the experimental frame specification.

2

Basic Principles of Continuous System Modeling

Preview

In this chapter, we shall introduce some basic concepts of continuous system modeling. By the end of this chapter, the student should be able to code simple modeling problems in some of the currently used simulation languages. The languages ACSL, DARE–P, and DESIRE are introduced in order to demonstrate the similarities that exist between the various continuous system simulation languages.

2.1 Introduction

While the continuous system simulation languages that are available on the software market today differ somewhat in terms of syntactical details, they are all derivations of the Continuous System Simulation Language (CSSL) specification generated by the CSSL committee in 1967 [2.1]. Their basic principles are thus identical. In particular, they all are based on a *state–space description* of the system equations, i.e., on a set of first–order ordinary differential equations (ODEs) of the form:

$$\dot{\mathbf{x}} = \mathbf{f}(\mathbf{x}, \mathbf{u}, t) \tag{2.1}$$

A typical simulation language may allow us to express a representative model through the set of equations:

DYNAMIC
$$thrust = \mathrm{f}(t)$$
$$h. \quad = v$$
$$v. \quad = a$$
$$a \quad = (1/m) * (thrust - m * g)$$
$$m. \quad = -\mathrm{c1}*\mathrm{abs}(thrust)$$
$$g \quad = \mathrm{c2}/(h + r) * *2$$
END

which describes the vertical motion of a rocket that is just about to perform a soft landing on the surface of the moon. This model contains three *state variables*, namely, the altitude h, the vertical component of the velocity v, and the mass m of the rocket describing a *third–order model* of the rocket dynamics. The dot (.) denotes the first derivative with respect to the independent variable which, in simulation languages, is traditionally assumed to be time (t). This model also specifies three additional *auxiliary variables*, namely, the vertical component of the rocket's acceleration a, which encodes Newton's law applied to the rocket's rigid body; the gravity force g, which increases quadratically with decreasing altitude of the rocket; and the *thrust*, which is specified as an externally computed function of the simulation clock t. The model also references three constants, namely, the lunar radius r, the gravitational constant $c2$, and the fuel efficiency constant $c1$. The first two state equations (for the state variables h and v) denote the mechanical interrelation between position, velocity, and acceleration of a rigid body, while the third state equation (for the state variable m) describes the reduction of the rocket's mass due to fuel consumption, which is assumed to be proportional to the absolute value of the applied thrust.

This simple model teaches us a number of things. First, we notice that some of the variables, such as a, are referenced *before* they have been defined. In most programming environments, this would result in a run–time exception, but not in CSSLs. The underlying physical phenomena that are captured through the equations of the simulation model are all taking place *in parallel*. This is reflected in CSSLs by allowing the user to specify equations as *parallel code*. The sequence of the equations that make up the dynamic model of the system is entirely immaterial. An *equation sorter,* which is an intrinsic part of most CSSLs, will sort the equations at compilation time into an executable sequence.

An immediate consequence of this decision is the rule that no variable can be defined more than once (since otherwise the sorter wouldn't know which definition to use). CSSLs belong to the class of *single assignment languages* (SALs). All time–dependent variables (state variables and auxiliary variables) must be defined exactly once inside the dynamic model description section of the simulation program.

The dynamic model equations are really of a *declarative* rather than an *executive* nature. In CSSLs, the equal sign denotes *equality* rather than *assignment*. (Some programming languages, such as PASCAL, distinguish between these two meanings of the equal operator by using a separate operator symbol, namely, "=" to denote *equality* and ":=" to denote *assignment*. However, other programming languages, such as FORTRAN, don't make this distinction, which is bound to create a certain degree of confusion.) A statement such as:

$$i = i + 1 \qquad (2.2)$$

which is one of the most common statements in traditional programming languages, meaning that the value of the integer variable i is to be incremented by one, is entirely meaningless in a CSSL–type language. If we interpret Eq.(2.2) as a *mathematical formula*, the formula is incorrect because we can cancel the variable i from both sides of the equal sign, which would imply that 0 is equal to 1, which is obviously not true. On the other hand, if we interpret Eq.(2.2) as a *declarative statement* (which reflects a little more accurately what the simulation languages do) then we realize that we are confronted here with a *recursive declaration*, which makes the sorting algorithm as helpless as I am when I read in one of my TIME–LIFE cookbooks that the recipe for taratoor requires one tablespoon of tahini while the recipe for tahini calls for one tablespoon of taratoor (I meanwhile solved that problem — I buy tahini in the store).

2.2 The Algebraic Loop Problem

The preceding discussion unveils that not all problems related to equation sorting are solved by simply requesting that every variable be declared exactly once. The example can be expressed in terms of a CSSL notation as:

$$taratoor = \mathrm{f}(tahini) \qquad\qquad (2.3a)$$
$$tahini \;\;\; = \mathrm{g}(taratoor) \qquad\qquad (2.3b)$$

With this program segment, the equation sorter gets stuck in exactly the same way that I did. The typical response of most CSSLs in this case is to flag the program as nonexecutable and return an error message of the type: *Algebraic loop detected involving variables taratoor and tahini.* Algebraic loops can involve one or several variables.

2.3 Memory Functions

Let us look a little closer at some of the equations of our rocket model, for example:

$$vdot = a \qquad\qquad\qquad\qquad\qquad\qquad (2.4a)$$
$$a \;\;\;\; = (1/m) * (thrust - m * g) \qquad\quad (2.4b)$$
$$v \;\;\;\; = \mathrm{INTEG}(vdot, v0) \qquad\qquad\quad (2.4c)$$

I used here a slightly different notation, which is also commonly found in many CSSLs. Instead of specifying the model in a state–space representation (with the integration operation being implied), some CSSLs provide for an explicit integration operator (called IN-TEG in our example). This notation is a little more bulky, but it has the advantage that the initial condition v_0 can be specified as part of the dynamic model description rather than being treated separately together with the constants.

From this description, we could get the impression that the variable v is a function of the variable $vdot$ (through Eq.(2.4c)), that $vdot$ in turn is a function of the variable a (through Eq.(2.4a)), and that finally a is a function of v (through Eq.(2.4b)).

Did we just detect an algebraic loop involving the three variables v, $vdot$, and a? To answer this question, we need to look a little more closely into the process of *numerical integration*. As was mentioned in Chapter 1, it is necessary to discretize the continuous process of numerical integration to make it treatable on a digital computer.

Many schemes exist that demonstrate how this can be accomplished and we shall discuss those at length in the companion book on continuous system simulation. However, for now let us just look at the two simplest schemes that can be devised. Figure 2.1 depicts the process of numerical integration.

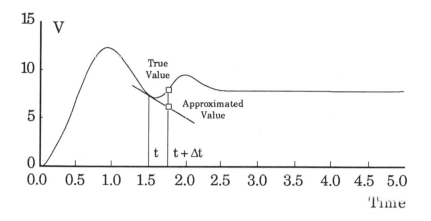

Figure 2.1. A simple scheme for numerical integration.

Given the value of our state variable v at any time t, we can approximate the value of v some Δt time units later through the formula:

$$v(t + \Delta t) \approx v(t) + \Delta t * \dot{v}(t) \tag{2.5}$$

which is commonly referred to as *Euler's integration rule*.

Since v at time $t = t_0$ is given as v_0, we can immediately evaluate Eq.(2.4b), followed by Eq.(2.4a). At this point, we have evaluated $\dot{v}(t_0)$. Therefore, we can now evaluate Eq.(2.5) to find $v(t_0 + \Delta t)$. Now, we can again evaluate Eq.(2.4b), followed by Eq.(2.4a) to find $\dot{v}(t_0 + \Delta t)$, a.s.f.

Obviously, the integration function has broken the algebraic loop since it depends on values of variables at past values of time only. A function that has this property is called a *memory function*. Any memory function will break algebraic loops, and thus the sorter must know whether or not a function that it comes across is a memory function. The two most prominent memory functions in continuous system simulation are the *integration function* and the *delay function*. Some CSSLs allow the user to declare his or her own additional memory functions.

2.4 Explicit Versus Implicit Integration

Let us repeat the previous discussion with a slightly modified integration scheme. Figure 2.2 demonstrates this scheme.

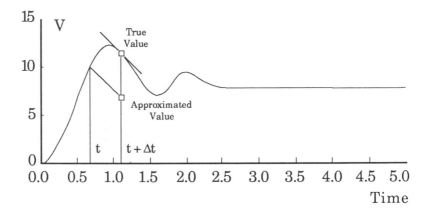

Figure 2.2. Numerical integration by backward Euler technique.

In this scheme, the solution $v(t + \Delta t)$ is approximated using the values of $v(t)$ and $\dot{v}(t + \Delta t)$ through the formula:

$$v(t + \Delta t) \approx v(t) + \Delta t * \dot{v}(t + \Delta t) \tag{2.6}$$

This scheme is commonly referred to as the backward Euler integration rule.

As can be seen, this integration formula depends on current as well as past values of variables. It is thus *not* a memory function. Consequently, it will not break up the algebraic loop. It can easily be verified that in order to compute $\dot{v}(t_0 + \Delta t)$ we require knowledge of $a(t_0 + \Delta t)$ (according to Eq.(2.4a)), which in turn requires knowledge of $v(t_0 + \Delta t)$ (according to Eq.(2.4b)). However, in order to compute $v(t_0 + \Delta t)$ (according to Eq.(2.6)), we need to know $\dot{v}(t_0 + \Delta t)$. In other words, the algebraic loop has not been broken.

Integration algorithms that are described by memory functions are therefore often referred to as *explicit integration techniques*, whereas algorithms that are not of the memory function type are called *implicit integration techniques*. Although the implicit integration techniques are advantageous from a numerical point of view (as we shall learn later), the additional computational load created by the necessity to solve simultaneously a set of nonlinear algebraic equations at least once every integration step makes them undesirable for use in CSSLs. Therefore, all integration techniques that are commonly available in CSSL–type languages are of the explicit type.

2.5 Implicit Loop Solvers

Let us return once more to the algebraic loop that we met earlier in this chapter. Clearly the best way to deal with algebraic loops is to solve them manually and replace the "delinquent" equations by a set of new equations that are now explicitly solvable.

Let us assume that the sorter found a set of two algebraically coupled equations:

$$y = -2 * x + 3 * u \qquad (2.7a)$$
$$x = -y/3 + 7 * u \qquad (2.7b)$$

We can easily satisfy the sorter's needs by manually solving these two equations as follows:

$$y = -33 * u \qquad (2.8a)$$
$$x = +18 * u \qquad (2.8b)$$

which can now be inserted in place of the two original equations.

However, this technique does not always work. Often, the set of equations does not have an analytical solution. In that case, we would like to be able to "buy tahini in the store." Some CSSLs, such as CSMP–III [2.4], provide a special mechanism for this purpose, which is called an *implicit loop solver*. CSMP–III enables us to reformulate the problem as follows:

```
tahini     = IMPL(store_tahini, errmax, next_tahini)
taratoor   = f(tahini)
next_tahini= g(taratoor)
```

which is a (somewhat clumsy) way to specify the following algorithm. First, we buy a can of *tahini* in the store (*store_tahini*). Then we produce one batch of *taratoor*. However, it could well be that the *store_tahini* is somewhat different from our favorite *tahini* recipe. Therefore, we now use the just–produced *taratoor* to make a new version of *tahini* (called *next_tahini*), from which we then can produce a new batch of *taratoor*, a.s.f., until two consecutive batches of *tahini* taste almost the same (their difference is smaller than *errmax*). (The third parameter of the IMPL function provides the compiler with the name of the assignment variable of the last statement that belongs to the algebraic loop.)

Now remember that this iteration must take place once per function evaluation, i.e., at least once every integration step. In other words, the next day, we have to go and buy a new can of *store_tahini* and the process starts all over again. After a short while, we may decide that this algorithm is wasteful. Surely, the final solution of the last day's *tahini* is a better starting value for today's *tahini* than *store_tahini*. In other words, I should save one tablespoon of the last day's *tahini* and put it in the freezer for reuse today rather than starting from scratch with a new can of *store_tahini*.

CSMP–III allows us to formulate this modified algorithm as follows:

```
INITIAL
   last_tahini = store_tahini
DYNAMIC
   . . .
   tahini      = IMPL(last_tahini, errmax, next_tahini)
   taratoor    = f(tahini)
   next_tahini = g(taratoor)
   . . .
   NOSORT
   last_tahini = tahini
END
```

2.6 Procedural Sections

Many CSSLs don't provide for an IMPL function. However, other mechanisms are available that can replace this construct easily. One such mechanism is the *procedural section*, which is offered by almost all CSSL–type languages.

A procedural section can be considered a "sandwich" equation, i.e., a set of regular *procedural statements* that are treated by the sorter like one equation to be sorted as a whole with all the other equations, while the assignment statements that form the procedural block stay together and are left unchanged. The header of the procedural section instructs the equation sorter about the place where the section needs to be inserted. The sorter *never* checks the inside of the procedural section itself.

For instance, the previous algorithm could be expressed in DARE–P [2.10] as follows:

```
$D1
      ...
      PROCED tahini = last_tahini
         dish_count = 0
   10    taratoor   = f(last_tahini)
         tahini     = g(taratoor)
         dish_count = dish_count + 1
         IF (dish_count.gt.max_dish) GO TO 20
         IF (abs(tahini - last_tahini).gt.errmax) GO TO 10
         GO TO 30
   20    WRITE(6, 21)
   21    FORMAT(' Iteration failed to converge')
   30    CONTINUE
      ENDPRO
      ...
      PROCED dummy = tahini
         last_tahini = tahini
      ENDPRO
   END
   last_tahini = store_tahini
   END
```

As can be seen, DARE–P employs FORTRAN statements to describe procedural sections. This is understandable since DARE–P is compiled (preprocessed) into FORTRAN. Most CSSLs use their *intermediate language* to express procedural sections.

Note that DARE–P procedures serve a completely different purpose than procedures in most of the other programming languages (such as PASCAL). They are used as structuring elements to declare sandwich equations. A DARE–P procedure is *not* a subprogram.

The outputs of the procedure are specified in the PROCED statement to the left of the equal sign (separated by comma), while its inputs are specified to the right of the equal sign. The sorter will place the procedural block such that all its inputs have been evaluated before the procedure is computed and none of its outputs are used before the procedure has been computed. In other words, the inputs and outputs of the procedure follow the SAL rule, while this is not true for assignments inside the procedure.

In our example, the first procedure is used to program out the iteration loop that was implied by the previous IMPL construct. The iteration loop ends when we can't tell subsequent batches of *tahini* apart any longer or when all dishes are dirty, whichever occurs first. The second procedure is used to break the algebraic loop. Since the sorter never looks inside the procedure, it will not detect that *last_tahini* is actually being redefined inside the second procedure. However, the header information of the two procedures will ensure that the second procedure is placed *after* the first since it needs *tahini* as an input, whereas the same variable *tahini* was declared as an output of the first procedure. Each PROCED is accompanied by an END statement to mark the end of the procedure.

The $D1 statement marks the beginning of the dynamic model description. The second–to–last END statement marks the end of the dynamic model description. Between this END and the final END of the code segment, DARE–P expects the declaration of constants and initial conditions.

2.7 The Basic Syntax of Current CSSLs

Let me formulate the simple lunar landing problem in terms of three current CSSLs, namely, ACSL [2.8], DARE–P [2.10], and DESIRE [2.5]. This is to demonstrate how similar the various CSSLs are in their basic formalisms, i.e., to show that, once we understand one, we know them all.

In ACSL [2.8], the problem can be formulated as follows:

PROGRAM *Lunar Landing Maneuver*
  INITIAL
    constant ...
      $r = 1738.0E3$, $c2 = 4.925E12$, $f1 = 36350.0$, ...
      $f2 = 1308.0$, $c11 = 0.000277$, $c12 = 0.000277$, ...
      $h0 = 59404.0$, $v0 = -2003.0$, $m0 = 1038.358$, ...
      $tmx = 230.0$, $tdec = 43.2$, $tend = 210.0$
    cinterval $cint = 0.2$
  END \$ *"of INITIAL"*
  DYNAMIC
    DERIVATIVE
      $thrust = (1.0- \text{ step}(tend)) * (f1 - (f1 - f2)*\text{step}(tdec))$
      $c1 \quad = (1.0- \text{ step}(tend)) * (c11 - (c11 - c12)*\text{step}(tdec))$
      $h \quad = \text{integ}(v, h0)$
      $v \quad = \text{integ}(a, v0)$
      $a \quad = (1.0/m) * (thrust - m * g)$
      $m \quad = \text{integ}(mdot, m0)$
      $mdot \quad = -c1*\text{abs}(thrust)$
      $g \quad = c2/(h + r) * *2$
    END \$ *"of DERIVATIVE"*
    termt $(t.\text{ge.}tmx \text{ .or. } h.\text{le.}0.0 \text{ .or. } v.\text{gt.}0.0)$
  END \$ *"of DYNAMIC"*
END \$ *"of PROGRAM"*

From the previous discussions, this program should be almost self-explanatory. "step" is an ACSL system function, the output of which is zero as long as the system variable t (the simulation clock) is smaller than the parameter (in our case *tdec* and *tend*). Therefore, the *thrust* takes a value of $f1$ for t smaller than *tdec*. It then takes a value of $f2$ for t between *tdec* and *tend*, and it takes a value of 0.0 thereafter. The fuel–efficiency constant $c1$ is computed by the same mechanism. This equation was introduced since the main retro motor which produces the thrust $f1$ and the three vernier engines which together produce the thrust $f2$ may have different fuel efficiencies. *Tmx* denotes the final time of the simulation. *Cint* denotes the *communication interval*, i.e., it tells ACSL how often results are to be stored in the simulation data base. Finally, *termt* is a *dynamic termination criterion*. Whenever the logical expression of the *termt* statement becomes true, the simulation will terminate. Notice that *tmx* is not a system constant and must be manually tested in the *termt* statement.

    Notice that this program does not contain any output statements. In the compilation process, the ACSL code is first preprocessed into FORTRAN by the *ACSL preprocessor*, then it is compiled further into machine code and thereafter it is linked with the *ACSL run–time library*. During execution of the resulting code, ACSL automatically switches to an *interactive mode* in which parameter values can be modified, simulation runs can be performed, and simulation outputs can be plotted.

The two–step compilation is standard practice in most CSSLs since it provides machine independence, as it can be assumed that all computers provide for a FORTRAN compiler.

Now let me write down the corresponding DARE–P [2.10] code:

```
$D1
    THRUST = (1.0 − STPE) * (F1 − (F1 − F2) * STPD)
    C1       = (1.0 − STPE) * (C11 − (C11 − C12) * STPD)
    H.       = V
    V.       = A
    A        = (1.0/XM) * (THRUST − XM * G)
    XM.      = −C1*ABS(THRUST)
    G        = C2/(H + R) * *2
    STPE     = STP(T,TEND)
    STPD     = STP(T,TDEC)
    TERMINATE −H * V
$F
    FUNCTION STP(T,TON)
    STP = 0.0
    IF (T.GE.TON) STP = 1.0
    RETURN
    END
END
R = 1738.0E3,  C2 = 4.925E12,  F1 = 36350.0
F2 = 1308.0,  C11 = 0.000277,  C12 = 0.000277
H = 59404.0,  V = −2003.0,  XM = 1038.358
TMAX = 230.0,  NPOINT = 301,  TDEC = 43.2,  TEND = 210.0
END
*LUNAR LANDING MANEUVER
GRAPH H
GRAPH V
END
```

DARE–P is a much older (and much more old–fashioned) language than ACSL. It was originally designed for CDC machines, which did not operate on a full ASCII character set, and consequently, even today, many versions of DARE–P don't support lower–case characters. The mass m had to be renamed into XM since any variable starting with I, J, K, L, M, or N is considered to be of type INTEGER and no mechanism exists in DARE–P to reassign the type of such variables. In the previous DARE–P program segment,

I had ignored these restrictions in order to improve the readability of the code and I shall do so again in the future since, contrary to many other areas of life, in programming languages age does not deserve reverence. Nevertheless, DARE–P still has its beauties and advantages as we shall see later. One of its true advantages is the fact that its syntax is easy to learn.

Other than that, the two programs are similar. DARE–P prefers the "dot" notation over the INTEG operator. The dynamic termination condition is here called TERMINATE and operates on a numerical rather than a logical expression. The simulation run terminates whenever the numerical expression associated with the TERMINATE statement becomes negative. This is a little less powerful than ACSL's technique since it could eventually happen that both the altitude h and the velocity v change their sign within the same integration step, and under those circumstances, the TERMINATE condition would not trigger. DARE–P does not allow the user to specify several termination conditions on separate TERMINATE statements. STP is not a DARE–P system function but can easily be created by use of the $F block in which arbitrary FORTRAN subprograms can be coded. (ACSL provides for the same mechanism by permitting the user to code her or his FORTRAN subprograms following the final END statement of the ACSL program.) The final time is here called *TMAX* rather than *tmx*, and is treated as a system constant, which is automatically tested. The communication interval is determined indirectly by specifying how many data points are to be stored during the simulation run. *NPOINT* = 301 does not give us a communication interval of 0.2 sec, but is the largest number allowed in DARE–P before some internal arrays overflow. (FORTRAN programming is so much fun!)

The model description is followed by a program section in which constants and initial conditions are specified. The final code section of any DARE–P program lists the output commands.

Notice that the segment separators in DARE–P are column sensitive. All $ block markers ($D1 and $F in our example) must be coded with the $ starting in column 2, and the three DARE–P END separators (before and after the constants and following the output declarations) must be placed in columns 1 to 3. (The DARE–P syntax is an excellent example of how *not* to design a language grammar, however, in the early 1970s when DARE–P was developed, software engineering was still in its infancy, and as a university product, DARE–P did not have the financial means behind it that would have permitted a constant upgrading of the product.)

Let us finally look at DESIRE [2.5]. The following DESIRE program is equivalent to the two previous programs.

```
- - - - - - - - - - - - - - - - - - - - - - - - - - - - - - - -
-- CONTINUOUS SYSTEM
-- Lunar Landing Maneuver
- - - - - - - - - - - - - - - - - - - - - - - - - - - - - - - -
-- Constants
r = 1738.0E+3  |  c2 = 4.925E+12  |  f1 = 36350.0
f2 = 1308.0  |  c11 = 0.000277  |  c12 = 0.000277
tdec = 43.2  |  tend = 210.0
-- Initial conditions
h = 59404.0  |  v = -2003.0  |  m = 1038.358
- - - - - - - - - - - - - - - - - - - - - - - - - - - - - - - -
TMAX = 230.0  |  DT = 0.1  |  NN = 1151
- - - - - - - - - - - - - - - - - - - - - - - - - - - - - - - -
-- auto scale
scale = 1
XCCC = 1
label TRY
drunr
if XCCC < 0 then XCCC = -XCCC
                 scale = 2 * scale
                 go to TRY
             else  proceed
- - - - - - - - - - - - - - - - - - - - - - - - - - - - - - - -
DYNAMIC
- - - - - - - - - - - - - - - - - - - - - - - - - - - - - - - -
thrust  = ((f1 - f2)*swtch(tdec - t) + f2)*swtch(tend - t)
c1      = ((c11 - c12)*swtch(tdec - t) + c12)*swtch(tend - t)
mdot    = -c1*abs(thrust)
g       = c2/(h + r) ∧ 2
a       = (1.0/m) * (thrust - m * g)
d/dt h  = v
d/dt v  = a
d/dt m  = mdot
- - - - - - - - - - - - - - - - - - - - - - - - - - - - - - - -
OUT
term -h
term v
hs = h * 1E-5  |  ms = m * 5E-4  |  vs = v * 5E-4
dispt hs, vs, ms
- - - - - - - - - - - - - - - - - - - - - - - - - - - - - - - -
/ - -
/PIC 'lunar.prc'
/ - -
```

While ACSL was designed for moderately sized industrial simulation problems and DARE–P was designed for easy learning (classroom environments), DESIRE has been designed for maximum interactivity and for ultrafast execution speed.

These goals have considerably influenced the software design. DESIRE is one of the few CSSLs that is not based upon a target language for improved portability since portability always goes at the expense of execution speed. The statements above the DYNAMIC declaration describe the *experiment* to be performed on the model, i.e., they constitute procedural code to be executed only once. These statements are coded in an interpreted enhanced BASIC and are therefore slow in execution. However, since this code is executed only once rather than constantly during the simulation run, speed is not the issue. The statements following the DYNAMIC declaration describe the *dynamic model*. They are microcompiled into threaded code when the simulation starts. A simulation run is performed whenever, during execution of the experiment section, a "drun" or "drunr" statement is met. Although the model description code is heavily optimized, the execution of the microcompiler is ultrafast. The preceding program compiles within less than 0.1 sec and thereafter, the program is ready to run. However, in order to keep the compilation time down, it was decided *not* to provide for an automated equation sorter, i.e., it is the user's responsibility to place the equations in an executable sequence.

DESIRE is modeled after the analog computers of the past. It provides for the ultimate in flexibility, interactivity, and responsiveness. The "dispt" statement, for instance, invokes a *run–time display*, i.e., the user can view the results of his or her simulation run as they develop. The price to be paid for this sportscar among the simulation languages is a certain reduction in programming comfort, program reliability (compile–time checks), and program robustness (run–time checks).

DESIRE is by far the most modern of the three languages. While a version exists that executes on VAX/VMS, the beauty of DESIRE lies really in its PC implementation. The execution speed of programs that consist of several hundred differential equations on a 486–class machine is breathtaking. This is due to the fact that the user doesn't share this powerful machine with anybody else. As the designer of the software, Granino Korn, always says, "Time–sharing is for the birds." DESIRE certainly goes with the trend of modern design software, which is away from mainframes onto ever–more-powerful engineering workstations such as the 486s, the Macintosh

II, and the MicroVAX and Sun workstations.

Other than that, the program listed earlier requires little additional explanation. DESIRE is *case–sensitive*, i.e., *TRY*, *Try*, and *try* are three perfectly good variable names, but they denote three different variables. This is important to remember in particular in the context of system variables (for example, the simulation clock is called *t* and not *T*). *NN* is a system variable that serves the same purpose as *NPOINT* in DARE–P. The integration step size *DT* must be specified since the default integration algorithm of DESIRE is a *fixed–step Heun algorithm* rather than the *variable–step fourth–order Runge–Kutta algorithm*, which is most commonly used as the default algorithm of CSSLs. *XCCC* (what a name!) is a system variable that is set negative by DESIRE when a run–time display variable exceeds the scaling bounds (defined by the system variable *scale*). Since DE-SIRE simulation runs are so fast, we can afford to simply repeat the entire simulation with a larger scale for the display, rather than bother to store the previous data and recover them from memory. "swtch" is a DESIRE system function that uses a slightly different syntax than ACSL's "step" function, but performs the same task. DESIRE uses a "dot" notation (as DARE–P), but a different syntax (d/dt).

DESIRE's OUT block is similar to ACSL's portion of the DY-NAMIC block which is outside the DERIVATIVE section. Statements in the OUT block are executed only once per communication interval rather than during every integration step. This is not important except for speeding up the execution of the simulation run. *hs*, *ms*, and *vs* are scaled variables. Scaling was necessary to make all three variables comparable in amplitude, since DESIRE's run–time display uses the same scaling factor for all dependent variables (again in order to speed up the execution). The "term" statement operates on a numerical condition (as in DARE–P), but this time, it triggers whenever the associated expression becomes positive. Moreover, DE-SIRE tolerates several separate "term" statements to be specified in a row.

Let us now discuss the results of this simulation study. Figure 2.3 shows the time trajectories of five of the simulation variables.

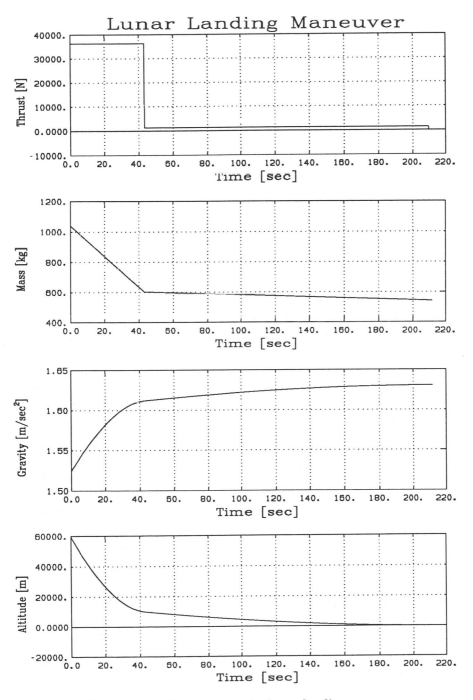

Figure 2.3. Results from the lunar landing maneuver.

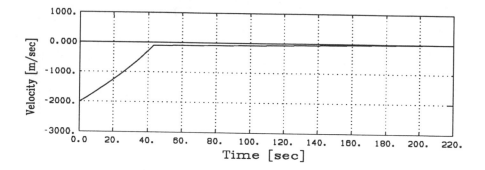

Figure 2.3. Results from the lunar landing maneuver (continued).

The main retro motor is fired at time zero, and remains active during the first 43 sec of the landing maneuver, slowing the landing module down from an initial vertical velocity of 1800 m/sec to about 150 m/sec. From then on, only the three vernier engines are used. At time 210 sec, the module has landed. (A soft landing requires a vertical velocity of less than 5 m/sec at impact.) During these 210 sec, the module has lost half of its initial weight due to fuel consumption.

The results shown on these graphs were produced by the ACSL program which, however, was executed under control of yet another program, called CTRL–C [2.9], a technique that we shall use frequently in this textbook. The graphs themselves were produced after the simulation run was completed by use of CTRL–C's graphic routines. A similar interface exists also between ACSL and Pro–MATLAB [2.7]. More and other simulation systems will be discussed later.

The main purpose of this section was to illustrate how similar the various CSSLs are to each other. Once we have mastered any one of them, we basically know them all. In this textbook, we shall mostly use ACSL for demonstration purposes since we believe it to be the most flexible, convenient, and widely used among the currently available CSSLs. However, for those who have another CSSL software installed in their computer, it should not prove overly difficult to transcribe the presented concepts in terms of their CSSL software.

2.8 Discontinuity Handling

Already the first simple example of a continuous system which we presented in this chapter showed the need to model *discontinuous systems*. During the course of the simulation, the thrust of the rocket changed abruptly twice. This was not an accident. Discrete changes within continuous systems are extremely common and any decent CSSL should offer means to code such discrete changes in a convenient and efficient way. Unfortunately, this is not the case. The need for such a language element had not been anticipated by the CSSL committee [2.1], and therefore, none of the subsequently developed CSSL systems offered such a feature. This is the drawback of too early a software standardization — it tends to freeze the state of the art and thereby hampers future developments. This issue had not been addressed to its full extent until 1979 [2.2,2.3]. Among the three systems presented in this chapter, only ACSL offers a decent discontinuity handling facility.

Some discrete changes are known in advance and can be scheduled to occur at a particular point in time. This type of discontinuity is called a *time-event*. The change in the rocket thrust belongs to this type of discontinuity. Time–events can be scheduled during the execution of the INITIAL block. At the time of occurrence, the DERIVATIVE block will be interrupted, and a special DISCRETE block is executed once. During the execution of the DISCRETE block, new time–events can be scheduled to happen at arbitrary time instances in the future. Once the DISCRETE block has been completed, the program uses its current state as a new "initial condition" and resumes execution of the DERIVATIVE block.

The DISCRETE block is used to describe a discrete change of one or several otherwise continuous variables. As the companion book on continuous system simulation explains, it is essential that the numerical integration is interrupted during discrete state changes since all numerical integration techniques are based upon polynomial extrapolations, which cannot properly be used to describe discontinuous functions.

Let me illustrate this concept by means of the previous example. A more elegant reformulation of the lunar landing maneuver in ACSL is now presented. This version of the code contains two time–events, one to shut down the main retro motor and one to shut down the vernier engines. The time of shut–down has been precomputed and the two events can therefore be scheduled from within the INITIAL block.

```
PROGRAM Lunar Landing Maneuver
  INITIAL
    constant ...
      r = 1738.0E3,  c2 = 4.925E12,  f1 = 36350.0, ...
      f2 = 1308.0,  c11 = 0.000277,  c12 = 0.000277, ...
      h0 = 59404.0,  v0 = −2003.0,  m0 = 1038.358, ...
      tmx = 230.0,  tdec = 43.2,  tend = 210.0
    cinterval cint = 0.2
    schedule shutlg .at. tdec
    schedule shutsm .at. tend
    ff = f1
    cc = c11
  END $ "of INITIAL"
  DYNAMIC
    DERIVATIVE
      thrust = ff
      c1     = cc
      h      = integ(v, h0)
      v      = integ(a, v0)
      a      = (1.0/m) * (thrust − m * g)
      m      = integ(mdot, m0)
      mdot   = −c1*abs(thrust)
      g      = c2/(h + r) * *2
    END $ "of DERIVATIVE"
    DISCRETE shutlg
      ff = f2
      cc = c12
    END $ "of DISCRETE shutlg"
    DISCRETE shutsm
      ff = 0.0
      cc = 0.0
    END $ "of DISCRETE shutsm"
    termt (t.ge.tmx .or. h.le.0.0 .or. v.gt.0.0)
  END $ "of DYNAMIC"
END $ "of PROGRAM"
```

At time *tdec*, the retro motor is shut down. This is being accomplished by the time–event *shutlg*. At time *tend*, the vernier engines are also shut down. This is being accomplished by the time–event *shutsm*. Notice that the two variables *ff* and *cc* must be assigned initial values. This solution is much more readable than the previous one. It also has numerical benefits as we shall see in the companion book on continuous system simulation.

Sometimes, time–events recur in constant time intervals. Although such events could be coded in exactly the same way by simply scheduling the next occurrence of the same event from within the DISCRETE block itself, this case is so common that ACSL offers a simplified way of scheduling recurring events. If a DISCRETE block contains an INTERVAL statement, it is implicitly scheduled to occur once every sampling interval. No "schedule" statement is necessary in that case. The following program shows how the set of difference equations:

$$x_1(k + 1) = (1 - a)x_1(k) + b\, x_1(k)x_2(k) \qquad (2.9a)$$
$$x_2(k + 1) = (1 + c)x_2(k) - d\, x_1(k)x_2(k) \qquad (2.9b)$$

can be coded in ACSL.

```
PROGRAM Difference Equations
INITIAL
  constant ...
    a = 0.07,  b = 0.02,  c = 0.05,  d = 0.02, ...
    tmx = 100.0
  cinterval cint = 1.0
  x1 = 2.5
  x2 = 1.7
END $ "of INITIAL"
DYNAMIC
  DISCRETE difrnc
    interval Ts = 1.0
    PROCEDURAL (x1, x2 =)
      x12 = x1 * x2
      x1new = (1.0 - a) * x1 + b * x12
      x2new = (1.0 + c) * x2 - d * x12
      x1 = x1new
      x2 = x2new
    END $ "of PROCEDURAL"
  END $ "of DISCRETE difrnc"
  termt (t.ge.tmx)
END $ "of DYNAMIC"
END $ "of PROGRAM"
```

The DISCRETE block *difrnc* is implicitly scheduled to occur once every *Ts* time unit. The communication interval *cint* is chosen to be identical with the sampling interval. The model is executed over 100 iterations of the set of difference equations. The PROCEDURAL block is ACSL's equivalent of DARE–P's PROCED block, which we met before. It is needed here to prevent the equation sorter from detecting algebraic loops between *x1* and *x1new*, and between *x2* and *x2new*.

Sometimes, discontinuities depend on the model states rather than on time. Such discontinuities are therefore called *state–events*. For example, we could reformulate our lunar lander example such that the main retro motor is shut down when the landing module has reached an altitude of 9934 m rather than at time 43.2 sec. This may, in fact, be a more robust formulation of our problem since it may be less sensitive to disturbances such as a somewhat inaccurately guessed initial mass value.

State–events cannot be scheduled ahead of time. ACSL "schedules" state–events from within the DERIVATIVE section of the model whenever a specified *state condition* is met. The following code shows the reformulated lunar landing maneuver.

```
PROGRAM Lunar Landing Maneuver
INITIAL
  constant ...
    r = 1738.0E3,  c2 = 4.925E12,  f1 = 36350.0, ...
    f2 = 1308.0,  c11 = 0.000277,  c12 = 0.000277, ...
    h0 = 59404.0,  v0 = −2003.0,  m0 = 1038.358, ...
    tmx = 230.0
  cinterval cint = 0.2
  ff = f1
  cc = c11
END $ "of INITIAL"
DYNAMIC
  DERIVATIVE
    thrust = ff
    c1    = cc
    h     = integ(v, h0)
    v     = integ(a, v0)
    a     = (1.0/m) * (thrust − m * g)
    m     = integ(mdot, m0)
    mdot  = −c1*abs(thrust)
    g     = c2/(h + r) * *2
    schedule shutlg .xp. 9934.0 − h
    schedule shutsm .xp. 15.0 − h
  END $ "of DERIVATIVE"
  DISCRETE shutlg
    ff = f2
    cc = c12
  END $ "of DISCRETE shutlg"
  DISCRETE shutsm
    ff = 0.0
    cc = 0.0
  END $ "of DISCRETE shutsm"
  termt (t.ge.tmx .or. h.le.0.0 .or. v.gt.0.0)
  END $ "of DYNAMIC"
END $ "of PROGRAM"
```

In ACSL, state–events are fired whenever the specified state condition crosses zero in the positive direction (using the ".xp." operator), in the negative direction (using the ".xn." operator), or in an arbitrary direction (using the ".xz." operator).

2.9 Model Validation

Validating a given model for a particular purpose is a very difficult issue that we are not yet ready to deal with in general. However, one particular validation technique exists that I would like to discuss at this point, namely, the issue of *dimensional consistency checking*.

The concept of dimensional consistency is a trivial one. In any equation, all terms must carry the same units. For example, in the equation:

$$a = \frac{1}{m}(thrust - m \cdot g) \tag{2.10}$$

the thrust is a force which is expressed in N or kg m sec^{-2}. We often shall write this as $thrust[N]$ or $[thrust] = N$. The square brackets denote "units of." Consequently, the term $m \cdot g$ must also be expressed in the same units, which checks out correctly since $[m] = $ kg, and g is an acceleration, and therefore $[g] = $ m sec^{-2}. Finally, we can check the dimensional consistency across the equal sign: $[a] = $ m sec^{-2} which indeed is the same as the dimension of a force divided by the dimension of a mass, since N kg$^{-1} \equiv$ m sec^{-2}. While this principle is surely trivial, I find that one of the most common errors in my students' papers is a lack of dimensional consistency across equations. The principle is obviously so trivial that the students simply don't bother to check it.

While it would be perfectly feasible to design simulation software that checks the dimensional consistency for us, this is not done in the currently used simulation languages for the following reason. Let us look at the simplest stable state–space equation:

$$\dot{x} = -x \tag{2.11}$$

Physically, this equation does not make sense since:

$$[\dot{x}] = [\frac{dx}{dt}] = \frac{[x]}{[t]} = [x] \cdot \sec^{-1} \tag{2.12}$$

i.e., different units are on the left and right of the equal sign. A correct differential equation would be:

$$\dot{x} = a \cdot x \qquad\qquad (2.13)$$

where the parameter (or constant) a assumes a value of $a = -1.0 \ \text{sec}^{-1}$.

If the simulation software would check the dimensional consistency for us, this would force us to code the preceding differential equation either using a constant or a parameter properly declared with its units, rather than being able to plug in a number directly. Eventually, it would be useful to have a modeling tool that can check dimensional consistency upon request. In any event, I strongly urge any potential modeler to perform this consistency check regularly.

2.10 Summary

We have introduced the basic syntax of a few current CSSLs, and discussed the major problems in formulating continuous system models, such as the algebraic loop problem. By now, the student should be able to code simple problems in any current CSSL. Please notice that it is *not* the aim of this text to provide the student with the details of any particular language syntax. For that purpose, the textbook should be accompanied by language reference manuals for the chosen simulation languages. We feel that a student should be perfectly capable of studying such a language reference manual on her or his own, and that it is much more important to provide insight into the basic principles of modeling and simulation mechanisms.

References

[2.1] Donald C. Augustin, Mark S. Fineberg, Bruce B. Johnson, Robert N. Linebarger, F. John Sansom, and Jon C. Strauss (1967), "The SCi Continuous System Simulation Language (CSSL)," *Simulation*, **9**, pp. 281–303.

[2.2] François E. Cellier (1979), *Combined Continuous/Discrete System Simulation by Use of Digital Computers: Techniques and Tools*, Ph.D. Dissertation, Diss ETH No 6483, ETH Zürich, CH–8092 Zürich, Switzerland.

[2.3] François E. Cellier (1986), "Combined Continuous/Discrete Simulation — Applications, Techniques and Tools," *Proceedings 1986 Winter Simulation Conference*, Washington, D.C., pp. 24–33.

[2.4] IBM Canada Ltd. (1972), *Continuous System Modeling Program III (CSMP-III) — Program Reference Manual*, Program Number: 5734–XS9, Form: SH19–7001–2, IBM Canada Ltd., Program Produce Centre, Don Mills, Ontario, Canada.

[2.5] Granino A. Korn (1989), *Interactive Dynamic-System Simulation*, McGraw–Hill, New York.

[2.6] Granino A. Korn and John V. Wait (1978), *Digital Continuous-System Simulation*, Prentice–Hall, Englewood Cliffs, N.J.

[2.7] Mathworks, Inc. (1987), *Pro-MATLAB with System Identification Toolbox and Control System Toolbox — User Manual*, South Natick, Mass.

[2.8] Edward E. L. Mitchell and Joseph S. Gauthier (1986), *ACSL: Advanced Continuous Simulation Language — User Guide and Reference Manual*, Mitchell & Gauthier Assoc., Concord, Mass.

[2.9] Systems Control Technology, Inc. (1985), *CTRL-C, A Language for the Computer-Aided Design of Multivariable Control Systems, User's Guide*, Palo Alto, Calif.

[2.10] John V. Wait and DeFrance Clarke III (1976), *DARE-P User's Manual*, Version 4.1, Dept. of Electrical & Computer Engineering, University of Arizona, Tucson, Ariz.

Bibliography

[B2.1] YaoHan Chu (1969), *Digital Simulation of Continuous Systems*, McGraw–Hill, New York.

[B2.2] Charles M. Close and Dean K. Frederick (1978), *Modeling and Analysis of Dynamic Systems*, Houghton Mifflin, Boston, Mass.

[B2.3] Wolfgang K. Giloi (1975), *Principles of Continuous System Simulation*, Teubner Verlag, Stuttgart, FRG.

[B2.4] Granino A. Korn and John V. Wait (1978), *Digital Continuous-System Simulation*, Prentice–Hall, Englewood Cliffs, N.J.

Homework Problems

[H2.1] Limit Cycle

The following two equations describe a so–called limit cycle:

$$\dot{x} = +y + \frac{k \cdot x \cdot (1 - x^2 - y^2)}{\sqrt{x^2 + y^2}} \qquad (H2.1a)$$

$$\dot{y} = -x + \frac{k \cdot y \cdot (1 - x^2 - y^2)}{\sqrt{x^2 + y^2}} \qquad (H2.1b)$$

The initial conditions are given as:

$$x(t = 0) = x0, \quad y(t = 0) = y0 \qquad (H2.1c)$$

Simulate this system over a period of 10 sec with the following four sets of initial conditions:

$$
\begin{aligned}
&x0 = 0.4, \quad y0 = 1.0 && (H2.1d) \\
&x0 = 0.0, \quad y0 = 0.5 && (H2.1e) \\
&x0 = 0.0, \quad y0 = 0.001 && (H2.1f) \\
&x0 = 1.0, \quad y0 = 1.5 && (H2.1g)
\end{aligned}
$$

The constant k takes a value of $k = 2.0$. Plot all four trajectories in the $[x, y]$–plane on the same graph. An example of what such a graph may look like is given in Fig.8.18 (for a different limit cycle though). For this purpose, I recommend using ACSL with either Pro–MATLAB or CTRL–C.

[H2.2] Cannon Ball [2.6]

A famous cannon ball used by the United States forces during the Independence War can be described by the following set of differential equations:

$$\ddot{x} = -R \cdot v \cdot \dot{x} \qquad (H2.2a)$$

$$\ddot{y} = -R \cdot v \cdot \dot{y} - g \qquad (H2.2b)$$

where x denotes the horizontal position, y denotes the vertical position, and v denotes the absolute velocity of the cannon ball. The following constants are needed in the model: $v_0 = 900$ ft/sec is the initial velocity, $R = 7.5 \times 10^{-5}$ ft^{-1} is the air friction constant, and $g = 32.2$ ft/sec^2 is the gravity constant.

The cannon was conquered by the Redcoats when they temporarily took Philadelphia in the late fall of 1777. This, of course, created a major security breach. As a consequence, the model was officially declassified by Congress in 1910.

Simulate the system for various shooting angles (10°, 15°, ... 65°). Each simulation is to be carried out for the duration needed until the cannon ball hits the ground again. Plot all trajectories as functions of time on one sheet of paper.

[H2.3] Hysteresis

Use two differential equations to generate the sine wave function

$$x(t) = 2 * \sin(t) \qquad (H2.3a)$$

Use two state–events to code a vertical hysteresis as shown in Fig.H2.3. Use ACSL's *LOGD* subroutine to record the corners of the discontinuity immediately before and after the event. Choose $a = 1.0$, $b = -1.0$, $p = 1.0$, and $n = -1.0$. Simulate the system during 10 sec, and plot $y(x)$. The resulting curve should look similar to Fig.H2.3.

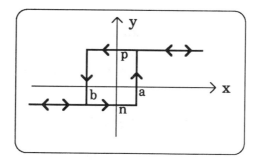

Figure H2.3. Vertical hysteresis function.

[H2.4] Difference Equations

Recode the lunar landing maneuver (the version with scheduled time–events) into a set of difference equations using the forward Euler integration algorithm presented in this chapter with a fixed step size of 0.1 sec.

Code the resulting set of nonlinear difference equations in ACSL and simulate. Compare the results with those presented in the chapter.

[H2.5] Bouncing Ball

A ball is dropped from an altitude of 2 m. No initial velocity is applied. The ball falls simply due to the gravitational force. When the ball hits the floor, it bounces. It is assumed that the coefficient of restitution is $k = 0.7$, i.e., the velocity immediately after the impact is exactly 70% of what it was just before the impact, but with reversed sign.

Simulate the system over 10 bounces and plot the altitude of the ball versus time. Recompute the communication interval during each bouncing event such that 10 communication intervals lie between any two subsequent bounces (this problem is so simple that the time of the next bounce can be computed analytically once the new initial velocity is known). Use ACSL's *LOGD* function to record the times of impact.

[H2.6] Model Validation

Determine a set of consistent measurement units for all constants of the lunar lander model. Verify the dimensional consistency across all of the equations contained in the lunar lander model.

Projects

[P2.1] Language Quality Assessment

Read the user manuals of ACSL [2.8], DARE–P [2.10], and DESIRE [2.5] carefully and write down a list of language features for each of them. Compile a table with three columns, one for each language, and many rows, one for each feature and evaluate the availability and implementation quality of each feature in each of the three languages by assigning a value between 0 (feature not available) and 2 (feature nicely implemented) to it. Add the accumulated figures of merit for each column and assess the overall quality of the three products.

Research

[R2.1] Model Validation

Compile a list of language features which could be implemented in a simulation language in order to partially automate the process of model validation. One of these features is the dimensional consistency check that was proposed in this chapter. Implement these features by means of compiler switches in a simulation language for which you have access to the source code of the compiler.

3

Principles of Passive Electrical Circuit Modeling

Preview

In this chapter, we shall discuss issues relating to the modeling of simple passive electrical circuits consisting of sources, resistors, capacitors, and inductors only. The traditional approach to this type of system is through either mesh equations or node equations. However, the resulting models are not in a state–space form and they cannot easily be converted into a state–space form thereafter. We shall also discuss another technique that allows us to derive a state–space model directly and we shall see why this approach is not commonly used. Very often, the resulting equations contain either algebraic loops or structural singularities.

3.1 Introduction

A good selection of textbooks deal with passive electrical circuits and simulations thereof [3.1,3.2,3.4,3.5]. The most commonly used modeling principles are to express the circuit equations either through a special selection of *mesh equations* (expressed in terms of so–called loop currents using Kirchhoff's voltage law) or through a special selection of *node equations* (expressed in terms of so–called cutset voltages using Kirchhoff's current law). Let me explain the basic idea behind these two methods by means of the example shown in Fig.3.1.

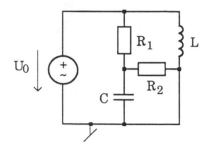

Figure 3.1. Example of a passive circuit.

We have two elements that can store energy (the capacitor C and the inductor L), and we thus expect to obtain two state equations in the end.

3.2 Mesh Equations

Let me first discuss how the loop current approach (mesh equations, Kirchhoff's voltage law) can be used to generate a mathematical model for this circuit. Figure 3.2 shows the same circuit after the circuit has been "colored" by introducing a "tree."

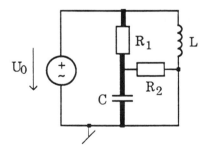

Figure 3.2. Passive circuit after selection of a tree.

The "tree_branches" of the tree are those branches of the circuit that have been marked by bold lines (i.e., the branches containing the resistor R_1 and the capacitor C). Let me define what a tree is.

A tree consists of a set of *connected tree_branches* such that the tree_branches alone don't form closed loops and such that any addition of another branch to the tree would create a closed loop consisting of tree-branches only. The remaining branches of the circuit structure are called the *links* of the circuit.

A considerable freedom exists in the selection of tree-branches. However, some rules must be observed.

(1) Mesh equations cannot tolerate any independent current sources. Node equations equation; cannot tolerate any independent voltage sources. If the circuit contains the wrong type of sources, they must be converted to equivalent sources of the other type. Figure 3.3 shows the conversion of independent sources.

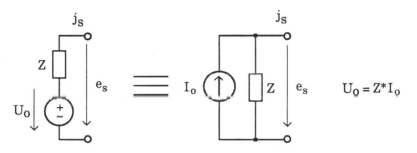

Figure 3.3. Conversion of independent sources.

Furthermore, if the "wrong" sources are ideal sources (i.e., they have zero impedance associated with them), they must first be moved into other branches until the problem disappears.

(2) In the case of mesh equations, all voltage sources should be placed in *links*. In the case of node equations, all current sources should be placed in *branches*. In this way, they will appear only once in the resulting set of equations.

Using the following notation:

$$
\begin{aligned}
n_n &::= \text{number of circuit nodes} \\
n_b &::= \text{number of circuit branches} \\
n_l &::= \text{number of links} \\
n_{tb} &::= \text{number of tree-branches} \\
\sigma_i &::= \text{number of ideal current sources} \\
\sigma_u &::= \text{number of ideal voltage sources} \\
n_{ej} &::= \text{number of mesh equations} \\
n_{ee} &::= \text{number of node equations}
\end{aligned}
$$

we can compute the number of tree-branches n_{tb} and the number of links n_l as follows:

$$n_{tb} = n_n - 1 \qquad (3.1a)$$
$$n_l = n_b - n_{tb} \qquad (3.1b)$$

and therefore, we can compute the number of equations that are needed for the two methods as:

$$n_{ee} = n_{tb} - \sigma_u \qquad (3.2a)$$
$$n_{ej} = n_l - \sigma_i \qquad (3.2b)$$

We usually select the technique that lets us get away with the smaller number of equations.

In our example, we have an ideal independent voltage source, thus mesh equations may be more convenient, i.e., we operate on Kirchhoff's voltage law rather than using Kirchhoff's current law.

It is useful to replace all passive circuit elements by impedances as shown in Fig.3.4 (i.e., we convert the circuit from the time domain to the frequency domain).

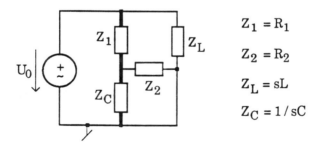

$$Z_1 = R_1$$
$$Z_2 = R_2$$
$$Z_L = sL$$
$$Z_C = 1/sC$$

Figure 3.4. Frequency–domain representation using impedances.

Now, we introduce so–called *loop currents*, one for each link of the circuit. A *loop* is a generalized mesh. Except for the one link that it represents, it consists of tree_branches only. Figure 3.5 depicts the three loops of our circuit. Once the tree has been selected, the loops are fully determined. Notice that the short–circuit at the lower right corner of Fig.3.5 is drawn for convenience only and does not qualify as a link. The loop currents j_1, j_2, and j_3 are identical to the link currents i_1, i_2, and i_3 of Fig.3.6. The tree_branch currents i_4 and i_5 are the directed sums of the loop currents that traverse the two tree_branches.

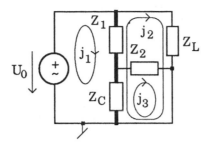

Figure 3.5. Circuit with tree and loop currents.

$$U_0 = Z_1 * (j_1 - j_2) + Z_C * (j_1 - j_2 - j_3) \tag{3.3a}$$
$$0 = Z_L * j_2 + Z_1 * (j_2 - j_1) + Z_C * (j_2 + j_3 - j_1) \tag{3.3b}$$
$$0 = Z_2 * j_3 + Z_C * (j_3 + j_2 - j_1) \tag{3.3c}$$

The terms can be reordered as follows:

$$U_0 = (Z_1 + Z_C) * j_1 - (Z_1 + Z_C) * j_2 - Z_C * j_3 \tag{3.4a}$$
$$0 = -(Z_1 + Z_C) * j_1 + (Z_1 + Z_C + Z_L) * j_2 + Z_C * j_3 \tag{3.4b}$$
$$0 = -Z_C * j_1 + Z_C * j_2 + (Z_2 + Z_C) * j_3 \tag{3.4c}$$

which can be expressed using a matrix notation as:

$$\begin{pmatrix} U_0 \\ 0 \\ 0 \end{pmatrix} = \begin{pmatrix} Z_1 + Z_C & -(Z_1 + Z_C) & -Z_C \\ -(Z_1 + Z_C) & Z_1 + Z_C + Z_L & Z_C \\ -Z_C & Z_C & Z_2 + Z_C \end{pmatrix} \cdot \begin{pmatrix} j_1 \\ j_2 \\ j_3 \end{pmatrix} \tag{3.5}$$

which can be abbreviated as:

$$\mathbf{e}_\sigma = \mathbf{Z}_m * \mathbf{j}_l \tag{3.6}$$

$\mathbf{e}_\sigma$ denotes the *source voltage vector*, $\mathbf{Z}_m$ denotes the *mesh impedance matrix*, and $\mathbf{j}_l$ denotes the *loop current vector*.

Somewhat more systematically, we can achieve the same result by starting off with two other matrices, namely, the mesh–incidence matrix and the branch–impedance matrix. The *mesh–incidence matrix* $\boldsymbol{\Phi}$, which in some texts is also called the *fundamental loop matrix*, is defined as a matrix that describes the circuit topology by coding the direction of the loop currents in the branches. Figure 3.6 illustrates the procedure.

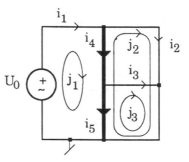

Figure 3.6. Circuit topology used for the mesh–incidence matrix.

This allows us to generate the following mesh–incidence matrix:

$$\boldsymbol{\Phi} = \begin{matrix} & \begin{matrix} i1 & i2 & i3 & i4 & i5 \end{matrix} \\ \begin{matrix} j1 \\ j2 \\ j3 \end{matrix} & \begin{pmatrix} 1 & 0 & 0 & 1 & 1 \\ 0 & 1 & 0 & -1 & -1 \\ 0 & 0 & 1 & 0 & -1 \end{pmatrix} \end{matrix} \tag{3.7}$$

which contains $+1$ entries where the direction of a loop current corresponds with the direction of the branch current, it contains -1 entries where the loop current and the branch current have opposite directions, and it contains 0 entries for branches in which the loop current is not present.

The *branch–impedance matrix* $\mathbf{Z}_b$ is defined as a diagonal matrix containing the individual branch impedances along the main diagonal:

$$\mathbf{Z}_b = \begin{matrix} & \begin{matrix} i1 & i2 & i3 & i4 & i5 \end{matrix} \\ \begin{matrix} i1 \\ i2 \\ i3 \\ i4 \\ i5 \end{matrix} & \begin{pmatrix} 0 & 0 & 0 & 0 & 0 \\ 0 & sL & 0 & 0 & 0 \\ 0 & 0 & R_2 & 0 & 0 \\ 0 & 0 & 0 & R_1 & 0 \\ 0 & 0 & 0 & 0 & 1/sC \end{pmatrix} \end{matrix} \tag{3.8}$$

which we sometimes abbreviate as:

$$\mathbf{Z}_b = \mathrm{diag}(0, sL, R_2, R_1, 1/sC) \tag{3.9}$$

We can now write all equations in a compact matrix form. Let us start with Kirchhoff's voltage law:

$$\boldsymbol{\Phi} \cdot \mathbf{u}_b = 0 \qquad (3.10)$$

where $\mathbf{u}_b$ denotes the vector of voltages across each of the circuit branches. This can then be expressed as:

$$\mathbf{u}_b = \mathbf{Z}_b \cdot \mathbf{i}_b + \mathbf{u}_\sigma \qquad (3.11)$$

where $\mathbf{i}_b$ denotes the vector of currents through each of the circuit branches and $\mathbf{u}_\sigma$ denotes the vector of voltage sources in the circuit branches. We can now transform the vector of branch currents into the vector of loop currents as follows:

$$\mathbf{i}_b = \boldsymbol{\Phi}^T \cdot \mathbf{j}_l \qquad (3.12)$$

Plugging the last three equations into each other, we find:

$$\boldsymbol{\Phi} \cdot \mathbf{Z}_b \cdot \boldsymbol{\Phi}^T \cdot \mathbf{j}_l = -\boldsymbol{\Phi} \cdot \mathbf{u}_\sigma \qquad (3.13)$$

A comparison to Eq.(3.6) yields:

$$\mathbf{Z}_m = \boldsymbol{\Phi} \cdot \mathbf{Z}_b \cdot \boldsymbol{\Phi}^T \qquad (3.14a)$$

$$\mathbf{e}_\sigma = -\boldsymbol{\Phi} \cdot \mathbf{u}_\sigma \qquad (3.14b)$$

We can now evaluate all loop currents at once by computing:

$$\mathbf{j}_l = \mathbf{Z}_m^{-1} \cdot \mathbf{e}_\sigma \qquad (3.15)$$

which we shall often abbreviate as:

$$\mathbf{j}_l = \mathbf{Z}_m \backslash \mathbf{e}_\sigma \qquad (3.16)$$

using the slash operator ("/") to denote matrix division from the right and the backslash operator ("\") to denote matrix division from the left. This is the notation used in MATLAB and CTRL–C. We can then immediately find all branch currents using Eq.(3.12) and finally we can find all branch voltages using Eq.(3.11). Notice, however, that the evaluation of Eq.(3.16) is more tricky than it seems at first sight since it involves the symbolic inversion of a polynomial matrix. Neither MATLAB nor CTRL–C can handle this type of matrix inversion.

3.3 Node Equations

Let me next discuss the alternative approach using node equations and Kirchhoff's current law. Since we now have a source of the "wrong" type, we first need to convert the circuit. Figure 3.7 shows how this is done.

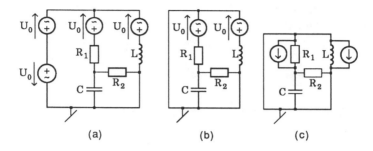

Figure 3.7. Conversion of the voltage source.

Since the "wrong" source is ideal, we start by moving the source into other branches. This is easily accomplished by compensating the source with an equivalent source of reverse polarity as shown in Figs.3.7a–b. Figure 3.7b is equivalent to the original circuit in every respect except for the potential at the additional top node. Now, we can convert the voltage sources to equivalent current sources as shown in Fig.3.7c. This circuit is again equivalent to the previous ones *except for the internal characteristics of the sources*. Consequently, the voltage across and the current through the inductor L and the resistor R_1 are no longer the same as before. In fact, the inductor has been short–circuited altogether. Since these "modifications" affect about half of our original circuit, this approach may not be sensible for the given problem. However, if we wish to determine the voltage across the capacitor only, this approach works perfectly well.

Instead of continuing with this example, let me demonstrate this technique by means of a slightly different example. Figure 3.8 shows another passive circuit.

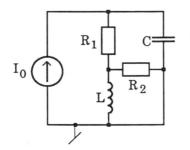

Figure 3.8. Another passive circuit.

Figure 3.9 demonstrates the steps needed to prepare the circuit for the formulation of node equations using Kirchhoff's current law.

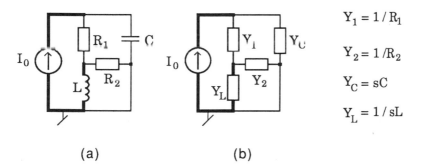

$$Y_1 = 1 / R_1$$

$$Y_2 = 1 / R_2$$

$$Y_C = sC$$

$$Y_L = 1 / sL$$

(a)   **(b)**

Figure 3.9. Preparation of the circuit for node equations.

Figure 3.9a shows the selection of the tree, which now should contain the current source. Every node of the circuit must be reached by the tree. It is usually a good idea to build the tree as a star with the center at the ground node (reference node). For this purpose, it is often necessary to introduce additional fictitious tree-branches (tree-branches with zero admittance). Figure 3.9b shows the conversion of the circuit from the time domain to the frequency domain, now using *admittances* rather than *impedances*.

Then we introduce so–called *cutset potentials*, one for each tree-branch of the circuit. A *cutset* is a generalized node. Except for the one tree-branch that it represents, it cuts through links only. Figure 3.10a depicts the two cutsets of our circuit. Once the tree has been selected, the cutsets are fully determined. The cutset potentials

e_1 and e_2 are identical to the node potentials at the nodes in which the tree_branches end. If every tree_branch connects one node of the circuit with the reference node (as in our example), the cutset potentials are also identical to the voltages across the tree_branches u_1 and u_2 of Fig.3.10b. The link voltages u_3, u_4, and u_5 are the directed sums of the cutset potentials that cut through the three links.

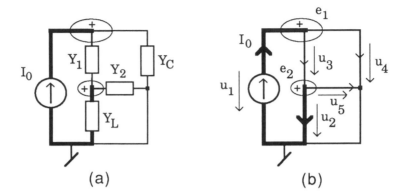

Figure 3.10. Introducing cutset voltages.

Figure 3.10a shows the introduction of cutset voltages and their polarities. Figure 3.10b places direction conventions on all branch voltages.

Using the circuit as shown in Fig.3.10a, we can immediately proceed to generate circuit equations by applying Kirchhoff's current law to every cutset of the tree:

$$I_0 = Y_1 * (e_1 - e_2) + Y_C * e_1 \qquad (3.17a)$$
$$0 = Y_L * e_2 + Y_1 * (e_2 - e_1) + Y_2 * e_2 \qquad (3.17b)$$

which can be reordered as:

$$I_0 = (Y_1 + Y_C) * e_1 - Y_1 * e_2 \qquad (3.18a)$$
$$0 = -Y_1 * e_1 + (Y_1 + Y_2 + Y_L) * e_2 \qquad (3.18b)$$

This can further be written in a matrix notation as:

$$\begin{pmatrix} I_0 \\ 0 \end{pmatrix} = \begin{pmatrix} Y_1 + Y_C & -Y_1 \\ -Y_1 & Y_1 + Y_2 + Y_L \end{pmatrix} \cdot \begin{pmatrix} e_1 \\ e_2 \end{pmatrix} \qquad (3.19)$$

which can be abbreviated as:

$$\mathbf{j}_\sigma = \mathbf{Y}_n * \mathbf{e}_l \tag{3.20}$$

where $\mathbf{j}_\sigma$ denotes the *source current vector*, $\mathbf{Y}_n$ denotes the *node admittance matrix*, and $\mathbf{e}_l$ denotes the *cutset potential vector*.

As before, we can achieve the same result more systematically by starting off with two other matrices, namely, the node–incidence matrix and the branch–admittance matrix. The *node–incidence matrix* $\mathbf{\Psi}$, which is sometimes also called the *fundamental cutset matrix*, is defined as a matrix that describes the circuit topology by recording the direction of the cutset voltages relative to the direction of the branch voltages. This procedure is illustrated in Fig.3.10b, which allows us to generate the following node–incidence matrix:

$$\mathbf{\Psi} = \begin{array}{c} \\ e1 \\ e2 \end{array} \begin{array}{cccccc} u1 & u2 & u3 & u4 & u5 \\ \left(\begin{array}{ccccc} 1 & 0 & 1 & 1 & 0 \\ 0 & 1 & -1 & 0 & 1 \end{array} \right) \end{array} \tag{3.21}$$

The node–incidence matrix contains $+1$ entries where the direction of a cutset voltage corresponds with the direction of the branch voltage, it contains -1 elements where the cutset voltage and the branch voltage have opposite directions and it contains 0 entries for branches in which the cutset is not present.

The *branch–admittance matrix* $\mathbf{Y}_b$ is defined as a diagonal matrix containing the individual branch admittances along the main diagonal:

$$\mathbf{Y}_b = \mathrm{diag}(0, 1/sL, 1/R_1, sC, 1/R_2) \tag{3.22}$$

We can again write all equations in a compact matrix form. Let us start with Kirchhoff's current law:

$$\mathbf{\Psi} \cdot \mathbf{i}_b = 0 \tag{3.23}$$

where $\mathbf{i}_b$ denotes the vector of currents through each of the circuit branches. This can then be expressed as:

$$\mathbf{i}_b = \mathbf{Y}_b \cdot \mathbf{u}_b + \mathbf{i}_\sigma \tag{3.24}$$

where $\mathbf{u}_b$ denotes the vector of voltages across each of the circuit branches and $\mathbf{i}_\sigma$ denotes the vector of current sources in the circuit branches. We can now transform the vector of branch voltages into the vector of cutset potentials as follows:

$$\mathbf{u}_b = \mathbf{\Psi}^T \cdot \mathbf{e}_l \tag{3.25}$$

Plugging the last three equations into each other, we find:

$$\mathbf{\Psi} \cdot \mathbf{Y}_b \cdot \mathbf{\Psi}^T \cdot \mathbf{e}_l = -\mathbf{\Psi} \cdot \mathbf{i}_\sigma \tag{3.26}$$

A comparison to Eq.(3.20) yields:

$$\mathbf{Y}_n = \mathbf{\Psi} \cdot \mathbf{Y}_b \cdot \mathbf{\Psi}^T \tag{3.27a}$$

$$\mathbf{j}_\sigma = -\mathbf{\Psi} \cdot \mathbf{i}_\sigma \tag{3.27b}$$

We can now evaluate all cutset potentials at once by computing:

$$\mathbf{e}_l = \mathbf{Y}_n \backslash \mathbf{j}_\sigma \tag{3.28}$$

We can then immediately find all branch voltages using Eq.(3.25) and finally we can find all branch currents using Eq.(3.24). Notice, however, that the evaluation of Eq.(3.28) is more tricky than it seems since it again involves the symbolic inversion of a polynomial matrix.

3.4 Disadvantages of Mesh and Node Equations

We have not yet answered the question how these techniques can help us to derive a set of first–order differential equations, i.e., our *state–space model*. Let us return once more to the original circuit example and the set of equations as formulated in Eqs.(3.3a–c). In order to derive a state–space description, we need to transform these equations back to the time domain:

$$U_0 = R_1(j_1 - j_2) + \frac{1}{C} \int_0^t (j_1 - j_2 - j_3)d\tau \tag{3.29a}$$

$$0 = L\frac{dj_2}{dt} + R_1(j_2 - j_1) + \frac{1}{C} \int_0^t (j_2 + j_3 - j_1)d\tau \tag{3.29b}$$

$$0 = R_2 j_3 + \frac{1}{C} \int_0^t (j_3 + j_2 - j_1)d\tau \tag{3.29c}$$

In order to obtain state equations, we need to get rid of the integral terms. This can be achieved by differentiating all three equations once:

$$\frac{dU_0}{dt} = R_1\left(\frac{dj_1}{dt} - \frac{dj_2}{dt}\right) + \frac{1}{C}(j_1 - j_2 - j_3) \tag{3.30a}$$

$$0 = L\frac{d^2 j_2}{dt^2} + R_1\left(\frac{dj_2}{dt} - \frac{dj_1}{dt}\right) + \frac{1}{C}(j_2 + j_3 - j_1) \tag{3.30b}$$

$$0 = R_2\frac{dj_3}{dt} + \frac{1}{C}(j_3 + j_2 - j_1) \tag{3.30c}$$

We realize that several bad things have happened.

(1) In our equations, the term dU_0/dt suddenly appears. This is a derivative of an input signal. We certainly don't want to operate on such a signal and yet we shall have a hard time getting rid of it again

(2) A second–derivative term appeared in our equations, which does not fit into our state–space description and which needs to be reduced to two first–order terms (by means of a technique that we shall discuss in Chapter 5).

(3) These equations seem to describe a *fourth–order system,* while we know that the order of our system can certainly not be higher than two. Consequently, linear dependencies must exist between some of the derivative terms in these equations.

This discussion clearly demonstrates that circuit equations are not geared toward the generation of a state–space model. We can draw two possible conclusions:

(1) The demonstrated methodology is not adequate to generate a state–space model, and thus, we need to develop a different modeling methodology that will allow us to generate the requested state–space model directly or

(2) State–space models are not the right approach to describe electrical circuits, and thus, we need to develop a different simulation methodology that will allow us to perform simulation runs using the previously generated circuit equations directly.

Both arguments have their pros and cons, and thus, we shall proceed along both avenues.

3.5 State–Space Models

Let me begin with the first alternative. A good technique to find a set of state equations directly is the following. We start by introducing variables for every single current and voltage in the circuit. This is shown in Fig.3.11 for our original circuit example.

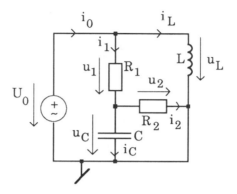

Figure 3.11. Passive circuit with all variables named.

In our example, we have introduced the following nine unknowns: u_1, u_2, u_C, u_L, i_1, i_2, i_C, i_L, and i_0.

We can now go ahead and write four branch equations and five mesh and/or node equations. In general, we add linearly independent equations until we have as many equations as we have variables in the circuit. For our example, the following set of equations is found:

$$u_1 = R_1 * i_1 \tag{3.31a}$$

$$u_2 = R_2 * i_2 \tag{3.31b}$$

$$u_L = L * \frac{di_L}{dt} \tag{3.31c}$$

$$i_C = C * \frac{du_C}{dt} \tag{3.31d}$$

$$U_0 = u_1 + u_C \tag{3.31e}$$

$$u_C = u_2 \tag{3.31f}$$

$$u_L = u_1 + u_2 \tag{3.31g}$$

$$i_0 = i_1 + i_L \tag{3.31h}$$

$$i_1 = i_2 + i_C \tag{3.31i}$$

Now, we need to solve these equations for the appropriate variables. We start by remembering that our goal is to derive a set of first–order differential equations. Therefore, we solve Eqs.(3.31c–d) for the derivative terms. Consequently, our two state variables will be the current through the inductor i_L and the voltage across the capacitor u_C. We can mark these two variables "solved" in our list of variables by crossing them out from the list of unknowns. We place the two derivative terms in Eqs.(3.31c–d) in square brackets (meaning that we want to solve for these variables) and we underline the two state variables in all other equations wherever they occur.

Now we can proceed with either of two philosophies:

(1) We can look for equations that have only one unknown left. We need to solve that equation for this one unknown or we won't use the equation at all.

(2) We can look for variables that occur in one equation only. We must use that equation to evaluate the unknown, otherwise we won't evaluate that variable at all.

With each unknown found, we proceed the same way. We cross it out from the list of unknowns, place it in square brackets in the equation that we plan to use for its evaluation, and underline it in all other equations. (Of course, all input variables, such as U_0 in our example, are known right away, and can thus be underlined in all equations from the beginning.) We proceed until all variables have been crossed out and all equations have been used up. In our example, this algorithm leads to the following *unique* solution:

$$u_1 = R_1 * [i_1] \tag{3.32a}$$

$$u_2 = R_2 * [i_2] \tag{3.32b}$$

$$u_L = L * [\frac{di_L}{dt}] \tag{3.32c}$$

$$i_C = C * [\frac{du_C}{dt}] \tag{3.32d}$$

$$U_0 = [u_1] + u_C \tag{3.32e}$$

$$u_C = [u_2] \tag{3.32f}$$

$$[u_L] = u_1 + u_2 \tag{3.32g}$$

$$[i_0] = i_1 + i_L \tag{3.32h}$$

$$i_1 = i_2 + [i_C] \tag{3.32i}$$

which can now be rearranged as follows:

$$i_1 = u_1/R_1 \tag{3.33a}$$

$$i_2 = u_2/R_2 \tag{3.33b}$$

$$\frac{di_L}{dt} = u_L/L \tag{3.33c}$$

$$\frac{du_C}{dt} = i_C/C \tag{3.33d}$$

$$u_1 = U_0 - u_C \tag{3.33e}$$

$$u_2 = u_C \tag{3.33f}$$

$$u_L = u_1 + u_2 \tag{3.33g}$$

$$i_0 = i_1 + i_L \tag{3.33h}$$

$$i_C = i_1 - i_2 \tag{3.33i}$$

Since most CSSLs allow us to specify auxiliary algebraic equations in addition to the state equations, and since they usually provide for an equation sorter, this set of equations represents a perfectly good CSSL dynamic model description.

3.6 Algebraic Loops

Let us now consider the slightly modified circuit depicted in Fig.3.12:

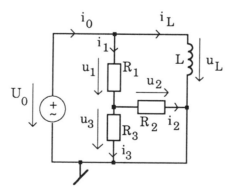

Figure 3.12. Another passive circuit with all variables named.

In this example, we have introduced the following nine unknowns: u_1, u_2, u_3, u_L, i_1, i_2, i_3, i_L, and i_0.

We can go ahead and write four branch equations and five mesh and/or node equations as before.

$$u_1 = R_1 * i_1 \qquad (3.34a)$$
$$u_2 = R_2 * i_2 \qquad (3.34b)$$
$$u_3 = R_3 * i_3 \qquad (3.34c)$$
$$u_L = L * \frac{di_L}{dt} \qquad (3.34d)$$
$$U_0 = u_1 + u_3 \qquad (3.34e)$$
$$u_3 = u_2 \qquad (3.34f)$$
$$u_L = u_1 + u_2 \qquad (3.34g)$$
$$i_0 = i_1 + i_L \qquad (3.34h)$$
$$i_1 = i_2 + i_3 \qquad (3.34i)$$

We try to solve these equations for the appropriate variables using the previously introduced procedure. However, this time the solution is not unique. After marking the first three equations, we are stuck:

$$u_1 = R_1 * i_1 \qquad (3.35a)$$
$$u_2 = R_2 * i_2 \qquad (3.35b)$$
$$u_3 = R_3 * i_3 \qquad (3.35c)$$
$$u_L = L * [\frac{di_L}{dt}] \qquad (3.35d)$$
$$U_0 = u_1 + u_3 \qquad (3.35e)$$
$$u_3 = u_2 \qquad (3.35f)$$
$$[u_L] = u_1 + u_2 \qquad (3.35g)$$
$$[i_0] = i_1 + i_L \qquad (3.35h)$$
$$i_1 = i_2 + i_3 \qquad (3.35i)$$

At this point, all the remaining equations contain at least two unknowns and all the remaining unknowns appear in at least two equations. We now have to make a choice. We can do this in an arbitrary fashion. For example, we could decide to solve Eq.(3.35e) for variable u_3. From then on, everything else will follow as before and we obtain the following set of equations:

$$[u_1] = R_1 * i_1 \qquad (3.36a)$$
$$u_2 = R_2 * [i_2] \qquad (3.36b)$$
$$u_3 = R_3 * [i_3] \qquad (3.36c)$$
$$u_L = L * [\frac{di_L}{dt}] \qquad (3.36d)$$
$$U_0 = u_1 + [u_3] \qquad (3.36e)$$
$$u_3 = [u_2] \qquad (3.36f)$$
$$[u_L] = u_1 + u_2 \qquad (3.36g)$$
$$[i_0] = i_1 + i_L \qquad (3.36h)$$
$$[i_1] = i_2 + i_3 \qquad (3.36i)$$

However, the fact that we had to make a choice invariably results in an *algebraic loop*. Let us rearrange the equations and try to recognize the resulting algebraic loop.

$$u_1 = R_1 * i_1 \qquad (3.37a)$$
$$i_2 = u_2/R_2 \qquad (3.37b)$$
$$i_3 = u_3/R_3 \qquad (3.37c)$$
$$\frac{di_L}{dt} = u_L/L \qquad (3.37d)$$
$$u_3 = U_0 - u_1 \qquad (3.37e)$$
$$u_2 = u_3 \qquad (3.37f)$$
$$u_L = u_1 + u_2 \qquad (3.37g)$$
$$i_0 = i_1 + i_L \qquad (3.37h)$$
$$i_1 = i_2 + i_3 \qquad (3.37i)$$

In order to compute u_3 from Eq.(3.37e), we need knowledge of u_1. However, in order to compute u_1 from Eq.(3.37a), we need knowledge of i_1. In order to compute i_1 from Eq.(3.37i), we need knowledge of i_3. Finally, in order to compute i_3 from Eq.(3.37c), we need knowledge of u_3, which closes the algebraic loop.

Unfortunately, algebraic loops are extremely common in electrical circuits, and this is the most serious drawback of the outlined technique.

3.7 Structural Singularities

Let us now look at yet another circuit as depicted in Fig.3.13.

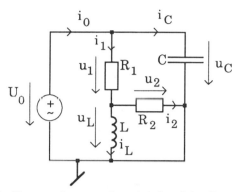

Figure 3.13. Yet another passive circuit with all variables named.

In this example, we have introduced the following nine unknowns: u_1, u_2, u_C, u_L, i_1, i_2, i_C, i_L, and i_0.

We go ahead and write four branch equations and five mesh and/or node equations as before.

$$u_1 = R_1 * i_1 \tag{3.38a}$$
$$u_2 = R_2 * i_2 \tag{3.38b}$$
$$u_L = L * \frac{di_L}{dt} \tag{3.38c}$$
$$i_C = C * \frac{du_C}{dt} \tag{3.38d}$$
$$U_0 = u_1 + u_L \tag{3.38e}$$
$$u_L = u_2 \tag{3.38f}$$
$$u_C = u_1 + u_2 \tag{3.38g}$$
$$i_0 = i_1 + i_C \tag{3.38h}$$
$$i_1 = i_2 + i_L \tag{3.38i}$$

We try to solve these equations for the appropriate variables using the same procedure. However, this time the problem is *overspecified*. This situation is frequently referred to as a *structural singularity*. The term stems from the equivalent situation as it occurs in mechanical system modeling.

We are stuck after the first three equations have been used:

$$u_1 = R_1 * i_1 \qquad\qquad (3.39a)$$

$$u_2 = R_2 * i_2 \qquad\qquad (3.39b)$$

$$u_L = L * \left[\frac{di_L}{dt}\right] \qquad\qquad (3.39c)$$

$$i_C = C * \left[\frac{du_C}{dt}\right] \qquad\qquad (3.39d)$$

$$U_0 = u_1 + u_L \qquad\qquad (3.39e)$$

$$u_L = u_2 \qquad\qquad (3.39f)$$

$$u_C = u_1 + u_2 \qquad\qquad (3.39g)$$

$$[i_0] = i_1 + i_C \qquad\qquad (3.39h)$$

$$i_1 = i_2 + i_L \qquad\qquad (3.39i)$$

At this point, we have no equation left to compute i_C. Obviously, we cannot use u_C as a state variable since we need that equation to compute i_C. We thus must revise our strategy, solve Eq.(3.39d) for i_C rather than for u_C, and continue from there. Let us see what happens now:

$$u_1 = R_1 * i_1 \qquad\qquad (3.40a)$$

$$u_2 = R_2 * i_2 \qquad\qquad (3.40b)$$

$$u_L = L * \left[\frac{di_L}{dt}\right] \qquad\qquad (3.40c)$$

$$[i_C] = C * \frac{du_C}{dt} \qquad\qquad (3.40d)$$

$$U_0 = u_1 + u_L \qquad\qquad (3.40e)$$

$$u_L = u_2 \qquad\qquad (3.40f)$$

$$[u_C] = u_1 + u_2 \qquad\qquad (3.40g)$$

$$[i_0] = i_1 + i_C \qquad\qquad (3.40h)$$

$$i_1 = i_2 + i_L \qquad\qquad (3.40i)$$

At this point, we are left with an algebraic loop as before.

3.8 Disadvantages of State–Space Models

In the last example, we could not avoid leaving one of the differential terms on the right–hand side of the equal sign. Whenever we face the situation of having no equation left to solve for a partic-ular unknown or being left with an unused equation that contains only known variables (assuming that we didn't choose linearly dependent equations right from the beginning), we are confronted with a *degenerate system* (a so–called structural singularity). In our example, the system really is of first order and not of second order, and the additional differentiator is a *true differentiator,* which cannot be eliminated from the circuit. Such systems do not have a state–space model. The best that we can hope for (and in a linear system, we can always achieve this) is to move the differentiation operator out of the integration loop into the output equation. We then end up with a *generalized state–space model* of the form:

$$\dot{\mathbf{x}} = \mathbf{A}\mathbf{x} + \mathbf{B}\mathbf{u} \qquad (3.41a)$$

$$\mathbf{y} = \mathbf{C}\mathbf{x} + \mathbf{D}(s)\mathbf{u} \qquad (3.41b)$$

where s denotes the Laplacian operator, and $\mathbf{D}(s)$ is a polynomial in s. Consequently, the output vector $\mathbf{y}$ depends explicitly on derivatives of the input vector $\mathbf{u}$.

Algebraic loops and structural singularities are serious problems that make the derivation of state–space models for electrical circuits difficult, if not impossible. For this reason, the approach is not commonly used in today's conventional circuit simulators. However, the approach has its beauties and, if successful, can speed up the run–time execution of the resulting model quite dramatically. Several techniques exist to reduce algebraic loops and structural singularities automatically and we shall discuss some of these techniques in Chapter 5 of this text. However, whether or not algebraic loops and/or structural singularities can always be removed in an automated (i.e., algorithmic) fashion is still an unanswered question. Therefore, the development of such algorithms must be considered open research.

3.9 Summary

We have introduced the two standard techniques used in the analysis of electrical circuitry, the mesh–equation approach and the node–equation approach. More details on the implementation of these techniques in modern circuit simulators (such as SPICE) will be presented in Chapter 6 of this text. Details of the numerical techniques required to simulate these types of models (using *implicit differentiation*) are discussed in the companion book on continuous system simulation. We have also shown an alternative approach to circuit analysis, an approach that leads directly to a state–space description. However, we have shown that this route is quite problematic due to algebraic loops and structural singularities that occur frequently in electrical circuits. More details about these problems and how we deal with them are presented in Chapter 5 of this text, together with a tool (DYMOLA) that helps us automate the discussed algorithms.

There are many books on the market which describe electrical circuit models. The books referenced in this chapter are exemplary. However, all these books (and also many others that I have seen) concentrate on *analytical modeling techniques* and not on *simulation modeling techniques*, i.e., their goal is to derive a system description that is amenable to analytical techniques rather than to numerical simulation. In that respect, this chapter in my book is different from other presentations of the same material.

References

[3.1] P. R. Bélanger, E. L. Adler, and N. C. Rumin (1985), *Introduction to Circuits with Electronics: An Integrated Approach*, Holt, Rinehart and Winston, New York.

[3.2] Leonard S. Bobrow (1981), *Elementary Linear Circuit Analysis*, Holt, Rinehart and Winston, New York.

[3.3] Gene H. Hostetter, Clement J. Savant, Jr., and Raymond T. Stefani (1989), *Design of Feedback Control Systems*, second edition, Saunders College Publishing, New York.

[3.4] Lawrence P. Huelsman (1984), *Basic Circuit Theory*, second edition, Prentice–Hall, Englewood Cliffs, N.J.

[3.5] David E. Johnson, John L. Hilburn, and John R. Johnson (1978), *Basic Electric Circuit Analysis*, Prentice–Hall, Englewood Cliffs, N.J.

Homework Problems

[H3.1] Choosing Between Mesh and Node Equations

Figure H3.1 shows a simple passive circuit. The circuit contains one *dependent current source*. The current i_4 is at all times proportional to the voltage v_3. The proportionality factor is 4 A V^{-1}.

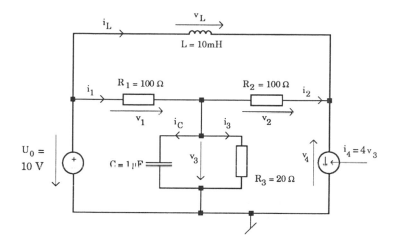

Figure H3.1. Circuit diagram of a simple passive circuit.

Determine how many mesh equations (node equations) will be needed to describe this circuit.

[H3.2] Mesh Equations

For the circuit of Fig.H3.1, replace all storage elements by equivalent impedances, choose an appropriate tree, and determine a consistent set of mesh equations. Present these equations in a matrix form.

[H3.3] Node Equations

For the circuit of Fig.H3.1, reduce the independent voltage source to a set of current sources, replace all storage elements by equivalent admittances, choose an appropriate tree, and determine a consistent set of node equations. Present these equations in a matrix form.

[H3.4] CSSL Model

For the circuit of Fig.H3.1, use the state–space modeling approach to determine a consistent set of simulation equations. Code these equations in any CSSL and simulate the system over 50 μsec.

[H3.5] Linear State–Space Model

From the simulation equations of Hw.[H3.4], eliminate all auxiliary variables and write the resulting state equations in a matrix form of the type

$$\dot{\mathbf{x}} = \mathbf{A} \cdot \mathbf{x} + \mathbf{b} \cdot u \qquad\qquad (H3.5a)$$
$$y = \mathbf{C} \cdot \mathbf{x} + \mathbf{d} \cdot u \qquad\qquad (H3.5b)$$

where the output vector consists of the variables v_3 and i_C and where the single input is the independent voltage source U_0.

Use CTRL–C (or MATLAB) to determine the eigenmodi of this system (which are the eigenvalues of the $\mathbf{A}$ matrix). Choose the final time of the simulation to be minus three times the inverse of the slower of the two eigenmodi.

Simulate a step response of this system using CTRL–C (MATLAB) directly. For this purpose, it is necessary to create a *time base*, i.e., a vector of time values at which we wish to sample the simulated trajectories. This is accomplished with the statement:

$$[> \quad t = 0 : tmx/1000 : tmx;$$

which generates a vector of length 1001 containing numbers that are equidistantly spaced between 0.0 and tmx. This vector contains the communication points. Next, we need to generate the input signal sampled at the communication points. Since we want to simulate a step response, we can create a vector of length 1001 each element of which is 1.0. This can be achieved with the statement:

$$[> \quad u = U0 * \text{ONES}(t);$$

Finally, we can apply an initial condition and simulate the system over time, using the CTRL–C statements:

$$[> \quad x0 = [v30; iL0];$$
$$[> \quad \text{SIMU}('IC', x0)$$
$$[> \quad y = \text{SIMU}(a, b, c, d, u, t);$$

The MATLAB solution looks very similar.

Determine analytically the steady–state value of the output vector to step input. At steady–state, all derivatives have died out, and thus, $\dot{\mathbf{x}}_{ss} = 0$. Compare the analytically found value with the numerically found value as a means to validate your simulation.

[H3.6] Structural Singularity [3.3]

Figure H3.6 shows another simple passive circuit. Use the state–space modeling technique to determine a set of simulation equations. You will notice that this circuit exhibits a structural singularity.

Prove that the two inductive currents are linearly dependent on each other, and thus, do not qualify as two separate state variables.

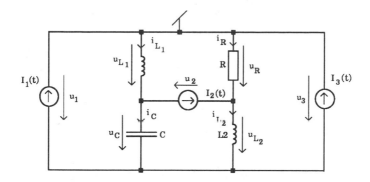

Figure H3.6. Circuit diagram of another simple passive circuit.[†]

[H3.7] Passive Filter
Figure H3.7 shows a simple passive filter.

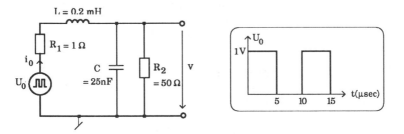

Figure H3.7. Circuit diagram of a simple passive filter.

Use the state–space modeling approach to derive a set of first–order differential equations to describe this system. Use any CSSL to simulate the system over 50 μsec, and display the resulting trajectories for u_0, i_0, and u_C. Use the step–function approach to generate the square–wave voltage source U_0.

[H3.8] Connecting Power to an Unloaded Power Line
Figure H3.8a shows an unloaded and initially unenergized power line that must be brought on–line.

[†] Adapted figure from DESIGN OF FEEDBACK CONTROL SYSTEMS, Second Edition by Gene Hostetter *et al.*, ©1989 by Saunders College Publishing, a division of Holt, Rinehart and Winston, Inc., reprinted by permission of the publisher.

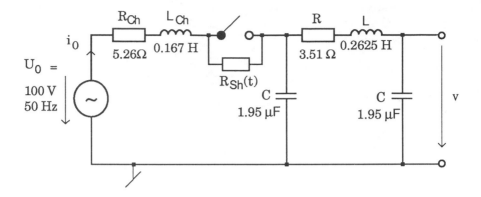

Figure H3.8a. Circuit diagram of a power line.

The AC voltage source is connected to the power line, which has a characteristic impedance of $R_{Ch} + j \cdot L_{Ch}$. The power line itself is represented through a single π–element.

The aim of this experiment is to determine how we can minimize the overshoot on the power line. For this purpose, we shall connect the voltage source to the power line (by closing the switch) at different phase angles. We shall try seven different phase angles equidistantly spaced between $0°$ and $180°$.

Also, we want to check whether a shunt resistor may help suppress the overshoot on the power line. For this purpose, we introduce a time–varying shunt resistor R_{Sh} as shown in Fig.H3.8b.

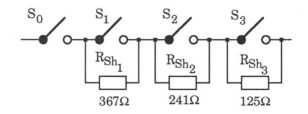

Figure H3.8b. Circuit diagram of the shunt resistor.

The shunt resistor R_{Sh} is placed in parallel with the main switch. It contains four switches, which are being closed at various time instants: the switch S_0 closes at time zero, S_1 closes after 3.5 msec, S_2 closes after 9 msec, and S_3 closes after 18 msec.

Simulate the system with and without shunt resistor for all seven phase angles and plot v on two separate graphs (seven curves per graph). Use the state–space approach for modeling. The shunt resistor is easiest coded as a FORTRAN subroutine.

[H3.9]* Resonance Circuit

Figure H3.9 shows a simple resonance circuit.

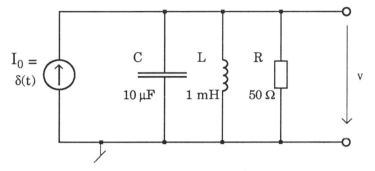

Figure H3.9. Resonance circuit.

We want to analyze the response of this circuit to a Dirac current impulse:

$$i_0(t) = \delta(t) \qquad\qquad (H9.3a)$$

For this purpose, we perform two separate simulation experiments.

In the first experiment, we approximate the Dirac impulse by a rectangular pulse of length δt. Choose the height of the pulse such that the identity:

$$\int_{-\infty}^{\infty} \delta(t)\, dt = 1 \qquad\qquad (H3.9b)$$

is preserved. Repeat this experiment for three different values of δt, namely, $\delta t = 6$ μsec, $\delta t = 0.6$ μsec, and $\delta t = 0.06$ μsec. Simulate the system for a duration of $t_{max} = 2$ msec. Use a fixed–step integration algorithm with the step size $\Delta t = 0.01 \cdot \delta t$ up to the time δt, and a much increased step size of $\Delta t = 60$ μsec thereafter. Plot $v(t)$ on one graph for all three simulation runs. Use the state–space approach and formulate the model in any of the CSSLs. Some languages (such as ACSL) don't provide for fixed–step algorithms. In that case, simply set the communication interval accordingly. In ACSL, you can set the communication interval initially in a CINTERVAL declaration:

cinterval $cint = 2.0E\text{-}5$

This declares the variable *cint*. Modify *cint* immediately in the INITIAL section to:

$$cint = 0.01 * width$$

and declare a discrete event of type *change* to happen at time *width* $(= \delta t)$:

schedule *change* **.at.** *width*

This statement will force ACSL to enter a DISCRETE block (to be coded as part of the DYNAMIC block but outside the DERIVATIVE block) by the name of *change* in which you can modify the communication interval:

DISCRETE *change*
cint = 60.0*E*-6
END $ *"of DISCRETE change"*

The current I_0 can, of course, be modified simultaneously.

In the second experiment, we shall notice that this is a linear system that can be written in the form:

$$\dot{\mathbf{x}} = \mathbf{A} \cdot \mathbf{x} + \mathbf{b} \cdot u \qquad (H3.9c)$$

$$y = \mathbf{c}' \cdot \mathbf{x} + d \cdot u \qquad (H3.9d)$$

where the single input u is the current source $i_0(t)$ and the single output y is the voltage $v(t)$. We can further notice that any such system can be analytically solved. The analytical solution (convolution integral) is:

$$y(t) = \mathbf{c}' \exp(\mathbf{A}t)\mathbf{x}_0 + \mathbf{c}' \int_0^t \exp(\mathbf{A}(t - \tau))\mathbf{b}u(\tau)d\tau \qquad (H3.9e)$$

and since $u(t) = \delta(t)$, we can use the sifting property of the Dirac distribution to evaluate the integral. This allows us to reformulate the given problem, which has a Dirac input and no initial condition as another equivalent problem that has no input but a nonvanishing initial condition. Determine what the equivalent initial condition has to be in terms of matrices $\mathbf{A}$, $\mathbf{b}$, $\mathbf{c}'$, and d. Simulate this modified problem and compare the results to those obtained from the previous simulations. It may be easiest to simulate this linear time–invariant problem in CTRL–C (or MATLAB) directly.

4

Principles of Planar Mechanical System Modeling

Preview

In this chapter, we shall deal with the dynamic behavior of translational and rotational planar motions. The basic physical law governing this type of systems can be expressed in terms of either Newton's law for translational and rotational motions or the d'Alembert principle. The concepts will be demonstrated by means of a number of practical examples, such as a crane crab system and an inverted pendulum. Toward the end of the chapter, we shall discuss electromechanical transducers.

4.1 Introduction

Mechanical systems are quite similar to electrical systems. Some of the basic principles are the same for both types of systems. In electrical circuit modeling, we have learned that the sum of all currents in a node must always be zero. In modeling rigid mechanical systems, we shall see that all internal forces and torques at any point of the body must add up to zero. In electrical circuit modeling, we have learned that the potentials of all connecting branches at a node must be the same. In mechanical systems, it is true that the positions, velocities, and accelerations, both translational and rotational must be the same at any connecting point in the system.

In fact, some electrical engineers like to convert mechanical systems to equivalent electrical circuits, and thereafter treat those using the techniques that were discussed in Chapter 3. We shall not do this, but we shall see in Chapter 7 that a deeper truth lies behind these similarities.

However, differences also exist that make it a little harder to deal with mechanical systems than with electrical circuits. Let me summarize these dissimilarities:

(1) Geometry plays an important role in mechanical systems. This is not so in electrical systems unless one goes to very high frequencies (microwave frequencies) or to very small dimensions (integrated circuits). Mechanical systems operate in a three–dimensional space that is difficult to capture on two–dimensional drawings. Therefore, it is a little more difficult to grasp the functioning of most mechanical devices. However, this is compensated by the fact that electrical circuits often contain thousands of circuit elements. Mechanical systems are always rather simple devices.

(2) To model electrical circuits, only two types of signals were required: *voltages* and *currents*. Every facet of an electrical circuit can be described in terms of these two quantities. In mechanical systems, each body can be exerted simultaneously by a *force* (which usually has an x, a y, and a z component), and by a *torque* (again with three components). As a consequence, we need to operate on a larger number of variables in order to capture all mechanical properties of a system.

(3) Mechanical systems are always subjected to *constraints*. Masses can bump into each other or can fall down; springs cannot be compressed or pulled to an arbitrary extent; bodies are not infinitely stiff in reality, but can sag or be otherwise deformed (elastically or even plastically). This is, of course, also true for electrical systems, but to a much lesser extent. Electrical systems are much "cleaner" than mechanical systems.

What this really means is that electrical systems exhibit a more crisp separation of the various governing physical phenomena in both the *time* and *space* dimensions. Most electrical circuits operate in the kHz–to–MHz range. In this frequency range, the electrical phenomena as studied in Chapter 3 are strongly dominant. Mechanical and thermal side effects are much slower, usually too slow to be considered in the model except for DC analysis. Quantum mechanical effects are much faster, usually so fast that they can be ignored altogether or at least, their influence can be aggregated to a statistical description (noise analysis). Also, in the kHz–to–MHz range, the geometry of the circuit layout can still be ignored (except inside an integrated circuit chip). On the other hand, mechanical systems operate in the Hz range. Therefore, it may be necessary to consider thermal side effects. More important, however, are the geometrical (spatial) influences. A simulation system that does not allow us to formulate the geometrical constraints inherent in the model is therefore virtually worthless.

However, let us start with the most simple principle that governs the mechanical behavior of rigid bodies. This principle was formulated first by Sir Isaac Newton. It describes the dynamics of a rigid body in both translational and rotational terms.

4.2 Newton's Law for Translational Motions

Newton's law is often quoted as follows: The sum of all forces exerted on a rigid body equals the mass of the body multiplied by its acceleration, i.e.:

$$m \cdot a = \sum_{\forall i} f_i \qquad (4.1)$$

However, a little more precisely, the law should be written as:

$$\frac{d(m \cdot v)}{dt} = \sum_{\forall i} f_i \qquad (4.2)$$

where the term $m \cdot v$ (the mass multiplied by the velocity) is sometimes called the *momentum* or *impulse* $\mathcal{I}$ of the rigid body, i.e.:

$$\frac{d\mathcal{I}}{dt} = \sum_{\forall i} f_i \qquad (4.3)$$

This distinction becomes important if the body is not all that "rigid" after all, but loses mass on the way. However, we must be a little cautious when we apply this formula. Figure 4.1 illustrates Newton's law as applied to our lunar landing module from Chapter 2.

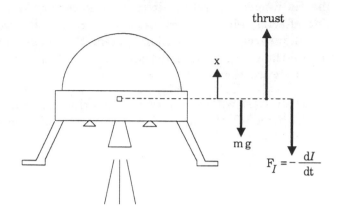

Figure 4.1. Lunar landing module.

Since the mass of the rocket changes with time, we are inclined to believe that, for this system, Newton's law can be written as:

$$\frac{d(m \cdot v)}{dt} = \frac{dm}{dt} \cdot v + m \cdot \frac{dv}{dt} = thrust - m \cdot g \qquad (4.4)$$

which can be rewritten as:

$$m \cdot a = thrust - m \cdot g - \frac{dm}{dt} \cdot v \qquad (4.5)$$

which is different from the equation that we used in Chapter 2.

Let us try to *validate* our model. A good validation technique is the following: We make simplifying assumptions until the problem is reduced to a much simpler problem for which we can check the plausibility of the results obtained. Applied to our rocket: We shall assume that our space craft is far away from any planetary mass. Consequently, we may ignore the gravity term. Moreover, we shall assume that the thrust is always nonnegative, i.e.:

$$thrust \geq 0.0 \qquad (4.6)$$

Thus, we obtain the following set of equations:

$$a = \frac{1}{m} \cdot (thrust - \dot{m} \cdot v) \qquad (4.7a)$$

$$\dot{m} = -c \cdot thrust \qquad (4.7b)$$

where Eq.(4.7a) is the simplified Newton equation and Eq.(4.7b) is the simplified fuel consumption equation. If we now plug Eq.(4.7b) into Eq.(4.7a), we find:

$$a = \frac{1}{m} \cdot thrust \cdot (1.0 + c \cdot v) \tag{4.8}$$

If we assume that we travel initially with a constant velocity of

$$v = -\frac{1}{c} \tag{4.9}$$

backward through space, the last factor of Eq.(4.8) cancels out and we shall never again be able to accelerate or decelerate our space craft. What a fate!

Quite obviously, something has gone awry. The problem is the following: Newton's law is not truly a "law of physics." It is a derived law, i.e., a "law of mathematics." The real "law of physics" states that the total momentum of a closed system must be conserved, or more generally:

$$\mathcal{I}(t + \Delta t) = \mathcal{I}(t) + \Delta \mathcal{I}(t \rightarrow t + \Delta t) \tag{4.10}$$

The total momentum $\mathcal{I}$ of a system at time $t+\Delta t$ equals the total momentum at time t plus the (positive or negative) momentum added to (subtracted from) the system between time t and time $t + \Delta t$. Let us apply this law to our space craft:

$$(m - \Delta m) \cdot (v + \Delta v) + \Delta m \cdot v = m \cdot v + thrust \cdot \Delta t \tag{4.11}$$

The first term on the left–hand side of Eq.(4.11) denotes the momentum of the space craft at time $t + \Delta t$. The second term denotes the momentum of the cloud of exhaust at the same time. The first term on the right–hand side of Eq.(4.11) denotes the momentum of the space craft at time t, and the second term denotes the added momentum due to the drive of the space craft. Notice that we must somehow include the exhaust. Either we consider the cloud of exhaust a part of our system by adding it to the left–hand side of Eq.(4.11) or we must consider that the exhaust leaves the system between time t and time $t + \Delta t$ and subtract this term from the right–hand side of Eq.(4.11).

Neglecting terms in Eq.(4.11) which are of second order small, we find:

$$m \cdot \Delta v = thrust \cdot \Delta t \tag{4.12}$$

or by dividing through Δt and by letting Δt go to zero:

$$m \cdot \frac{dv}{dt} = thrust \tag{4.13}$$

Thus, we must use the more familiar form of Newton's law, although the mass of the space craft is undeniably changing with time. Initially, we had simply forgotten to take the cloud of exhaust into account. We could have arrived at the same conclusion by considering the total *kinetic energy* of the system instead of its momentum, since the energy must also be conserved, but the momentum was easier to use in this example.

Let us return now to the discussion of Newton's law. In an alternative approach, we can introduce a fictitious "mass force":

$$F_{\mathcal{I}} = -\frac{d\mathcal{I}}{dt} \tag{4.14}$$

and, adding this force to our set of forces acting on the rigid body, we can reformulate Newton's law as follows:

$$\sum_{\forall i} f_i = 0 \tag{4.15}$$

In this modified form, Newton's law is known as the *d'Alembert principle*. The two formulations are equivalent.

Let us exercise Newton's law by means of the simple mechanical system depicted in Fig.4.2.

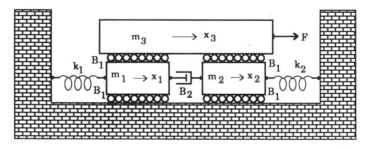

Figure 4.2. Simple translational problem [4.11].[†]

Let us assume that all elements of this mechanical system operate in their linear range. The top body m_3 does neither fall down, nor does

---

[†] Adapted figure from CONTROL SYSTEMS ENGINEERING by Norman Nise, ©1991 by Benjamin/Cummings, Inc., printed by permission of the publisher.

it lose any of its wheels. Also, the top body does not sag and always covers all the wheels that separate it from the two lower bodies m_1 and m_2. The springs k_1 and k_2 do not overexpand or overcontract, and the same is true for the hydraulic cylinder B_2. Let us assume furthermore that, for all times $t < 0$, the driving force F is zero and the system is in an equilibrium state in which all three positions x_1, x_2, and x_3 are defined as zero.

This problem is very simple. One approach to tackling such problems is to freeze all bodies but one and see what happens to that body when we try to move it. Let us start with body m_3. If we apply a force F at time $t = 0$ pulling m_3 to the right, the two frictions between m_3 and the two other bodies m_1 and m_2, which are told not to move, will oppose our attempts. This settles the question of the signs and we find immediately the equation:

$$m_3 \frac{d^2 x_3}{dt^2} = F - B_1 \left(\frac{dx_3}{dt} - \frac{dx_2}{dt} \right) - B_1 \left(\frac{dx_3}{dt} - \frac{dx_1}{dt} \right) \qquad (4.16a)$$

The *reaction* to these friction forces (i.e., the same forces but with opposite signs) are responsible for getting the bodies m_1 and m_2 moving. Thus, we can, for instance, write the equation for body m_2:

$$m_2 \frac{d^2 x_2}{dt^2} = B_1 \left(\frac{dx_3}{dt} - \frac{dx_2}{dt} \right) - B_1 \frac{dx_2}{dt} - B_2 \left(\frac{dx_2}{dt} - \frac{dx_1}{dt} \right) - k_2 \, x_2 \quad (4.16b)$$

and finally for body m_1:

$$m_1 \frac{d^2 x_1}{dt^2} = B_1 \left(\frac{dx_3}{dt} - \frac{dx_1}{dt} \right) - B_1 \frac{dx_1}{dt} + B_2 \left(\frac{dx_2}{dt} - \frac{dx_1}{dt} \right) - k_1 \, x_1 \quad (4.16c)$$

In order to obtain a set of state–space equations, we need to solve Eqs.(4.16a–c) for their highest derivatives and to reduce the second–order differential equations to sets of first–order differential equations. The resulting state equations are as follows:

$$\dot{x}_1 = v_1 \qquad (4.17a)$$

$$\dot{v}_1 = \frac{1}{m_1} [-k_1 \, x_1 - (2B_1 + B_2)v_1 + B_2 \, v_2 + B_1 \, v_3] \qquad (4.17b)$$

$$\dot{x}_2 = v_2 \qquad (4.17c)$$

$$\dot{v}_2 = \frac{1}{m_2} [B_2 \, v_1 - k_2 \, x_2 - (2B_1 + B_2)v_2 + B_1 \, v_3] \qquad (4.17d)$$

$$\dot{x}_3 = v_3 \qquad (4.17e)$$

$$\dot{v}_3 = \frac{1}{m_3} [B_1 \, v_1 + B_1 \, v_2 - 2B_1 \, v_3 + F(t)] \qquad (4.17f)$$

which can be coded directly in any of the simulation systems that were introduced in Chapter 2.

The preceding approach works well for simple problems, but it can become confusing when more parts are involved that move in all directions and rotate at the same time. In those more complicated cases, another approach works better. This will be illustrated next.

In the second approach, we start again by identifying parts of the system that can be moved without the rest of the system moving with them. We now cut the system open at the interfaces between the moving subsystem and the frozen subsystems and replace the influence of the frozen subsystems on the moving subsystem by an equivalent force acting on the moving subsystem and the *reaction* (i.e., the influence of the moving subsystem on the frozen subsystems) by an equivalent force acting on the frozen subsystems. These two *internal forces* are always of the same size but of opposite direction (i.e., they annihilate each other when the system is recombined). Figure 4.3 demonstrates this approach for the case of our simple mechanical system.

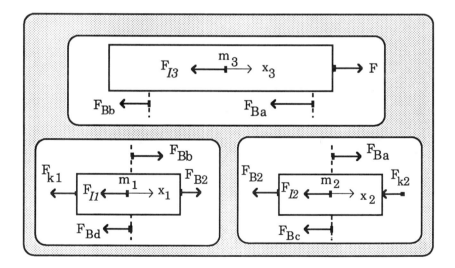

Figure 4.3. Simple translational problem cut open.

Now, we can apply the d'Alembert principle to the three bodies separately and write those equations down together with the equations governing the behavior of the individual forces:

$$F(t) = F_{I3} + F_{Ba} + F_{Bb} \tag{4.18a}$$

$$F_{Ba} = F_{I2} + F_{Bc} + F_{B2} + F_{k2} \tag{4.18b}$$

$$F_{Bb} + F_{B2} = F_{I1} + F_{Bd} + F_{k1} \tag{4.18c}$$

$$F_{I1} = m_1 \frac{dv_1}{dt} \tag{4.18d}$$

$$\frac{dx_1}{dt} = v_1 \tag{4.18e}$$

$$F_{I2} = m_2 \frac{dv_2}{dt} \tag{4.18f}$$

$$\frac{dx_2}{dt} = v_2 \tag{4.18g}$$

$$F_{I3} = m_3 \frac{dv_3}{dt} \tag{4.18h}$$

$$\frac{dx_3}{dt} = v_3 \tag{4.18i}$$

$$F_{Ba} = B_1(v_3 - v_2) \tag{4.18j}$$

$$F_{Bb} = B_1(v_3 - v_1) \tag{4.18k}$$

$$F_{Bc} = B_1 \, v_2 \tag{4.18l}$$

$$F_{Bd} = B_1 \, v_1 \tag{4.18m}$$

$$F_{B2} = B_2(v_2 - v_1) \tag{4.18n}$$

$$F_{k1} = k_1 \, x_1 \tag{4.18o}$$

$$F_{k2} = k_2 \, x_2 \tag{4.18p}$$

Notice that the directions of the arrows of the internal forces are arbitrary (as were the directions of currents and voltages in electrical circuits). However, we must adjust the equations to our conventions. This is demonstrated in Fig.4.4. If friction forces and/or spring forces have the opposite direction to the position (and velocity and acceleration) of a rigid body, the contribution of the body itself is counted positively in the force equation while the contribution of the environment is counted negatively. If the directions are the same, the contribution of the environment is counted positively, while the contribution of the body itself is counted negatively. If the direction of the inertial force is opposite to the direction of the position, the equation is entered with a plus sign, otherwise with a minus sign.

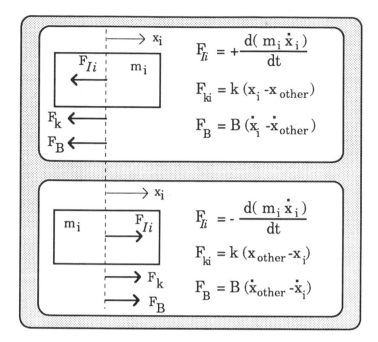

Figure 4.4. Convention for direction of forces.

Now we can try to solve these equations using the methodology advocated in Chapter 3. As in Chapter 3, we start by solving all differential equations for their derivative terms and proceed until all equations and unknowns have been used up.

$$F(t) = [F_{I3}] + F_{Ba} + F_{Bb} \tag{4.19a}$$

$$F_{Ba} = [F_{I2}] + F_{Bc} + F_{B2} + F_{k2} \tag{4.19b}$$

$$F_{Bb} + F_{B2} = [F_{I1}] + F_{Bd} + F_{k1} \tag{4.19c}$$

$$F_{I1} = m_1 [\frac{dv_1}{dt}] \tag{4.19d}$$

$$[\frac{dx_1}{dt}] = v_1 \tag{4.19e}$$

$$F_{I2} = m_2 [\frac{dv_2}{dt}] \tag{4.19f}$$

$$[\frac{dx_2}{dt}] = v_2 \tag{4.19g}$$

$$F_{I3} = m_3 \left[\frac{dv_3}{dt}\right] \qquad (4.19h)$$

$$\left[\frac{dx_3}{dt}\right] = v_3 \qquad (4.19i)$$

$$[F_{Ba}] = B_1(v_3 - v_2) \qquad (4.19j)$$

$$[F_{Bb}] = B_1(v_3 - v_1) \qquad (4.19k)$$

$$[F_{Bc}] = B_1\, v_2 \qquad (4.19l)$$

$$[F_{Bd}] = B_1\, v_1 \qquad (4.19m)$$

$$[F_{B2}] = B_2(v_2 - v_1) \qquad (4.19n)$$

$$[F_{k1}] = k_1\, x_1 \qquad (4.19o)$$

$$[F_{k2}] = k_2\, x_2 \qquad (4.19p)$$

which can then be rewritten as:

$$F_{I3} = F(t) - F_{Ba} - F_{Bb} \qquad (4.20a)$$

$$F_{I2} = F_{Ba} - F_{Bc} - F_{B2} - F_{k2} \qquad (4.20b)$$

$$F_{I1} = F_{Bb} + F_{B2} - F_{Bd} - F_{k1} \qquad (4.20c)$$

$$\frac{dv_1}{dt} = F_{I1}/m_1 \qquad (4.20d)$$

$$\frac{dx_1}{dt} = v_1 \qquad (4.20e)$$

$$\frac{dv_2}{dt} = F_{I2}/m_2 \qquad (4.20f)$$

$$\frac{dx_2}{dt} = v_2 \qquad (4.20g)$$

$$\frac{dv_3}{dt} = F_{I3}/m_3 \qquad (4.20h)$$

$$\frac{dx_3}{dt} = v_3 \qquad (4.20i)$$

$$F_{Ba} = B_1(v_3 - v_2) \qquad (4.20j)$$

$$F_{Bb} = B_1(v_3 - v_1) \qquad (4.20k)$$

$$F_{Bc} = B_1\, v_2 \qquad (4.20l)$$

$$F_{Bd} = B_1\, v_1 \qquad (4.20m)$$

$$F_{B2} = B_2(v_2 - v_1) \qquad (4.20n)$$

$$F_{k1} = k_1\, x_1 \qquad (4.20o)$$

$$F_{k2} = k_2\, x_2 \qquad (4.20p)$$

which again can be programmed immediately using any of the previously introduced simulation languages.

4.3 Newton's Law for Rotational Motions

This version of Newton's law is often quoted as follows: The sum of all torques exerted on a rigid body equals the inertia of the body multiplied by its angular acceleration, i.e.:

$$J\dot{\omega} = \sum_{\forall i} \tau_i \qquad (4.21)$$

However, a little more precisely, the law should be written as:

$$\frac{d(J\omega)}{dt} = \sum_{\forall i} \tau_i \qquad (4.22)$$

where the term $J\omega$ (the inertia multiplied by the angular velocity) is the *twist* $\mathcal{T}$ of the rigid body, i.e.:

$$\frac{d\mathcal{T}}{dt} = \sum_{\forall i} \tau_i \qquad (4.23)$$

which is sometimes also called the *angular momentum*. As before, we can introduce a fictitious "inertial torque" $\tau_\mathcal{T}$:

$$\tau_\mathcal{T} = -\frac{d\mathcal{T}}{dt} \qquad (4.24)$$

and reformulate Newton's law using the d'Alembert principle as:

$$\sum_{\forall i} \tau_i = 0 \qquad (4.25)$$

Let me illustrate the modeling of rotational motions by means of another simple mechanical system as illustrated in Fig.4.5.

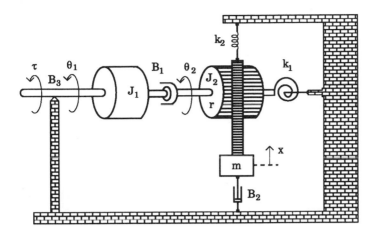

Figure 4.5. A simple rotational system [4.11].[†]

Since the system is sufficiently simple, we can proceed along the first route, and write second–order differential equations right away. However, we first need to understand what the gear is doing to our system. This is illustrated in Fig.4.6, which describes the transformation of the rotational subsystem to the translational subsystem. The equations that govern the gear are the same irrespective of whether the cause is a torque applied to the pinion or a force applied to the rack.

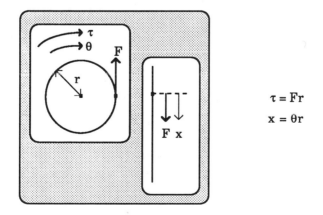

$$\tau = Fr$$
$$x = \theta r$$

Figure 4.6. Modeling mechanical gears.

[†] Adapted figure from CONTROL SYSTEMS ENGINEERING by Norman Nise, ©1991 by Benjamin/Cummings, Inc., printed by permission of the publisher.

Now we are ready to write down the equations. Let us begin with the subsystem J_1:

$$J_1 \frac{d^2\theta_1}{dt^2} = \tau - B_1\left(\frac{d\theta_1}{dt} - \frac{d\theta_2}{dt}\right) - B_3\frac{d\theta_1}{dt} \qquad (4.26a)$$

Let me introduce two additional variables, namely, a torque τ_G denoting the influence of the rack on the pinion J_2 and a force F_G denoting the influence of the pinion J_2 on the rack. This allows us to write an equation for the second subsystem (J_2):

$$J_2 \frac{d^2\theta_2}{dt^2} = B_1\left(\frac{d\theta_1}{dt} - \frac{d\theta_2}{dt}\right) - k_1\,\theta_2 - \tau_G \qquad (4.26b)$$

and for the third subsystem:

$$m\frac{d^2x}{dt^2} = F_G - m\,g - B_2\frac{dx}{dt} - k_2\,x \qquad (4.26c)$$

Now we only need to describe the gear:

$$\tau_G = r\,F_G \qquad (4.26d)$$

$$x = r\,\theta_2 \qquad (4.26e)$$

Of course, from Eq.(4.26e) we can immediately derive two more equations:

$$\frac{dx}{dt} = r\frac{d\theta_2}{dt}, \qquad \frac{d^2x}{dt^2} = r\frac{d^2\theta_2}{dt^2} \qquad (4.27)$$

Let us eliminate x and F_G from Eq.(4.26c) by replacing these terms with θ_2 and τ_G:

$$m\,r\frac{d^2\theta_2}{dt^2} = \frac{1}{r}\tau_G - m\,g - B_2\,r\frac{d\theta_2}{dt} - k_2\,r\,\theta_2 \qquad (4.28)$$

which can be solved for τ_G:

$$\tau_G = m\,r^2\frac{d^2\theta_2}{dt^2} + B_2\,r^2\frac{d\theta_2}{dt} + k_2\,r^2\,\theta_2 + m\,g\,r \qquad (4.29)$$

Plugging Eq.(4.29) into Eq.(4.26b) and rearranging the terms, we find:

$$[J_2 + m\,r^2]\frac{d^2\theta_2}{dt^2} = B_1\frac{d\theta_1}{dt} - [B_1 + B_2\,r^2]\frac{d\theta_2}{dt} - [k_1 + k_2\,r^2]\theta_2 - m\,g\,r \qquad (4.30)$$

The term $[J_2 + m\ r^2]$ is the *apparent inertia* of the body J_2, i.e., the inertia that is visible when we measure the inertia from the rotational end of the gear. Similarly, the terms $[B_1 + B_2\ r^2]$ and $[k_1 + k_2\ r^2]$ denote *apparent friction* and *apparent spring coefficients*.

We are now ready to generate a set of state equations:

$$\dot{\theta}_1 = \omega_1 \tag{4.31a}$$

$$\dot{\omega}_1 = \frac{1}{J_1}[-(B_1 + B_3)\omega_1 + B_1\omega_2 + \tau(t)] \tag{4.31b}$$

$$\dot{\theta}_2 = \omega_2 \tag{4.31c}$$

$$\dot{\omega}_2 = \frac{1}{J_2 + m\ r^2}[B_1\omega_1 - (k_1 + k_2\ r^2)\theta_2$$
$$- (B_1 + B_2\ r^2)\omega_2 - m\ g\ r] \tag{4.31d}$$

$$x = r\theta_2 \tag{4.31e}$$

which can directly be coded in any of the simulation languages. Equation (4.31e) denotes an *output equation*. It is not needed in order to solve the set of differential equations, but is computed only for the purpose of display on output. Consequently, it does not make sense to code this equation inside the DERIVATIVE section of the program. To demonstrate this new concept, let me write down an excerpt of an ACSL [4.9] program implementing this model:

```
PROGRAM Rotational Mechanical System
  INITIAL
    constant
      "Place values for all constants here"
    J1inv  = 1.0/J1    $    J2apin = 1.0/(J2 + m * r * *2)
    kap    = k1 + k2 * r * *2    $    Bap = B1 + B2 * r * *2
    B13    = -(B1 + B3)    $    mgr = m * g * r
  END $ "of INITIAL"
  DYNAMIC
    DERIVATIVE
      th1dot  = om1
      om1dot = J1inv * (B13 * om1 + B1 * om2 + tau)
      th2dot  = om2
      om2dot = J2apin * (B1 * om1 - kap * th2 - Bap * om2 - mgr)
      th1    = integ(th1dot, th10)
      om1    = integ(om1dot, om10)
      th2    = integ(th2dot, th20)
      om2    = integ(om2dot, om20)
    END $ "of DERIVATIVE"
    x = r * th2
    termt (t.ge.tmx)
  END $ "of DYNAMIC"
END $ "of PROGRAM"
```

This program demonstrates new concepts. Since the DERIVATIVE segment of the program is being executed over and over again, it is important to keep this segment as short as possible by throwing out all computations that are not necessary for the solution of the differential equations in order to make the execution of the simulation program fast. In this context, all constant expressions should be moved into the INITIAL segment and all output equations should be moved out of the DERIVATIVE block. Output equations will then be evaluated once per communication interval only, namely, immediately before the output variables are stored in the data base.

DESIRE [4.8] and DARE–P [4.17] offer equivalent features. In DESIRE, output equations can be placed below the OUT statement and in DARE–P, they can be coded in a separate $D2 block.

Let me now demonstrate the other approach. Figure 4.7 shows the same system after it has been decomposed into three subsystems and after all internal and fictitious forces/torques have been introduced.

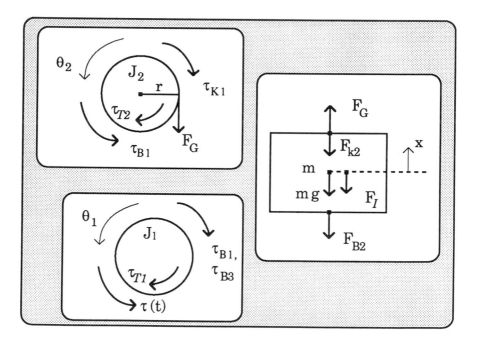

Figure 4.7. Rotational system after decomposition into parts.

Previously, it was stated that subsystems should be selected such that each subsystem contains exactly one independently movable body. In this example, we did not adhere to this rule. The subsystems J_2 and m cannot be moved independently from each other. It will be demonstrated how this decision will affect our model.

We are now ready to write simulation equations directly. As before, the simulation equations comprise the equations resulting from the d'Alembert principle, as well as the equations describing the individual forces/torques (equivalent to the "branch equations" of electrical circuits).

$$\tau(t) = \tau_{T1} + \tau_{B1} + \tau_{B3} \tag{4.32a}$$

$$\tau_{B1} = \tau_{T2} + \tau_{k1} + \tau_G \tag{4.32b}$$

$$F_G = F_{\mathcal{I}} + F_{k2} + F_{B2} + m\,g \tag{4.32c}$$

$$\tau_{T1} = J_1 \frac{d\omega_1}{dt} \tag{4.32d}$$

$$\frac{d\theta_1}{dt} = \omega_1 \tag{4.32e}$$

$$\tau_{T2} = J_2 \frac{d\omega_2}{dt} \tag{4.32f}$$

$$\frac{d\theta_2}{dt} = \omega_2 \tag{4.32g}$$

$$F_{\mathcal{I}} = m \frac{dv}{dt} \tag{4.32h}$$

$$\frac{dx}{dt} = v \tag{4.32i}$$

$$\tau_G = r\,F_G \tag{4.32j}$$

$$x = r\theta_2 \tag{4.32k}$$

$$\tau_{B1} = B_1(\omega_1 - \omega_2) \tag{4.32l}$$

$$\tau_{B3} = B_3\omega_1 \tag{4.32m}$$

$$F_{B2} = B_2\,v \tag{4.32n}$$

$$\tau_{k1} = k_1\theta_2 \tag{4.32o}$$

$$F_{k2} = k_2\,x \tag{4.32p}$$

Among these equations, we find six differential equations. However, we know already that this is a fourth–order system. The reason for this discrepancy is easily understood by looking at Eq.(4.32k). The variables x and θ_2 are related to each other in a linear fashion, i.e., they do not qualify as separate state variables. The problem was caused by the fact that we ignored the rule that systems that cannot be moved independently should not be split in two. Such a decision will always create structural singularities. Nothing is wrong with this approach though, we must only be prepared to do some extra work in the end in order to formulate an executable simulation model.

Rather than proceeding with this example, I prefer to demonstrate the learned concepts by means of a realistically complex problem, namely, the analysis of a crane crab system.

4.4 The Crane Crab Example

Figure 4.8 shows a crane crab system that is used in a mechanical shop to move heavy loads from one place to another. A cart moves horizontally on a bridge. The cart is pulled with a nonelastic rope. The rope is moved by the motor $M1$. The load hangs on another rope the length of which can be controlled by motor $M2$. It is assumed that the masses of the ropes are negligible, that both ropes are ideally stiff (they don't exhibit either an elastic or a plastic deformation), and the bridge is ideally stiff (no sag).

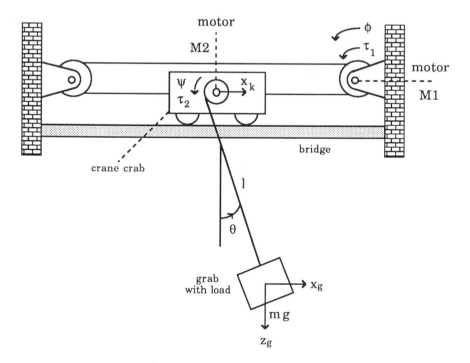

Figure 4.8. Crane crab system.

Since the system is fairly complex, we decided to use the decomposition technique. We decompose the system into four separate subsystems describing (a) the crane crab, (b) the motor $M1$, (c) the motor $M2$, and (d) the grab with the load. This is demonstrated in Fig.4.9.

The first subsystem (the crane crab) exhibits a translational movement in horizontal direction only (the vertical forces were also drawn, but they must add up to zero at all times). The second subsystem exhibits a rotational movement only, as does the third subsystem. The fourth subsystem exhibits translational movements in two directions. This seems to indicate that the system is of 10th order. However, linear dependencies exist between the subsystems (we again cut the system into smaller portions than can be moved independently). The position x_k determines the angle ϕ completely. Also, the angle ψ influences both the x_g and the z_g coordinates of the load (for a constant value of ψ, the load has only one circular path along which it can move). Consequently, the system order will have to be reduced to six and structural singularities will pop up between our initially chosen model variables.

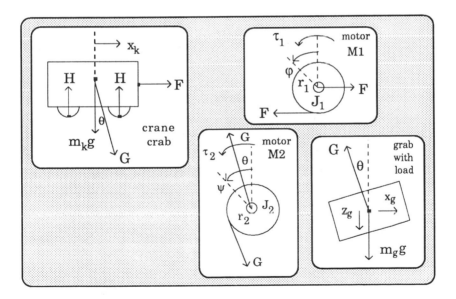

Figure 4.9. Crane crab system decomposed.

Let us write an initial set of simulation equations now. For a change, we shall not introduce any fictitious forces/torques and shall operate on Newton's law directly.

$$m_k \, \ddot{x}_k = F + G \, \sin(\theta) \tag{4.33a}$$

$$J_1 \, \ddot{\phi} = \tau_1(t) - r_1 \, F \tag{4.33b}$$

$$J_2 \, \ddot{\psi} = \tau_2(t) + r_2 \, G \tag{4.33c}$$

$$m_g \, \ddot{x}_g = -G \, \sin(\theta) \tag{4.33d}$$

$$m_g \, \ddot{z}_g = m_g \, g - G \, \cos(\theta) \tag{4.33e}$$

$$x_k = r_1 \phi \tag{4.33f}$$

$$\ell = r_2 \psi \tag{4.33g}$$

$$x_g = x_k + \ell \, \sin(\theta) \tag{4.33h}$$

$$z_g = \ell \, \cos(\theta) \tag{4.33i}$$

It was easy so far. We ended up with nine highly nonlinear equations in the nine unknowns x_k, x_g, z_g, ϕ, ψ, θ, ℓ, F, and G. In order to formulate a set of simulation equations, we shall need to analytically compute the second derivatives of Eqs.(4.33f–i). This leads to:

$$\ddot{x}_k = r_1 \ddot{\phi} \tag{4.34a}$$

$$\ddot{\ell} = r_2 \ddot{\psi} \tag{4.34b}$$

$$\ddot{x}_g = \ddot{x}_k + \ell\ddot{\theta}\cos(\theta) - \ell\dot{\theta}^2\sin(\theta) + 2\dot{\ell}\dot{\theta}\cos(\theta) + \ddot{\ell}\sin(\theta) \tag{4.34c}$$

$$\ddot{z}_g = -\ell\ddot{\theta}\sin(\theta) - \ell\dot{\theta}^2\cos(\theta) - 2\dot{\ell}\dot{\theta}\sin(\theta) + \ddot{\ell}\cos(\theta) \tag{4.34d}$$

Now let us eliminate the variables x_g and z_g by plugging Eq.(4.33d) into Eq.(4.34c) and by plugging Eq.(4.33e) into Eq.(4.34d).

$$-G\sin(\theta) = m_g[\ddot{x}_k + \ell\ddot{\theta}\cos(\theta) - \ell\dot{\theta}^2\sin(\theta) + 2\dot{\ell}\dot{\theta}\cos(\theta) + \ddot{\ell}\sin(\theta)] \tag{4.35a}$$

$$m_g g - G\cos(\theta) = m_g[-\ell\ddot{\theta}\sin(\theta) - \ell\dot{\theta}^2\cos(\theta) - 2\dot{\ell}\dot{\theta}\sin(\theta) + \ddot{\ell}\cos(\theta)] \tag{4.35b}$$

These equations can be simplified by the following operation:

$$\text{Eq.(4.35}a) \cdot \cos(\theta) - \text{Eq.(4.35}b) \cdot \sin(\theta) \Rightarrow \text{Eq.(4.36}a)$$
$$\text{Eq.(4.35}a) \cdot \sin(\theta) + \text{Eq.(4.35}b) \cdot \cos(\theta) \Rightarrow \text{Eq.(4.36}a)$$

This generates the equations:

$$-g \, \sin(\theta) = \ddot{x}_k \, \cos(\theta) + \ell\ddot{\theta} + 2\dot{\ell}\dot{\theta} \tag{4.36a}$$

$$m_g \, g \, \cos(\theta) - G = m_g[\ddot{x}_k \, \sin(\theta) - \ell\dot{\theta}^2 + \ddot{\ell}] \tag{4.36b}$$

At this point, we have seven equations in seven unknowns, namely, Eqs.(4.33a–c), Eqs.(4.34a–b), and Eqs.(4.36a–b). This set of equations is solvable, except for the fact that it contains algebraic loops.

We must either continue to eliminate variables (we can eliminate F and G, for example) until the algebraic loops disappear or we must place the entire set of equations into an IMPL construct as described in Chapter 2. In the given example, the latter approach may be more feasible.

4.5 Modeling Pulleys

Pulleys are frequently used elements to enable human operators to lift heavy loads. Figure 4.10 shows a four–pulley hoist that may serve as an example.

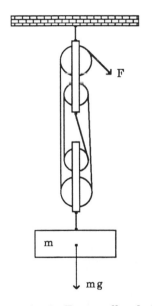

Figure 4.10. Four–pulley hoist.

The question of interest is the following: Which force F is necessary in order to keep the system in an equilibrium state? The answer is trivial. If we cut the system in the middle, we realize that the internal forces (tensions) in the four ropes must add up to mg, otherwise the lower two wheels would move either up or down. Furthermore, if we cut the system above the lowermost wheel, we see that the tensions in the two outermost ropes must be the same, otherwise the lowermost wheel would rotate. We can thus conclude that the tensions in all four ropes must be equal, i.e., $mg/4$. Consequently, in order to prevent the uppermost wheel from rotating, we need to apply a force $F = mg/4$, i.e., we require only one–fourth of the force to lift the heavy load m as compared to a direct lift.

4.6 The Inverse Pendulum Problem

Let us look at one more problem. A double pendulum is balanced on a cart. We would like to simulate what happens to the pendulum as a function of time if various types of control inputs F are applied to the system. The system is shown in Fig.4.11.

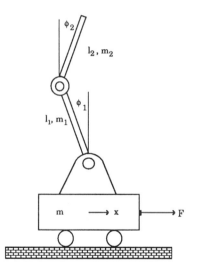

Figure 4.11. Double pendulum balanced on a cart.

This system can be used to study a number of interesting control problems. The most interesting question is the following: Assume that we start out with both sticks in the upright position. Assume that a small disturbance moves the sticks away from the unstable equilibrium point. Can we balance the two sticks to return to the upright position simply by moving the cart back and forth? In other words, can we find a control strategy that stabilizes the system around this steady–state point? Amazingly enough, the answer is yes. In fact, it has been proven that an infinitely large number of sticks can (at least theoretically) be balanced in this way. By making these infinitely many sticks infinitely short, we just reinvented the Indian magician's rope trick (no flute though).

In order to tackle the modeling task, we must first realize that we can replace the two sticks of lengths ℓ_1 and ℓ_2 and homogeneously distributed masses m_1 and m_2 by two other sticks with their masses concentrated in their centers of gravity. We can then decompose the system into three parts by introducing the internal forces at the cutting points. This is shown in Fig.4.12.

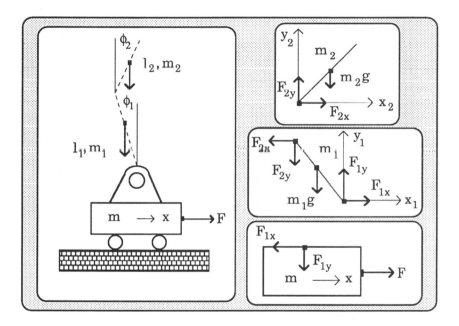

Figure 4.12. Double pendulum decomposed.

Now we are ready to write the differential equations for each of the three subsystems. Notice that the two angles ϕ_1 and ϕ_2 are counted positively clockwise from the vertical.

$$m\,\ddot{x} = F - F_{1x} \tag{4.37a}$$

$$m_1\,\ddot{x}_1 = F_{1x} - F_{2x} \tag{4.37b}$$

$$m_2\,\ddot{x}_2 = F_{2x} \tag{4.37c}$$

$$m_1\,\ddot{y}_1 = F_{1y} - F_{2y} - m_1\,g \tag{4.37d}$$

$$m_2\,\ddot{y}_2 = F_{2y} - m_2\,g \tag{4.37e}$$

$$J_1\ddot{\phi}_1 = -(F_{1x} + F_{2x})\frac{\ell_1}{2}\cos(\phi_1) + (F_{1y} + F_{2y})\frac{\ell_1}{2}\sin(\phi_1) \quad (4.37f)$$

$$J_2\ddot{\phi}_2 = -F_{2x}\frac{\ell_2}{2}\cos(\phi_2) + F_{2y}\frac{\ell_2}{2}\sin(\phi_2) \quad (4.37g)$$

$$x_1 = x + \frac{\ell_1}{2}\sin(\phi_1) \quad (4.37h)$$

$$y_1 = \frac{\ell_1}{2}\cos(\phi_1) \quad (4.37i)$$

$$x_2 = x + \ell_1\sin(\phi_1) + \frac{\ell_2}{2}\sin(\phi_2) \quad (4.37j)$$

$$y_2 = \ell_1\cos(\phi_1) + \frac{\ell_2}{2}\cos(\phi_2) \quad (4.37k)$$

The model consists of 11 highly nonlinear equations in the 11 un-
knowns x, x_1, x_2, y_1, y_2, ϕ_1, ϕ_2, F_{1x}, F_{1y}, F_{2x}, and F_{2y}. As in the
case of the crane crab system, we are plagued by structural singu-
larities. Looking at the degrees of freedom of the system, we realize
that we can move the cart in one direction, the first stick in one
direction relative to the cart and the second stick in one direction
relative to the first stick, i.e., we have three independently movable
bodies with one direction each, that is, the system must be of sixth
order. However, looking into our set of equations, we seem to have
a 14th–order model here. This discrepancy can be explained (as before) by the four linear constraints expressed in Eqs.(4.37h–k). In
order to formulate a simulation model, we would again have to compute second derivatives for these four equations, detect the resulting
algebraic loops, and eliminate variables until they disappear, or solve
the equations in an IMPL block.

4.7 Modeling Electromechanical Systems

We are now ready to model electromechanical devices. Electrical and
mechanical systems can interact in several ways, the most prominent
of which is through magnetic fields. This is how all electrical motors
work. Let us look at one such motor in a little more detail, namely,
the DC motor. Figure 4.13 shows an electromechanical diagram of
this motor.

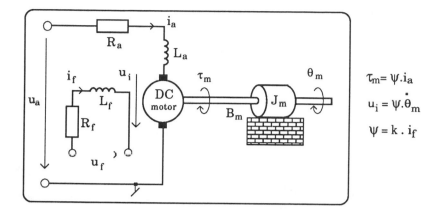

Figure 4.13. Electromechanical diagram of a DC motor.

The motor has two separate coils, the *armature coil*, which is mounted on the rotating part of the motor, and the *field coil*, which is mounted on the stationary part of the motor. The current flowing through the field coil generates a magnetic field. If current flows through the armature coil as well, a force is generated in the armature coil which is responsible for the rotation of the cylinder that is anchored to the armature coil. The resulting torque τ_m is proportional to the applied field current and to the applied armature current:

$$\tau_m = k \cdot i_f \cdot i_a \tag{4.38}$$

Very often, the DC motor is operated with a *constant field*, i.e., the angular velocity of the motor is controlled through a variation of the applied armature current. Such a configuration is called *armature control*. In that case, Eq.(4.38) can also be written as:

$$\tau_m = \psi \cdot i_a \tag{4.39}$$

where ψ is sometimes called the *torque constant* and sometimes the *Back EMF constant*, since the same constant appears in a second equation:

$$u_i = \psi \cdot \omega_m \tag{4.40}$$

which describes the voltage induced in the armature coil under the influence of the rotation.

With these two equations, we can now model the DC motor as a whole since it consists of the two electrical subsystems describing the field and the armature, the mechanical subsystem describing the inertia and friction of the rotating cylinder, plus the two coupling equations that connect the mechanical subsystem to the electrical subsystem. The equations should be self–explanatory by now.

$$u_f = R_f \, i_f + L_f \frac{di_f}{dt} \tag{4.41a}$$

$$u_a = R_a \, i_a + L_a \frac{di_a}{dt} + u_i \tag{4.41b}$$

$$J_m \frac{d^2\theta_m}{dt^2} = \tau_m - \tau_L - B_m \frac{d\theta_m}{dt} \tag{4.41c}$$

$$\tau_m = k \, i_f \, i_a \tag{4.41d}$$

$$u_i = k \, i_f \, \omega_m \tag{4.41e}$$

Figure 4.14 shows a block diagram of the DC motor.

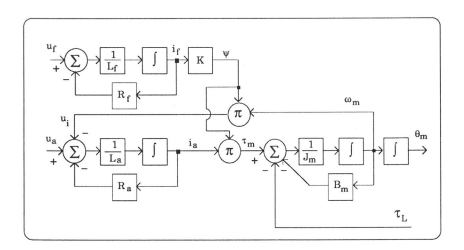

Figure 4.14. Block diagram of a DC motor.

The motor has three different inputs, namely, the voltage applied to the field coil u_f, the voltage applied to the armature coil u_a, and the torque load τ_L that results from the machinery that is being driven by the motor. The block diagram also explains the popularity of armature control. If the field current is constant, the device degenerates to a *linear system*, which is better amenable to an analytical treatment (although, for the purpose of simulation, we couldn't care less).

4.8 Summary

In this chapter, we dealt with the problem of modeling planar mechanical systems in both translational and rotational coordinates. More can be said about mechanical systems than we were able to cover in this chapter. Unfortunately, we lack the necessary space in this text for an enhanced discussion. Notice that one of our major goals is to bridge the gap between the various application areas of differential equation models and to develop a consistent terminology and methodology to deal with such models. It was not the aim of this chapter to duplicate the tremendous effort that went into the design of textbooks in mechanics.

What are the topics that were left out of this chapter and where are they discussed?

This chapter discussed planar systems only. A more general discussion of the subject matter should include three–dimensional mechanical problems. A free–moving rigid body can translate along and rotate around the three spatial axes independently. Consequently, Newton's law (or the d'Alembert principle) must be formulated six times, once for each degree of freedom. Therefore, a free–moving rigid body is described through a 12th–order state–space model. Examples of three–dimensional motions are presented in the two projects of this chapter.

A number of textbooks deal with the modeling of mechanical systems, textbooks that are geared more toward the needs of mechanical engineering students [4.1,4.12,4.13]. However, all these texts are junior–level, not senior–level textbooks. Consequently, while they cover the modeling of mechanical systems on a larger number of pages, they simply proceed at a somewhat slower pace and do not really extend their coverage beyond our discussion. In particular, none of those textbooks covers general motions of mechanical systems in three space dimensions. Moreover, all these texts are geared toward *analytical modeling* rather than *simulation modeling*. Consequently, they stop with the derivation of the differential equations themselves and don't bother to translate these differential equations into state–space models. Consequently, they don't discuss the concept of structural singularities and algebraic loops at all. However, since these texts are meant to be used by mechanical engineers, they provide nice chapters on hydraulic system modeling and pneumatic system modeling, topics for which we lack the space in this textbook, which has been written more with electrical engineering and systems engineering students in mind.

A good selection of general Newtonian mechanics textbooks exists that indeed go far beyond our coverage of the topic [4.5,4.18,4.21]. All of these texts discuss three–dimensional motions in great detail. However, these textbooks deal with the physics of mechanical systems only and are not really meant to be modeling textbooks.

Secondly, when modeling moving bodies (such as an aircraft), it is quite common to describe the motion of the body relative to a coordinate system that moves along with the body. The origin of the moving coordinate system is then often assumed to be the center of gravity of that body. Typical examples are bodies that move within an Earth–fixed coordinate system, but relative to the movement of planet Earth, or a robot's end–effector, which moves relative to the position of its wrist. The total movement of a rigid body is thus decomposed into a movement relative to a moving coordinate frame and an absolute movement of that coordinate frame itself. In this case, Newton's law (or the d'Alembert principle) must be slightly modified by including two additional fictitious forces, namely, the *centripetal force* F_z, and the *Coriolis force* F_C. Again, this topic is carefully discussed in all classical Newtonian mechanics textbooks. An example of a relative motion is presented in Pr.[P4.2].

Finally, a more modern approach to dealing with mechanical systems is through the use of the *Euler equation*, which replaces and generalizes Newton's law. Newton's law, as discussed in this chapter, assumes that the equations of motion are described in Cartesian coordinates. This is not always practical. We can overcome this limitation by formulating the total kinetic energy of all bodies in the system in as many different velocities as the system contains independently moving bodies. For example, for the system in Fig.4.5, we find:

$$E_k = \frac{1}{2}J_1\dot{\theta}_1^2 + \frac{1}{2}J_2\dot{\theta}_2^2 + \frac{1}{2}m\dot{x}^2 = \frac{1}{2}J_1\dot{\theta}_1^2 + \frac{1}{2}(J_2 + mr^2)\dot{\theta}_2^2 \qquad (4.42)$$

We call the θ_i and x_i variables our *generalized displacements* q_i. In these variables, we can then reformulate Newton's law as follows:

$$\frac{d}{dt}\left(\frac{\partial E_k}{\partial \dot{q}_i}\right) - \frac{\partial E_k}{\partial q_i} = Q_i \qquad (4.43)$$

where Q_i stands for the sum of all generalized forces in the direction of the generalized displacements q_i. This formulation of Newton's law is more powerful than the previously used formulation since it is valid independent of the coordinate frame being used. Furthermore, with this approach, it is no longer necessary to cut the system into individually moving pieces by introducing coupling forces that we must later eliminate again. Therefore, this approach is often more economical than the direct application of Newton's law.

A special case is the set of *conservative systems*, i.e., systems without energy dissipation (i.e., the frictionless systems). For those systems, yet another formulation of the equations of motion can be found. This time, we consider also the potential energy E_p of the system. Thereafter, we build the so-called *Lagrangian* of the system, which is the difference between the total kinetic energy and the total potential energy

$$L(q_1, \ldots, q_n, \dot{q}_1, \ldots, \dot{q}_n) = E_k - E_p \qquad (4.44)$$

Using the Lagrangian, we can reformulate Newton's law as follows:

$$\frac{d}{dt}\left(\frac{\partial L}{\partial \dot{q}_i}\right) - \frac{\partial L}{\partial q_i} = 0 \qquad (4.45)$$

Equation (4.45) is referred to as the *Lagrange equation*.

Another formulation of Eq.(4.45) is through the use of the so-called *Hamiltonian* of the system, which is equivalent to the total free energy of the system

$$H(q_1, \ldots, q_n, \dot{q}_1, \ldots, \dot{q}_n) = E_k + E_p \qquad (4.46)$$

We can now replace the derivatives of the generalized displacements by generalized momenta. In translational coordinates, the generalized momentum of a displacement x_i is its momentum $p_i = \mathcal{I}_i = m_i \dot{x}_i$, whereas in rotational coordinates, the generalized momentum of an angle θ_i (a generalized displacement) is its twist $p_i = \mathcal{T}_i = J_i \dot{\theta}_i$. We can thus reformulate the Hamiltonian in terms of the new variables q_i and p_i. Using this version of the Hamiltonian, we can reformulate Newton's law as follows:

$$\dot{p}_i = -\frac{\partial H}{\partial q_i}, \quad \dot{q}_i = \frac{\partial H}{\partial p_i} \qquad (4.47)$$

Equations (4.47) are referred to as the *Hamilton equations*. We shall resume this discussion in Chapters 7 and 8.

In many mechanical systems, the kinematic constraints are the dominating factors that determine the motion of the system. The system is so rigid and moves so slowly that the dynamics of the system are no longer considered important and are therefore simply being ignored. Inputs to these systems are no longer forces and torques, but rather positions and angles. The goal of the investigation is to determine the positions and angles of all parts of the system in response to the applied inputs. The responses are considered instantaneous. Time appears only through the input functions themselves, which are often assumed to be functions of time. Consequently, these models are not differential equation models and are therefore outside the realm of this textbook.

Very little was said in this chapter about software for dealing with mechanical problems. Indeed, very little has been accomplished in this respect. A nicely written senior–level textbook exists that describes in detail both the kinematics and the kinetics of two–dimensional (planar) and three–dimensional mechanical systems [4.10]. However, while this textbook describes clearly how such systems are being modeled, and while it is also computer–oriented, it is somewhat disappointing with respect to the maturity of software concepts and user interfaces of the programs introduced to deal with the simulation of these systems. It contains no more than a bunch of FORTRAN–coded subroutines that the student is encouraged to adapt to his or her needs. Unfortunately, this seems to be the state of the art in mechanical system simulation.

Three hot research topics in mechanical system modeling can be named. One topic is related to full digital flight simulators. Such simulators are used for three separate purposes: (1) as parts of autopilots, (2) for pilot training, and (3) for system troubleshooting. The equations of flight motion are straightforward, and a number of good textbooks exist that deal with those in great detail [4.3,4.6]. The major problem with flight simulators relates to their execution speed. Flight simulators must execute in real time, which either calls for very fast computers (which are still fairly expensive) or special parallel–processor architectures for which no good distributed real–time operating systems exist yet. Consequently, flight simulators still rely heavily on assembly programming and special tricks to reduce the execution time required.

A second hot topic is robot modeling for the purpose of robot control. Again, good textbooks can be found that deal with robot modeling in great detail [4.4,4.14,4.15,4.18]. The major problem here is with the kinematic constraints. Robots are highly nonlinear systems and the algebraic loops created by the kinematic constraints often cannot be solved analytically. In many cases, the robot dynamics are not considered at all and the movement of the robot is dictated entirely by its kinematic constraints. Newer papers, in particular those dealing with dexterous hand movements and other sorts of fine motion planning, include the dynamic equations of motion. The problems here are mostly concerned with the highly nonlinear and nonmeasurable friction and backlash coefficients. Clearly, more research must be devoted to the development of user–friendly general–purpose robotics software. A good amount of work went recently into the development of three–dimensional graphics engines, which allow display of the three–dimensional motion of robots on a computer screen.

A third hot topic is related to the modeling of mechanical limbs for the purpose of prosthesis design. To my knowledge, no modern engineering textbook exists yet that deals with this topic, but a series of very good research papers have been devoted to this topic recently.

References

[4.1] Charles M. Close and Dean K. Frederick (1978), *Modeling and Analysis of Dynamic Systems*, Houghton Mifflin, Boston, Mass.

[4.2] J. Denavit and R. S. Hartenberg (1955), "A Kinematic Notation for Lower–Pair Mechanisms Based on Matrices," *ASME J. of Applied Mechanics*, 22(2), pp. 215–221.

[4.3] Bernard Etkin (1982), *Dynamics of Flight*, John Wiley & Sons, New York.

[4.4] King–Sun Fu, Rafael C. Gonzales, and C. S. George Lee (1987), *Robotics: Control, Sensing, Vision, and Intelligence*, McGraw–Hill, New York.

[4.5] Herbert Goldstein (1980), *Classical Mechanics*, second edition, Addison–Wesley, Reading, Mass.

[4.6] Robert K. Heffley and Wayne F. Jewell (1972), *Aircraft Handling Qualities Data*, Report NASA–CR–2144, Systems Technology, Inc., Washington, D.C.

[4.7] Gene H. Hostetter, Clement J. Savant, Jr., and Raymond T. Stefani (1989), *Design of Feedback Control Systems*, second edition, Saunders College Publishing, New York.

[4.8] Granino A. Korn (1989), *Interactive Dynamic–System Simulation*, McGraw–Hill, New York.

[4.9]   Edward E. L. Mitchell and Joseph S. Gauthier (1986), *ACSL: Advanced Continuous Simulation Language — User Guide and Reference Manual*, Mitchell & Gauthier Assoc., Concord, Mass.

[4.10] Parviz E. Nikravesh (1988), *Computer–Aided Analysis of Mechanical Systems*, Prentice–Hall, Englewood Cliffs, N.J.

[4.11] Norman S. Nise (1991), *Control Systems Engineering*, Benjamin/Cummings, Ft. Collins, Colo.

[4.12] Katsuhiko Ogata (1978), *System Dynamics*, Prentice–Hall, Englewood Cliffs, N.J.

[4.13] William J. Palm III (1983), *Modeling, Analysis, and Control of Dynamic Systems*, John Wiley, New York.

[4.14] Richard P. Paul (1981), *Robot Manipulators: Mathematics, Programming, and Control — The Computer Control of Robot Manipulators*, MIT Press, Boston, Mass.

[4.15] Mark W. Spong and M. Vidyasagar (1989), *Robot Dynamics and Control*, John Wiley, New York.

[4.16] Pentti J. Vesanterä, and François E. Cellier (1989), "Building Intelligence Into an Autopilot — Using Qualitative Simulation to Support Global Decision Making," *Simulation*, **52**(3), pp. 111–121.

[4.17] John V. Wait and DeFrance Clarke III (1976), *DARE–P User's Manual*, Version 4.1, Dept. of Electrical & Computer Engineering, University of Arizona, Tucson, Ariz.

[4.18] Jens Wittenburg (1977), *Dynamics of Systems of Rigid Bodies*, Teubner Verlag, Stuttgart, F.R.G.

[4.19] William A. Wolovich (1987), *Robotics: Basic Analysis and Design*, Holt, Rinehart and Winston, Inc., New York.

[4.20] Bernard P. Zeigler, François E. Cellier, and Jerzy W. Rozenblit (1988), "Design of a Simulation Environment for Laboratory Management by Robot Organizations," *J. of Intelligent and Robotic Systems*, **1**, pp. 299–309.

[4.21] Hans Ziegler (1965), *Mechanics*, Addison–Wesley, Reading, Mass.

Homework Problems

[H4.1] Lagrangian

In the system of Fig.4.5, replace all frictions (dissipative terms) temporarily by springs. Call the spring constants β_1, β_2, and β_3. In this way, we have transformed our previously dissipative system into another system that is conservative. Its potential energy can be computed as:

$$E_p = \frac{1}{2}\beta_3\theta_1^2 + \frac{1}{2}\beta_1(\theta_1 - \theta_2)^2 + \frac{1}{2}k_1\theta_2^2 + \frac{1}{2}k_2x^2 + \frac{1}{2}\beta_2x^2 + mgx - \tau\theta_1 \quad (H4.1)$$

Find the Lagrangian of this system and plug it into Eq.(4.37). Derive a set of differential equations describing this conservative system. In the very end, replace each term of type $\beta_i q_j$ by the original friction term $B_i\dot{q}_j$ and show that you end up with the same set of differential equations as with our original approach. This cheap engineering trick allows us to use the Lagrangian and/or the Hamiltonian to analyze dissipative systems.

[H4.2] Hamiltonian

Find a Hamiltonian for the modified system of Hw.[H4.1], replace the derivatives of the generalized displacements $\dot{q}_i$ by the generalized momenta p_i and write down the modified expression for the Hamiltonian in the q_i and p_i variables. Then apply Eq.(4.39) and show that you end up with the same set of differential equations as in Hw.[H4.1] prior to replacing the dissipative terms back into the system equations.

[H4.3] Cervical Syndrome [4.7]

Some people (such as my sister) suffer from a so–called cervical syndrome. Their neck is not sufficiently stiff to connect their head solidly with their upper torso. Therefore, if their upper torso is exposed to vibrations, such as when riding in a car, these people often react with severe headaches.

A car manufacturer wants to design a new car in which these problems are minimized. Resonance phenomena are to be studied with the purpose of avoiding resonance frequencies of the human body to appear as eigen-frequencies of the car.

Figure H4.3 shows a mechanical model of a sitting human body. The legs are left out since they do not contribute to potential oscillations of the upper body. The data are average data for a human adult.

Derive a state–space model for this system. Since this is a linear time–invariant system, put it in linear state–space form and simulate the system directly in CTRL–C (or MATLAB) using an AC force input of 1.5 Hz. The output of interest is the distance between the head and the upper torso.

In order to analyze the resonance phenomena, we wish to obtain a Bode diagram of this system. Create a frequency base logarithmically spaced between 0.01 and 100 Hz using CTRL–C's (MATLAB's) "logspace" function. Then compute a Bode diagram using the "bode" function. Convert the amplitude into decibels and plot, on two graphs, the magnitude and the phase in a semilogarithmic scale using CTRL–C's (MATLAB's) "plot" function. Both CTRL–C and MATLAB offer interactive help on all these functions (in CTRL–C, you need also the "window," "title," "xlabel," and "ylabel" functions; MATLAB offers similar facilities). Determine all resonance frequencies together with the maximum overshoot at these frequencies.

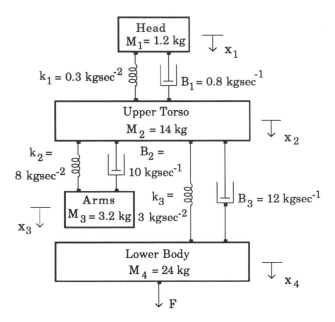

Figure H4.3. Mechanical model of a sitting human body [4.7].[†]

Finally, we wish to perform a sensitivity analysis. We want to study the variability of the spring constant and the damper between the head and the upper torso. For this purpose, we assume a variability of k_1 and B_1 of ±50%. Repeat the frequency analysis for the four worst–case combinations and determine the range of resonance frequencies to be avoided. Determine also the maximum overshoot to be expected in the worst case. I suggest that you create a CTRL–C (MATLAB) function which computes the **A** matrix and **b** vector as a function of the two model parameters k_1 and B_1. This will allow you to generate the four models more easily. While you execute these simulations, keep a diary of what you are doing (using CTRL–C's or MATLAB's "diary" function).

[H4.4] Electromechanical System

The electromechanical system shown in Fig.H4.4 can represent either a microphone, a loudspeaker, or a vibrating table.

[†] Adapted figure from DESIGN OF FEEDBACK CONTROL SYSTEMS, Second Edition by Gene Hostetter *et al.*, ©1989 by Saunders College Publishing, a division of Holt, Rinehart and Winston, Inc., reprinted by permission of the publisher.

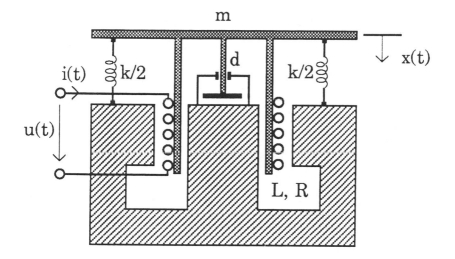

Figure H4.4. Electromechanical system.

A moving induction coil is placed in the circular gap of a permanent magnet with the magnetic induction B. It can oscillate in the axial direction. The coil has the inductance L and resistance R. It consists of w windings with a radius r. The length of the coil is ℓ. Coupled to the coil is a mechanical system with a mass m, a spring constant k, and a damping factor d. The coupling between the electrical and mechanical system can be described by the following two equations:

$$F = B \cdot i \cdot \ell \qquad (H4.4a)$$
$$u_i = B \cdot \dot{x} \cdot \ell \qquad (H4.4b)$$

where F denotes the force exerted on the mechanical system as a function of the current i that flows through the coil and u_i denotes the voltage induced in the coil as a result of the mechanical movement.

Determine a state–space model for this electromechanical system and draw a block diagram with $u(t)$ as input and $x(t)$ as output.

[H4.5] Translational System [4.11]

The translational mechanical system shown in Fig.H4.5 is to be modeled.

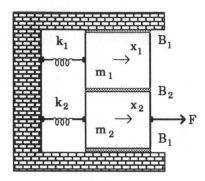

Figure H4.5. Translational mechanical system [4.11].[†]

Find a linear state–space model for this system with $F(t)$ as input and $x_1(t)$ as output.

[H4.6]* Mixed Translational and Rotational System

The mixed translational and rotational system shown in Fig.H4.6 is to be modeled and simulated. This is a simplified version of a slipping clutch. A force F pulls (or pushes) a mass $m = 3.2$ kg along a floor. Viscous friction exists between the mass m and the floor. The friction force is:

$$F_{Fr} = B_1 \dot{x} \qquad (H4.6a)$$

where $B_1 = 0.8$ kg sec^{-1}. The mass m is also attached to the rear wall with a spring. The spring constant is $k_1 = 5$ kg sec^{-2}. A cylinder with the inertia $J = 0.001$ kg m^2 and the radius $r = 0.05$ m sits on top of the mass. It can either roll on the mass or slip over the mass. Coulomb friction exists between the mass m and the cylinder. The friction force is:

$$F_{Coul} \leq B_2 \, \text{sgn}(\dot{x}) \qquad (H4.6b)$$

where $B_2 = 0.4$ kg m sec^{-2}. As long as the internal force at the contact point between the mass m and the cylinder is smaller than the maximum Coulomb friction, the cylinder will roll and behaves exactly like a gear. However, as soon as the internal force becomes larger than the maximum Coulomb friction, the cylinder will start to slip. Thereafter, the cylinder will continue to slip until the two velocities at the contact point have equalized again. At this point in time, the cylinder will return to its rolling

---

[†] Adapted figure from CONTROL SYSTEMS ENGINEERING by Norman Nise, ©1991 by Benjamin/Cummings, Inc., printed by permission of the publisher.

mode. The cylinder is attached to the two side walls with two rotational springs with the spring constants $k_2 = k_3 = 0.001$ kg m^2 sec^{-2}.

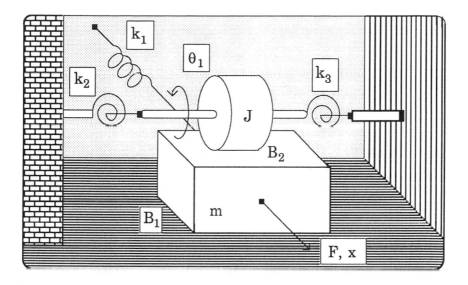

Figure H4.6. Mixed translational and rotational system.

Model this nonlinear system in ACSL using $F(t)$ as input and both $x(t)$ and $\theta_1(t)$ as outputs. Apply a force:

$$F(t) = 0.1 \cdot t \cdot \sin(t) \qquad (H4.6c)$$

and simulate this system during 75 sec.

This simulation is not so simple. We need to toggle between two different models. The overall model is of second order whenever the cylinder rolls, but of fourth order when the cylinder slips. This is therefore a so–called *variable structure model*. In order to model the switching between the two modes correctly, we require two state–events. ACSL's state–event scheduler is called from within the DERIVATIVE segment. The two statements:

> **schedule** *goslip* **.xn.** *slpcon*
> **schedule** *goroll* **.xn.** *rolcon*

can be placed within the DERIVATIVE section. *Slpcon* is a real expression that triggers the execution of a DISCRETE section by the name of *goslip* to be executed whenever *slpcon* changes its sign from positive to negative. *Rolcon* works accordingly.

Projects

[P4.1] Aircraft Modeling [4.16]

Flight stability can be studied through two independent models: longitudinal and lateral. Longitudinal motions can be modeled independently from the lateral ones if the following simplifying assumptions are valid:

(1) The airplane is perfectly symmetrical with respect to its median longitudinal plane.

(2) No gyroscopic effects of spinning masses (engine rotors, airscrews, etc.) act on the aircraft.

In this project, we want to adopt these assumptions and consider the longitudinal model of a B747 aircraft in cruise flight at high altitude.

A longitudinal flight is characterized by the absence of forces and moments that would cause its lateral motion. (Notice the terminological confusion: the term "moment" is here used as a synonym for "torque," and not in the sense in which we have introduced that term before. However, I decided to stick to the conventional terminology since this is the one that you will commonly find when you scan through aerodynamics literature.) Furthermore, the aeroelastic nature of the airplane's structure is neglected as well, so that the rigid body equations of motion apply to the model.

The mathematical model described in this project reflects an essentially longitudinal flight restricted to longitudinal deviations from a trimmed reference flight condition. This reference flight is characterized by the requirement that the resultant force and torque acting on the aircraft's center of mass are zero.

We define a reference flight condition as being characterized by a steady longitudinal and horizontal flight where the resultant force and moment acting on the plane are zero. The headwind is assumed to be constant and horizontal.

The theory presented in this project is developed with respect to a set of body–fixed axes named the *stability axes*. The origin of this coordinate system is the center of gravity of the airplane: the x–axis points in the direction of the motion of the airplane in the reference flight condition, the z–axis points "downward," and the y–axis runs spanwise and points to the right.

Three angles are defined to describe the relative position of the velocity vector of the center of gravity of the airplane with respect to an Earth–fixed reference frame and a fuselage–fixed reference frame.

α is the angle of attack (or incidence) of the airplane, which describes the inclination of the resultant velocity vector **v** to the x–axis of the body–fixed coordinate system. The usual notation of the velocity components in the stability axes is u for the x–axis component, and w for the z–axis component. Hence

$$\alpha = \tan^{-1}\left(\frac{w}{u}\right) \qquad (P4.1a)$$

γ is the flight–path angle of the aircraft, representing the inclination of the velocity vector to the horizontal, i.e., to the x–component of the Earth–fixed reference frame.

θ is the pitch angle, being the one that is best sensed by a human pilot, for it represents the relative position between the two reference frames. It is defined in terms of the previous angles as follows:

$$\theta - \gamma + \alpha \qquad (P4.1b)$$

The relationship between the different variables is graphically depicted in Fig.P4.1a.

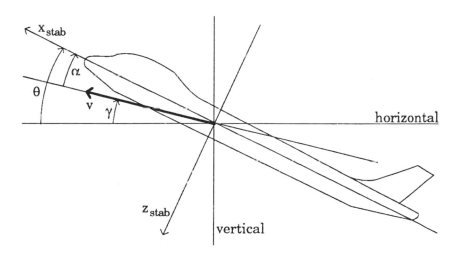

Figure P4.1a. The reference angles [4.16].

The tangential and normal components of the resultant force and the moment about the center of gravity of the airplane considered as a rigid body whose mass is constant over time can be written in terms of the reference angle γ and θ as:

$$F_t = m\frac{d\mathbf{v}}{dt} \qquad (P4.1c)$$

$$F_n = m\mathbf{v}\frac{d\gamma}{dt} \qquad (P4.1d)$$

$$M_y = I_y\frac{d^2\theta}{dt^2} \qquad (P4.1e)$$

The quantities affecting the airplane in flight are its weight W, the thrust T developed by the engines, the aerodynamic forces lift L and drag D, and the aerodynamic pitching moment M.

The *weight* of the aircraft will be considered constant (thus the weight of the fuel consumed during the flight is neglected).

The *thrust* developed by the propulsive system will be considered a function of the flight velocity and of its own control variable δ_T, the throttle opening. For reasons of simplicity, the thrust line will be assumed to coincide with the x–axis of the stability axes. The center of gravity, by definition, is in this axis, and therefore, the thrust does not affect the moment directly.

The aerodynamic forces L and D compose the force response of the aircraft to the motion. They act in the mean aerodynamic center of the wing, causing the aerodynamic moment M about the center of gravity, which is defined to be positive for a nose–down effect. The *lift* is defined as being the normal component of the aerodynamic force with respect to the flight path and the *drag* is its tangential component.

The forces and moments acting on the aircraft are graphically depicted in Fig.P4.1b.

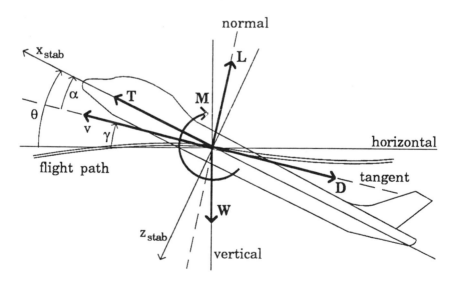

Figure P4.1b. Forces and moments acting on the airplane [4.16].

The standard way of expressing the aerodynamic forces D and L and the longitudinal aerodynamic moment M is through their nondimensional aerodynamic coefficients C_D, C_L, and C_M:

$$D = \frac{1}{2}\rho v^2 S C_D \qquad (P4.1f)$$

$$L = \frac{1}{2}\rho v^2 S C_L \qquad (P4.1g)$$

$$M = \frac{1}{2}\rho v^2 S \frac{\bar{c}}{2} C_M \qquad (P4.1h)$$

which shows their direct dependence on the local air density ρ, the square of the cruising speed v, and the size of the aerodynamic surface S of the airplane.

Parameter $\bar{c}/2$ in the expression for the moment stands for the characteristic length (for the nondimensional coefficients), taken as half of the mean aerodynamic chord $\bar{c}$ of the wing.

The three nondimensional coefficients C_D, C_L, and C_M express the aerodynamic response of the airplane to variations in the following aerodynamic variables:

(1) α, the angle of attack.
(2) δ_e, the elevator deflexion.
(3) $\dot{\alpha}$, the angle of attack rate.
(4) q, the pitch rate.

Equations (P4.1i–k) describe the nondimensional aerodynamic coefficients expressed by a Taylor series expansion around an initial value (subscript 0) for which α, δ_e, $\dot{\alpha}$, and q are zero:

$$C_L = C_{L_0} + \frac{\partial C_L}{\partial \alpha}\alpha + \frac{\partial C_L}{\partial \delta_e}\delta_e + \frac{\partial C_L}{\partial \dot{\alpha}}\dot{\alpha} + \frac{\partial C_L}{\partial q}q \qquad (P4.1i)$$

$$C_D = C_{D_0} + \frac{\partial C_D}{\partial \alpha}\alpha \qquad (P4.1j)$$

$$C_M = C_{M_0} + \frac{\partial C_M}{\partial \alpha}\alpha + \frac{\partial C_M}{\partial \delta_e}\delta_e + \frac{\partial C_M}{\partial \dot{\alpha}}\dot{\alpha} + \frac{\partial C_M}{\partial q}q \qquad (P4.1k)$$

The aerodynamic reactions of the airplane can be represented approximately by means of stability derivatives, i.e., the coefficients of the Taylor series expansion.

Note that, as C_D is strongly influenced by the angle of attack, all other influences can be neglected.

The α derivatives C_{L_α}, C_{D_α}, and C_{M_α} describe how changes in the angle of attack α affect the aerodynamic forces and moments. An increase in the angle of attack generally induces an increase in the lift, an increase in the drag, and a decrease in the pitching moment.

The δ_e derivatives, $C_{L_{\delta_e}}$ and $C_{M_{\delta_e}}$, describe the effect that a deflexion of the elevator has on the lift and pitching moment. A positive elevator deflexion is defined as being *elevator down*, which causes an increase in the lift and a negative pitching moment increment.

The $\dot{\alpha}$ derivatives, $C_{L_{\dot{\alpha}}}$ and $C_{M_{\dot{\alpha}}}$, represent the adjustment of the pressure distribution on the aerodynamic surfaces to sudden changes in the angle of attack, as, for example, when sudden changes in the incidence of the headwind occur.

The q derivatives, C_{L_q} and C_{M_q}, represent the aerodynamic effects induced by a rotation of the airplane about its spanwise axis when the angle of attack is kept constant, for example, keeping the fuselage horizontal in an arbitrarily varying flight path.

These two rotational effects can be visualized considering a flight along an arbitrary flight path: first with the fuselage of the plane always tangential to the flight path (angle of attack kept zero) and second with the fuselage always horizontal (pitch angle kept zero).

Finally, the nondimensional aerodynamic coefficients can be expressed by the set of equations:

$$C_L = C_{L_0} + C_{L_\alpha}\alpha + C_{L_{\delta_e}}\delta_e + \frac{\bar{c}/2}{v}\left[C_{L_{\dot{\alpha}}}\dot{\alpha} + C_{L_q}q\right] \qquad (P4.1l)$$

$$C_D = C_{D_0} + C_{D_\alpha}\alpha \qquad (P4.1m)$$

$$C_M = C_{M_0} + C_{M_\alpha}\alpha + C_{M_{\delta_e}}\delta_e + \frac{\bar{c}/2}{v}\left[C_{M_{\dot{\alpha}}}\dot{\alpha} + C_{M_q}q\right] \qquad (P4.1n)$$

Note that the rotational derivatives, $C_{L_{\dot{\alpha}}}$, C_{L_q}, $C_{M_{\dot{\alpha}}}$, and C_{M_q}, are multiplied by $\bar{c}/(2 \cdot v)$. This comes from the fact these derivatives are, in fact, with respect to the quantity $\dot{\alpha}\bar{c}/(2 \cdot v)$ or $q\bar{c}/(2 \cdot v)$ where $\bar{c}$ is the mean aerodynamic chord of the wing and v is the cruising velocity.

For example, consider $C_{L_{\dot{\alpha}}}$:

$$C_{L_{\dot{\alpha}}} = \frac{\partial C_L}{\partial \frac{\dot{\alpha}\bar{c}/2}{v}} = \frac{\partial C_L}{\frac{\bar{c}/2}{v}\partial\dot{\alpha}} \qquad (P4.1o)$$

and therefore:

$$\frac{\partial C_L}{\partial\dot{\alpha}} = \frac{\bar{c}/2}{v}C_{L_{\dot{\alpha}}} \qquad (P4.1p)$$

Longitudinal flight control means control of the velocity vector **v** acting on the center of gravity of the airplane. The two available control elements are δ_e for the elevator deflexion and δ_T for the throttle control of the thrust.

The immediate response of the aircraft to a $\Delta\delta_e$ at constant throttle is a brisk rotation in pitch and a consequent change in both angle of attack and lift, followed by a curvature $\dot{\gamma}$ of the flight path. After this first fast transient, the new steady–state flight is characterized by the new values of γ_{ss} and u_{ss}. The steady–state speed u_{ss} is fixed by the value of $C_{L_{ss}}$, which, in turn, is determined by δ_e.

The immediate effect of a positive $\Delta\delta_T$ with fixed δ_e is essentially a change in the velocity followed by a change in the flight–path angle γ.

However, as a given δ_e fixes a constant steady–state velocity, the final effect of opening the throttle will be a change in the flight–path angle without changing the speed.

We are now able to write down the equations of motion. This will not be done in the stability axes, but in the tangential and normal axes with respect to the flight path, because this simplifies the equations:

$$m\dot{v} = T \, \cos(\alpha) - D - W \, \sin(\gamma) \tag{P4.1q}$$
$$mv\dot{\gamma} = T \, \sin(\alpha) + L - W \, \cos(\gamma) \tag{P4.1r}$$
$$I_y\dot{q} = M \tag{P4.1s}$$
$$\dot{\theta} = q \tag{P4.1t}$$

Equation (P4.1b) reflects the relationship between the reference angles, while the position of the airplane with respect to the ground is given by Eqs.(P4.1u–v):

$$\dot{h} = v \, \sin(\gamma) \tag{P4.1u}$$
$$\dot{x} = v \, \cos(\gamma) \tag{P4.1v}$$

Equations (P4.1f–h) specify the aerodynamic quantities L, D, and M, and the nondimensional coefficients C_L, C_D, and C_M are given by Eqs.(P4.1l–n).

The two control laws implemented in the model are standard procedure in the control of flight stability. Feedback of the pitch angle deviation from its trimmed value (for which the airplane is in steady horizontal reference flight) into the elevator deflexion suppresses effectively the phugoid mode of the airplane, which is slow and very lightly damped. The second control loop was similarly implemented feeding back the velocity into the thrust.

$$\delta_e = \delta_{e_{trim}} + K_\theta(\theta - \theta_{trim}) \tag{P4.1w}$$
$$T = T_{trim} + K_u(u - u_{trim}) \tag{P4.1x}$$

The subscript *trim* refers to the trimmed value of the variable and u is the x–component of the velocity in the stability axes, or:

$$u = v \cdot \cos(\alpha) \tag{P4.1y}$$

In tables P4.1a–e, all the values used for the flight–related constants are listed. The airplane–related physical data have been chosen for a large commercial/cargo Boeing 747 jet plane in cruise flight at an altitude of $20,000$ ft and a speed of Mach .5 (≈ 500 ft/sec).

The aerodynamic coefficients are adapted for a trimmed reference flight with a given set of initial conditions, which is characterized by a horizontal steady flight at 500 ft/sec, an altitude of 20,000 ft, a zero angle of attack, an elevator deflexion of 1.6° (0.0279 rad), and a constant thrust of 33,000 lb. The initial conditions are set such that the flight will start perfectly trimmed since approximation errors in the aerodynamic constants must still be corrected. The data for this model were taken from Heffley and Jewell [4.6].

Table P4.1a. Airplane Constants.

I_y	$= 27,000,000.0$	$[\text{lb ft}^2]$
W	$= 500,000.0$	$[\text{slug}]$
$\bar{c}$	$= 27.3$	$[\text{ft}]$
S	$= 6000.0$	$[\text{ft}^2]$

Table P4.1b. Physical Constants.

g	$= 32.2$	$[\text{ft/sec}^2]$
ρ	$= 0.0012$	$[\text{lb/ft}^3]$, at 20,000 ft

Table P4.1c. Aerodynamic Constants.

C_{D_0}	$= 0.036667$	$[\ ]$
C_{D_α}	$= 0.26$	$[1/\text{rad}]$
C_{L_0}	$= 0.5455$	$[\ ]$
C_{L_α}	$= 5.2$	$[1/\text{rad}]$
$C_{L_{\delta_e}}$	$= 0.36$	$[1/\text{rad}]$
$C_{L_{\dot{\alpha}}}$	$= 2.0$	$[1/\text{rad}]$
C_{L_q}	$= 5.5$	$[1/\text{rad}]$
C_{M_0}	$= 0.039$	$[\ ]$
C_{M_α}	$= -0.74$	$[1/\text{rad}]$
$C_{M_{\delta_e}}$	$= -1.4$	$[1/\text{rad}]$
$C_{M_{\dot{\alpha}}}$	$= -8.0$	$[1/\text{rad}]$
C_{M_q}	$= -22.0$	$[1/\text{rad}]$

Table P4.1d. Feedback Gains.

K_θ	$= 0.25$	$[\ ]$
K_u	$= 40.0$	$[\text{lb/sec}]$

Table P4.1e. Initial Conditions.

v_0	$= 500.1375$	$[\text{ft/sec}]$
h_0	$= 20,000.0$	$[\text{ft}]$
x_0	$= 0.0$	$[\text{ft}]$
q_0	$= 0.0$	$[\text{rad/sec}]$
α_0	$= -0.000055$	$[\text{rad}]$
θ_0	$= -0.000055$	$[\text{rad}]$
γ_0	$= 0.0$	$[\text{rad}]$
δ_{e_0}	$= 0.027886$	$[\text{rad}]$
T_0	$= 33,005.5$	$[\text{slug}]$

Develop a state–space model for this highly nonlinear system. Write a simulation program in any CSSL that describes this state–space model. The inputs and outputs of this aircraft are shown in Fig.P4.1c.

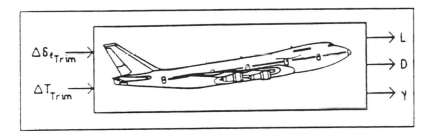

Figure P4.1c. Inputs and outputs of the B747 airplane [4.16].

Simulate step responses of the aircraft to a step of ±0.001 rad for the elevator deflexion and to a step of ±3000 slug for the thrust.

Convert all measurement units to Georgi system units and repeat the simulation in these modified units. Show that the results are the same as before.

[P4.2] Robot Modeling

The problem of modeling the motion of robots can be decomposed into the problem of determining statically the position of the end–effector in Cartesian coordinates as a function of the generalized positions of the individual joints (robot kinematics) and the problem of determining dynamically the generalized positions of the joints as a function of the applied generalized forces (robot kinetics).

If the generalized joint positions are given, *forward kinematics* will determine the position of the end–effector in Cartesian coordinates. If, on the other hand, the desired end–effector location is given, the *inverse kinematics* problem must be solved in order to determine the generalized joint positions that are necessary to bring the end–effector to the desired location.

Since the state variables in robot arm models are usually the joint variables but tasks are usually described in Cartesian coordinates, it is necessary to determine values for the joint variables that lead the end–effector to its desired position and orientation. Therefore, the inverse kinematics representation is the more important of the two.

A systematic method developed by Denavit and Hartenberg, called the 4×4 D-H homogeneous transformation matrix, can be used to describe the kinematic relationship of one coordinate system relative to another coordinate system [4.2]. The homogeneous transformation matrix for frame i with respect to frame k is:

$$\mathbf{T_k^i} = \begin{bmatrix} \mathbf{R_k^i} & \mathbf{d_k^i} \\ \mathbf{p'} & s \end{bmatrix} \qquad (P4.2a)$$

where:

$\mathbf{R_k^i}$ = 3×3 rotation matrix for frame i relative to frame k

$\mathbf{d_k^i}$ = 3×1 translation column vector for the origin of frame i relative to frame k

$\mathbf{p'}$ = 1×3 perspective transformation row vector. In robotics, this is unimportant, and therefore, it is set equal to zero

s = Scaling factor. In robotics, it is always set equal to 1.

Since any robot arm can be characterized by a series of movable links, every two adjacent links are connected by a joint, except for the end–effector. We can associate a fixed Cartesian coordinate system with each joint of the robot arm and define the joint i as the intersection of the link $i-1$ with the link i. Let us consider the joint i as depicted in Fig.P4.2a

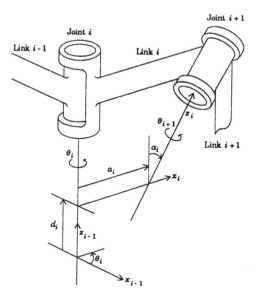

Figure P4.2a. Joint coordinate system for joint i [4.4].[†]

[†] Adapted figure from ROBOTICS: CONTROL, SENSING, VISION, AND INTEL-LIGENCE by King–Sun Fu *et al.*, ©1987 by McGraw–Hill, Inc., reprinted by permission of the publisher.

and define the coordinate system as follows

> z_{i-1} = An attached axis at the end of link $i-1$, directed along joint i about which or along which link i moves
> x_{i-1} = Axis directed from z_{i-1} to z_i, perpendicular to both
> y_{i-1} = The unique axis that completes a right–hand Cartesian coordinate system with x_{i-1} and z_{i-1}
> o_{i-1} = The origin of the frame $i-1$ coordinate system.

With so–defined coordinate systems attached to the joints i and $i+1$, four link parameters can be defined to specify the configuration of the frame i coordinate system relative to the frame $i-1$ coordinate system.

> θ_i = Rotation angle about z_{i-1} of x_{i-1} into x_i, taken in the right–hand sense about z_{i-1}
> d_i = Distance from x_{i-1} to x_i, measured along z_{i-1}
> a_i = Distance from z_{i-1} to z_i, measured along x_i
> α_i = Rotation angle about x_i of z_{i-1} into z_i, taken in the right–hand sense about x_i.

By following the order of these four operations, the overall D-H coordinate transformation matrix for frame i relative to frame $i-1$ can be expressed as:

$$A_{i-1}^{i} = \begin{bmatrix} \text{rotation} \\ \text{about} \\ z_{i-1}\ \text{axis} \end{bmatrix} \cdot \begin{bmatrix} \text{translation} \\ \text{along} \\ z_{i-1}\ \text{axis} \end{bmatrix} \cdot \begin{bmatrix} \text{translation} \\ \text{along} \\ x_i\ \text{axis} \end{bmatrix} \cdot \begin{bmatrix} \text{rotation} \\ \text{about} \\ x_i\ \text{axis} \end{bmatrix}$$

$$= \begin{bmatrix} C\theta_i & -S\theta_i & 0 & 0 \\ S\theta_i & C\theta_i & 0 & 0 \\ 0 & 0 & 1 & 0 \\ 0 & 0 & 0 & 1 \end{bmatrix} \cdot \begin{bmatrix} 1 & 0 & 0 & 0 \\ 0 & 1 & 0 & 0 \\ 0 & 0 & 1 & d_i \\ 0 & 0 & 0 & 1 \end{bmatrix} \cdot \begin{bmatrix} 1 & 0 & 0 & a_i \\ 0 & 1 & 0 & 0 \\ 0 & 0 & 1 & 0 \\ 0 & 0 & 0 & 1 \end{bmatrix} \cdot \begin{bmatrix} 1 & 0 & 0 & 0 \\ 0 & C\alpha_i & -S\alpha_i & 0 \\ 0 & S\alpha_i & C\alpha_i & 0 \\ 0 & 0 & 0 & 1 \end{bmatrix}$$

$$= \begin{bmatrix} C\theta_i & -S\theta_i C\alpha_i & S\theta_i S\alpha_i & a_i C\theta_i \\ S\theta_i & C\theta_i C\alpha_i & -C\theta_i S\alpha_i & a_i S\theta_i \\ 0 & S\alpha_i & C\alpha_i & d_i \\ 0 & 0 & 0 & 1 \end{bmatrix} \qquad (P4.2b)$$

where $C\theta_i$ denotes $\cos(\theta_i)$ and $S\theta_i$ denotes $\sin(\theta_i)$.

Therefore, if a point P is fixed in the i frame and is expressed in coordinates with respect to the i frame coordinate system as

$$r_i = \begin{bmatrix} p_{x_i} \\ p_{y_i} \\ p_{z_i} \\ 1 \end{bmatrix} \qquad (P4.2c)$$

then

$$\mathbf{r}_{i-1} = \mathbf{A}^i_{i-1} \cdot \mathbf{r}_i \qquad (P4.2d)$$

is the coordinate vector of point P with respect to the $i-1$ frame coordinate system.

The joint coordinate system and the four link parameters for each link of the Stanford arm are depicted in Fig.P4.2b.

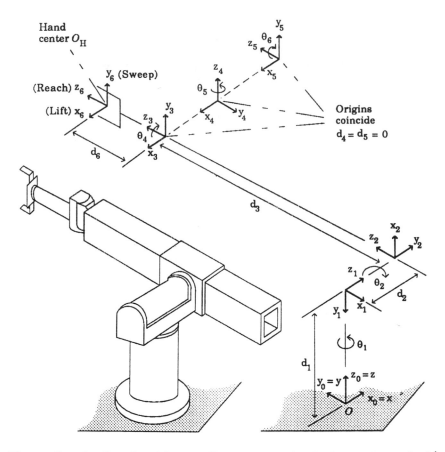

Figure P4.2b. Complete joint coordinate systems for the Stanford arm [4.4].[†]

The link parameters are listed in Table P4.2a.

[†] Adapted figure from ROBOTICS: CONTROL, SENSING, VISION, AND INTELLIGENCE by King–Sun Fu *et al.*, ©1987 by McGraw–Hill, Inc., reprinted by permission of the publisher.

Table P4.2a. Link Parameters for the Stanford Arm.

Joint i	θ_i	d_i	a_i	α_i
1	θ_1-90	d_1	0	-90
2	θ_2-90	d_2	0	90
3	-90	d_3	0	0
4	θ_4	0	0	-90
5	θ_5	0	0	90
6	θ_6	d_6	0	0

The position and orientation for the coordinate system of link i (frame i) with respect to the base coordinate system (frame 0) can be obtained via a chain product of the A matrices and can be expressed as:

$$\mathbf{T}_0^i = \mathbf{A}_0^1 \cdot \mathbf{A}_1^2 \ldots \mathbf{A}_{i-1}^i \qquad (P4.2e)$$

$\mathbf{T}_0^6$ describes the position and orientation of the end–effector with respect to the base coordinate system. Equation (P4.2e) is therefore referred to as the *kinematic equation* of the robot arm. It describes the fundamental kinematic behavior of the arm.

By assuming d_1 and d_6 to be zero, Paul derived the complete $\mathbf{A}_{i-1}^i$ and $\mathbf{T}_0^i$ matrices of the Stanford arm [4.14]. These matrices are:

$$\mathbf{A}_0^1 = \begin{bmatrix} C_1 & 0 & -S_1 & 0 \\ S_1 & 0 & C_1 & 0 \\ 0 & -1 & 0 & 0 \\ 0 & 0 & 0 & 1 \end{bmatrix} \qquad \mathbf{A}_1^2 = \begin{bmatrix} C_2 & 0 & S_2 & 0 \\ S_2 & 0 & -C_2 & 0 \\ 0 & 1 & 0 & d_2 \\ 0 & 0 & 0 & 1 \end{bmatrix}$$

$$\mathbf{A}_2^3 = \begin{bmatrix} 1 & 0 & 0 & 0 \\ 0 & 1 & 0 & 0 \\ 0 & 0 & 1 & d_3 \\ 0 & 0 & 0 & 1 \end{bmatrix} \qquad \mathbf{A}_3^4 = \begin{bmatrix} C_4 & 0 & -S_4 & 0 \\ S_4 & 0 & C_4 & 0 \\ 0 & -1 & 0 & 0 \\ 0 & 0 & 0 & 1 \end{bmatrix}$$

$$\mathbf{A}_4^5 = \begin{bmatrix} C_5 & 0 & S_5 & 0 \\ S_5 & 0 & -C_5 & 0 \\ 0 & 1 & 0 & 0 \\ 0 & 0 & 0 & 1 \end{bmatrix} \qquad \mathbf{A}_5^6 = \begin{bmatrix} C_6 & -S_6 & 0 & 0 \\ S_6 & C_6 & 0 & 0 \\ 0 & 0 & 1 & 0 \\ 0 & 0 & 0 & 1 \end{bmatrix} \qquad (P4.2f)$$

where C_i now denotes $\cos(\theta_i)$ and S_i stands for $\sin(\theta_i)$, and therefore

$$\mathbf{T}_0^6 = \begin{bmatrix} r_{11} & r_{12} & r_{13} & d_x \\ r_{21} & r_{22} & r_{23} & d_y \\ r_{31} & r_{32} & r_{33} & d_z \\ 0 & 0 & 0 & 1 \end{bmatrix} \qquad (P4.2g)$$

where:

$$r_{11} = C_1 \left[C_2 \left(C_4 C_5 C_6 - S_4 S_6 \right) - S_2 S_5 C_6 \right] - S_1 \left(S_4 C_5 C_6 + C_4 S_6 \right)$$

$$r_{21} = S_1 \left[C_2 \left(C_4 C_5 C_6 - S_4 S_6 \right) - S_2 S_5 C_6 \right] + C_1 \left(S_4 C_5 C_6 + C_4 S_6 \right)$$

$$r_{31} = -S_2 \left(C_4 C_5 C_6 - S_4 S_6 \right) - C_2 S_5 C_6$$

$$r_{21} = C_1 \left[-C_2 \left(C_4 C_5 S_6 + S_4 C_6 \right) + S_2 S_5 S_6 \right] - S_1 \left(-S_4 C_5 S_6 + C_4 C_6 \right)$$

$$r_{22} = S_1 \left[-C_2 \left(C_4 C_5 S_6 + S_4 C_6 \right) + S_2 S_5 S_6 \right] + C_1 \left(-S_4 C_5 S_6 + C_4 C_6 \right)$$

$$r_{23} = S_2 \left(C_4 C_5 S_6 + S_4 C_6 \right) + C_2 S_5 S_6$$

$$r_{31} = C_1 \left(C_2 C_4 S_5 + S_2 C_5 \right) - S_1 S_4 S_5$$

$$r_{32} = S_1 \left(C_2 C_4 S_5 + S_2 C_5 \right) + C_1 S_4 S_5$$

$$r_{33} = -S_2 C_4 S_5 + C_2 C_5$$

$$d_x = C_1 S_2 d_3 - S_1 d_2$$

$$d_y = S_1 S_2 d_3 + C_1 d_2$$

$$d_z = C_2 d_3$$

Given any relative orientation of the end–effector specified as r_6 (from Eq.(P4.2c), and given any set of joint coordinates θ_i and d_j, we can compute the momentary value of the $\mathbf{T}_0^6$ matrix, and therefore determine the absolute position and orientation of the end–effector. This technique solves the forward kinematics problem. Notice that these equations are highly nonlinear in the joint angles, and therefore, most mechanical manipulators don't have a closed–form solution to the inverse kinematics problem.

Develop a static simulation program that allows you to specify all joint coordinates of the Stanford arm as arbitrary functions of time and that will compute the absolute position and orientation of its end–effector over time.

The kinetic equations of a robot arm are primarily a function of the generalized external forces that are applied to the individual joints. We are going to use the Lagrangian to describe the dynamic motion of the Stanford arm under the (unfortunately very unrealistic) assumption of negligible friction. The derivation of the Lagrangian is systematic and the resulting equations are compact and explicit in terms of joint torques and joint variables. The resulting equations can be easily expressed in terms of state equations and can be used for the design of control strategies.

For an n–degree–of–freedom robot arm, the Lagrangian can be written as:

$$\frac{d}{dt} \left(\frac{\partial L}{\partial \dot{q}_i} \right) - \frac{\partial L}{\partial q_i} = 0 \qquad i = 1, 2, \ldots, n \tag{P4.2h}$$

where:

$$L = E_k - E_p \tag{P4.2i}$$

is the difference between the total kinetic energy E_k and the total potential energy E_p of the robot arm. The q_i are the generalized coordinates of the robot arm.

Based on the transformation matrices A_i, and following a tedious but straightforward procedure, the Lagrangian can be converted to the following form:

$$\tau_i = \sum_{j=1}^{n} D_{ij}(\mathbf{q})\ddot{q}_i + I_{a_i}\ddot{q}_i + \sum_{j=1}^{n}\sum_{k=1}^{n} C_{ijk}(\mathbf{q})\dot{q}_j\dot{q}_k + G_i(\mathbf{q}) \qquad i = 1, 2, \ldots, n \quad (P4.2j)$$

where:

$$
\begin{aligned}
D_{ii} &= \text{Effective inertia at joint } i \\
D_{ij} &= \text{Coupling inertia between joints } i \text{ and } j \\
I_{a_i} &= \text{Actuator inertia at joint } i \\
C_{ijj} &= \text{Centripetal forces at joint } i \text{ due to velocity at joint } j \\
C_{ijk} &= \text{Coriolis forces at joint } i \text{ due to velocities at joints } j \text{ and } k \\
G_i &= \text{Gravity load at joint } i \\
\mathbf{q} &= \text{Vector of generalized coordinates of the robot arm.}
\end{aligned}
$$

These equations are highly nonlinear and time–dependent. However, from the servo stability and position control point of view, the inertia terms and gravity terms are much more important than the centripetal and the Coriolis forces. The centripetal and Coriolis forces are important only when the robot arm is moving at high speed. Therefore, we can simplify the equations by ignoring these forces. We shall also ignore the coupling inertias between links 4, 5, and 6 and consider all these masses to be concentrated at the end of link 3. Paul compared the values in each of the inertia and gravity terms, neglected small value terms, and thereby obtained simplified equations [4.14]. The remaining nonzero terms are as follows:

$$
\begin{aligned}
b_{60}^L &= m_6^L k_{6zz}^2 \\
b_{50}^L &= m_5 k_{5yy}^2 + m_6^L k_{6xx}^2 \\
b_{40} &= m_4 k_{4yy}^2 \\
b_{30}^L &= m_3 + m_4 + m_5 + m_6^L \\
b_{20} &= m_2 k_{2yy}^2 + m_3 k_{3yy}^2 \\
b_{21} &= 2m_3 \bar{z}_3 \\
b_{22}^L &= 2(m_5 \bar{z}_5 + m_6^L \bar{z}_6) \\
b_{10}^L &= m_1 k_{1yy}^2 + m_2 k_{2xx}^2 + 2m_2 \bar{y}_2 d_2 + d_2^2(m_2 + m_3 + m_4 + m_5 + m_6^L) \\
b_{11} &= m_3 k_{3xx}^2
\end{aligned}
$$

$$D_{66}^L = b_{60}^L + I_{a_6}$$
$$D_{55}^L = b_{50}^L + I_{a_5}$$
$$D_{44}^L = b_{40} + b_{50}^L S_5^2 + I_{a_4}$$
$$D_{33}^L = b_{30}^L + I_{a_3}$$
$$D_{22}^L = b_{20} + b_{21} q_3 + b_{30}^L q_3^2 + b_{22}^L q_3 C_5 + I_{a_2}$$
$$D_{11}^L = b_{10}^L + b_{11} S_2^2 + b_{21} S_2^2 q_3 + b_{30}^L S_2^2 q_3^2 + b_{22}^L S_2^2 q_3 C_5 + b_{22}^L d_2 S_4 S_5 + I_{a_1}$$
$$D_{12}^L = D_{21}^L = -b_{30}^L C_2 d_2 q_3$$
$$D_{13}^L = D_{31}^L = -b_{30}^L S_2 d_2$$

and:

$$c_{50}^L = g(m_5 \bar{z}_5 + m_6^L \bar{z}_6)$$
$$c_{40}^L = c_{50}^L$$
$$c_{30}^L = g(m_3 + m_4 + m_5 + m_6^L)$$
$$c_{20} = -g m_3 \bar{z}_3$$
$$c_{21}^L = -c_{30}^L$$
$$c_{22}^L = -c_{50}^L$$
$$G_5^L = c_{50}^L(-C_2 S_5 + S_2 C_4 C_5)$$
$$G_4^L = c_{40}^L S_2 S_4 S_5$$
$$G_3^L = c_{30}^L C_2$$
$$G_2^L = c_{20} S_2 + c_{21} S_2 q_3 + c_{22}(C_2 C_4 S_5 + S_2 C_5)$$

where:

$$m_i \quad = \text{Mass of link } i$$
$$m_i k_{izz}^2 = \text{Inertia of link } i \text{ with respect to the } z_i\text{–axis}$$
$$\bar{z}_i \quad = \text{Center of mass of link } i \text{ in the } z_i\text{–direction}$$
$$I_{a_i} \quad = \text{Inertia of actuator } i$$
$$d_2 \quad = \text{Offset of link 3 (equal to 16.2 cm.)}$$

The superscript L indicates that this constant term is load–dependent.

In order to determine numerical values for the dynamic constants of the Stanford arm, we assume that the center of mass of the load is identical to that of the end–effector itself, which is located at 24.76 cm in the z_6–direction. The other dynamic constants are listed in Tables P4.2b–c.

Table P4.2b. Translational Dynamic Constants.

Link	Mass [kg]	$\bar{x}$ [cm]	$\bar{y}$ [cm]	$\bar{z}$ [cm]
1	9.29	0	1.75	-11.05
2	5.01	0	-10.54	0
3	4.25	0	0	-64.47
4	1.08	0	0.92	-0.54
5	0.63	0	0	5.66
6	0.51	0	0	15.54

Table P4.2c. Rotational Dynamic Constants.

Link	Ia [kgm^2]	mk_{xx}^2 [kgm^2]	mk_{yy}^2 [kgm^2]	mk_{zz}^2 [kgm^2]
1	0.953	0.276	0.255	0.071
2	2.193	0.108	0.018	0.100
3	0.782	2.51	2.51	0.006
4	0.106	0.002	0.001	0.001
5	0.097	0.003	0.003	0.0004
6	0.020	0.013	0.013	0.0003

Equation (P4.2j) can be expressed in a much more compact matrix form:

$$\vec{\tau}(t) = \mathbf{D}(\mathbf{q}(t))\,\ddot{\mathbf{q}}(t) + \mathbf{C}(\mathbf{q}(t),\dot{\mathbf{q}}(t)) + \mathbf{G}(\mathbf{q}(t)) \qquad (P4.2k)$$

By defining:

$$\mathbf{x}(t) = [\,\mathbf{q}'(t),\dot{\mathbf{q}}'(t)\,]' \qquad (P4.2l)$$

as the $2n \times 1$ state vector, Eq.(P4.2k) can be expressed in state–space form as

$$\dot{\mathbf{x}}(t) = \begin{bmatrix} \mathbf{0} & \mathbf{I} \\ \mathbf{0} & \mathbf{0} \end{bmatrix} \mathbf{x}(t) - \begin{bmatrix} \mathbf{0} \\ \mathbf{D}(\mathbf{x})^{-1}(\mathbf{G}(\mathbf{x}) + \mathbf{C}(\mathbf{x})) \end{bmatrix} + \begin{bmatrix} \mathbf{0} \\ \mathbf{D}(\mathbf{x})^{-1} \end{bmatrix} \vec{\tau}(t) \qquad (P4.2m)$$

This state–space model allows us to compute the joint coordinates of the robot arm as a function of the applied generalized external forces. We can then use the solution to the forward kinematics problem as an output equation that converts the joint coordinates into Cartesian coordinates of the end–effector.

Write a simulation program that computes the dynamic (kinetic) motion of the Stanford arm as a function of arbitrary generalized external forces applied to the robot.

Research

[R4.1] Robot Modeling and Simulation

Develop a robot modeling language that allows the user to specify the robot geometry in a user–friendly manner and that is able to solve the forward kinematics and forward kinetics problems in a convenient manner for an arbitrary robot geometry. Provide a library of preprogrammed robot models for the most common robots such as the Stanford arm and the Puma 560.

Develop a robot simulation language that allows you to access the previously defined robot model and simulate robot motions. It should provide for an input protocol that allows the user to "program" the robot model in quite the same manner in which one would program a real robot, i.e., the language should provide for a "teach pendant" mode and a mode to execute previously taught command sequences. The robot simulator should have an internal display feature that allows the user to display the position of the simulated robot on–line, but it should also be able to return joint coordinate telemetry data to the calling program that might eventually be executed from another computer. It may be advantageous from a point of view of flexibility if the commands to the robot model can also be sent to the robot simulator in the form of command telemetry packages rather than being entered in an interactive command mode. This will allow us to simulate several robots residing on several separate computers cooperating with each other by exchanging messages through command and telemetry packages.

One of our current research efforts (sponsored by NASA) involves the modeling and simulation of a hierarchy of cooperating robots for Space applications [4.20]. For this project, the availability of a program as described would be highly beneficial.

5

Hierarchical Modular Modeling of Continuous Systems

Preview

To this point, we have dealt with very simple and small problems. In this chapter, we shall cover some of the techniques necessary for modeling larger systems. Very often, systems consist of subsystems that may be described in quite different ways. Besides state–space representations and topological descriptions (which we have met previously), subsystems may also be described in the frequency domain in terms of transfer functions or may simply be given as a static characteristic relating one output variable to one or several input variables. It is therefore important that models can be structured. Modular modeling enables us to encapsulate subsystem descriptions and treat them as unseparable entities that can be incorporated in a hierarchical fashion within ever–more–complex system descriptions.

5.1 Modeling Transfer Functions

Let us assume a system is described by the following transfer function:

$$G(s) = 200 \frac{(s+1)}{(s+10)(s+20)} \tag{5.1}$$

In order to make this system amenable to simulation, we need to convert the specification back from the frequency domain into the time domain. The easiest way to do this is the following.

$$G(s) = \frac{200 + 200s}{200 + 30s + s^2} = \frac{P(s)}{Q(s)} = \frac{Y(s)}{U(s)} \tag{5.2}$$

where $G(s)$ denotes the transfer function, $P(s)$ denotes its numerator polynomial, $Q(s)$ denotes its denominator polynomial, $Y(s)$ denotes the output signal, and $U(s)$ denotes the input signal. We introduce an additional signal $X(s)$

$$G(s) = \frac{Y(s)}{U(s)} = \frac{Y(s)}{X(s)} \cdot \frac{X(s)}{U(s)} \tag{5.3}$$

such that

$$\frac{X(s)}{U(s)} = \frac{1}{Q(s)} \tag{5.4a}$$

$$\frac{Y(s)}{X(s)} = P(s) \tag{5.4b}$$

We shall look at Eq.(5.4a) first. We can rewrite this for our example as:

$$[200 + 30s + s^2]X(s) = U(s) \tag{5.5}$$

which can be transformed back into the time domain as:

$$200x(t) + 30\dot{x}(t) + \ddot{x}(t) = u(t) \tag{5.6}$$

assuming that all initial conditions are zero, which is standard practice when operating on transfer functions. We now solve Eq.(5.6) for its highest derivative:

$$\ddot{x}(t) = -200x(t) - 30\dot{x}(t) + u(t) \tag{5.7}$$

Finally, we introduce the following state variables:

$$\xi_1 = x \tag{5.8a}$$
$$\xi_2 = \dot{x} \tag{5.8b}$$

which leads us to the following state–space model:

$$\dot{\xi}_1 = \xi_2 \tag{5.9a}$$
$$\dot{\xi}_2 = -200\xi_1 - 30\xi_2 + u \tag{5.9b}$$

We now look at Eq.(5.4b), which can be written for our example as:

$$Y(s) = [200 + 200s]X(s) \qquad (5.10)$$

or in the time domain:

$$y(t) = 200x(t) + 200\dot{x}(t) \qquad (5.11)$$

and using our state variables:

$$y = 200\xi_1 + 200\xi_2 \qquad (5.12)$$

We can rewrite Eqs.(5.9a–b) and Eq.(5.12) in a matrix form as:

$$\dot{\xi} = \begin{pmatrix} 0 & 1 \\ -200 & -30 \end{pmatrix} \xi + \begin{pmatrix} 0 \\ 1 \end{pmatrix} u \qquad (5.13a)$$

$$y = \begin{pmatrix} 200 & 200 \end{pmatrix} \xi \qquad (5.13b)$$

In general, if a system is specified through the transfer function:

$$G(s) = \frac{b_0 + b_1 s + b_2 s^2 + \ldots + b_{n-1} s^{n-1}}{a_0 + a_1 s + a_2 s^2 + \ldots + a_{n-1} s^{n-1} + s^n} \qquad (5.14)$$

we can immediately convert this to the following state–space description:

$$\dot{\mathbf{x}} = \begin{pmatrix} 0 & 1 & 0 & \cdots & 0 & 0 \\ 0 & 0 & 1 & \cdots & 0 & 0 \\ \vdots & \vdots & \vdots & \ddots & \vdots & \vdots \\ 0 & 0 & 0 & \cdots & 1 & 0 \\ 0 & 0 & 0 & \cdots & 0 & 1 \\ -a_0 & -a_1 & -a_2 & \cdots & -a_{n-2} & -a_{n-1} \end{pmatrix} \mathbf{x} + \begin{pmatrix} 0 \\ 0 \\ \vdots \\ 0 \\ 0 \\ 1 \end{pmatrix} u \qquad (5.15a)$$

$$y = \begin{pmatrix} b_0 & b_1 & b_2 & \cdots & b_{n-2} & b_{n-1} \end{pmatrix} \mathbf{x} \qquad (5.15b)$$

In other words, the *state matrix* consists of all zero elements except for a superdiagonal of one elements and the last row in which the negative coefficients of the denominator polynomial are stored. The *input vector* consists of zero elements except for the last element, which is one, and the *output vector* contains the positive coefficients of the numerator polynomial.

This technique will work fine as long as the numerator polynomial is of lower degree than the denominator polynomial. If this is not the case, we need to divide the numerator by the denominator first and separate in this way the direct input/output coupling from the remainder of the system. This procedure will be illustrated by means of another simple example:

$$G(s) = \frac{6s^3 + 32s^2 + 10s + 2}{2s^2 + 8s + 4} \tag{5.16}$$

We always start by normalizing the highest–degree coefficient of the denominator polynomial to one, i.e.:

$$G(s) = \frac{3s^3 + 16s^2 + 5s + 1}{s^2 + 4s + 2} \tag{5.17}$$

The division of polynomials works the same way as the division of regular numbers:

$$
\begin{aligned}
&(3s^3 + 16s^2 + \quad 5s + \ 1\ ) : (s^2 + 4s + 2) = 3s + 4 \\
-\ &\ \ 3s^3 + 12s^2 + \quad 6s \\[4pt]
\hline
&\backslash \qquad 4s^2 - \qquad s + \ 1 \\
-\ &\qquad\ \ 4s^2 + \quad 16s + \ 8 \\[4pt]
\hline
&\backslash \qquad\quad -17s - \ 7
\end{aligned}
$$

i.e., $G(s)$ can also be written as:

$$G(s) = (3s + 4) + \frac{-17s - 7}{s^2 + 4s + 2} \tag{5.18}$$

which can be interpreted as a parallel connection of two subsystems, as depicted in Fig.5.1.

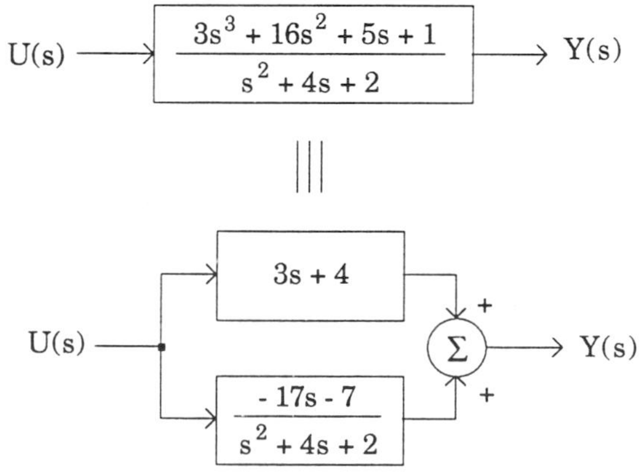

Figure 5.1. Separation of the direct input/output coupling.

The transfer function has been split into a polynomial that contains the *input/output coupling* of the system and a remainder transfer function, the numerator of which is now guaranteed to be of lower degree than the denominator polynomial. This contains the so–called *strictly proper* portion of the system.

In the case of our example, we end up with the following simulation model:

$$\dot{x}_1 = x_2 \tag{5.19a}$$
$$\dot{x}_2 = -2x_1 - 4x_2 + u \tag{5.19b}$$
$$y = -7x_1 - 17x_2 + 4u + 3\dot{u} \tag{5.19c}$$

As can be seen, a true differentiation of the input signal u was unavoidable in this case. This is always true when the numerator polynomial of a transfer function is of a higher degree than the denominator polynomial. As a small consolation, the necessary numerical differentiation is performed as part of the evaluation of output equations and has thereby been removed from the simulation loop. Numerical errors made in the process of numerical differentiation will not grow by being passed around the integration loop many times. We came upon this situation in Chapter 3, and at that time, I had mentioned (without a proof) that essential differentiators can, in linear systems, always be moved out of the simulation loop into the output equations. It has now become clear why this is the case and how this can be accomplished in practice.

5.2 Modeling Static Characteristics

Often, static but nonlinear functional relationships exist between input variables and output variables of a subsystem. Often, mathematical equations describing these relationships are not available. Instead, these relationships have been found through experimentation with a real system.

Let us demonstrate this concept by means of our lunar landing module, which now should be equipped to land on Earth instead. Of course, this wouldn't work with the rockets designed in the previous model, but let us be forbearing with these lesser details. However, it will be important to modify our mechanical equations to take the air density into consideration. This is proportional to the active surface S, the air density ρ, and the square of the velocity v^2

$$F_{air} = k \ S \ \rho(h) \ v^2 \tag{5.20}$$

The air density is an experimentally determined function of the altitude, as depicted in Fig.5.2.

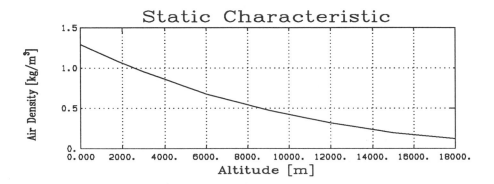

Figure 5.2. Earth's air density as a function of altitude.

Most CSSLs provide for mechanisms to describe such functions in a tabular form. For instance, DARE–P [5.21] provides for a separate tabular function block in which static characteristics of one and two variables can be coded. The preceding example could be coded in DARE–P as follows:

```
$ T
RHO, 12
0.0,      1.293
300.0,    1.256
600.0,    1.22
1200.0,   1.152
1800.0,   1.082
3000.0,   0.955
4500.0,   0.815
6000.0,   0.676
9000.0,   0.476
12000.0,  0.319
15000.0,  0.196
18000.0,  0.122
END
```

where the first column denotes the independent variable, here altitude expressed in m, and the second column denotes the dependent variable, here air density expressed in kg m^{-3}. The first line specifies the name of the table and the number of recordings collected. This table can be used in the model like a FORTRAN function, i.e.

$$A = (THRUST - XM * G - XK * S * \mathrm{RHO}(H) * V * *2)/XM \quad (5.21)$$

ACSL [5.16] uses a different syntax, but the idea is the same. In ACSL, the preceding problem would be formulated as follows:

```
table RHO, 1, 12/ ...
       0.0,    300.0,   600.0, 1200.0,   1800.0, ...
    3000.0,  4500.0,  6000.0, 9000.0,  12000.0, ...
   15000.0, 18000.0,   ...
      1.293,   1.256,   1.22 ,  1.152,    1.082, ...
      0.955,   0.815,   0.676,  0.476,    0.319, ...
      0.196,   0.122/
```

which can also be used like a FORTRAN function. DESIRE [5.13] offers a similar mechanism, and extensions to several independent variables exist.

5.3 Dynamic Table Load

The tabular function as presented earlier is not all that useful when we are confronted with huge amounts of data, such as wind tunnel data of an aircraft wing. We certainly don't want to recode (or even reedit) the data into the format of such tables. For such purpose, it is mandatory that tables can be loaded dynamically from a data base. Amazingly, this feature was offered 15 years ago in the software system CSMP–III [5.9] (*call tvload*), and yet, none of the currently advocated systems offers such a function as a standard feature. However, this function is easy to implement when needed.

5.4 Modular and Hierarchical Modeling

Another requirement that comes immediately to mind is the need to model systems in a modular and hierarchical manner. It should be possible to create reusable modules that can be grouped hierarchically. Modules should be groupable in exactly the same way as real equipment is, i.e., if one wishes to model a cupboard full of electronics, one should be able to model the cupboard as a rack filled with individual instruments, each instrument as a box filled with printed circuit boards, each printed circuit board as a collection of chips, each chip as consisting of a set of transistors, and each transistor through a set of individual discrete elements.

In practice, nobody in his or her right mind would ever attempt to model such a system in all details. Models are always goal–driven. The goal dictates the level at which the highest hierarchical layer is placed. Usually, the details of lower hierarchical layers are less important to the specified goal and should be aggregated into atomic units that are not further decomposed. However, it is important that the conceptual mechanism of hierarchical structuring be preserved in the modeling environment. Two hierarchical levels are extremely common in models of practical systems, three to four hierarchical levels can still be found. In Chapter 8, we shall present a model of a solar–heated house that contains a five–layer hierarchy.

Let us discuss some major mechanisms for hierarchical modeling as they are provided in continuous system modeling environments. By far the most popular mechanism is the *macro facility*.

5.5 The Macro Facility

Most CSSLs allow us to formulate subsystem descriptions as *macros*. Superficially seen, macros look very similar to subprograms in traditional programming languages. This concept will be demonstrated by means of the electromechanical system depicted in Fig.5.3.

A new lightweight fiber–optic deep–sea communication cable is to be laid through the British Channel between Calais in France and Dover in the United Kingdom. The cable comes on a huge reel that is placed on a ship. The ship moves slowly from one coast to the other while constantly leaving cable behind. A large DC motor unrolls the cable from the reel. A speedometer detects the speed of the cable as it comes off the reel. A simple proportional and integral (PI) controller is used to keep the cable speed v at its preset value V_{set}.

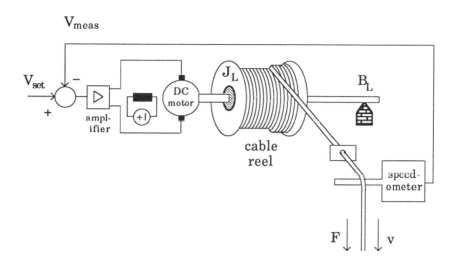

Figure 5.3. Functional diagram of the cable reel system.

In modeling this system, we must first realize that we cannot use the block diagram of the DC motor as it was developed in the last chapter. The reason is that the inertia J_L of the cable reel changes with time. Consequently, we must use the modified version of Newton's law by operating on the twist T of the motor. Figure 5.4 shows a modified block diagram of the DC motor that will work correctly for our application.

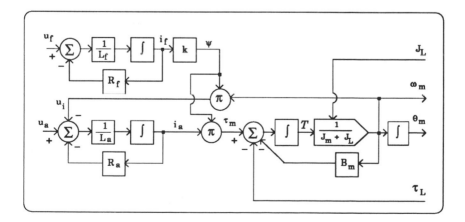

Figure 5.4. Modified block diagram of a DC motor.

The modified version of Newton's law manifests itself through a swapping of the inertial box and one of the integrator boxes in the mechanical subsystem. Notice also that it was necessary to introduce a fourth input variable, namely, the inertia of the load, J_L. It wouldn't have been a good idea to combine this term with the torque load (by treating it as a fictitious "inertial torque") since, under those circumstances, an algebraic loop would have resulted.

Let us now analyze the dynamics of the cable reel itself. The length of one winding of the cable is obviously:

$$\ell_w = 2\pi R \tag{5.22}$$

where R is the current radius of the cable reel. With the width of the cable reel being W, and the diameter of the cable being D, the length of one cable layer can obviously be computed to:

$$\ell_l = \ell_w \frac{W}{D} = \frac{2\pi R \; W}{D} \tag{5.23}$$

The velocity of the cable can be computed to be:

$$v = R\frac{d\theta_m}{dt} \tag{5.24}$$

and since the velocity was assumed to be approximately constant, we can approximate the time to unroll one layer of cable as:

$$t_l = \frac{\ell_l}{v} = \frac{2\pi W}{D\omega_m} \tag{5.25}$$

During this time, the radius of the cable reel is reduced by one cable diameter:

$$\Delta R = -D \tag{5.26}$$

and therefore, we can find a differential equation describing the change of the cable reel radius as follows:

$$\frac{dR}{dt} \approx \frac{\Delta R}{t_l} = -\frac{D^2}{2\pi W}\omega_m \tag{5.27}$$

The inertia of the cable reel is a function of the radius R. It can be computed by the following formula:

$$J_L = 0.5\pi W\rho(R^4 - R^4_{empty}) + J_0 \tag{5.28}$$

where ρ denotes the density of the cable material and J_0 denotes the inertia of the empty cable reel. The torque load τ_L consists of the friction torque and the torque produced by the force F, which is a result of the weight of the already–laid cable. However, this force supports the DC motor rather than impedes it. Therefore, this term is entered with the opposite sign

$$\tau_L = B_L \omega_m - F\,R \tag{5.29}$$

Figure 5.5 shows a block diagram of the cable reel dynamics.

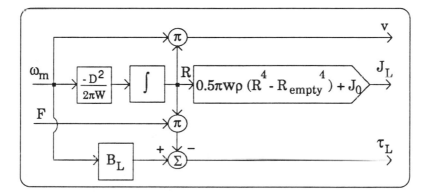

Figure 5.5. Block diagram of cable reel dynamics.

Finally, we need to look into the internal dynamics of the speedometer. These were specified by the manufacturer in the frequency domain as:

$$G(s) = \frac{3}{s+3} \tag{5.30}$$

which can immediately be transformed into the state–space representation:

$$\dot{\xi} = -3\xi + u \tag{5.31a}$$
$$y = 3\xi \tag{5.31b}$$

We could now go ahead and simply concatenate all these equations to a monolithic program. However, in a sufficiently large model, this approach is certainly prone to error. It seems desirable to be able to formulate the program through modules that represent the structural components of the system, as depicted in Fig.5.6.

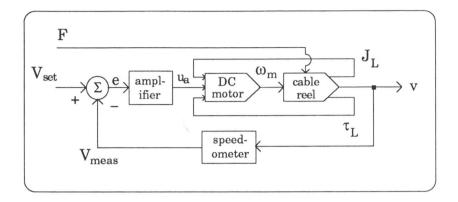

Figure 5.6. Block diagram of the overall cable reel system.

This can be achieved by using the *macro facility*. Let me demonstrate this facility by means of the ACSL [5.16] language (DARE–P
[5.21] does not offer a macro facility, and while DESIRE [5.13] provides for such a facility, it is much less powerful than the one offered
in ACSL). The following macro describes the speedometer:

> **MACRO** $TACHO(out, in, x0)$
> **MACRO redefine** x, $xdot$
> **MACRO standval** $x0 = 0.0$
> $xdot = -3 * x + in$
> x $= \text{integ}(xdot, x0)$
> out $= 3 * x$
> **MACRO END**

This macro can then be used in the ACSL program by use of the
statement:

$$TACHO(vmeas = v)$$

the **standval** directive allows us to assign default values to parameters that can then be omitted in the macro call. The **redefine**
directive is necessary due to the SAL rule. If the model would contain several speedometers, several "$xdot = ...$" statements would
otherwise appear in the resulting program after the macro call has
been replaced by the macro definition and the equation sorter would
complain. The **redefine** directive instructs the macro handler to
replace each occurrence of the names x and $xdot$ during macro replacement by a unique identifier (in ACSL, a Z followed by five

digits). In other words, if it is intended to use a macro several times within a program, all local variables of the macro must be declared in a **redefine** directive.

Let us now write a macro to describe the cable reel dynamics:

$$\textbf{MACRO } CABREL(v, tauL, JL, R, omega, F, D, W, rho, \ldots$$
$$Rfull, Rempty, J0, BL)$$

> **MACRO redefine** $Rdot$
> **constant** $pi = 3.14159$
> $Rdot = -((D * D)/(2.0 * pi * W)) * omega$
> $R \quad = \text{integ}(Rdot, Rfull)$
> $v \quad = R * omega$
> $JL \quad = 0.5 * pi * W * rho * (R ** 4 - Rempty ** 4) + J0$
> $tauL = BL * omega - F * R$

MACRO END

From this macro, we can learn several new lessons. First, we may notice the long parameter list. While it would be perfectly acceptable *not* to list all the constants among the input parameters of the macro, this would prevent us from simulating several cable reels at once with different values for these parameters. Unfortunately, while the macro facility provides us with a mechanism to hierarchically decompose program structures, it does not allow us to hierarchically decompose the data structures along with the program structures. We shall discuss later in this chapter another mechanism that will provide us with such a feature. Clearly, when using the macro facility, all parameter values must be passed on to higher and higher hierarchical levels until the calling sequences become totally unmanageable. This is a serious drawback of the macro concept. Also, we have learned before that it helps the efficiency of the program execution if all constant computations, such as $-D^2/(2\pi W)$ and R^4_{empty}, are moved out of the DERIVATIVE section into the INITIAL section of the program. We cannot do so in this case without destroying the integrity of the macro. Unfortunately, few CSSL macro handlers have been devised to contain an INITIAL section (although this would be quite easy to implement). One simulation language that offers such a facility is SYSMOD [5.19].

Now, let us look at the macro describing the dynamics of the DC motor.

MACRO *DCMOT(theta, omega, ua, uf, tauL, JL, ...*
  Ra, La, Rf, Lf, k, Jm, Bm, flag, if0, ia0, T0, th0)
  MACRO redefine *ia, iadot, if, ifdot, ui, psi*
  MACRO redefine *taum, Twist, Tdot*
  MACRO standval *if0* = 0.0, *ia0* = 0.0, *T0* = 0.0
  MACRO standval *th0* = 0.0
  MACRO if ($flag = IND$) *labind*
    if   = uf/Rf
    ia   = $(ua - ui)/Ra$
  MACRO goto *goon*
  MACRO *labind*..**continue**
    ifdot = $(uf - Rf * if)/Lf$
    if   = $integ(ifdot, if0)$
    iadot = $(ua - ui - Ra * ia)/La$
    ia   = $integ(iadot, ia0)$
  MACRO *goon*..**continue**
    psi   = $k * if$
    taum   = $psi * ia$
    ui   = $psi * omega$
    Tdot   = $taum - tauL - Bm * omega$
    Twist = $integ(Tdot, T0)$
    omega = $Twist/(Jm + JL)$
    theta = $integ(omega, th0)$
  MACRO END

Also this macro exhibits a number of additional features. Let us look once more at the armature equation:

$$u_a = u_i + R_a \, i_a + L_a \frac{di_a}{dt} \tag{5.32}$$

Usually, Eq.(5.32) will be implemented in the form of a differential equation in state–space form, i.e.:

$$\frac{di_a}{dt} = \frac{u_a - u_i - R_a \, i_a}{L_a} \tag{5.33}$$

This is done since we always wish to transform continuous models into a state–space form such that all differential equations can be numerically integrated instead of numerically differentiated. However, the electrical time constants of the DC motor are often so much smaller than the mechanical time constant that the effect of the armature inductance on the overall system behavior can be neglected. In this case, we cannot operate on Eq.(5.33) since, if we set $L_a = 0.0$, this results in a division by zero. Instead, we must return to Eq.(5.32), delete the term in L_a from the equation, and rewrite it as:

$$i_a = \frac{u_a - u_i}{R_a} \tag{5.34}$$

Notice that this example confronts us with two different versions of DC motor models. Instead of creating two separate macros for these two cases, it was decided to code them as two variants within the same macro. The constant parameter *flag* can assume either the value *IND* or *NOIND* in the macro call. Depending on the setting of this compile–time parameter, the macro replacer will generate code in the form of either Eq.(5.33) or Eq.(5.34).

We are now ready to code the entire cable reel simulation program.

```
PROGRAM Cable Reel Dynamics
INITIAL
  MACRO TACHO(out, in, x0)
    . . .
  MACRO END
  MACRO CABREL(...)
    . . .
  MACRO END
  MACRO DCMOT(...)
    . . .
  MACRO END
  constant ...
    Rfull = 1.2, Rempty = 0.6, W = 1.5, D = 0.0127, ...
    rho = 1350.0, J0 = 150.0, Jm = 5.0, Bm = 0.2, ...
    BL = 6.5, Ra = 0.25, La = 0.5E-3, Rf = 1.0, ...
    Lf = 0.002, kmot = 1.5, kprop = 6.0, kint = 0.2, ...
    vset = 15.0, kship = 10.0, F0 = 100.0, tmx = 3600.0
  cinterval cint = 1.0
  algorithm ialg = 2
  nsteps nstp = 1000
  uf = 25.0
END $ "of INITIAL"
DYNAMIC
  DERIVATIVE
    err   = vset − vmeas
    ua    = kprop * err + kint*integ(err, 0.0)
    DCMOT(theta, omega = ua, uf, tauL, JL, ...
          Ra, La, Rf, Lf, kmot, Jm, Bm, "NOIND")
    CABREL(v, tauL, JL, R = omega, F, ...
          D, W, rho, Rfull, Rempty, J0, BL)
    F     = amax1(kship * (vset − v), F0)
    vmeas = TACHO(v)
  END $ "of DERIVATIVE"
  termt (t.ge.tmx .or. R.lt.Rempty)
END $ "of DYNAMIC"
END $ "of PROGRAM"
```

As the TACHO call demonstrates, it is allowable to call macros that produce only a single output as a *function* instead of as a *procedure*. The **algorithm** and **nsteps** instructions were necessary to keep the numerical integration happy. The meaning of these instructions will be discussed later.

The results of this simulation are shown in Fig.5.7.

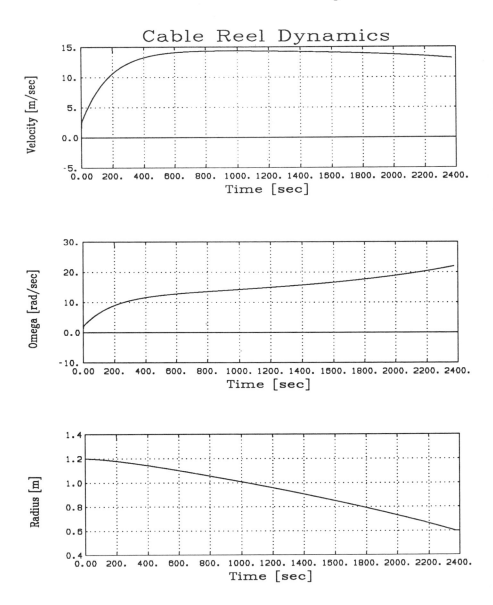

Figure 5.7. Simulation results of cable reel dynamics.

As can be seen, it took almost 10 min to accelerate the cable reel to its steady–state speed. Obviously, the DC motor is a little weak for this application and it might have made sense to replace it with a more sturdy hydraulic motor. However, the control worked fine. The motor accelerates constantly in order to keep the speed of the unrolling cable at a constant value.

It made little sense to model the electrical time constant of the DC motor. This time constant is so much smaller than the mechanical time constants that the results would look the same if the inductances were included, but the simulation would execute much more slowly since the step size of the integration algorithm must adapt itself to the fastest time constant in the system.

Let us now look at the equations that the macro handler generates during the macro expansion. For illustration, the inductances were included this time.

main:	err	$= vset - vmeas$
	ua	$= kprop * err + kint * Z09987$
	$Z09987$	$= \text{integ}(err, 0.0)$
dcmot:	$Z09996$	$= (uf - Rf * Z09997)/Lf$
	$Z09997$	$= \text{integ}(Z09996, 0.0)$
	$Z09998$	$= (ua - Z09995 - Ra * Z09999)/La$
	$Z09999$	$= \text{integ}(Z09998, 0.0)$
	$Z09994$	$= kmot * Z09997$
	$Z09993$	$= Z09994 * Z09999$
	$Z09995$	$= Z09994 * omega$
	$Z09991$	$= Z09993 - tauL - Bm * omega$
	$Z09992$	$= \text{integ}(Z09991, 0.0)$
	$omega$	$= Z09992/(Jm + JL)$
	$theta$	$= \text{integ}(omega, 0.0)$
cabrel:	$Z09990$	$= -((D * D)/(2.0 * 3.14159 * W)) * omega$
	R	$= \text{integ}(Z09990, Rfull)$
	v	$= R * omega$
	JL	$= 0.5 * 3.14159 * W * rho * (R ** 4 - Rempty ** 4) + J0$
	$tauL$	$= BL * omega - F * R$
main:	F	$= \text{amax1}(kship * (vset - v), F0)$
tacho:	$Z09988$	$= -3 * Z09989 + v$
	$Z09989$	$= \text{integ}(Z09988, 0.0)$
	$vmeas$	$= 3 * Z09989$

We recognize that, while the **redefine** declaration is necessary for proper functioning of the macro mechanism, it does not exactly contribute to the readability of the generated code.

Let us go one step further, and sort these equations into an executable sequence. The result is as follows.

dcmot: $Z09996 = (uf - Rf * Z09997)/Lf$
$Z09994 = kmot * Z09997$
$Z09993 = Z09994 * Z09999$
cabrel: $JL\quad = 0.5 * 3.14159 * W * rho * (R * *4 - Rempty * *4) + J0$
tacho: $vmeas = 3 * Z09989$
main: $err\quad = vset - vmeas$
$ua\quad = kprop * err + kint * Z09987$
dcmot: $omega = Z09992/(Jm + JL)$
cabrel: $Z09990 = -((D * D)/(2.0 * 3.14159 * W)) * omega$
$v\quad = R * omega$
$tauL\quad = BL * omega - F * R$
main: $F\quad = \mathrm{amax1}(kship * (vset - v), F0)$
tacho: $Z09988 = -3 * Z09989 + v$
dcmot: $Z09995 = Z09994 * omega$
$Z09991 = Z09993 - tauL - Bm * omega$
$Z09998 = (ua - Z09995 - Ra * Z09999)/La$
main: $Z09987 = \mathrm{integ}(err, 0.0)$
dcmot: $Z09997 = \mathrm{integ}(Z09996, 0.0)$
$Z09999 = \mathrm{integ}(Z09998, 0.0)$
$Z09992 = \mathrm{integ}(Z09991, 0.0)$
$theta\quad = \mathrm{integ}(omega, 0.0)$
cabrel: $R\quad = \mathrm{integ}(Z09990, Rfull)$
tacho: $Z09989 = \mathrm{integ}(Z09988, 0.0)$

A number of different algorithms exist that the equation sorter can use. One algorithm is the following. We start by assuming all outputs of memory functions (i.e., the outputs of integrators) to be known. We then skim through the equations and try to find one that does not define a memory function and has only known variables to the right of the equal sign. If we find one, we write it to the output file and add the defined variable to the set of known variables. When we reach the end of the input file, we check whether any equations were written to the output file during this pass. If not, we obviously have one or several algebraic loops and another algorithm is being activated to detect algebraic loops among the remaining equations. On the other hand, if we have written one or several equations to the output file, we check whether the set of remaining equations is now empty or contains only memory function definitions, otherwise we go back and start again with the remaining equations. We continue until the set of unsorted equations contains only memory function definitions, which can then be added to the output file or until no more equations can be written to the output file due to the presence of algebraic loops.

We learn from the preceding example that during the sorting process, equations from the various macros are being completely interspersed. As a consequence, the macro replacement must be performed before the equation sorter can be activated, and consequently, macros cannot be separately compiled, but must always be stored as source code. In this respect, macros are totally different from the subprograms of general–purpose programming languages.

This last observation explains why DESIRE's [5.13] macro facility is not very powerful. Since DESIRE does not provide for an equation sorter, macros can only be coded for subsystems that do not require sorting of the equations after the macro replacement has taken place. This makes DESIRE's built–in macro facility virtually worthless. However, in the next section, we shall introduce an alternative utility that can replace the macro handler altogether and that will work well in connection with DESIRE.

Let us now look at another example that will unravel more of the shortcomings of the macro facility as it is offered in today's CSSLs. One of the requests that we may have when using a DC motor is that the angular velocity of the motor ω_m is insensitive to changes in the torque load T_L. Unfortunately, this is not the case in the standard armature control configuration. Let us look once more at the equations describing the dynamics of the DC motor:

$$\frac{di_a}{dt} = \frac{u_a - u_i - R_a\, i_a}{L_a} \qquad (5.35a)$$

$$\tau_m = \psi\, i_a \qquad (5.35b)$$

$$u_i = \psi\, \omega_m \qquad (5.35c)$$

$$\frac{d\omega_m}{dt} = \frac{\tau_m - \tau_L}{J_m + J_L} \qquad (5.35d)$$

In *steady–state*, all derivatives are zero, i.e., Eqs.(5.35a–d) can be rewritten as:

$$0.0 = u_a - u_i - R_a\, i_a \qquad (5.36a)$$

$$\tau_m = \psi\, i_a \qquad (5.36b)$$

$$u_i = \psi\, \omega_m \qquad (5.36c)$$

$$0.0 = \tau_m - \tau_L \qquad (5.36d)$$

which can be reduced to:

$$\omega_m = \frac{1}{\psi} u_a - \frac{R_a}{\psi^2} \tau_L \qquad (5.37)$$

Equation (5.37) can be graphically depicted as shown in Fig.5.8.

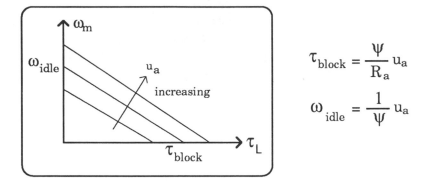

$$\tau_{block} = \frac{\psi}{R_a} u_a$$

$$\omega_{idle} = \frac{1}{\psi} u_a$$

Figure 5.8. Angular velocity versus torque load in DC motor.

In order to keep the angular velocity constant under the influence of a changing load, it is necessary to change the armature voltage along with the load. However, the load is usually considered an unpredictable *disturbance* of the system. For this reason, armature control may not be the best of all choices. However, if we are able to feed the armature circuit with a constant current source and apply field control, the dependency of the angular velocity from the torque load vanishes. Unfortunately, constant current sources for high–power applications aren't so easy to come by. A more practical solution to the problem is to use a Ward–Leonard group as depicted in Fig.5.9.

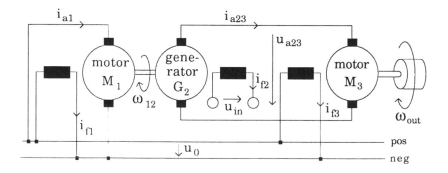

Figure 5.9. Ward–Leonard group.

The DC motor M_1 is driven with constant armature and constant field and thereby produces an angular velocity that is used to drive the generator G_2. The generator is a machine of the same type as the motor. The electromechanical coupling works both ways. We can either generate a rotation by having currents flow through both the field and the armature circuits (motor) or we can induce a voltage in the armature circuit by rotating the machine externally if we feed current through the field circuit at the same time (generator). Figure 5.10 depicts a functional diagram of the DC generator.

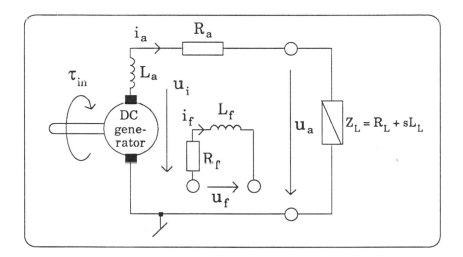

Figure 5.10. Functional diagram of a DC generator.

When the machinery is operated in its generator mode, the main input to the system is the driving torque. It causes the motor to rotate. Once an angular velocity ω has been built up, it induces a voltage u_i on the electrical side, which causes a current i_a to flow through the armature coil. The armature current i_a causes a mechanical torque τ_L to be built up back on the mechanical side, which opposes the driving torque. The armature current i_a is also responsible for building up an armature voltage u_a across the two armature terminals. The armature voltage u_a is subtracted from the induced voltage u_i, thereby weakening the armature current i_a. This process continues until an equilibrium is reached. The load is now electrical, symbolized in our model by a resistive load R_L and an inductive load L_L, which, in themselves, are not part of the DC generator and are therefore additional inputs to the DC generator model. Figure 5.11 depicts the block diagram of the DC generator.

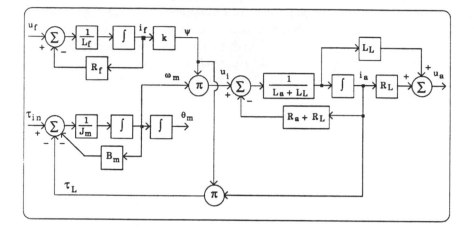

Figure 5.11. Block diagram of a DC generator.

The following ACSL macro can be used to code the DC generator equations.

```
MACRO DCGEN(theta, omega, ua, ia, tauin, uf, RL, LL, ...
    Ra, La, Rf, Lf, k, Jm, Bm, if0, ia0, om0, th0)
    MACRO redefine iadot, if, ifdot, ui, psi
    MACRO redefine tauL, omdot
    MACRO standval if0 = 0.0, ia0 = 0.0, om0 = 0.0
    MACRO standval th0 = 0.0
    ifdot  = (uf − Rf * if)/Lf
    if     = integ(ifdot, if0)
    iadot  = (ui − (Ra + RL) * ia)/(La + LL)
    ia     = integ(iadot, ia0)
    ua     = RL * ia + LL * iadot
    psi    = k * if
    tauL   = psi * ia
    ui     = psi * omega
    Tdot   = tauin − tauL − Bm * omega
    Twist  = integ(Tdot, T0)
    omega  = Twist/Jm
    theta  = integ(omega, th0)
MACRO END
```

As in the case of the electrical circuits of Chapter 3, we see that the same physical device with emphasis on the same physical phenomena can lead to quite different model equations depending on the environment in which the model is being used. We learned before that macros aren't really modular with respect to the incorporated data structures. Now, we learn that macros aren't even modular with respect to the represented program structures. The same physical device calls for quite different macros depending on the environments in which it is supposed to operate. The simplest "macro" representing an electrical resistor, for instance, must be stored in the macro library in two different versions, one modeling the equation:

$$u_R = R \cdot i_R \qquad (5.38a)$$

and the other modeling the equation:

$$i_R = \frac{u_R}{R} \qquad (5.38b)$$

If the resistor is placed over a current source, the current i_R through the resistor is known and we need to use the macro that reflects the model according to Eq.(5.38a), whereas if we place the resistor over a voltage source, the voltage u_R across the resistor is known and we need to use the macro that reflects the model according to Eq.(5.38b). Obviously, an equation sorter is insufficient. We require an *equation solver*, which accepts general equalities of the type:

$$< \text{expression} > \; = \; < \text{expression} > \qquad (5.39a)$$

or:

$$< \text{expression} > \; = \; 0.0 \qquad (5.39b)$$

and which can solve these equalities for arbitrary variables. Later in this chapter, we shall discuss a software that satisfies this requirement.

However, let us return once more to the Ward–Leonard example. We now have two separate macros to describe the two DC motors and the DC generator. Let us try to call them from an ACSL program and see what happens. Unfortunately, this won't work. According to our models, ω_{12} is supposed to be a state variable of the DC motor M_1 and at the same time a state variable of the DC generator G_2. This obviously can't be true. We just detected a degeneracy (structural singularity) of our system. Similarly, the armature current i_{a23} is supposedly a state variable of the DC generator G_2 and at the same time a state variable of the DC motor M_3. So, we have found a second degeneracy of our system.

What can we learn from this example? While we can propose the implementation of an *automated single–equation solver* as an extension to the commonly used *equation sorter* to take care of the problem illustrated in Eqs.(5.38a–b), even such a mechanism won't suffice.

Algebraic loops are quite common and they cut right across the borders between individual macros, i.e., the IMPL block as presented in Chapter 2 is totally incompatible with the demand of modular modeling. For this purpose, we should request a *compiler option*, which would instruct the CSSL preprocessor to automatically generate the necessary IMPL block structures for us *after* the macro replacement and a partial equation sorting have already taken place. Unfortunately, even this feature won't help us with system degeneracies that occur as a result of subsystem coupling.

Let me repeat my conclusion from Chapter 3. We have exactly two choices. Either we put state equations as a mechanism to describe simulation models to the sword once and for all and use the topological system description directly for the simulation or we find a *much* more powerful mechanism to generate appropriate state equations out of the topological system description than our simple macro facility represents. Let me develop this second path a little further in the remainder of this chapter.

5.6 Modular State–Space Models

As we have seen, it is necessary to store macros at all times in source form and it is essential that the macro handler be executed *before* anything else happens to the simulation program. Therefore, it is not really essential that a CSSL provides for a macro handler of its own. It would be equally acceptable to employ a totally independent general–purpose macro handler as a separate program to be called before the simulation compiler is entered. In this section, a new tool, DYMOLA [5.5], will be presented, which is a stand–alone program that can be used as a front end to several different simulation languages. Two different versions of DYMOLA have been written. One is coded in PASCAL, the other is coded in SIMULA. The SIMULA version runs on UNIVAC computers, the PASCAL version runs on VAX/VMS and PC compatibles. DYMOLA is a *program generator* since a compiler switch decides for what simulation language code is to be generated. DYMOLA currently supports the simulation languages DESIRE [5.13] and SIMNON [5.4] (another direct executing language) and plain FORTRAN. An interface to ACSL [5.16] is currently under development. DYMOLA is not a *simulation language* in its own right since it does not provide for a simulation engine of its own. Instead, DYMOLA is a *modeling language* since it supports the user in coding more–readable and better–modularized hierarchically structured model descriptions. We shall dwell more on this topic in Chapter 15 of this text. On first glance, DYMOLA looks like a powerful macro handler.

Let us discuss how the preceding DC motor example can be coded in DYMOLA:

```
model type DCMOT
    terminal theta, omega, ua, uf, tauL, JL
    local ia, if, ui, psi, taum, Twist
    parameter Ra, Rf, kmot, Jm
    parameter La = 0.0, Lf = 0.0, Bm = 0.0
    default ua = 25.0, uf = 25.0
        Lf*der(if) = uf - Rf * if
        La*der(ia) = ua - ui - Ra * ia
        psi        = kmot * if
        taum       = psi * ia
        ui         = psi * omega
        der(Twist) = taum - tauL - Bm * omega
        Twist      = (Jm + JL) * omega
        der(theta) = omega
end
```

This code is fairly self–explanatory. However, let us discuss some of the special properties of DYMOLA model descriptions.

(1) DYMOLA variables belong to either the type **terminal** or **local**. They are of type **terminal** if they are supposed to be connected to something outside the model. They are of type **local** if they are totally connected inside the model.

(2) Terminals can be either inputs or outputs. What they are often depends on the environment to which they are connected. However, the user can specify what he or she wants them to be by explicitly declaring them as **input** or **output** rather than as **terminal**.

(3) Terminals can have default values. In this case, they don't need to be externally connected.

(4) DYMOLA constants can be declared to be of type **parame-**

ter. For parameters, values can be assigned from outside the model. Parameters can have default values, in which case it is not necessary to assign a value to them from outside the model.

(5) Derivatives are either expressed using the "der(.)" operator or a prime ('). It is also allowed to use a "der2(.)" operator or a double prime (") to denote a second derivative and even higher derivatives are admissible. Contrary to most CSSLs, DYMOLA allows us to use these operators anywhere in the equation, to both the left and right of the equal sign.

(6) Consequently, it is not possible to set initial conditions for the integrators inside a model, which is clearly a disadvantage of DYMOLA.

(7) DYMOLA equations use the syntax of Eq.(5.39a). During the process of *model expansion*, equations are solved for the appropriate variable. For this reason, the SAL rule no longer applies. It is perfectly acceptable to have der($Twist$) on the left–hand side of one equation and $Twist$ on the left–hand side of another.

(8) Terms that are multiplied by a zero parameter are automatically eliminated during the model expansion. Consequently, if $La = 0.0$, the model equation $La * \text{der}(ia) = ua - ui - Ra * ia$ is first replaced by the modified model equation $0.0 = ua - ui - Ra * ia$, which then results in one of three simulation equations, namely, (i) $ua = ui + Ra * ia$, (ii) $ui = ua - Ra * ia$, or (iii) $ia = (ua - ui)/Ra$, depending on the environment in which the model is used. However, if $La \neq 0.0$, the model equation is always transformed into the simulation equation $\text{der}(ia) = (ua - ui - Ra * ia)/La$. This is an elegant way to solve the "variant macro" problem of ACSL.

(9) Rule (8) indicates that parameters with value 0.0 are treated in a completely different manner from all other parameters. This decision has a side effect. Parameters that are not set equal to zero are preserved in the generated simulation code and can be interactively altered through the simulation program directly without a need to return to DYMOLA. Parameters with value 0.0 are optimized away by the DYMOLA compiler and are not

represented in the simulation code. However, the advantages of this decision are overwhelming, since this does away with an entire class of structural singularities.

The preceding DYMOLA model can then be called in the following way:

submodel (DCMOT) $dcm1(Ra = 2.0, Rf = 5.5, kmot = 1.0, Jm = 15.0)$

It can be *connected* to the outside world using a dot notation:

$$dcm1.ua = kalph * err$$
$$dcm1.uf = 12.0$$
$$dcm1.JL = crl1.JL$$
$$dcm1.tauL = crl1.tauL$$
$$crl1.omega = dcm1.omega$$

where *crl1* is the name of a model of the cable reel type.

DYMOLA models are much more modular than ACSL macros since equations are automatically solved during model expansion for the variable that is appropriate in the context of the model call environment. The use of *named parameters* instead of *positional parameters* upon invocation of a DYMOLA model helps with long parameter lists. Default values can and should be assigned to many parameters, and with the named parameter convention the user can selectively specify values only for those parameters for which the default values are not appropriate. The connection mechanism as presented so far is very general, although a little clumsy. Each connection corresponds to connecting two points of a circuit with a wire.

It can be noticed that wires are frequently grouped into cables or buses. For example, consider an RS–232 connector. The RS–232 male connector has 25 pins, while the corresponding RS–232 female connector has 25 holes. It seems natural that a modeling language should provide for an equivalent mechanism. DYMOLA does this by providing so–called cuts.

Let us look at the cable reel example once more. It can be noticed that the cable reel and DC motor have three variables in common, namely, *omega*, *tauL*, and *JL*. We can therefore go ahead and declare those three variables in a **cut** rather than as simple **terminal** variables. The modified model type DCMOT looks now as follows:

```
model type DCMOT
   terminal theta, ua, uf
   cut mech(omega, tauL, JL)
   local ia, if, ui, psi, taum, Twist
   parameter Ra, Rf, kmot, Jm
   parameter La = 0.0, Lf = 0.0, Bm = 0.0
   default ua = 25.0, uf = 25.0
      Lf*der(if) = uf − Rf * if
      La*der(ia) = ua − ui − Ra * ia
      psi        = kmot * if
      taum       = psi * ia
      ui         = psi * omega
      der(Twist) = taum − tauL − Bm * omega
      Twist      = (Jm + JL) * omega
      der(theta) = omega
end
```

If we declare a similar cut in the model type CABREL, we can invoke in the main program a DC motor *dcm1* of type DCMOT and a cable reel *crl1* of type CABREL, and connect the cut *mech* of *dcm1* at the cut *mech* of *crl1*. This is coded as follows:

```
submodel (DCMOT) dcm1(Ra = ...)
submodel (CABREL) crl1(Bl = ...)
connect dcm1:mech at crl1:mech
```

The **connect** statement automatically generates the three model equations:

$$dcm1.omega = crl1.omega$$
$$dcm1.tauL = crl1.tauL$$
$$dcm1.JL = crl1.JL$$

Cuts can be hierarchically structured. For example, we could modify the model type DCMOT once more:

model type *DCMOT*
    terminal *theta*
    cut *mech(omega, tauL, JL)*
    cut *elect(ua, uf)*
    cut *both[mech, elect]*
    local *ia, if, ui, psi, taum, Twist*
    parameter *Ra, Rf, kmot, Jm*
    parameter $La = 0.0$, $Lf = 0.0$, $Bm = 0.0$
    default $ua = 25.0$, $uf = 25.0$
      $Lf * \mathrm{der}(if) = uf - Rf * if$
      $La * \mathrm{der}(ia) = ua - ui - Ra * ia$
      $psi \qquad\quad = kmot * if$
      $taum \qquad\: = psi * ia$
      $ui \qquad\quad\;\; = psi * omega$
      $\mathrm{der}(Twist) = taum - tauL - Bm * omega$
      $Twist \qquad\: = (Jm + JL) * omega$
      $\mathrm{der}(theta) \;\: = omega$
  end

in which case we can either connect the cut *mech* and the cut *elect* separately, or we can connect *both* together. During expansion of the **connect** statement, DYMOLA checks that the connected cuts are structurally compatible with each other.

However, in many cases, even this won't suffice. We may notice that, by connecting a wire between two points in an electrical circuit, we actually connect *two* variables, namely, the *potential* at the two points and the *current* that flows through the new wire. However, the two connections work differently. This is illustrated in Fig.5.12.

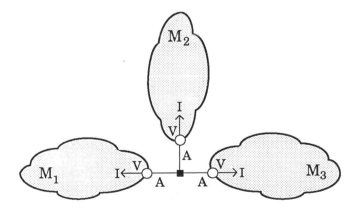

Figure 5.12. Connection conventions in an electrical circuit.

While the potentials of all cuts that are connected at a point must be equal, the currents must add up to zero. Variables of type potential are called *across variables*, while variables of type current are called *through variables*. DYMOLA also provides for this second type of connection. The generalized form of a DYMOLA cut looks as follows:

cut $< $ cut_name $ > (<$ across_variables $ > / <$ through_variables $ >)$

If the three models $M1$, $M2$, and $M3$ of Fig.5.12 each have a cut A declared as:

cut $A(V/I)$

we can use the connect statement:

connect $M1{:}A$ **at** $M2{:}A$ **at** $M3{:}A$

which will generate the following model equations:

$$M1.V = M2.V$$
$$M2.V = M3.V$$
$$M1.I + M2.I + M3.I = 0.0$$

Notice that currents at cuts are normalized to point *into* their subsystems. If a current is directed the opposite way, it must take a negative sign on the cut definition.

This concept is more generally useful than only for electrical circuits. In a mechanical system, all positions, velocities, and accelerations are across variables, while all forces and torques are through variables. In a hydraulic system, water level and pressure are across variables, while water flow is a through variable. In a thermic system, temperature is an across variable, while heat flow is a through variable, etc. These similarities between different types of physical systems are particularly emphasized in the bond graph modeling methodology, which will be discussed in Chapter 7 of this text.

One cut can be declared as the **main cut**. The main cut is the *default* cut in a connection, i.e., it suffices to specify the model name to connect the main cut of a submodel.

Sometimes it is useful to allow connections to take place *inside* a model instead of across model boundaries. For this purpose, DYMOLA provides a **node** declaration. Nodes are named and cuts can be connected to nodes. Nodes are hierarchically structured the same way cuts are.

```
model M
  cut A(v1, v2), B(v3, v4), C(v5, v6)
  main cut D[A, B, C]
    ...
end
...
node N
connect M at (N, N, N)
```

The connect statement is equivalent to:

$$\text{connect } M{:}A \text{ at } N, \ M{:}B \text{ at } N, \ M{:}C \text{ at } N$$

which is identical to saying:

$$\text{connect } M{:}A \text{ at } M{:}B \text{ at } M{:}C$$

which will result in the following set of equations:

$$M.v1 = M.v3$$
$$M.v3 = M.v5$$
$$M.v2 = M.v4$$
$$M.v4 = M.v6$$

Sometimes, it is also useful to connect a type of variable through from a source to a destination. For this purpose, DYMOLA allows us to declare a directed **path** from an input cut to an output cut.

Let us assume we have a model describing a pump that is declared as follows:

```
model pump
  cut inwater(w1), outwater(w2)
  path water < inwater − outwater >
    ...
end
```

Let us assume we have two more models describing a pipe and a tank with compatibly declared cuts and paths, then we can connect the water flow from the pump through the pipe to the tank with the statement:

$$\text{connect } (water) \ pump \ \textbf{to} \ pipe \ \textbf{to} \ tank$$

One path can always be declared as the **main path**. If the main path is to be connected, the path name can be omitted in the connect statement.

Besides the **at** and **to** operators, DYMOLA provides some additional connection mechanisms, which are sometimes useful. The **reversed** operator allows us to connect a path in the opposite direction. The **par** operator allows for a parallel connection of two paths, and the **loop** operator allows us to connect paths in a loop. Also, DYMOLA provides for abbreviations of some of these operators. The "=" symbol can be used as an alternative to **at**, the "−" operator can be used instead of **to**, the "//" operator can be used as an alternative to **par**, and the "\" operator denotes **reversed**. I shall present examples of the use of these elements in Chapter 6.

5.7 The Equation Solver

DYMOLA can solve equations for any variable that appears *linearly* in the equation. This does not mean that the equation as a whole must be linear. For instance, DYMOLA is able to handle the following equation:

$$7 * x + y * y - 3 * x * y = 25 \qquad (5.40)$$

if the variable it wants to solve this equation for is x. In this case, DYMOLA will transform Eq.(5.40) into:

$$x = (25 - y * y)/(7 - 3 * y) \qquad (5.41)$$

However, it cannot solve Eq.(5.40) for the variable y.

For some simple cases, it would be very easy to implement the appropriate transformation rules to handle even nonlinear equations, but most nonlinear equations don't provide for unique solutions. For example, the problem:

$$x^2 + y^2 = 1 \qquad (5.42)$$

when solved for y has the two solutions:

$$y = +\sqrt{1 - x^2} \qquad (5.43a)$$
$$y = -\sqrt{1 - x^2} \qquad (5.43b)$$

DYMOLA would have no way of knowing which of the two solutions to use. The same is true when the nonlinear equation is solved numerically by automatically generating an IMPL block around the equation. The numerical algorithm will simply approach one of the two solutions, often depending on the chosen initial value, and that may be the wrong one. This will be illustrated by means of the following equation:

$$x^2 - 5x + 2 = 0 \qquad (5.44)$$

We could reformulate this problem as follows:

$$x = \sqrt{5x - 2} \qquad (5.45)$$

We now choose an initial value for x, and plug it iteratively into Eq.(5.45) until convergence. Figure 5.13 illustrates what happens in this case.

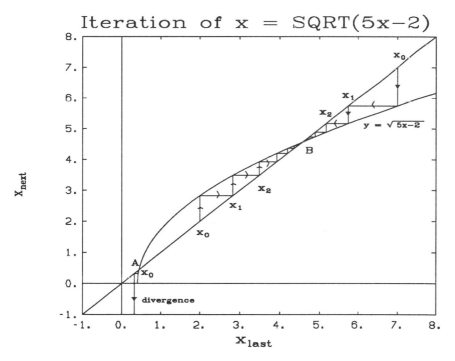

Figure 5.13. Iteration of nonlinear equation.

As can be seen, two solutions to this nonlinear equation exist. Depending on the starting value, we either approach solution B or the algorithm diverges. Solution A is an *unstable solution* of this iteration process.

However, we could have decided to formulate the problem in a different way:

$$x = 0.2x^2 + 0.4 \qquad (5.46)$$

and iterate this equation instead. Figure 5.14 shows what happens in this case.

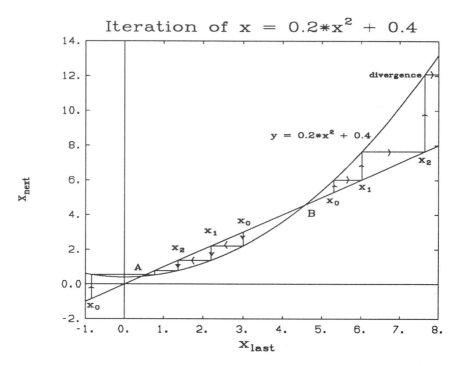

Figure 5.14. Iteration of nonlinear equation.

Obviously, this problem must have the same two solutions A and B as the previous one. However, as can be seen, this time, solution A is the stable solution, while solution B is unstable. Also, the *range of attraction* (i.e., the set of stable initial conditions) is different from the previous situation.

No generally applicable algorithm can be found for the automated solution of nonlinear equations. The best that can probably be achieved is that the DYMOLA preprocessor stops when it comes across a nonlinear equation and requests help from the user. It may then store this information away for later reuse in another compilation of the same model. One possible answer that the user may provide is to request the system to build an IMPL block around the equation and tell it which initial value to use for the iteration.

5.8 Code Optimization

DYMOLA already provides for a feature to eliminate trivial equations of the type $a = b$, by eliminating one of the two variables and replacing other occurrences of this variable in the program by the retained variable. This can significantly speed up the execution of the simulation program. However, much more could be done. It is foreseen to enhance the code optimizer by a fully automated algorithm to move all constant computations from the simulation language's DYNAMIC section into its program control section, and, in the case of DESIRE, all output computations from the DYNAMIC section into the OUT section. The code optimization is somewhat target–language–specific.

5.9 Linear Algebraic Loops

Let us revisit the electrical circuit as presented in Fig.3.12. The resulting set of equations was given in Eqs.(3.34a–i). Let us discuss what DYMOLA would do with these equations.

Let us start from the initial set of equations:

$$u1 = R1 * i1 \tag{5.47a}$$

$$u2 = R2 * i2 \tag{5.47b}$$

$$u3 = R3 * i3 \tag{5.47c}$$

$$uL = L * \text{der}(iL) \tag{5.47d}$$

$$U0 = u1 + u2 \tag{5.47e}$$

$$u3 = u2 \tag{5.47f}$$

$$uL = u1 + u2 \tag{5.47g}$$

$$i0 = i1 + iL \tag{5.47h}$$

$$i1 = i2 + i3 \tag{5.47i}$$

which are nine equations in the nine unknowns: $u1$, $u2$, $u3$, uL, $i0$, $i1$, $i2$, $i3$, and iL. $U0$ is not an unknown since this is the input to the system.

DYMOLA will start by recognizing that Eq.(5.47d) contains a "der(.)" operation, which determines that this equation must be solved for the derivative and which moves iL from the list of unknowns to the list of knowns. It further recognizes that uL and $i0$ appear only in one of the remaining equations each, namely, Eq.(5.47g) and Eq.(5.47h). This moves the variables uL and $i0$ from the list of unknowns to the list of knowns. Finally, it is recognized that Eq.(5.47f) is a trivial equation. We solve it for $u3$, and simultaneously replace all other occurrences of $u3$ by $u2$. At this point, we have the following set of solved equations:

$$\text{der}(iL) = uL/L \tag{5.48a}$$
$$u3 = u2 \tag{5.48b}$$
$$uL = u1 + u2 \tag{5.48c}$$
$$i0 = i1 + iL \tag{5.48d}$$

Five equations remain unsolved:

$$u1 = R1 * i1 \tag{5.49a}$$
$$u2 = R2 * i2 \tag{5.49b}$$
$$u2 = R3 * i3 \tag{5.49c}$$
$$i1 = i2 + i3 \tag{5.49d}$$
$$U0 = u1 + u2 \tag{5.49e}$$

They must be used to evaluate the five remaining unknowns $u1$, $u2$, $i1$, $i2$, and $i3$. Each remaining equation contains at least two unknowns, and each of the unknowns appears in at least two equations. Thus, we have an algebraic loop.

At this moment, DYMOLA is stuck. However, this will change soon. DYMOLA could easily recognize that all unknowns appear linearly in all the remaining equations and thus rewrite the system of equations in a matrix form as:

$$
\begin{pmatrix}
1 & 0 & -R1 & 0 & 0 \\
0 & 1 & 0 & -R2 & 0 \\
0 & 1 & 0 & 0 & -R3 \\
0 & 0 & 1 & -1 & -1 \\
1 & 1 & 0 & 0 & 0
\end{pmatrix}
\cdot
\begin{pmatrix}
u1 \\ u2 \\ i1 \\ i2 \\ i3
\end{pmatrix}
=
\begin{pmatrix}
0 \\ 0 \\ 0 \\ 0 \\ U0
\end{pmatrix}
\tag{5.50}
$$

which could be coded in the following way:

$$Z09999 = [\; 1, 0, -R1, \quad 0, \quad 0; \quad \ldots$$
$$0, 1, \quad 0, \; -R2, \; 0; \quad \ldots$$
$$0, 1, \quad 0, \quad 0, -R3; \quad \ldots$$
$$0, 0, \quad 1, \; -1, \; -1; \quad \ldots$$
$$1, 1, \quad 0, \quad 0, \quad 0]$$
$$[u1; u2; i1; i2; i3] = \mathrm{inv}(Z09999) * [0; 0; 0; 0; U0]$$

Notice the notation. The square bracket denotes the beginning of a matrix definition. The "," operator separates elements in neighboring columns, while the ";" operator separates elements in neighboring rows.

If all we have here is a *linear algebraic loop*, matrix $Z09999$ will be nonsingular and the set of matrix equations has a unique solution.

At this point, the *code optimizer* can become active and recognize that all elements within $Z09999$ are constants, i.e., that the evaluation of the matrix can be moved out of the DYNAMIC section into the INITIAL block (DESIRE's control section), and that even $\mathrm{inv}(Z09999)$ is a constant expression, which can be moved out into the control section where the matrix inversion will be performed exactly once prior to the execution of the simulation run.

Since DESIRE [5.13] is able to handle matrix expressions elegantly and very efficiently, this will be an easy task to implement. Currently, the regular version of DESIRE handles matrices only within the interpreted control section and not within the compiled DYNAMIC block. However, a modified version DESIRE/NEUNET [5.14] exists already, which was designed particularly for the simulation of neural networks. This version handles matrix expressions within the DYNAMIC block in a very efficient manner.

5.10 Nonlinear Algebraic Loops

The problem with nonlinear algebraic loops is the same as with the solutions of single nonlinear equations. Depending on how we iterate the set of equations, we may end up with one solution, or another, or none at all. Unfortunately, no generally applicable method can be found that would deal with this problem once and for all. As before, the best that DYMOLA may be able to do is interrupt the compilation, display the set of coupled algebraic equations on the screen together with the set of unknowns contained in these equations, and ask for help. Proper help may not always be easy to provide.

One way to tackle this problem is to get away from state–space models altogether. Instead of solving the set of first–order differential equations:

$$\dot{x} = f(x, u, t) \qquad (5.51)$$

we could try to find integration algorithms that can solve the more general set of equations:

$$A \cdot \dot{x} = f(x, u, t) \qquad (5.52)$$

directly, where A is allowed to be a singular matrix. This formulation takes care of all linear algebraic loops. Integration algorithms for this type of problems have been known for quite a while. Non-linear algebraic loops can be handled by integration algorithms that are able to solve the following set of implicit differential equations directly:

$$f(x, \dot{x}, u, t) = 0.0 \qquad (5.53)$$

The price to be paid for this generality is a reduction in execution speed and solution robustness. It can no longer be guaranteed that these equations have exactly one correct solution for each set of initial conditions. MODEL [5.17] is a language in which this approach was implemented. MODEL is another experimental language with a user interface that is quite similar to that of DYMOLA.

5.11 Structural Singularities

Usually, each component of a system that can store energy is represented by one or more differential equations. Capacitors and inductors of electrical circuits can store energy. Each capacitor (inductor) normally gives cause to one first–order differential equation. Mechanical masses can store two forms of energy, potential energy and kinetic energy. Each separately movable mass in a mechanical system usually gives rise to a second–order differential equation, which is equivalent to two first–order differential equations. However, sometimes this is not so. If we take two capacitors and connect them in parallel, the resulting system order is still one. This is due to the fact that a linear dependence exists between the two voltages over the two capacitors (they are the same), and thus, they do not both qualify as state variables.

Such situations are called *system degeneracies* or *structural singularities*. Usually, subsystems will be designed such that no such singularities occur. The two parallel capacitors are simply represented in the model by one equivalent capacitor with the value:

$$C_{eq} = C_1 + C_2 \qquad (5.54)$$

However, difficulties occur when subsystems are connected together and when the structural singularity is a direct result of the coupling of the two subsystems. Let us assume the two subsystem orders of subsystems S_1 and S_2 are n_1 and n_2. If the coupled system S_c has a system order n_c smaller than the sum of n_1 and n_2, a structural singularity exists that is a result of the subsystem coupling.

These problems must be carefully analyzed and they are often quite difficult to circumvent. For this purpose, let us study once more our Ward–Leonard group from before. The goal is to design a model of a DC motor/generator that is powerful enough to be used under all circumstances, i.e., it must be possible to model the Ward–Leonard group by coupling together three submodels of the same type DCMOT.

Let us start by looking at the coupling between the motor M_1 and the generator G_2. Under this coupling, the two angular velocities are forced to be the same except for their signs, which are opposite since the two machines are coupled back to back:

$$\omega_2 = -\omega_1 \qquad (5.55)$$

Since the angular velocities of both submodels are essentially outputs of integrators, a structural singularity has occurred. We can now realize that the torque produced by each of the two rotating machines represents the torque load of the other machine and that the inertia of each of the two machines is seen as an inertial load by the other machine. Therefore, each machine can contain the equations:

$$\frac{dT}{dt} = \tau_m - \tau_{load} - B_m\,\omega \qquad (5.56a)$$

$$\omega = T/(J_m + J_{load}) \qquad (5.56b)$$

and the couplings can be expressed as follows:

$$\tau_{load\_1} = \tau_{m\_2} - B_{m\_2}\,\omega_2 \qquad (5.57a)$$

$$\tau_{load\_2} = \tau_{m\_1} - B_{m\_1}\,\omega_1 \qquad (5.57b)$$

$$J_{load\_1} = J_{m\_2} \qquad (5.57c)$$

$$J_{load\_2} = J_{m\_1} \qquad (5.57d)$$

Now, we realize that we don't even need to specify the equality of the angular velocities any longer since they will be guaranteed automatically. In the coupled model, we waste computing time since we integrate the same angular velocity twice, but the structural singularity has been avoided.

Let us now look at the coupling between the generator G_2 and the motor M_3. We could express the coupling through a connection of the two armature cuts (u_a/i_a), but we immediately realize that the two currents, which are now forced to be equal except for their signs, are outputs of two integrators.

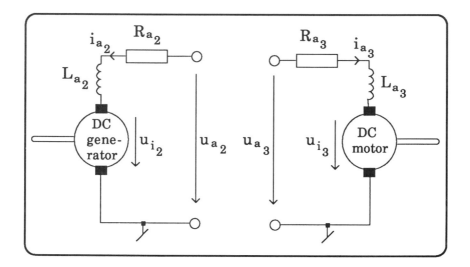

Figure 5.15. Electrical coupling in Ward–Leonard group.

Thus, this coupling represents a second structural singularity. Figure 5.15 depicts the conventions for voltage and current directions. Therefore, we can write down the following equation:

$$ui_2 + Ra_2\, ia_2 + La_2\frac{dia_2}{dt} = ui_3 + Ra_3\, ia_3 + La_3\frac{dia_3}{dt} \qquad (5.58)$$

which can be written in terms of ia_2 as:

$$\frac{dia_2}{dt} = (u_{load\_2} - ui_2 - Ra_2\, ia_2)/(La_2 + L_{load\_2}) \qquad (5.59a)$$

$$u_{load\_2} = ui_3 + Ra_3 \, ia_3 \qquad (5.59b)$$

$$L_{load\_2} = La_3 \qquad (5.59c)$$

or in terms of ia_3:

$$\frac{dia_3}{dt} = (u_{load\_3} - ui_3 - Ra_3 \, ia_3)/(La_3 + L_{load\_3}) \qquad (5.60a)$$

$$u_{load\_3} = ui_2 + Ra_2 \, ia_2 \qquad (5.60b)$$

$$L_{load\_3} = La_2 \qquad (5.60c)$$

In other words, we can solve the structural singularity in exactly the same way as in the case of the mechanical coupling, again at the expense of some extra computation since Eqs.(5.59a–c) and Eqs.(5.60a–c) represent the same physical variables. Therefore, we can write the following equations into our generic DCMOT model:

$$\frac{di_a}{dt} = (u_{load} - u_i - R_a \, i_a)/(L_a + L_{load}) \qquad (5.61a)$$

$$u_a = u_i + R_a \, i_a + L_a \frac{di_a}{dt} \qquad (5.61b)$$

The coupling equations are written as follows:

$$u_{load\_2} = ui_3 + Ra_3 \, ia_3 \qquad (5.62a)$$

$$L_{load\_2} = La_3 \qquad (5.62b)$$

$$u_{load\_3} = ui_2 + Ra_2 \, ia_2 \qquad (5.62c)$$

$$L_{load\_3} = La_2 \qquad (5.62d)$$

Let us now look at how these equations must be modified if we have decided to ignore the armature inductances. In that case, in order to avoid an algebraic loop, the resistance of the other machine cannot be included in the load voltage u_{load}, but must instead be treated as a resistive load. The generic model equations now look as follows:

$$i_a = (u_{load} - u_i)/(R_a + R_{load}) \qquad (5.63a)$$

$$u_a = u_i + R_a \, i_a \qquad (5.63b)$$

and the coupling equations can be written as follows:

$$u_{load\_2} = ui_3 \qquad (5.64a)$$

$$R_{load\_2} = Ra_3 \qquad (5.64b)$$

$$u_{load\_3} = ui_2 \qquad (5.64c)$$

$$R_{load\_3} = Ra_2 \qquad (5.64d)$$

The following ACSL [5.16] macro is a general macro that can be used both as a DC motor and as a DC generator.

```
MACRO DCMOT(theta, omega, taum, Jm, ua, RLa, ia, ui, uf, uld, RLld, ...
    tauld, Jld, Ra, La, Rf, Lf, kmot, Jm0, Bm, flag, if0, ia0, T0, th0)
    MACRO redefine iadot, if, ifdot, psi
    MACRO redefine Twist, Tdot
    MACRO standval if0 = 0.0, ia0 = 0.0, T0 = 0.0
    MACRO standval th0 = 0.0
    MACRO if (flag = IND) labind
    if      = uf/Rf
    ia      = (uld − ui)/(Ra + RLld)
    ua      = ui + Ra * ia
    RLa     = Ra
    MACRO goto goon
    MACRO labind..continue
    ifdot   = (uf − Rf * if)/Lf
    if      = integ(ifdot, if0)
    iadot   = (uld − ui − Ra * ia)/(La + RLld)
    ia      = integ(iadot, ia0)
    ua      = ui + Ra * ia + La * iadot
    RLa     = La
    MACRO goon..continue
    psi     = kmot * if
    taum    = psi * ia
    ui      = psi * omega
    Tdot    = taum − tauld − Bm * omega
    Twist   = integ(Tdot, T0)
    omega   = Twist/(Jm + Jld)
    theta   = integ(omega, th0)
    Jm      = Jm0
MACRO END
```

With this macro, we can simulate the Ward–Leonard group in the control configuration depicted in Fig.5.16.

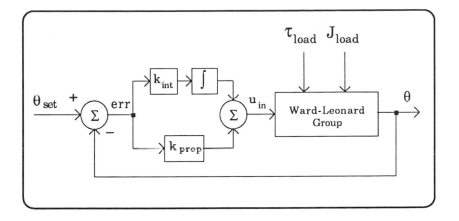

Figure 5.16. Block diagram of controlled Ward–Leonard group.

The Ward–Leonard group is embedded in a position control circuit with a PI controller. The angular position θ_{set} is set to 10.0 initially and is reduced to 5.0 at time 50.0. The Ward–Leonard group is originally idle, but it is loaded both with a torque τ_{load} and an inertia J_{load} at time 100.0. The simulation extends over 200.0 time units. The following program implements this control problem.

```
PROGRAM Ward–Leonard group
  INITIAL
    MACRO DCMOT(...)
      ...
    MACRO END
    constant ...
      Jm0 = 0.05, Bm = 2.0E-4, Ra = 10.0, La = 0.5E-3, ...
      Rf = 25.0, Lf = 2.2E-3, kmot = 0.5, kampl = 0.06, ...
      kint = 0.002, uf1 = 25.0, uf3 = 25.0, uld1 = 25.0, ...
      RLld1 = 0.0, tmx = 200.0
    cinterval cint = 0.5
    thset = 10.0
    Jld  = 0.0
    tauld = 0.0
    schedule angle .at. 50.0
    schedule load .at. 100.0
  END $ "of INITIAL"
```

```
DYNAMIC
DERIVATIVE
  err     = thset − theta
  uin     = kampl * err + kint*integ(err, 0.0)
  DCMOT(th1, om1, tm1, Jm1, ua1, RLa1, ia1, ui1 = ...
        uf1, uld1, RLld1, tld1, Jld1, ...
        Ra, La, Rf, Lf, kmot, Jm0, Bm, "NOIND")
  DCMOT(th2, om2, tm2, Jm2, ua2, RLa2, ia2, ui2 = ...
        uf2, uld2, RLld2, tld2, Jld2, ...
        Ra, La, Rf, Lf, kmot, Jm0, Bm, "NOIND")
  DCMOT(th3, om3, tm3, Jm3, ua3, RLa3, ia3, ui3 = ...
        uf3, uld3, RLld3, tld3, Jld3, ...
        Ra, La, Rf, Lf, kmot, Jm0, Bm, "NOIND")
  tld1    = tm2 − Bm * om2
  Jld1    = Jm2
  uf2     = −uin
  uld2    = ui3
  RLld2   = RLa3
  tld2    = tm1 − Bm * om1
  Jld2    = Jm1
  uld3    = ui2
  RLld3   = RLa2
  tld3    = tauld
  Jld3    = Jld
  theta   = th3
END $ "of DERIVATIVE"
DISCRETE angle
  thset   = 5.0
END $ "of DISCRETE angle"
DISCRETE load
  Jld     = 0.05
  tauld   = 0.01
END $ "of DISCRETE load"
termt (t.gt.tmx)
END $ "of DYNAMIC"
END $ "of PROGRAM"
```

The discontinuous driving functions were implemented using time–events.

The results of this simulation are shown in Fig.5.17.

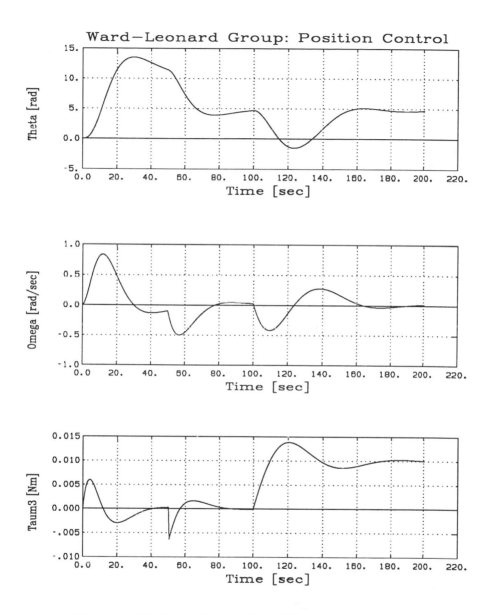

Figure 5.17. Simulation results of Ward–Leonard group.

The output position θ_3 follows the input rather well. The overshoot behavior is due to the integral portion of the control circuit. The torque produced in motor M_3 is responsible for the rotation of the motor. Since the motor is idle, the torque returns to zero as soon as the desired position has been reached. However, after time $t = 100.0$, the motor is loaded with a torque load. Now, in order to keep the position stationary, the motor needs to produce a constant motor torque that compensates for the torque load. The integral portion of the controller is responsible for returning the angular position of the motor bias–free back to its desired value after each disturbance. Nevertheless, the group is not able to keep the position constant under the varying load. In order to improve the performance of the system, we must guarantee a constant rotation of the driving shaft ω_1. This can be achieved by replacing the driving motor M_1 with a stronger machine, or eventually, by replacing it with a synchronous AC motor.

The modeling approach did work since the DCMOT macro does not call for many equations to be rearranged with respect to their outputs. However, quite a bit of insight was needed in order to get all equations into an adequate form and to solve all problems related to system degeneracies and algebraic loops. The approach is therefore not very convenient or user–friendly and it may be quite difficult to apply this solution to a more intricate problem than the one presented.

We are convinced that, in principle, the DYMOLA approach is the better answer to the problem. However, DYMOLA won't do the job yet. A number of extensions will be needed before DYMOLA can take care of such problems in a completely automated manner:

(1) DYMOLA should be able to eliminate variables not only from equations of the type $a = b$, but also from equations of the type $a + b = 0$.

(2) DYMOLA should be able to recognize equations that have been specified twice and eliminate the duplicate automatically.

(3) DYMOLA should be able to handle superfluous connections, i.e., if we specify that $\theta_2 = -\theta_1$, it is obviously true that also $\omega_2 = -\omega_1$. However, DYMOLA won't let us specify this additional connection at the current time. Superfluous connections should simply be eliminated during the model expansion.

(4) DYMOLA should recognize that connections of outputs of integrators can always be converted into connections of inputs of these integrators, i.e., if we have specified that $ia_3 = -ia_2$, it is obviously true that $iadot_3 = -iadot_2$. This reformulation can help to eliminate structural singularities, usually at the cost of generating additional algebraic loops.

(5) DYMOLA should be enabled to handle linear algebraic loops in the manner previously suggested.

A fair amount of program development and research is still needed before DYMOLA can be turned into a production code, but we are convinced that this is a good way to go. We shall return to this problem in Chapter 15 of this text.

5.12 Large–Scale System Modeling

A true disadvantage of the previously proposed methodology is the fact that, as with macros, DYMOLA models need to be stored as source code. This may be quite impractical if models of maybe 20,000 lines of code are to be simulated. A small structural modification of one single equation within one single submodel will force us to recompile the entire code, which may take quite a long time and consume an undue amount of computing resources.

Separate compilation of submodels is difficult to achieve. One way to solve this problem is to generate the target code such that each single equation is preceded by a label and followed by a "goto" statement with an address that is not static but is stored in a large connection table. If a submodel is to be modified, it suffices to recompile that submodel and to correct the connection table accordingly. None of the systems presented so far offers this capability. One system that does offer this facility is SYSMOD [5.19]. The SYSMOD language is a superset of PASCAL. The SYSMOD system consists of a PASCAL–coded preprocessor that compiles (sub–)models into FORTRAN subroutines, a FORTRAN–coded simulation run–time system containing the integration routines and output routines, and a special–purpose "linker" that is responsible for updating the connection table after a recompilation of a submodel has occurred.

The price for this separate submodel compilation capability is a reduction in run–time efficiency. According to information obtained from the producers of SYSMOD, the run–time overhead is about 20% which seems quite acceptable. We have not yet had a chance to verify or reject this information on the basis of significantly large sample programs.

5.13 Graphical Modeling

With the advent of increasingly powerful engineering workstations, the demand has risen to model systems graphically on the screen. Submodels are maintained in a *model library*. Such submodels could be either DYMOLA models, regular CSSL macros, transfer functions, linear state–space descriptions written in matrix form, or static characteristics. Each of these models is associated with an *icon*, which is stored in an icon library. Invoking a submodel simply means to place the corresponding item on the screen. Connections between submodels are done by drawing a line between two terminals (cuts) of two icons.

Several such systems exist already on the software market. Some of these systems are *generic program generators* in that they allow the user to specify what code she or he wants to generate as a result of the *graphical compilation* (i.e., the evaluation of the graph). One such system, EASE+ [5.8], has been successfully employed as a graphical preprocessor to ACSL. Others are either stand–alone systems, such as Easy5 [5.1], or they are integral parts of particular software systems, such as SYSTEM–BUILD [5.11], which has been designed as a modeling tool for MATRIX$_X$ [5.10], or MODEL–C, which is a modeling tool for CTRL–C [5.18].

Most of these systems are based on the concept of block diagram modeling. The basic building blocks are those used in block diagrams, i.e., single–input/single–output (SISO) system descriptions, summers, and branching points. The disadvantage of these systems is obvious. They do not provide for a hierarchical decomposition of data structures (cuts) or for the representation of through variables.

Some systems are specialized tools for particular types of models. For example, WORKVIEW [5.20] is a graphical modeling system for electronic circuits. As a result of the graphical compilation, WORKVIEW generates a PSpice [5.15] program. WORKVIEW will be presented in Chapter 6.

Another system, HIBLIZ [5.6], has been designed specifically as a graphical preprocessor to DYMOLA. HIBLIZ supports all concepts that DYMOLA does and can therefore be used for generic modeling of arbitrarily coupled systems with hierarchical cuts and across as well as through variables. The result of the graphical compilation is a DYMOLA program. Unfortunately, while most of the available graphical systems have been developed for PC compatibles, HIBLIZ currently runs on Silicon Graphics (IRIS) machines only, which makes the software much less accessible (though faster executing).

One problem with the graphical approach is the fact that the screen is not large enough to depict reasonably complex systems. One typical solution to this problem is to resort to a *virtual screen*. The virtual screen can be made arbitrarily large and the actual screen covers a *window* out of the virtual screen, i.e., the physical screen can be moved over the virtual screen much like a short–sighted person may move a magnifying glass over a page of a book. Another approach is the *zoom in* facility, which allows us to modify the size of the icons on the screen. Virtual screens and zooming are often combined. When zooming in on a portion of the virtual screen, the window of the physical screen is made smaller, i.e., it covers a smaller portion of the virtual screen.

Also in this respect, HIBLIZ offers a rather unique feature called a *breakpoint*. Breakpoints allow the programmer to alter the drawing as a function of the magnification (zooming). As an example, it is possible to start out with a box as large as the virtual screen, which is empty except for the name of the problem that appears in large letters. The physical screen at this moment coincides with the virtual screen. As soon as we start zooming in on the graph, a breakpoint is passed, and suddenly, the title disappears and is replaced by some text that describes the purpose of the model. When zooming in further, another breakpoint is passed and the text is replaced by a set of smaller boxes with interconnections. Each box contains the name of the submodel represented by this box. Now we can zoom in on any of these boxes. Meanwhile, the physical screen has become considerably smaller than the virtual screen and we no longer see the entire picture at once. Again, a new breakpoint is passed and now we see a text that describes the purpose of the submodel on which we are currently focusing. When zooming in further, the text is replaced by the internal structure of the submodel, which may consist of some more submodels with interconnections. Ultimately, we come to the layer of *atomic models*, in which case the description may be replaced by a graph denoting a static characteristic, a state–space model, a transfer function, or a set of DYMOLA statements. If we zoom in on the point where an interconnection meets a box, we can notice that each such point in reality is represented by a little box itself. Zooming in further on that little box, we can determine the nature of the cut that is represented by the interconnection, i.e., we can learn about the variables represented in the cut.

The graphical representation can also be used later, i.e., during or after the simulation phase. For example, HIBLIZ allows pointing to a particular connection after the simulation has been executed. As a result of this action, a new window is opened in which the trajectories of all variables contained in the cut are displayed as functions of time.

In the long run, this is clearly the right approach. However, in order to enhance the accessibility of the HIBLIZ code, it is hoped that the developers of the software (the Technical University of Lund, Sweden) will port the code to *X Windows*. We shall talk more about HIBLIZ in Chapter 15 of this text.

It was demonstrated how hierarchical modeling has become the key issue to coping with the increasing demands of modern large–scale continuous system simulation. I would like to acknowledge in particular the important research results obtained by Hilding Elmqvist of the Technical University at Lund. In 1975, Hilding Elmqvist produced the first direct executing CSSL language, SIM-NON, which paved the way for the work that resulted much later in the DESIRE software, which has been advocated in this text. SIMNON was developed by Hilding Elmqvist as his master's thesis. Later in 1978, Hilding developed DYMOLA for his Ph.D. dissertation, a software system that is still, 13 years later, very much state of the art. In 1982, he developed the first prototype of HIBLIZ, at least three years before any competitor's products came on the market. HIBLIZ is even today considerably more powerful than all of its competitors, since it implements hierarchical cuts and across as well as through variables.

5.14 Summary

In this chapter, we discussed the problems associated with large–scale system modeling: modularity and hierarchical decomposition of submodels. We introduced a new language, DYMOLA, which is particularly well suited to support the process of modeling large–scale systems. In subsequent chapters, we shall present many examples of the concepts introduced here.

References

[5.1] Boeing Computer Services (1988), *EASY5/W — User's Manual*, Engineering Technology Applications (ETA) Division, Seattle, Wash.

[5.2] François E. Cellier (1979), *Combined Continuous/Discrete System Simulation by Use of Digital Computers: Techniques and Tools*, Ph.D. dissertation, Diss ETH No 6483, Swiss Federal Institute of Technology, Zürich, Switzerland.

[5.3] Olle I. Elgerd (1971), *Electric Energy Systems Theory: An Introduction*, McGraw-Hill, New York.

[5.4] Hilding Elmqvist (1975), *SIMNON — An Interactive Simulation Program for Nonlinear Systems — User's Manual*, Report CODEN: LUTFD2/(TFRT-7502), Dept. of Automatic Control, Lund Institute of Technology, Lund, Sweden.

[5.5] Hilding Elmqvist (1978), *A Structured Model Language for Large Continuous Systems*, Ph.D. dissertation, Report CODEN: LUTFD2/(TRFT-1015), Dept. of Automatic Control, Lund Institute of Technology, Lund, Sweden.

[5.6] Hilding Elmqvist (1982), "A Graphical Approach to Documentation and Implementation of Control Systems," *Proceedings Third IFAC/IFIP Symposium on Software for Computer Control (SOCOCO'82)*, Madrid, Spain, Pergamon Press, Oxford, U.K.

[5.7] EPRI (1983), *Modular Modeling System (MMS): A Code for the Dynamic Simulation of Fossil and Nuclear Power Plants*, Report: CS/NP-3016-CCM, Electric Power Research Institute, Palo Alto, Calif.

[5.8] Expert-EASE Systems, Inc. (1988), *EASE+ — User's Manual*, Belmont, Calif.

[5.9] IBM Canada Ltd. (1972), *Continuous System Modeling Program III (CSMP-III) — Program Reference Manual*, Program Number: 5734-XS9, Form: SH19-7001-2, IBM Canada Ltd., Program Produce Centre, Don Mills, Ontario, Canada.

[5.10] Integrated Systems, Inc. (1984), *MATRIX$_X$ User's Guide, MATRIX$_X$ Reference Guide, MATRIX$_X$ Training Guide, Command Summary and On-Line Help*, Santa Clara, Calif.

[5.11] Integrated Systems, Inc. (1985), *SYSTEM-BUILD User's Guide*, Santa Clara, Calif.

[5.12] Dirk L. Kettenis (1988), *COSMOS — Reference Manual*, Dept. of Computer Science, Agricultural University Wageningen, Wageningen, The Netherlands.

[5.13] Granino A. Korn (1989), *Interactive Dynamic-System Simulation*, McGraw-Hill, New York.

[5.14] Granino A. Korn (1991), *Neural-Network Experiments on Personal Computers*, MIT Press, Cambridge, Mass.

[5.15] MicroSim Corp. (1987), *PSpice User's Manual*, Irvine, Calif.

[5.16] Edward E. L. Mitchell and Joseph S. Gauthier (1986), *ACSL: Advanced Continuous Simulation Language — User Guide and Reference Manual*, Mitchell & Gauthier Assoc., Concord, Mass.

[5.17] Thomas F. Runge (1977), *A Universal Language for Continuous Network Simulation*, Technical Report UIUCDCS–R–77–866, Dept. of Computer Science, University of Illinois, Urbana–Champaign.

[5.18] Systems Control Technology, Inc. (1985), *CTRL–C, A Language for the Computer-Aided Design of Multivariable Control Systems, User's Guide*, Palo Alto, Calif.

[5.19] Systems Designers plc (1986), *SYSMOD User Manual*, Release 1.0, D05448/14/UM, Ferneberga House, Farnborough, Hampshire, U.K.

[5.20] Viewlogic Systems, Inc. (1988) *WORKVIEW Reference Guide*, Release 3.0, and: *VIEWDRAW Reference Guide*, Version 3.0, Marlboro, Mass.

[5.21] John V. Wait and DeFrance Clarke III (1976), *DARE–P User's Manual*, Version 4.1, Dept. of Electrical & Computer Engineering, University of Arizona, Tucson, Ariz.

Homework Problems

[H5.1]* Control System

Figure H5.1 shows a typical single–input/single–output (SISO) control system designed in the frequency domain. A second–order plant is controlled by use of a lead compensator. The output variable x is measured using measurement equipment with dynamic behavior of its own.

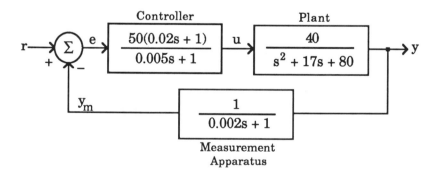

Figure H5.1. Block diagram of a SISO control system.

Many CSSLs (such as ACSL) offer built–in macros to model systems that are either totally or partially described in the frequency domain directly without need to transform the model back into the time domain. Model this system in ACSL using the LEDLAG macro to model the controller, the TRAN macro to model the plant, and the REALPL macro to model the measurement equipment. Simulate a step response of this system. Use a dynamic termination condition to bring the simulation to an end as soon as three values of y that are separated in time by $\Delta t = 0.03$sec differ less than 0.001 from each other. Use ACSL's DELAY operator to construct the signals:

$$\Delta y_1 - y(t) \quad y(t \quad \Delta t) \tag{$H5.1a$}$$
$$\Delta y_2 = y(t - \Delta t) - y(t - 2\Delta t) \tag{$H5.1b$}$$

which can then be used in a **termt** condition.

Since this is a linear system, we could also simulate it in CTRL–C (or MATLAB) directly. Use CTRL–C's (MATLAB's) TF2SS function to transform each of the three transfer functions into state–space models. Thereafter use CTRL–C's INTERC function to obtain a state–space model for the total interconnected system. Thereafter, use CTRL–C's SIMU or STEP functions to simulate the system with step input. Remember that CTRL–C offers interactive help for all its functions. If you use MATLAB, consult the manual for the names of these functions. All CTRL–C capabilities are also offered in MATLAB.

[H5.2]* Sampled–Data Control System [5.16]

Figure H5.2 shows a sampled–data control system. The plant to be controlled is a second order "Type 1" system (indicating that the plant has a pole at the origin). This model represents a DC motor in which the armature inductance has been neglected. The output variable x is the angular position of the motor. The system is to be controlled by a microprocessor. An optimal controller was designed in the s–domain and it was found that a lead compensator of the form:

$$G_c = \frac{1 + 2.5s}{1 + 0.5s} \tag{$H5.2a$}$$

would be optimal for this system using unit feedback (i.e., assuming an ideal measurement apparatus).

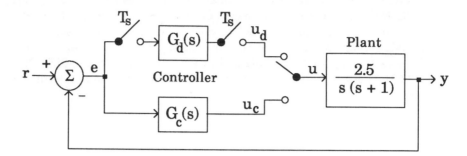

Figure H5.2. Block diagram of a sampled–data control system.

We want to simulate the continuous control system in ACSL using the built–in LEDLAG macro to describe the controller and the TRAN macro to describe the plant. Simulate the system over a duration of 5 sec.

Unfortunately, the preceding simulation is not very realistic since the microprocessor cannot properly represent the lead compensator dynamics. We can compute an equivalent discrete controller in CTRL–C (or MATLAB). Use the TF2SS function to get a state–space representation of your controller. Thereafter, use the C2D function to transform the continuous–time controller model to an equivalent discrete–time controller model. Use a sampling interval of $T_s = 0.1$ sec. The transformation affects only the **A** matrix and the **b** vector (in our example, they are both scalars), while the output equation remains unchanged. Augment your previous ACSL program by including the discrete controller using the design that was computed in CTRL–C (MATLAB). Use a DISCRETE block with an **interval** specification to model the discrete controller. This can be achieved using the following language construct:

> **DISCRETE** *controller*
>   **interval** $Ts = 0.1$
>   **PROCEDURAL**
> $xinew = a * xi + b * e$
> $udis\ \ = c * xi + d * e$
> $xi\ \ \ \ = xinew$
>   **END**
> **END**

The **interval** specification will ensure that this block is being executed exactly once every T_s time units. The PROCEDURAL declaration is necessary to break the algebraic loop between ξ and ξ_{new}. Upon execution of the DISCRETE block, a new value of u_{dis} is computed, which thereafter stays constant for the duration of one sampling period (sample and hold). Model the switch using ACSL's FCNSW function. Simulate the step response of the sampled–data system and compare the results to the idealized continuous–time simulation performed earlier.

Even the sampled–data model is not truly realistic since it takes the computer some time to perform the computations expressed in the DISCRETE block. Assume that the time needed for the computation is $\Delta t = 0.003$ sec. This can be simulated by replacing the u_{dis} signal in the DISCRETE block by another signal, say u_{st}, and schedule from within the DISCRETE block another DISCRETE block called DAC to be executed Δt time units in the future using ACSL's time–event scheduling facility. All that the DAC block needs to do is pass the current value of the u_{st} variable on to u_{dis}. Simulate the step response of the once more refined model, and compare the results with those of the previous simulations. If you run ACSL through CTRL–C or MATLAB, you can easily plot all simulation results on top of each other on one graph.

[H5.3] Water Flow Through a Reservoir

A community uses a small reservoir for irrigation purposes, but also to prevent damage otherwise produced by flooding during storms. When the reservoir contains a water volume of $V = V_D = 4000$ m^3, water will begin to overflow from the reservoir into a series of drainage channels. Assume that at $t = 0$, the reservoir contains $V_0 = 3900$ m^3 of water. A storm that lasts for 25 hours adds water to the reservoir as specified in Table H5.3a.

Table H5.3a. Water Inflow $\dot{q}_{in}$ During a Storm.

Time t [h]	Inflow $\dot{q}_{in}$ [m^3/h]
0.0	10.0
1.0	35.0
2.0	58.0
3.0	70.0
4.0	75.0
5.0	68.0
6.0	55.0
7.0	38.0
8.0	28.0
9.0	21.0
10.0	18.0
11.0	16.0
12.0	14.0
13.0	13.0
14.0	12.0
15.0	11.0
20.0	10.5
25.0	10.0

The overflow is described by the function:

$$\dot{q}_{out} = 0.02 \cdot f(\dot{q}_{out}) \cdot [\max(V - V_D, 0.0)]^{1.5} \qquad (H5.3)$$

where the overflow characteristic $f(\dot{q}_{out})$ is specified in Table H5.3b.

Table H5.3b. Water Outflow Characteristics $f(\dot{q}_{out})$.

Outflow $\dot{q}_{out}$ [m^3/h]	$f(\dot{q}_{out})$
0.0	0.80
25.0	0.85
50.0	0.95
75.0	1.00
100.0	1.00

Code this problem in ACSL using tabular functions and an IMPL block to describe the implicitly defined overflow characteristic.

Simulate this system over the duration of the storm and plot on separate graphs the inflow rate, outflow rate, and volume content of the reservoir.

[H5.4] Surge Tank Simulation

A water turbine is fed from a reservoir through a pressurized pipe. The turbine in turn feeds an electrical generator (a synchronous AC machine). In case of a short–circuit on the electrical circuit, it may be necessary to shut down the generator very fast (within a few seconds), which may force us to close the valve in front of the turbine within a very short time period. When the valve is open, water flows from the reservoir into the turbine. This water flow contains a potentially destructive amount of kinetic energy. If we are forced to close the valve quickly, the kinetic energy could destroy the rear end of the pressure tunnel. In order to prevent such damage from happening, we need to add a surge tank to the system into which the water can escape. The surge tank will help to convert the kinetic energy into potential energy, thereby reducing the maximum pressure in the pressure tunnel. The system is shown in Fig.H5.4a.

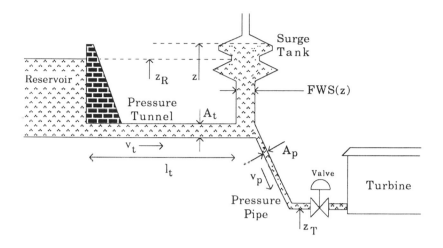

Figure H5.4a. Power generation using a water turbine.

The reservoir surface is at an altitude of $z_R = 3168$ m. It is assumed that the reservoir is sufficiently large so that the outflow of water from the reservoir does not noticeably change the water level in the reservoir.

We can compute the amount of water that flows from the reservoir into the pressure tunnel by formulating Newton's law for the pressure tunnel:

$$m\dot{v}_t = F_{in} - F_{out} - F_{Fr} \qquad (H5.4a)$$

F_{in} and F_{out} are the forces produced by the water pressure at the inflow and outflow of the pressure tunnel. These forces are equal to the water pressure multiplied by the active surface, i.e.,

$$F_{in} = A_t \cdot p_{in} \qquad (H5.4b)$$
$$F_{out} = A_t \cdot p_{out} \qquad (H5.4c)$$

where the cross section of the pressure tunnel A_t has a value of 12 m². The water pressures can be computed to be

$$p_{in} = \rho \cdot g \cdot h_{in} \qquad (H5.4d)$$
$$p_{out} = \rho \cdot g \cdot h_{out} \qquad (H5.4e)$$

where ρ is the water density ($\rho = 1000$ kg m⁻³), g is the gravity ($g = 9.81$ m sec⁻²), and h_{in} and h_{out} denote the momentary values of the water columns (i.e., h_{in} is the depth of the reservoir, and h_{out} is the current water level in the surge tank minus the altitude of the pressure tunnel). The mass of the water can be specified as the product of density and volume:

$$m = \rho \cdot V = \rho \cdot A_t \cdot \ell_t \qquad (H5.4f)$$

where ℓ_t is the length of the pressure tunnel ($\ell_t = 13,580$ m). In turbulent flow, the inner friction of the water is proportional to the square of the velocity:

$$F_{Fr} = k \cdot A_t \cdot \rho \cdot g \cdot v_t |v_t| \qquad (H5.4g)$$

where k is the friction constant ($k = 4.1$ m^{-1} sec^2). This information suffices to generate a differential equation for the tunnel velocity v_t.

At the rear end of the pressure tunnel, we can formulate the mass continuity equation (what comes in must go out), assuming that the water is ideally incompressible:

$$\dot{q}_t = \dot{q}_C + \dot{q}_p \qquad (H5.4h)$$

where the mass flow rate in the pressure tunnel $\dot{q}_t$ is the product of the tunnel velocity and the tunnel cross section:

$$\dot{q}_t = A_t \cdot v_t \qquad (H5.4i)$$

the mass flow rate in the pipe is the product of the pipe velocity v_p, the pipe cross section ($A_p = 0.6$ m^2), and the current percentage of valve opening $S(t)$:

$$\dot{q}_p = A_p \cdot v_p \cdot S(t) \qquad (H5.4j)$$

Finally, the mass flow rate into the surge tank $\dot{q}_C$ is the product of the cross section of the surge tank and the time derivative of the water level z in the surge tank. In other words, Eq.(H5.4h) can be rewritten as:

$$A_t \ v_t \ dt = A_p \ v_p \ S(t) \ dt + \text{FSW}(z) \ dz \qquad (H5.4k)$$

where $S(t)$ stands for either the closing or opening characteristic of the valve chosen to follow one of the static characteristics shown in Fig.H5.4b–c:

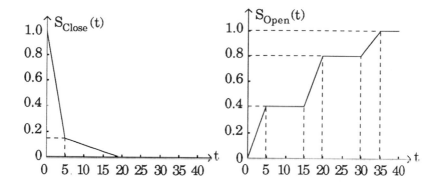

Figure H5.4b–c. Closing/opening characteristics of the valve.

$FWS(z)$ represents cross sections of the surge tank at different altitudes. This tabular function is given in Table H5.4.

Table H5.4. Topology of the Surge Tank.

Altitude z [m]	Surge Tank Surface FWS [m^2]
3100.0	20.0
3110.0	20.0
3120.0	100.0
3130.0	40.0
3160.0	40.0
3170.0	80.0
3180.0	80.0
3190.0	10.0
3210.0	10.0

The only quantity that we are still missing is the pipe velocity v_p. To compute this variable, we need to see what happens at the valve. Let me assume that the valve is initially open. Water flows into the turbine, which carries a kinetic energy of:

$$E_k = \frac{1}{2}mv_p^2 = \frac{1}{2}\rho \, V \, v_p^2 \qquad (H5.4l)$$

The water also carries a potential energy of:

$$E_p = mgh = \rho \, V \, g \, h \qquad (H5.4m)$$

If we now start to close the valve, the total free energy doesn't change, i.e., the reduction in kinetic energy must be equal to the increase in potential energy:

$$\frac{1}{2} \rho \, \Delta V \, v_p^2 = \rho \, \Delta V \, g \, h \qquad (H5.4n)$$

and therefore:

$$v_p = \sqrt{2gh} \qquad (H5.4o)$$

Under the assumption that the change in the water level can be neglected, we can write this as:

$$v_p = \sqrt{2g(z_R - z_T)} \qquad (H5.4p)$$

The altitude of the turbine z_T is 3072 m.

The critical quantity that we are interested in is the inertial pressure at the rear end of the pressure tunnel during valve openings and closings. The inertial pressure can be computed as:

$$p_I = \frac{F_I}{A_t} = \frac{m \dot{v}_t}{A_t} = \frac{\rho \ell_t A_t \dot{v}_t}{A_t} = \rho \ell_t \dot{v}_t \qquad (H5.4q)$$

Simulate separately one opening and one closing of the valve over a duration of 2000 sec each, starting from steady–state initial conditions. Plot on separate graphs the tunnel velocity $v_t(t)$, the water level $z(t)$ in the surge tank, and the inertial pressure $p_I(t)$ as functions of time, and determine a numerical value for the maximum absolute inertial pressure.

For a configuration without a surge tank, the equations become so simple that the inertial pressure can be computed analytically. Determine the maximum value of the inertial pressure if the surge tank is being removed and compare this value to the one found by simulation before.

[H5.5] Macros

Four bicyclists start at the coordinates indicated in Fig.H5.5. They travel with constant velocities $v_1 = 17$ km/h, $v_2 = 14$ km/h, $v_3 = 12$ km/h, and $v_4 = 15$ km/h. Each bicyclist travels at all times straight into the momentary direction of her or his next neighbor, i.e., bicyclist #1 tries to catch bicyclist #2, etc.

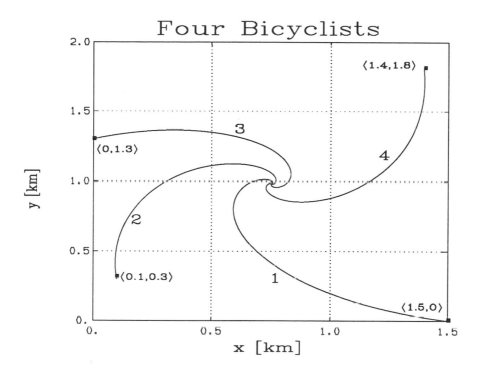

Figure H5.5. Phase plane plot of four bicyclists trying to catch each other.

Compute the positional coordinates of each of the four bicyclists as functions of time. Terminate the simulation as soon as the distance between any two bicyclists has decreased to below 10 m. Develop a macro that describes the motion of any one of the four bicyclists and call that macro four times in your simulation program.

[H5.6] Electric Power Generation [5.5]

Figure H5.6 shows the configuration of a small electric power system consisting of two synchronous generators, three transmission lines, and three different loads.

It is assumed that all voltages and currents in this system are basically sinusoidal with slowly varying amplitudes and phases. We can therefore conveniently model the system with complex voltages and currents.

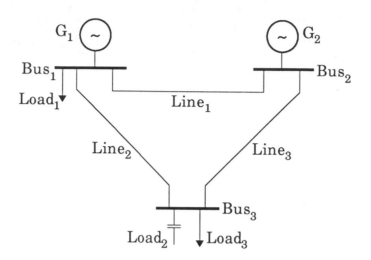

Figure H5.6. Topology of a small electric power system [5.5].

The generators are represented by AC voltage sources series connected with a pure reactance:

$$V_g = E_g - j\, X_d\, I_g \qquad (H5.6a)$$

where E_g is the ideal voltage source, V_g is the effective voltage that is connected to the net, I_g is the current that flows through the generator, and X_d is the (inductive) reactance of the generator. Both generators have a reactance of $X_d = 0.054\Omega$. The electric power produced by the generator is the real part of the product of the generator voltage E_g and the generator current I_g:

$$P_g = \mathrm{Re}\{E_g \cdot I_g\} \qquad (H5.6b)$$

For reasons of energy conservation, the electric power P_g must be equal to the mechanical power P_t of the water turbine that is responsible for rotating the generator. If this is temporarily not true, the phase angle between current and voltage will change to make it true. A model for this adaptation is the so–called *swing equation* [5.3]:

$$\frac{H}{\pi f_0} \ddot{\varphi} + D\dot{\varphi} = P_t - P_g \qquad (H5.6c)$$

where f_0 is the net frequency (60 Hz in the United States and 50 Hz in Europe) and H and D are generator parameters. For our two generators, we want to assume that $D_1 = D_2 = 0$, $H_1 = 30$ J, and $H_2 = 300$ J.

The transmission lines are modeled by (inductive) reactances:

$$V_{out} = V_{in} - j \, X_L \, I_L \qquad (H5.6d)$$

where X_L is the reactance of the transmission line. In our example, we want to assume that all three transmission lines are equal and have reactances of 0.05Ω.

The loads are modeled as impedances:

$$V_{Load} = Z \cdot I_{Load} \qquad (H5.6e)$$

We want to analyze the reaction of the power system to various load functions. Therefore, the loads are not *a priori* specified. All we know is that the second load is purely capacitive, while the other two loads can be anything.

Create three separate DYMOLA model types for the three types of components: generators, transmission lines, and loads. Then connect these components to the topology shown in Fig.H5.6. While we made the task of modeling easy and convenient, we now have to pay the price: the model does not contain enough differential equations and therefore the system contains several algebraic loops. Use the DYMOLA preprocessor to detect those algebraic loops and determine how many variables are involved in each of them.

Projects

[P5.1] The Domino Game

The domino game consists of 55 stones with the dimensions $x_D = 8$ mm, $y_D = 2.4$ cm, and $z_D = 4.6$ cm. The mass of each stone is $m = 10$ g. We place these 55 stones in a series at a distance d from each other. We push the first stone and the entire series of stones falls flat [5.2]. This is shown in Fig.P5.1a.

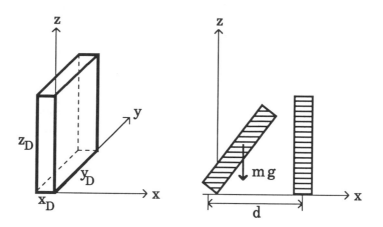

Figure P5.1a. Domino game.

The question that we wish to answer is the following: At what distance d between two consecutive stones is the chain velocity v_{Ch} maximized. The chain velocity can be defined as follows:

$$v_{Ch}(k) = \frac{d}{t_{Imp}(k+1) - t_{Imp}(k)} \qquad (P5.1a)$$

which is a discrete function. k stands for stone $^\#k$, and T_{Imp} is the time of impact, i.e., $t_{Imp}(k+1)$ is the time when stone $^\#k$ bumps into stone $^\#(k+1)$. This problem must have a nontrivial answer since, for very small values of d, the impacting stone has not yet gained a sufficiently large momentum at impact, while, for large values of d, most of the momentum is directed toward the floor already.

This is a very difficult problem to model accurately. We want to make some simplifying assumptions. First, we assume the impact to be totally *elastic*, i.e., just before impact, the impacting stone has a certain momentum $\mathcal{I} = m \cdot v$ which can be decomposed into its horizontal component $\mathcal{I}_x$ and its vertical component $\mathcal{I}_z$. At impact, the horizontal component of the momentum is passed on to the next stone, i.e., immediately after impact, the horizontal momentum of the impacting stone is zero, while the horizontal momentum of the impacted stone has taken over the momentum from the impacting stone. Second, we shall neglect the interaction between the stones after impact, i.e., the interaction is assumed to be momentary only, and thereafter, both stones are treated separately again. Each stone is simulated until it hits the ground, i.e., we shall neglect the kinematic constraints of two different stones occupying the same point in space at the same time.

At time zero, the first stone is pushed at an altitude $h = 0.75 \cdot z_D$. It thereby receives an initial horizontal momentum of $\mathcal{I} = 0.002$ kg m sec^{-1}. This produces an initial velocity of:

$$v_0 = \frac{\mathcal{I}}{m} \qquad (P5.1b)$$

Due to Coulomb friction between the stone and the ground, the stone will not slip along the floor, but starts to roll. The initial angular velocity is therefore:

$$\omega_0 = \frac{v_0}{h} \qquad (P5.1c)$$

We can now formulate Newton's law in rotational coordinates:

$$I_y \dot{\omega} = \tau \qquad (P5.1d)$$

where I_y is the inertia of the stone to rotation around the y-axis. The inertia can be computed easily:

$$I_y = \frac{1}{3} m (x_D^2 + z_D^2) \qquad (P5.1e)$$

τ is the gravitational torque. Figure P5.1b shows what happens while the stone falls.

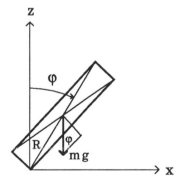

Figure P5.1b. Newton's law for the domino game.

The gravitational force can be decomposed into a normal and a tangential component. Only the tangential component produces a torque that can easily be computed:

$$\tau = mg \ R \ \sin(\varphi) \qquad (P5.1f)$$

where

$$R = \frac{1}{2}\sqrt{x_D^2 + z_D^2} \qquad (P5.1g)$$

Notice that the initial angle φ_0 is negative, namely,

$$\varphi_0 = -\tan^{-1}\left(\frac{x_D}{z_D}\right) \qquad (P5.1h)$$

and therefore, the initial torque is also negative. If the momentum $\mathcal{I}$ is chosen too small, the stone will never fall, but only rock for a while and then return to its initial upright position.

Generate an ACSL macro that describes the behavior of a stone. The differential equations are multiplied by a constant c, which is initially set equal to zero. When the stone is pushed (modeled using ACSL's state–event scheduling feature), it obtains its initial angular velocity ω_0 and the constant c is altered from zero to one. The stone starts moving and a new state–event is scheduled to occur at the time of next impact. At that moment, the angular velocity of the impacting stone changes abruptly and the stone continues to fall until $\varphi = 90°$. At that time, another state–event is scheduled, which will set the constant c back to zero. We shall assume the impact with the ground to be totally *plastic*, i.e., the fallen stone does not bounce.

Call this macro 55 times for the 55 stones of the domino game and simulate the overall system. Plan a strategy that will optimize the chain velocity v_{Ch}. Obviously, the distance d must be chosen in the range:

$$d \in [x_d \, , \ x_D + z_D] \qquad (P5.1i)$$

ACSL permits you to branch from the TERMINAL section of the program (which is executed after a simulation run has been completed) back to the INITIAL section using a GOTO statement. By doing so, a next simulation run is initiated. This feature enables you to program your strategy.

Each stone is described by a second–order differential equation. Consequently, the entire system order is 110 at all times, even though only a few stones move simultaneously. This is most unfortunate since this costs a lot of execution time in vain. Very few languages permit you to dynamically create/destroy processes involving differential equations, i.e., invoke a new instantiation of a macro at run–time during a discrete event. One such language is COSMOS [5.12].

Try to outwit ACSL by assuming that never more than eight stones will simultaneously move at any one time. Consequently, when you push stone #9, you can simply revive the already–fallen stone #1, etc. Thereby, the overall system order can be reduced from 110 to 16, which will speed up the simulation dramatically.

For the optimal distance d, plot the chain velocity over the stone number. For each simulation run, store away the last value of the chain velocity

$v_{Chl} = v_{Ch}(54)$ and the currently used value of d. After all simulations have been completed, plot $v_{Chl}(d)$.

[P5.2] Robot Modeling

We wish to analyze once more the behavior of the Stanford arm (for a detailed description, cf. Pr.[P4.2]). This time, we wish to model each limb separately. Describe a general limb through a DYMOLA model type. Each two consecutive limbs are connected with a joint. Describe in another set of DYMOLA model types the different types of joints (revolute and prismatic). Build a model of the entire Stanford arm by connecting the limbs and joints together.

Employ the DYMOLA preprocessor to obtain a total set of equations describing the Stanford arm. Obviously, the kinematic constraints imposed by the connections will result in structural singularities. Use the differentiation algorithm outlined in this chapter to manually get rid of these singularities.

Research

[R5.1] The Modular Modeling System

The *Modular Modeling System (MMS)* [5.7] was developed by the Electric Power Research Institute (EPRI) for the simulation of various types of electric power distribution systems. Currently, two implementations of MMS exist, one in the form of a macro library for ACSL [5.16], the other as an application of Easy5 [5.1]. Both implementations suffer from the fact that the parent languages do not provide for truly modular modeling facilities. Consequently, the macros that make up for the MMS library are rather involved, somewhat clumsy, unnecessarily slow in execution, and not truly flexible. It would therefore make a lot of sense to reimplement MMS in DYMOLA, which is ideally suited to provide for a flexible and totally modular parent language environment.

Reimplement MMS in DYMOLA. Since power system models can be quite large, neither SIMNON [5.4] nor DESIRE [5.13] are well suited as simulation run–time environments. Enhance DYMOLA to alternatively generate ACSL code as well. A graphical pre/postprocessor for MMS exists that was written in EASE+ [5.8]. In the context of recoding MMS in DYMOLA, I suggest replacing this pre/postprocessor with HIBLIZ [5.6].

[R5.2] Structural Singularities

Design a general–purpose algorithm that reduces structural singularities by analytic differentiation. Implement this algorithm as part of the DYMOLA language. Project Pr.[P5.2] can serve as a test case for this algorithm.

6

Principles of Active Electrical Circuit Modeling

Preview

In this chapter, we shall discuss the tools and techniques of today's professional circuit designers. We shall discuss how SPICE works and what additional tools exist that support SPICE, such as the PROBE feature for viewing simulation trajectories graphically or WORKVIEW [6.11] for schematic capture. We shall then explore the pros and cons of using DYMOLA [6.3] for electronic circuit modeling.

6.1 Topological Modeling

As we have shown in Chapter 3, it is rather inconvenient to manually derive a state–space description for even fairly simple passive electrical circuits. We have seen that it may not be desirable to even attempt to generate such a state–space description due to the frequent algebraic loops and structural singularities inherent in most practical electrical circuits.

For this reason, circuit designers prefer to model their circuits by specifying a *topological description*. Let us revisit the simple passive circuit that was discussed in Chapter 3. It is shown again in Fig.6.1.

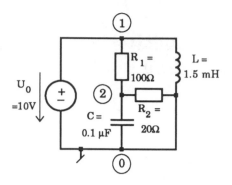

Figure 6.1. Example of a passive circuit.

Using the topological modeling approach, we simply number all nodes of the circuit and describe the circuit topology by specifying terminal nodes for each circuit element. This is the approach taken by all current circuit analysis programs. The most common among those is SPICE. Various SPICE dialects exist that have partly been coded in FORTRAN and partly in C. Among those, we currently recommend PSpice [6.9], which runs on both mainframes and PCs and works fairly well. The circuit in Fig.6.1 can be modeled in PSpice in the following manner:

```
Simple Passive Electrical Circuit
R1   1   2   100
R2   2   0   20
L    1   0   1.5m
C    2   0   0.1u
V0   1   0   10
.OP
.END
```

The **.OP** statement describes the simulation experiment, i.e., the type of analysis that is to be performed on the circuit (in this case, a computation of the DC steady–state value). The first character of the element names is semantically significant. It determines the type of circuit element being used. Node numbers are positive integers except for the ground node, which is always declared to be node 0. A character immediately following a numeric constant (an element value) denotes a scaling factor. For example, "m" stands for "milli," "u" for "micro." Notice that SPICE is a fairly old–fashioned language. Many SPICE dialects don't understand lower–case characters, but, as in the case of DARE–P, we shall ignore this deficiency for improved readability of the code.

SPICE uses the *branch–admittance matrix* approach to convert the topological circuit description into an implicit matrix description in the way described in Chapter 3.

6.2 Models of Active Devices in SPICE

Circuits that are of any practical interest today are always heavily nonlinear. This is due to the fact that most circuits today are built as *integrated circuits* and the electrical phenomena that occur in a p–n junction are governed by heavily nonlinear equations.

Let us look at a simple p–n junction first. Such a junction is shown in Fig.6.2.

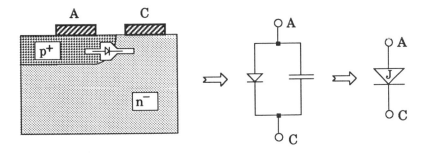

Figure 6.2. Model of a p–n junction.

Every p–n junction can be described as a *diode* that has its anode on the p–side and its cathode on the n–side of the junction, i.e., the junction is forward–biased if the potential on the p–side of the junction is higher than that on the n–side. In this case, current flows through the junction. Otherwise, the junction is reverse–biased. The relation between current through and voltage across a diode can be described by the following equation:

$$i_d = I_s[\exp(\frac{u_d}{V_T N_d}) - 1] + G_{min} u_d \qquad (6.1)$$

I_s is the saturation current of the diode, V_T is the thermal voltage, $V_T = kT/q$, and N_d is the emission coefficient. The second term denotes a resistive leakage current.

Especially for reverse–biased junctions, the electrical charge will build up along the junction since no current can flow. This charge can be described as a nonlinear capacitance that is connected in parallel to the diode. In Chapter 3, we saw that the relation between current through and voltage across a capacitor is:

$$i_c = C \frac{du_c}{dt} \tag{6.2}$$

However, this equation is only correct for linear capacitors. If the capacitance value changes over time, this equation must be replaced by the more general equation:

$$i_c = \frac{d}{dt}(C \cdot u_c) = \frac{dq_c}{dt} \tag{6.3}$$

where $q_c = C \cdot u_c$ is the electrical charge stored in the capacitance. Even more generally, the relation between current and voltage in the junction capacitance can be described as:

$$i_c = \frac{dq_c}{dt}, \quad q_c = f(u_c) \tag{6.4}$$

The charge in the junction capacitance can be approximately described by the following equation:

$$q_c = \tau_d i_d + \frac{\phi_d C_d [1 - (1 - \frac{u_d}{\phi_d})^{1 - m_d}]}{1 - m_d} \tag{6.5}$$

where i_d stands for the current through the junction diode and is in itself a function of u_d as stated in Eq.(6.1), τ_d is the time constant of the capacitance, and the parameters ϕ_d, C_d, and m_d model second–order transient effects.

Let us now look at transistors. A bipolar junction transistor (BJT) basically consists of two p–n junctions: a base–collector junction and a base–emitter junction. Figures 6.3a–b show two ways that BJTs can be built. The two p–n junctions can either be embedded into each other as shown in Fig.6.3a or placed next to each other as shown in Fig.6.3b. In the former case, the transistors are commonly referred to as *vertically diffused transistors* and in the latter case, they are called *laterally diffused transistors*. Both transistor types shown in Figs.6.3a–b are NPN transistors (identifying the doping of the three major regions (N:Emitter, P:Base, and N:Collector). PNP transistors look the same except that all doping concentrations are reversed.

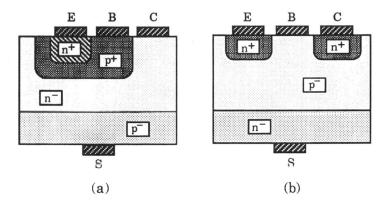

Figure 6.3a–b. Vertical and lateral NPN transistors.

NPN transistors can be modeled as shown in Fig.6.4.

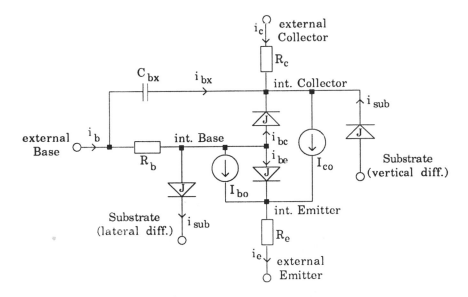

Figure 6.4. Model of vertical and lateral NPN transistors.

In the vertically diffused BJT, the substrate is connected to the collector, while in the laterally diffused BJT, it is connected to the base. The C_{bx} capacitance is actually a part of the junction capacitance of the base–collector junction diode. By splitting this capacitance between the external base and the internal base, the physically distributed junction charge can be represented a little more realistically. The model for PNP transistors looks the same except that all diode polarities are reversed.

The two dependent current sources are modeled in the following way:

$$I_{Co} = \frac{i_{be} - i_{bc}}{q_b} - \frac{i_{bc}}{B_R} - i_{bcn} \tag{6.6a}$$

$$I_{Bo} = \frac{i_{be}}{B_F} + i_{ben} + \frac{i_{bc}}{B_R} + i_{bcn} - \frac{i_{be}}{q_b} \tag{6.6b}$$

where i_{be} and i_{bc} denote the diode currents through the base–emitter and base–collector junction diodes; i_{ben} and i_{bcn} denote the same quantities, but this time using altered saturation currents and modified emission coefficients; B_F and B_R are the ideal forward and backward β coefficients, which denote the DC current gain factors from the base to the emitter and collector, i.e., I_E/I_B and I_C/I_B; and q_b is the base charge, in itself a computed quantity which, in SPICE, is approximated by the following equation:

$$q_b = \frac{q_1}{2}(1 + \sqrt{1 + 4q_2}) \tag{6.7}$$

where

$$q_1 = \frac{1}{1 - \frac{u_{bc}}{V_{AF}} - \frac{u_{be}}{V_{AR}}} \tag{6.8a}$$

$$q_2 = \frac{i_{bc}}{I_{KR}} + \frac{i_{be}}{I_{KF}} \tag{6.8b}$$

The base resistance is the most important resistance in the BJT model. Consequently, it is modeled more accurately than the other two resistances. The base resistance depends on the base current. The following equation is used to model the current dependence of the base resistance in SPICE:

$$r_{bb} = R_{BM} + 3(R_B - R_{BM})\frac{\tan(z) - z}{z(\tan(z))^2} \tag{6.9}$$

where:

$$z = \frac{-1 + \sqrt{\frac{1 + 144 i_b}{\pi^2 I_{RB}}}}{\frac{24}{\pi^2}\sqrt{\frac{i_b}{I_{RB}}}} \tag{6.10}$$

Not all SPICE dialects use the same equations. The equations presented in this text are those of the BBSPICE [6.1] dialect, an HSPICE [6.8] offspring, since this is the only version of SPICE for which I have source code available. Unfortunately, few SPICE manuals are explicit in these matters and many are inaccurate. It is therefore dangerous to rely in these matters on information provided in a user's manual rather than to extract it from the source code directly.

The equations presented in this chapter are in fact only a subset of those that are actually used in the SPICE model. For example, Eq.(6.10) won't work if the parameter I_{RB} takes on its default value of 0.0. In that case, SPICE automatically switches over to another simpler equation to approximate the base resistance, namely,

$$r_{bb} = R_{BM} + \frac{R_B - R_{BM}}{q_b} \qquad (6.9^{alt})$$

where q_b is the previously introduced base charge.

In fact, while the BJT model used by BBSPICE is the Gummel–Poon model, BBSPICE will automatically revert to the simpler Ebers–Moll model if default values are used for the secondary device parameters.

Also, it can easily happen that the denominator of Eq.(6.8a) becomes zero, in which case this equation would blow up. BBSPICE automatically flattens out denominators in the vicinity of zero and limits expressions to be exponentiated in size to prevent the model from blowing up.

Finally, many model parameters are temperature–dependent. Military–rated devices must operate correctly between $-55°C$ and $+120°C$. It is thus important that the circuit simulator can take effects of temperature variation into account. In BBSPICE, temperature variation is modeled by a quadratic approximation. Many of the device parameters have temperature coefficients associated with them. They all work basically in the same way. For instance, the temperature variation of the base resistance is modeled by:

$$\Delta Temp = Temp - T_{Room} \qquad (6.11a)$$
$$r_b = R_B + T_{RB1}\Delta Temp + T_{RB2}\Delta Temp^2 \qquad (6.11b)$$
$$r_{bm} = R_{BM} + T_{RM1}\Delta Temp + T_{RM2}\Delta Temp^2 \qquad (6.11c)$$

In reality, the temperature–dependent coefficients r_b and r_{bm} are plugged into equations (6.9) or (6.9^{alt}), and not the user–supplied constant coefficients R_B and R_{BM}.

BBSPICE offers an analysis statement of the type:

.TEMP − 55 − 30 − 5 20 45 70 95 120

which allows us to repeat whatever analysis was requested for a number of different temperature values. All temperature–dependent parameters are updated between simulation runs.

It would not be very practical if all model parameters would have to be user–specified for every transistor in the circuit separately. For example, in BBSPICE, BJTs contain 53 different model parameters that can be user–specified. Since typical circuits are fabricated as integrated circuits on one chip, and since for economic reasons not too many different processes can be involved in the fabrication of a single chip, chances are that most of the BJTs in a circuit are very similar, maybe except for the area that the emitter occupies (which, in turn, will influence many of the other parameters, i.e., the resistances will be divided by the area while the capacitances will be multiplied by the area; however, most SPICE dialects provide for one *area* parameter). To avoid this problem, SPICE allows us to group sets of device parameters into a **.MODEL** statement such as:

```
.MODEL  PROC35.N  NPN
+  IS = 1.1fA   BF = 190   BR = .1   EG = 1.205612   ISS = 0
+  ISE = 0   ISC = 0   NE = 1.5   NC = 2   BULK = SUB
+  VAF = 110   VAR = 0   IKF = 3.6mA   IKR = 0   SUBS = 1
+  RB = 1109.9   IRB = 5.63mA   RBM = 368.4   RE = 13.3   RC = 750
+  XTB = .006   XTI = 2.33   TRE1 = .0005   TRB1 = .005   TRC1 = .005
+  TRM1 = .005   TRE2 = 0   TRB2 = 0   TRC2 = 0   TRM2 = 0
+  CJE = .597pF   VJE = .77   MJE = .3
+  CJC = .36pF   VJC = .64   MJC = .425
+  CJS = 0   VJS = .75   MJS = 0   XCJC = 1   TLEV = 1
+  NF = 1   NR = 1   NS = 1   LEVEL = 2   VTF = 0   AF = 1
+  TF = 50psec   TR = 1usec   XTF = 1   ITF = 36mA   PTF = 0   KF = .16f
```

```
.MODEL  PROC35.P  PNP
+  IS = 1.1fA   BF = 125   BR = .1   EG = 1.205612   ISS = 0
+  ISE = 0   ISC = 0   NE = 1.5   NC = 2   BULK = SUB
+  VAF = 30   VAR = 0   IKF = 3.6mA   IKR = 0   SUBS = −1
+  RB = 778   IRB = 1.93mA   RBM = 57.6   RE = 23.3   RC = 450
+  XTB = .006   XTI = 2.33   TRE1 = .0005   TRB1 = .005   TRC1 = .005
+  TRM1 = .005   TRE2 = 0   TRB2 = 0   TRC2 = 0   TRM2 = 0
+  CJE = .652pF   VJE = .77   MJE = .3
+  CJC = .055pF   VJC = .64   MJC = .425
+  CJS = 0   VJS = .75   MJS = 0   XCJC = 1   TLEV = 1
+  NF = 1   NR = 1   NS = 1   LEVEL = 2   VTF = 0   AF = 1
+  TF = 100psec   TR = 1usec   XTF = 1   ITF = 90mA   PTF = 0   KF = .16F
```

In SPICE, lines starting with a "+" are continuation lines. Models can then be invoked by referring to their model identifier. For example, models of the two types *PROC35.N* and *PROC35.P* can be invoked through the statements:

$$Q120 \quad 7 \quad 21 \quad 15 \quad PROC35.N$$
$$Q121 \quad 9 \quad 42 \quad 31 \quad PROC35.P$$

where the character "Q" in the element name indicates the BJT. A small subset of the model parameters can also be specified during the element call itself in which case this value supersedes the value specified in the **.MODEL** statement, which in turn supersedes the default value. Typically, the *area* parameter will be specified during the element call since it varies from one transistor to the next while other process parameters remain the same.

How can we identify a decent set of model parameters for a given transistor? Two answers to this question can be given. On the one hand, some companies sell model libraries for various SPICE dialects that contain sets of model parameters for most of the commercially available semiconductor devices. One such product is ACCULIB [6.7], a model library for the SPICE dialect ACCUSIM [6.6]. On the other hand, it is possible to buy computerized data–acquisition systems that automatically produce test signals for a variety of semiconductor devices and quickly identify a set of model parameters for the given device. One such product is TECAP [6.4]. The TECAP system consists of an HP 9000 computer, an HP 4145 semiconductor analyzer, a network analyzer, and a capacitance meter. The chip to be modeled is inserted in the system. The system then generates a series of test signals for the chip and records the chip's responses. It then computes a rough first set of model parameters. Thereafter, it uses these model parameters as initial parameter values for a PSpice optimization study. Simulations are run and the parameters are adjusted until the PSpice simulation is in good agreement with the measured characteristics. When operated by an experienced user, TECAP requires roughly 20 min to identify the DC parameters of a BJT transistor and another 30 to 40 min to identify its AC parameters.

In this section, we have used the BJT to explain, by means of an example, how involved the active device models in modern circuit analysis programs are. We chose the BJT because its model is quite a bit simpler than the MOSFET model. However, the modeling principles are the same.

One disadvantage of the intrinsic "model" concept employed in SPICE is the fact that most device models are very complex and not transparent to the user. Furthermore, most SPICE versions (exceptions: BBSPICE [6.1] and HSPICE [6.8]), provide the engineer with insufficient access to the internal voltages, currents, and circuit element parameters of the models. Consequently, it is often quite difficult to interpret effects shown by the simulation since the equations used inside the models are not truly understood by and often are not known to the design engineers. Also, the SPICE models are rather difficult to maintain and upgrade. Recently, I added a four-terminal GaAs MESFET model to BBSPICE that we had received from another company. It took me a week and required changes to 25 subroutines to integrate this model with the BBSPICE software.

6.3 Hierarchical Modeling

In many circuit simulation studies, it is not practical to model each circuit down to the level of the available SPICE models (the BJTs, JFETs, and MOSFETs) since the same higher-level components (such as a logical inverter subcircuit) is reused in the circuit several times. It would, of course, be feasible to implement such a subcircuit as a new SPICE model and this is certainly the right approach if this same component is going to be used over and over again, such as a Zener diode. However, this approach is painfully slow and cumbersome at best.

Therefore, SPICE also offers the possibility to declare an ensemble of circuit elements as a *subcircuit* and reuse it thereafter as often as needed. Subcircuits are *macros*, as introduced in Chapter 5. However, at the abstraction level of a topological circuit description, macros *are* truly modular, which was not so in the case of state–space descriptions.

The following code shows a typical BBSPICE subcircuit that describes a generic gain stage of an operational amplifier:

```
.SUBCKT GAIN 100 101 1 2 RG = 500
    *Input Circuit
    RS    100   3    5 Ohm
    RIN   3     0    5 Ohm
    *Load Circuit
    RO    5     101 5 Ohm
    RL    101   0    5 Ohm
    *Gain Set Resistors
    RF1   5     26   500 Ohm
    RF2   26    25   500 Ohm
    RF3   4     25   500 Ohm
    CP1   25    0    1.4pF
    CP2   26    0    1.4pF
    RE    4     0    RG
    *Amplifier
    XAMP 3 4 5 1 2 SERSHNT
.ENDS GAIN
```

which then can be invoked through the statement:

$$XT1\ 50\ 51\ 7\ 8\ GAIN\ RG = 78.95$$

The node numbers of the subcircuit definition are formal positional parameters that are replaced by the actual node numbers during the invocation of the subcircuit. Other (internal) nodes of the subcircuit are local variables of the macro and are unaccessible from the outside. *RG* is a subcircuit parameter that can be assigned a default value in the subcircuit declaration (in our case 500). The default value can be overridden during the subcircuit invocation process. In our example, *RG* receives a value of 78.95. The "X" character in the element name (*XT1*) indicates that this is in fact not an element, but a subcircuit call. Subcircuits can be nested. In our example, the subcircuit GAIN calls upon another subcircuit of type SERSHNT.

In practice, the subcircuit concept in SPICE is a little clumsy and subcircuits often create problems. On the one hand, many SPICE versions give the user access to the top level of the hierarchy only, i.e., the node voltages and branch currents that are internal to a subcircuit cannot be displayed on output. On the other hand, most SPICE versions do not allow the user to specify an initial "nodeset" for internal nodes of subcircuits (cf. the subsequent discussion of ramping). This often creates serious problems with DC convergence. The cause of these difficulties is the fact that, in the original Berkeley SPICE, nodes had been *numbered* rather than *named*. Some of the newer versions of SPICE have solved this problem by introducing node names. In such versions, internal nodes of subcircuits can be accessed using a dot notation. For example, the voltage of the internal node *emitter* of the subcircuit *opamp* can be addressed as *V(opamp.emitter)*, and the current through the base resistance *rbb* of the subcircuit *opamp* can be addressed as *I(opamp.rbb)*.

6.4 Transient Analysis in SPICE

Transient analysis, in a circuit analysis program, corresponds to the previously introduced mechanism of system simulation. Starting from a given initial condition, the trajectory behavior of the system over time is determined. SPICE has a peculiar way of handling initial conditions. In SPICE, it is assumed that the initial values of all sources have been present for an infinite amount of time before the simulation starts. Let me explain this concept.

In a linear system of the type:

$$\dot{\mathbf{x}} = \mathbf{A}\mathbf{x} + \mathbf{B}\mathbf{u} \tag{6.12}$$

we can set $\mathbf{u}$ to its initial value $\mathbf{u_0}$ and compute a steady–state solution. In steady–state, all derivatives have died out, i.e., $\dot{\mathbf{x}} = 0.0$. Therefore, we find:

$$\mathbf{x_0} = -\mathbf{A}^{-1}\mathbf{B}\mathbf{u_0} \tag{6.13}$$

In a nonlinear system of the type:

$$\dot{\mathbf{x}} = \mathbf{f}(\mathbf{x}, \mathbf{u}, t) \tag{6.14}$$

we must solve an implicit set of nonlinear equations of the form

$$\mathbf{f}(\mathbf{x_0}, \mathbf{u_0}, t) = 0.0 \tag{6.15}$$

for the unknown vector $\mathbf{x_0}$. This is what SPICE attempts when it computes an "OP point."

SPICE tackles this problem by assuming an initial value for $\mathbf{x_0}$, linearizing the circuit around this assumed value, and then computing a better estimate for $\mathbf{x_0}$ using the approach of Eq.(6.13). This scheme is called *Newton–Raphson iteration*.

Let us look once more at the junction diode example. The current through the diode is a nonlinear function of the voltage across the diode. The algorithm starts with an initial voltage, which the user can specify in a so–called "nodeset," and which is assumed to be zero by default. Then, all nonlinear functions are linearized around the present working point, as shown in Fig.6.5.

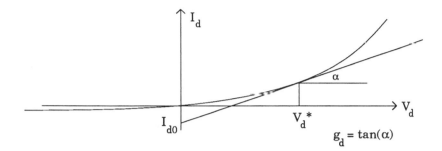

Figure 6.5. Linearization of nonlinear diode characteristic.

In the linearized model, the diode is replaced by a conductance $g_d = \partial i_d / \partial u_d$, and a current source I_{d0}. Now, we can use this model to solve the branch admittance matrix equation for a new, and hopefully improved, nodeset. The iteration continues until the difference between subsequent nodesets has become negligible.

Unfortunately, in a nonlinear circuit, no guarantee can be given that this algorithm will converge to the correct OP point, or that it will converge at all. DC convergence is one of the big problems in circuit analysis. The design engineer can help the program by specifying a good nodeset, i.e., by specifying good initial guesses for some or all of the node potentials in his or her circuit. The closer the nodeset is to the true value, the more likely it is that the iteration will converge to the desired value, and the faster the convergence will be.

Sometimes, this approach does not work since the design engineer is unable to guess a sufficiently close nodeset for her or his circuit. In these cases, design engineers have developed a technique of "ramping up all sources." This works in the following way: We first perform an experiment (i.e., a simulation) in which all (voltage and current) sources are initially set to zero and all active devices (transistors, diodes) have been switched off (one of the device parameters). In this setup, the OP point computation is trivial. All node voltages must obviously be zero. Now, we perform a *transient analysis* (i.e., a simulation) in which we apply a ramp to all sources, as shown in Fig.6.6.

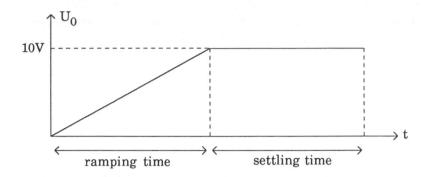

Figure 6.6. Ramping of sources in an electrical circuit.

The final value of each ramp is the desired initial value for the real simulation, which is then held constant for some time to allow the circuit to settle down. The final value of all node voltages can then be printed out and copied into a nodeset to be used by the subsequent simulation in which the real problem is being solved.

Some versions of SPICE (but not PSpice [6.9]) have automated this procedure. In BBSPICE [6.1], I have added a .RAMP statement of the form:

.RAMP < ramping time >< settling time >

Whenever the initial OP point fails to converge, BBSPICE will automatically modify the circuit internally to the previously described form, perform an invisible transient analysis, compute an initial nodeset, convert the circuit back to its initial specification, and then perform the desired analysis. This approach works very well, and the design engineers at Burr Brown are happy with this utility.

Once the initial condition has been determined, we can proceed to compute the trajectory behavior of the system over time. Here, we also face a number of awkward problems. In past chapters, we have learned that we always like to solve differential equations for their highest derivative terms and then integrate the derivatives into a new set of states via numerical integration. Unfortunately, this approach doesn't work in SPICE. Let us look once more at the junction capacitance equations, Eq.(6.4). We would like to solve the differential equation for $\dot{q}_c$. However, this would make q_c a "known" variable, and we would have to solve the equation $q_c = f(u_d)$, Eq.(6.5), for the variable u_d. However, this cannot be done since this equation does not have an analytical inverse, and moreover, since the initial value for u_d is user–specified (in the nodeset). Consequently, SPICE has no choice but to compute q_c from the nonlinear equation $q_c = f(u_d)$, and then evaluate i_c using *numerical differentiation*. As will be shown in the companion book on continuous system simulation, this approach is dubious from a numerical point of view unless we use an *implicit numerical differentiation scheme*.

Once all the currents have been determined, a new nodeset can be projected for Δt time units into the future.

6.5 Graphical Modeling

About three years ago, a new facet was added to industrial circuit design technology. This came with the advent of the new 386–based engineering workstations. For the first time, it was feasible to place a computer on every design engineer's desk, a computer that is not just a toy but is capable of solving even relatively large circuit analysis problems in a decent time frame. At Burr Brown, every design engineer has now a Compaq 386 on his or her desk, and most circuit analysis problems are being solved in PSpice, which runs beautifully on these machines. PSpice and BBSPICE offer powerful PROBE options that allow the design engineer to perform a transient analysis and then look at arbitrary node voltages or branch currents interactively on her or his screen. This feature has enhanced the efficiency of circuit design drastically. Previously, the engineer had to decide beforehand which variables he or she wanted to print out and if his or her analysis of the simulation results indicated that more output was needed, it became necessary to rerun the entire simulation, which often consumed several hours of execution time on a VAX 11/750.

Model input is now also being performed graphically using schematic capture programs. At Burr Brown, we use software called WORKVIEW [6.11], which allows us to draw electrical circuits on a virtual screen. WORKVIEW is one among roughly a dozen or so similar schematic capture programs currently on the market. In WORKVIEW, the physical screen can be moved over the virtual screen (zoom and pan). This is necessary for drawing complex circuits since it is often impossible to fit an entire circuit drawing on one physical screen. Figure 6.7 shows the WORKVIEW drawing of a simple (integrated) operational amplifier consisting of 12 BJTs, one resistor, and one capacitor.

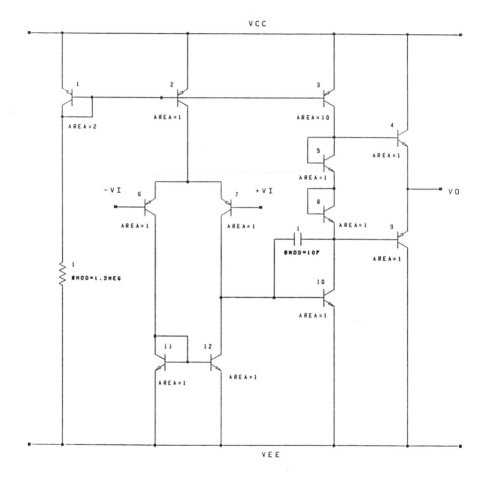

Figure 6.7. WORKVIEW drawing of an opamp.

WORKVIEW enables the user to automatically generate code for PSpice. The code that WORKVIEW generates for the simple opamp of Fig.6.7 is as follows:

```
* Project OPAMP
* WORKVIEW Wirelist Created with Version 3.0

Q12     1   2   3   3   PROC35.N   1
Q5      4   4   5   3   PROC35.N   1
Q3      4   6   7   3   PROC35.P   10
Q1      6   6   7   3   PROC35.P   2
Q2      8   6   7   3   PROC35.P   1
Q4      7   4   9   3   PROC35.N   1
Q6      2   10  8   3   PROC35.P   1
Q7      1   11  8   3   PROC35.P   1
Q8      5   5   12  3   PROC35.N   1
Q9      3   12  9   3   PROC35.P   1
Q10     12  1   3   3   PROC35.N   1
Q11     2   2   3   3   PROC35.N   1
C1      1   12  10p
R1      3   6   1.5Meg

* Dictionary 6
* + VI = 11
* − VI = 10
* V0 = 9
* VCC = 7
* VEE = 3
* GND = 0
```

.END

Notice that all BJTs are produced using the same process technology, i.e., the same photolithography and diffusion steps. They are all of either the *PROC35.N* type (for NPN transistors) or the *PROC35.P* type (for PNP transistors). The *area* parameter is specified on the element call statement. For example, the PNP transistor *Q3* occupies an area which is 10 times as large as the area that was used to determine the values of the model parameters on the .**MODEL** statement.

WORKVIEW also supports the concept of hierarchical modeling. A subcircuit can be drawn, its terminals can be identified, and a new symbol (icon) can then be sketched that must have as many terminals as the subcircuit has, and which, from now on, represents the subcircuit and can be used as a circuit element in the next higher level of the modeling hierarchy. In Fig.6.8, a new icon was constructed to denote the opamp and the opamp is being used as a modeling element in a higher–level circuit.

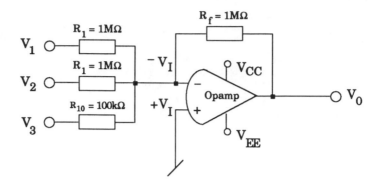

Figure 6.8. Drawing of an analog signal adder/inverter.

WORKVIEW can be requested to map its modeling hierarchies into PSpice's subcircuits. Alternatively, the graphical modeling hierarchy can be flattened out. We shall discuss the pros and cons of hierarchy flattening in Chapter 15 of this text.

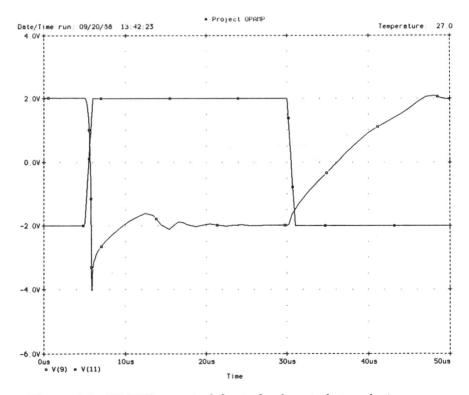

Figure 6.9. PROBE output of the analog inverter's transient response.

Figure 6.9 shows the transient behavior of this opamp as displayed by the PROBE feature. Signal $V(11)$ points to the input of the inverter while $V(9)$ shows the inverted output signal.

Obviously, this opamp is not all that great, but it suffices to illustrate the concept. After all, it is not the goal of this text to teach circuit design but to teach modeling concepts. Figure 6.10 shows the AC output (Bode diagram) of the opamp as shown by the PROBE feature. $VDB(9)$ depicts the gain of the opamp in decibels, while $VP(9)$ shows its phase in degrees. The opamp has a band width of approximately 400 kHz.

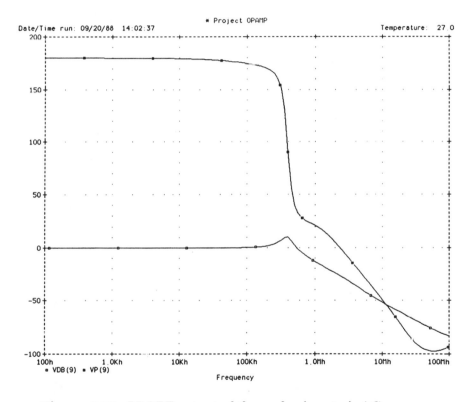

Figure 6.10. PROBE output of the analog inverter's AC response.

6.6 Circuit Design Using DYMOLA

In Chapter 5, we demonstrated that DYMOLA might provide us with an alternative approach to handling complex systems. Let us see how DYMOLA could be used as an alternative to SPICE for modeling electrical circuits. The advantages are obvious. If we are able to make DYMOLA powerful enough that it can handle arbitrarily complex circuits containing arbitrary algebraic loops and structural singularities, we can automatically generate a state–space model that will execute much more efficiently at run–time than the currently used SPICE code.

First, we need to model the basic circuit elements. This is straightforward. The following code shows the DYMOLA model types describing resistors, capacitors, inductors, voltage sources, and the ground. These models are equivalent to those used in BBSPICE.

```
model type resistor
  cut A(Va/I), B(Vb/-I)
  main cut C[A, B]
  main path P < A - B >
  local V, Rval
  parameter R = 0.0, TR1 = 0.0, TR2 = 0.0, area = 1.0
  external DTemp, DTempSq
  V = Va - Vb
  Rval = (R + TR1 * DTemp + TR2 * DTempSq)/area
  Rval * I = V
end

model type capacitor
  cut A(Va/I), B(Vb/-I)
  main cut C[A, B]
  main path P < A - B >
  local V, Cval
  parameter C = 0.0, TC1 = 0.0, TC2 = 0.0, area = 1.0
  external DTemp, DTempSq
  V = Va - Vb
  Cval = (C + TC1 * DTemp + TC2 * DTempSq) * area
  Cval*der(V) = I
end

model type inductor
  cut A(Va/I), B(Vb/-I)
  main cut C[A, B]
  main path P < A - B >
  local V, Lval
  parameter L = 0.0, TL1 = 0.0, TL2 = 0.0
  external DTemp, DTempSq
  V = Va - Vb
  Lval = L + TL1 * DTemp + TL2 * DTempSq
  Lval*der(I) = V
end
```

```
model type vsource
   cut A(Va/I), B(Vb/−I)
   main cut C[A, B]
   main path P < A − B >
   terminal V
   default V = 0.0
   V = Vb − Va
end

model type Common
   main cut A(V/.)
   V = 0.0
end
```

These DYMOLA models suffice to describe the simple passive circuit of Fig.6.1. The DYMOLA code needed to describe that circuit is as follows.

```
model RLC
   submodel(resistor)   R1(R = 100.0), R2(R = 20.0)
   submodel(capacitor) C(C = 0.1E-6)
   submodel(inductor)  L(L = 0.0015)
   submodel(vsource)   U0
   submodel Common
   parameter Temp = 300.0
   constant TRoom = 300.0
   local DTemp, DTempSq
   internal DTemp, DTempSq
   input u
   output y
   DTemp = Temp − TRoom
   DTempSq = DTemp * DTemp
   connect Common − U0 − ((R1 − (C//R2))//L) − Common
   U0.V = u
   y = R2.Va
end
```

In this model, several new language elements were introduced. The "−" sign is equivalent to the keyword **to** and denotes a series connection, whereas the "//" symbol is equivalent to the keyword **par** and denotes a parallel connection.

It is now time to analyze DYMOLA's declaration statements a little more closely. In DYMOLA, all variables must be declared.

Constants are variables that obtain once (upon declaration) a constant value and are never reassigned. **Parameters** are similar to constants. They never change their values during a simulation run. They can be reassigned, but only between simulation runs. This enables the compiler to extract all parameter computations from the dynamic loop. Parameters are one mechanism for data exchange between models. In this context, parameters are similar to formal read–only arguments of a subprogram call in a traditional programming language. Parameters cannot be reassigned within the model in which they are declared as parameters, only within the calling program. **Externals** are similar to parameters, but they provide for an implicit rather than an explicit data exchange mechanism. In this respect, they are similar to COMMON variables in a FORTRAN program. Externals are used to simplify the use of global constants or global parameters such as the temperature *Temp* or the thermal voltage *VT*. For security reasons, the calling program must acknowledge its awareness of the existence of these globals, by specifying them as **internal**. Notice, however, that "internal" is not a declaration, only a provision for redundancy, i.e., all internal variables must also be declared as something else.

Variables that may change their values during a simulation run are either **locals** or **terminals**, depending on whether they are connected to the outside world or local to the model in which they are used. Terminals can be assigned using a dot notation (as was done in the case of the voltage source: $U0.V = u$) or they can form part of one or several **cuts** in which case they can be connected. **Inputs** and **outputs** are special types of terminals. Variables declared as "terminals" or in "cuts" are undirected variables. For example, the statement $V = Vb - Va$ of the model *vsource* can be rearranged by the compiler into either $Vb = V + Va$ or $Va = Vb - V$ if needed. Had V been declared as "output," the compiler would be prevented from rearranging this equation.

Of course, not all circuits are conveniently described by sets of series and parallel connections. Alternatively, DYMOLA allows us to formulate the circuit in a topological manner similar to that used in SPICE. The following code shows an alternate way to formulate the circuit using DYMOLA.

```
model RLC
  submodel(resistor)  R1(R = 100.0),  R2(R = 20.0)
  submodel(capacitor) C(C = 0.1E-6)
  submodel(inductor)  L(L = 0.0015)
  submodel(vsource)   U0
  submodel Common
  node N0, N1, N2
  parameter Temp = 300.0
  constant TRoom = 300.0
  local DTemp, DTempSq
  internal DTemp, DTempSq
  input u
  output y

  DTemp = Temp − TRoom
  DTempSq = DTemp ∗ DTemp

  connect − >
    Common at      N0, − >
    U0        from N0 to N1, − >
    R1        from N1 to N2, − >
    R2        from N2 to N0, − >
    C         from N2 to N0, − >
    L         from N1 to N0

  U0.V = u
  y = R2.Va
end
```

The "− >" symbol denotes continuation lines in DYMOLA.

The generated code can be optimized by requesting DYMOLA to automatically throw out all terms that are multiplied by a parameter with value 0.0. Therefore, if the equation:

$$Cval = C + TC1 * DTemp + TC2 * DTemp * Dtemp \qquad (6.16)$$

uses its default values of zero for both $TC1$ and $TC2$, the equation would automatically degenerate into the equation

$$Cval = C \qquad (6.17)$$

at which time another DYMOLA rule will be activated that will throw out this equation altogether and replace the variable $Cval$ with C in the subsequent equation

$$Cval * \text{der}(V) = I \qquad (6.18)$$

since parameters take preference over local variables. In this way, DYMOLA can achieve indirectly the same run–time savings as SPICE does since only significant equations and terms will survive the translation. DYMOLA can also be requested to automatically extract all parameter computations from the dynamic portion of the simulation code into the initial portion of the code.

Let us now investigate what it would take to simulate a BJT in DYMOLA. As shown in Fig.6.4, the BJT model contains three junctions, and thus, we require a junction diode model. Its code is presented here.

```
model type jdiode
    cut Anode(Va/I), Cathode(Vb/-I)
    main cut C[Anode, Cathode]
    main path P < Anode - Cathode >
    local V, Ic, Qc, ISval, VDval, CDval
    terminal Id
    parameter ND = 1.0, IS = 1.0E-16, TD = 0.0, - >
            CD = 0.0, VD = 0.75, MD = 0.33, area = 1.0
    external DTemp, FTemp, Gmin, XTI, - >
            VT, EGval, VDfact

    {Electrical equations}
    V = Va - Vb
    I = Id + Ic
    Id = ISval * (exp(V/(VT * ND)) - 1.0) + Gmin * V
    Ic = der(Qc)
    Qc = TD * Id - >
            +VDval * CDval * (1 - (1 - V/VDval) * *(1 - MD))/(1 - MD)

    {Temperature adjustment equations}
    ISval = IS * area * exp((FTemp - 1.0) * EGval/VT) * FTemp * *XTI
    VDval = FTemp * (VD - VDfact) + VDfact
    CDval = CD * area/(1.0 + MD * (1.0 - VDval/VD + 4.0E-4 * DTemp))
end
```

This model contains one algebraic loop which, however, can be solved easily. Obviously, we wish to solve the differential equation for the derivative. Therefore, Qc is a state variable. Thus, we must solve the equation $Qc = \ldots$ for another variable, namely, either V or Id. However, the equation $Id = \ldots$ depends also on V, and thus, these two equations form an algebraic loop involving the variables V and Id. One way to solve this problem would be to replace the Id from the equation $Qc = \ldots$ by the other equation. In this way, the variable Id has been eliminated from the equation $Qc = \ldots$, and we can thus solve this equation for V. Thereafter, we can use the equation $Id = \ldots$ to determine Id. Unfortunately, the resulting equation:

$$Q_c = \tau_d I_s [\exp(\frac{V}{V_T N_d}) - 1] + \tau_d G_{min} V + \frac{\phi_d C_d [1 - (1 - \frac{V}{\phi_d})^{1-m_d}]}{1 - m_d} \quad (6.19)$$

is highly nonlinear in V and does not have an analytical inverse. Thus, we must either simplify the equation until it has an analytical inverse or employ a numerical iteration scheme to find V for any given value of Q_c from this equation.

The junction capacitance model shown earlier is not the one that is employed in BBSPICE since I recently implemented an improved model proposed by Van Halen [6.10]. This model avoids, in an elegant fashion, the singularity that occurred in the previously used model at $V = \phi_d$. However, since the new equation is even more bulky than the one shown in Eq.(6.19), I shall refrain from presenting it here. Moreover, the inversion problem that was demonstrated by means of Eq.(6.19) remains the same.

Next, we need a model for the variable base resistance of the BJT. Such a model is shown in the following code segment.

```
model type rbb
    cut A(Va/I), B(Vb/-I)
    main cut C[A, B]
    main path P < A - B >
    local V, Rval, RBval, RBMval, z, tz
    constant pi = 3.14159
    parameter RB = 0.0, RBM = 0.0, IRB = 0.0, area = 1.0, - >
                TRB1 = 0.0, TRB2 = 0.0, TRM1 = 0.0, TRM2 = 0.0
    external DTemp, DTempSq, qb

    V = Va - Vb
    RBval = (RB + TRB1 * DTemp + TRB2 * DTempSq)/area
    RBMval = (RBM + TRM1 * DTemp + TRM2 * DTempSq)/area
    Rval = if IRB = 0.0 - >
        then RBMval + (RBval - RBMval)/qb - >
        else RBMval + 3.0 * (RBval - RBMval) * (tz - z)/(z * tz * tz)
    z = if IRB = 0.0 - >
        then 0.0 - >
        else (-1 + sqrt(1 + 144 * I/(pi * pi * IRB * area))) - >
            /(24 * sqrt(I/(IRB * area))/(pi * pi))
    tz = if IRB = 0.0 then 0.0 else tan(z)
    Rval * I = V
end
```

This model demonstrates one of the weaknesses of DYMOLA. In order to be able to sort all equations properly, DYMOLA provides us with a funny–looking "if" statement of the form:

$$< var >= \text{if} < cond > \text{then} < expr > \text{else} < expr >$$

This becomes quite awkward when a condition propagates through a number of statements. Notice that in this text, we shall usually present the full models only and skip their degenerated versions in order to keep the models short and understandable. Notice further that this model contains an algebraic loop for the case $IRB \neq 0.0$ involving the variables $Rval$, z, and I. The simplified model (for $IRB = 0.0$) does not contain any algebraic loop.

Next, we require a model for the nonlinear external base junction capacitance $cbcx$. The following code describes this nonlinear capacitance.

```
model type cbcx
    cut A(./I), B(./–I)
    main cut C[A, B]
    main path P < A – B >
    local Qx, Vval, Cval
    terminal vbc
    parameter CJC = 0.0, MJC = 0.33, VJC = 0.75, XCJC = 1.0, – >
            area = 1.0
    external DTemp, FTemp, VDfact

    Qx = Vval * Cval * (1 – XCJC) – >
            *(1 – (1 – vbc/Vval) * *(1 – MJC))/(1 – MJC)
    der(Qx) = I

    {Temperature adjustment equations}
    Vval = FTemp * (VJC – VDfact) + VDfact
    Cval = CJC * area/(1.0 + MJC * (1.0 – VDval/VJC + 4.0E-4 * DTemp))
end
```

Finally, we need the BJT model itself. In SPICE, all four types (NPN and PNP, vertical and lateral) are coded in one single subroutine. In DYMOLA, this is hardly feasible at the moment. So let us look at one of the transistor types only, namely, the laterally diffused NPN transistor. The models for the other three types are similar.

model type *NPNlat*

    submodel (resistor) $rcc(R = RC, TR1 = TRC1, TR2 = TRC2, ->$
                 $area = area), ->$
                 $ree(R = RE, TR1 = TRE1, TR2 = TRE2, ->$
                 $area = area)$
    submodel $rbb(RB = RB, RBM = RBM, IRB = IRB, area = area, ->$
           $TRB1 = TRB1, TRB2 = TRB2, TRM1 = TRM1, ->$
           $TRM2 = TRM2)$
    submodel $cbcx(CJC = CJC, MJC = MJC, VJC = VJC, ->$
           $XCJC = XCJC, area = area)$
    submodel (jdiode) $dbc(ND = NR, IS = IS, TD = TR, area = area, ->$
              $CD = CJC * XCJC, VD = VJC, MD = MJC), ->$
              $dbe(ND = NF, IS = IS, TD = TF, area = area, ->$
                $CD = CJE, VD = VJE, MD = MJE), ->$
              $dbs(ND = NS, IS = ISS, TD = 0.0, area = area, ->$
                $CD = CJS, VD = VJS, MD = MJS)$
    submodel (csource) $ice0, ibe0$
    cut $Collector(VC/IC)$
    cut $Base(VB/IB)$
    cut $Emitter(VE/-IE)$
    cut $Substrate(VS/ISUB)$
    main cut $CBES \; [Collector, Base, Emitter, Substrate]$
    path $Basemitter \; < Base - Emitter >$
    path $Colemitter \; < Collector - Emitter >$
    node $IntCollector, IntBase, IntEmitter$

    constant $pi = 3.14159$
    parameter $IS = 1.0E\text{-}16, ISC = 0.0, ISE = 0.0, ISS = 0.0, ->$
          $BF = 100.0, BR = 1.0, TF = 0.0, TR = 0.0 ->$
          $NC = 2.0, NE = 1.5, NF = 1.0, NR = 1.0, NS = 1.0, ->$
          $VAF = 0.0, VAR = 0.0, IKF = 0.0, IKR = 0.0, ->$
          $VJC = 0.75, VJE = 0.75, VJS = 0.75, ->$
          $CJC = 0.0, CJE = 0.0, CJS = 0.0, ->$
          $MJC = 0.33, MJE = 0.33, MJS = 0.5, ->$
          $RB = 0.0, RBM = 0.0, RC = 0.0, RE = 0.0, ->$
          $TRB1 = 0.0, TRM1 = 0.0, TRC1 = 0.0, TRE1 = 0.0, ->$
          $TRB2 = 0.0, TRM2 = 0.0, TRC2 = 0.0, TRE2 = 0.0, ->$
          $XCJC = 1.0, EG = 1.11, XTI = 3.0, IRB = 0.0, area = 1.0$
local $q1, q2, qb, xti, vbc, vbe, ibc, ibe, ibcn, ->$
    $iben, IB0, IC0, EGval, EGroom, VDfact$
external $Temp, TRoom, DTemp, DTempSq, FTemp, ->$
    $Charge, Boltz, VT, VTroom, BT, BTroom, ->$
    $GapC1, GapC2, Gmin$
internal $DTemp, DTempSq, FTemp, VT, qb, EGval, ->$
    $VDfact, Gmin, xti$

> {*Compute frequently used internal voltages and currents*}
> $vbc = rbb.Vb - rcc.Vb$
> $vbe = rbb.Vb - ree.Va$
> $ibc = dbc.Id$
> $ibe = dbe.Id$
>
> {*Compute the base charge*}
> $q1 = 1.0/(1.0 - vbc/VAF - vbe/VAR)$
> $q2 = (ibc/IKR + ibe/IKF)/area$
> $qb = q1 * (1.0 + \text{sqrt}(1.0 + 4.0 * q2))/2.0$
> $xti = XTI$
>
> {*Compute the nonlinear current sources*}
> $ibcn = ISC * area * (\exp(vbc/(VT * NC)) - 1.0)$
> $iben = ISE * area * (\exp(vbe/(VT * NE)) - 1.0)$
> $IC0 = (ibe - ibc)/qb - ibc/BR - ibcn$
> $IB0 = ibe/BF + iben + ibc/BR + ibcn - ibe/qb$
>
> {*Compute the globals*}
> $EGroom = EG - GapC1 * TRoom * *2/(TRoom + GapC2)$
> $EGval = EG - GapC1 * Temp * *2/(Temp + GapC2)$
> $VDfact = -2.0 * VT * (1.5 * \log(FTemp) - >$
> $\qquad\qquad -0.5 * Charge * (EGval/BT - EGroom/BTroom))$
>
> {*Plug the internal circuit together*}
> **connect** rbb **from** *Base*   **to** *IntBase*
> **connect** rcc **from** *Collector*   **to** *IntCollector*
> **connect** ree **from** *IntEmitter*   **to** *Emitter*
> **connect** dbc **from** *IntBase*   **to** *IntCollector*
> **connect** dbe **from** *IntBase*   **to** *IntEmitter*
> **connect** dbs **from** *IntBase*   **to** *Substrate*
> **connect** $cbcx$ **from** *Base*   **to** *IntCollector*
> **connect** $ibe0$ **from** *IntBase*   **to** *IntEmitter*
> **connect** $ice0$ **from** *IntCollector* **to** *IntEmitter*
>
> $ice0.I0 = IC0$
> $ibe0.I0 = IB0$
> $cbcx.vbc = vbc$

    end

This model is currently not without its problems. It seems that the BJT is described as a fourth–order model since it contains four separate capacitances, each of which is described through a first–order differential equation. Unfortunately, this is an example of a degenerate system, as can be easily shown. Let us assume that we describe the capacitance of the base–collector junction diode through a differential equation. Then the charge over this capacitance $dbc.Qc$ is a state variable and we need to solve the "$dbc.Qc = \ldots$" equation for the variable $dbc.V$. However, this variable is the same variable as $NPNlat.vbc$, which is connected to the terminal variable of the external base collector junction capacitance, i.e., $cbcx.vbc = NPN$-$lat.vbc = dbc.V$. Therefore, we must solve the charge equation of the external capacitance for $cbcx.Qx$, which gets us into trouble since we would also like to solve this variable from the differential equation $der(cbcx.Qx) = cbcx.I$.

In order to overcome this model degeneracy, we would need to analytically compute the derivative of the equation $cbcx.Qx = \ldots$ and eliminate the differential equation.

Currently, the major benefit of these DYMOLA circuit descriptions is with their documentary value. We feel that our BJT model description is much more readable and understandable than the description given in most SPICE manuals and it is much easier to read than the SPICE source code listing. Currently, several of our graduate students are working on DYMOLA to make the software more powerful than it currently is and one of them is explicitly looking at DYMOLA as a tool for electrical circuit design.

6.7 How DYMOLA Works

Until now, we have only discussed how an input file for DYMOLA (i.e., a hierarchical model) is to be prepared. We have not yet seen what the DYMOLA preprocessor does with this model upon execution. This will be demonstrated now by means of a very simple example. Figure 6.11 shows an almost trivial electrical circuit consisting of one voltage source and two resistors.

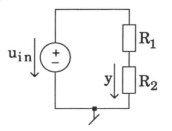

Figure 6.11. Schematic of a trivial electrical circuit.

Using the electrical component models that were introduced earlier in this chapter, this circuit can be described through the following DYMOLA program.

```
@vsource.elc
@resistor.elc
@common.elc

model circuit
   submodel (vsource) Uin
   submodel (resistor) R1(R = 10.0), R2(R = 20.0)
   submodel Common
   input u
   output y
   connect Common − Uin − R1 − R2 − Common
Uin.V0 = u
y = R1.Vb
end
```

The @ operator instructs the DYMOLA preprocessor to include an external file at this place. A DYMOLA program may contain definitions for (or inclusions of) an arbitrary number of model types followed by exactly one model that invokes the declared model types as submodels.

The following command sequence calls upon the DYMOLA preprocessor, which is requested to read in the model definition:

```
$ dymola
 > enter model
 − @circuit.dym
 >
```

At the operating system prompt ($), we call the DYMOLA preprocessor, which enters into an interactive mode and responds with its own prompt (>). The next statement instructs DYMOLA to read in a model. DYMOLA will present us with the next level prompt (−), until it has read in a complete model specification. We could enter equations here, but it is more practical to invoke them indirectly (@). At this point, DYMOLA is satisfied, since it found a model definition, and returns to its first–level interactive prompt (>).

Already at this point, DYMOLA has replaced all submodel references by their model definitions and has generated the additional equations that are a result of the submodel couplings (i.e., DYMOLA has replaced the "connect" statements by the coupling equations.

The result of this text replacement can be observed by issuing the command:

> **output equations**

which will result in the following display:

Common	$V = 0.0$
Uin	$V0 = Vb - Va$
R1	$V = Va - Vb$
	$R * I = V$
R2	$V = Va - Vb$
	$R * I = V$
circuit	$Uin.V0 = u$
	$y = R1.Vb$
	$R1.Va = Uin.Vb$
	$R1.I = Uin.I$
	$R2.Va = R1.Vb$
	$R2.I = R1.I$
	$R2.Vb = Uin.Va$
	$Common.V = R2.Vb$

assuming that a simplified resistor model that does not include the temperature variation effects or *area* parameter was used. The output can be redirected to a file (for printout) using the command:

> **outfile** *circuit.eq*1

to be issued prior to the "output equations" command.

At this point, we can try to determine which equation needs to be solved for what variable, and simultaneously, sort the equations into an executable sequence. This algorithm was thoroughly described in Chapter 3. In DYMOLA, the algorithm is invoked by issuing a "partition" command:

> **partition**

Thereafter, we may wish to look at the marked and sorted but not yet solved equations. This is achieved with the command:

> **outfile** *circuit.sr*1
> **output sorted equations**

which will write the following text to the file *circuit.sr1*:

Common	$[V] = 0.0$
circuit	$Common.V = [R2.Vb]$
	$R2.Vb = [Uin.Va]$
	$[Uin.V0] = u$
Uin	$V0 = [Vb] - Va$
circuit	$[R1.Va] = Uin.Vb$
−	$R2.I = [R1.I]$
−R2	$R * [I] = V$
−	$[V] = Va - Vb$
−circuit	$[R2.Va] = R1.Vb$
−R1	$V = Va - [Vb]$
−	$R * I = [V]$
circuit	$R1.I = [Uin.I]$
	$[y] = R1.Vb$

A first set of six equations was sorted correctly. This is followed by another set of six equations that form a linear algebraic loop. DYMOLA marks the equations belonging to an algebraic loop with "−." Finally, the last two equations can be properly sorted once the algebraic loop is solved. As this example demonstrates, algebraic loops occur rather commonly.

We can now proceed to optimize the code. The set of equations contains many *aliases*, i.e., the same physical quantity is stored several times under different variable names. This will slow down the execution of our simulation program. The command:

> **eliminate equations**

gets rid of equations of the type:

$$a = b$$

and replaces all occurrences of the variable a in all other equations by the symbol b. One exception to the rule must be stated: The eliminate operation never eliminates a variable that was declared as either input or output. If a is an output variable, DYMOLA will throw the equation away as well, but in this case, all occurrences of b are replaced by a. If both a and b are declared as output variables, the equation will not be eliminated.

This algorithm can be applied to either the original equations or the partitioned equations and will work equally well in both cases. The resulting code is shown here:

Common	$[Uin.Va] = 0.0$
Uin	$circuit.u = [Vb] - Va$
−R2	$R * [Uin.I] = V$
−	$[V] = circuit.y - Uin.Va$
−R1	$V = Uin.Vb - [circuit.y]$
−	$R * Uin.I = [V]$

In reality, this algorithm reduces all equations of the types:

$$+a = +b$$

and:

$$\pm a \pm b = 0$$

which are variants of the previously discussed case. The algorithm works also if either a or b is a constant. Consequently, the preceding set of equations is not yet the "final product." The truly reduced set of equations presents itself as follows:

−R2	$R * [Uin.I] = circuit.y$
−R1	$V = circuit.u - [circuit.y]$
−	$R * Uin.I = [V]$

We can now optimize the code further by requesting:

> **> eliminate parameters**

The algorithm that is now executed will perform the following tasks:

(1) All parameters with a numerical value of 0.0 or 1.0 are eliminated from the model and the numerical value is replaced directly into the equations.

(2) A numerical value of 1.0 that multiplies a term is eliminated from that term.

(3) A term that is multiplied by a numerical value of 0.0 is replaced as a whole by 0.0.

(4) Additive terms of 0.0 are eliminated altogether.

(5) If, in an equation, an expression consists of parameters and constants only, a new equation is generated that will evaluate this expression (assigned to a new generic variable) and the occurrence of the expression in the equation is replaced by the new generic variable.

(6) If an equation contains only one variable, it must be solved for that variable. This variable is then automatically redeclared as

a parameter and the equation is marked as a *parameter equation*, which can be moved from the DYNAMIC portion of the simulation program into the INITIAL portion of the simulation program.

In our example, nothing will happen if we apply this algorithm since none of the parameters has a value of either 1.0 or 0.0. However, in reality, we did not use a simplified resistor model but the full resistor model, in which the temperature variation coefficients defaulted to 0.0 and the *area* parameter defaulted to 1.0. Applying the algorithm to the full resistor model will reduce the equations to the format we found in our last display.

Notice that this algorithm may have undesirable side effects. Often, we may wish to start off with a simple model (by setting some parameters equal to 0.0) and successively make the model more realistic by assigning, in the simulation program, true values to the previously defaulted parameters. In this case, we shouldn't "eliminate parameters" since this algorithm will put the eliminated parameters to the sword once and for all. These parameters will no longer appear in the generated simulation program.

We can now proceed to "eliminate variables." This algorithm can only be applied after the equations have been partitioned. It affects only algebraic loops, and it affects each algebraic loop in the model separately. DYMOLA counts the times that each loop variable is referenced in an algebraic loop. Obviously, each loop variable must occur at least twice, otherwise, it would not be a loop variable. If we request DYMOLA to:

> eliminate variables

then DYMOLA will investigate all loop variables that occur exactly twice in a loop. If it found such a variable and if this variable appears linearly in at least one of the two equations, DYMOLA will solve the equation for that variable and replace the other occurrence of the variable by the evaluated expression, thereby eliminating this variable from the loop. If the eliminated variable is referenced anywhere after the loop, the equation defining this variable is not thrown away, but taken out of the loop and placed immediately after the loop. The same is true if the eliminated variable has been declared as an output variable. Although the same algorithm could be applied to variables that occur more than twice, this is not done since the algorithm tends to expand the code (the same, possibly long, expressions are duplicated several times).

If we apply this algorithm to our example, we first notice that our algebraic loop contains the three loop variables $Uin.I$, $circuit.y$, and $R1.V$. Each of them appears exactly twice in the loop. We investigate $Uin.I$ first and find that it occurs linearly in the first equation. Therefore, we solve that equation for $Uin.I$, replace the found expression in the third equation, and eliminate the variable since it was not declared as an output variable. The result of this operation is as follows:

$$-\textbf{R1} \qquad V = circuit.u - [circuit.y]$$
$$- \qquad R * (circuit.y / R2.R) = [V]$$

Since $circuit.y$ is an output variable, we prefer to analyze $R1.V$ next. We find that it occurs linearly in the second equation, solve for it, and replace the result in the first. The result is as follows:

$$\textbf{R1} \qquad R * ([circuit.y]/R2.R) = circuit.u - [circuit.y]$$

at which time the algebraic loop has disappeared. DYMOLA's equation solver is able to turn this equation into:

$$\textbf{R1} \qquad circuit.y = circuit.u/(1 + R/R2.R)$$

At this point, we can once more "eliminate parameters," which will lead to the following code:

Initial:
$D00001 = 1/(1 + R1.R/R2.R)$
Dynamic:
$circuit.y = D00001 * circuit.u$

DYMOLA offers yet another elimination algorithm. If we request DYMOLA to:

> **eliminate outputs**

DYMOLA will check for all variables in the DYNAMIC section of the code that appear only once in the set of equations. Obviously, the equations containing these variables must be used to evaluate them. Since these equations will not otherwise influence the behavior of the dynamic model, they can be marked as output equations and be taken out of the state–space model. If we apply this algorithm to our example, the following code results:

Initial:
$D00001 = 1/(1 + R1.R/R2.R)$
Dynamic:
Output:
$circuit.y = D00001 * circuit.u$

The command:

> > **partition eliminate**

will partition the equations and then automatically perform *all* types of elimination algorithms repetitively until the equations no longer change.

At the time of writing, not all of these code optimization techniques were fully implemented. However, the command "partition eliminate" can already be used. It simply skips those optimization tests that haven't been fully implemented.

We can receive a printout of the solved equations by issuing the command sequence:

> > **outfile** *circuit.sv*1
> > **output solved equations**

At this point, the circuit topology has been reduced to a (in this case trivial) state–space model. Now DYMOLA's code generator portion can be used to generate a simulation program for DESIRE [6.5], SIMNON [6.2], or FORTRAN. However, before we can do so, we need to add an experiment description to our DYMOLA program. The experiment may look as follows:

```
cmodel
   simutime 2E-5
   step 2E-7
   commupoints 101
   input 1, u(independ, 10.0)

ctblock
   scale = 1
   XCCC = 1
   label TRY
   drunr
if XCCC < 0 then XCCC = −XCCC | scale = 2 * scale | go to TRY
                 else proceed
ctend

outblock
   OUT
   dispt y
outend
end
```

This portion of code is specific for each of the target languages. The version shown here is the one required for DESIRE [6.5]. The

ctblock set of statements instructs DESIRE to automatically scale the run–time display. *XCCC* is a DESIRE variable that is set to −1 whenever the DESIRE program is interrupted with an "overflow." This happens when one of the displayed variables hits either the top or the bottom of the displayed window. At this time, the plot is simply rescaled and the simulation is rerun with a new *drunr* statement. Since DESIRE is so fast, it is not worth the effort to store the results of the previous attempt; instead, we simply rerun the entire simulation.

The experiment description is entered into DYMOLA using the statements:

```
> enter experiment
− @circuit.ctl
>
```

Now, we are ready to generate the simulation program:

```
> outfile circuit.des
> output desire program
```

To exit from DYMOLA and enter the DESIRE [6.5] program, we issue the command sequence:

```
> stop
$ desire 0
> load 'circuit.des'
> save
> run
> bye
$
```

The 0 parameter on the DESIRE call instructs DESIRE to not automatically load the last executed DESIRE program into memory, the **load** command loads the newly generated DESIRE program into memory, and the **save** command saves the program in binary form onto the file *circuit.prc* for a quicker reload at a later time using the command:

```
> old 'circuit'
```

The **run** command finally executes the DESIRE program and the **bye** command returns control to the operating system.

6.8 Summary

In this chapter, we discussed the tools that commercial circuit designers use to analyze and design their circuits. PSpice [6.9] and BBSPICE [6.1] are two of several available SPICE and ASTAP dialects. WORKVIEW [6.11] is one of a number of available schematic capture programs. I chose to mention those programs in the text, since I have the most familiarity with them. However, I truly believe that, among the programs I have had a chance to review, the selected ones are indeed the most powerful. With respect to circuit analysis programs, PSpice has probably today become the most popular among the SPICE dialects due to its availability and excellent implementation on PC–class machines. With respect to the schematic capture programs, some of the other commercially available programs don't provide an interface to PSpice, or their interface (the so–called netlisting feature) executes much slower, or they run only on more specialized hardware (such as Apollo workstations) that is not as readily available, or they are considerably more expensive. Finally, WORKVIEW offers a very convenient feature related to the simulation of digital circuits. It contains a *logic simulator* that allows the user to quickly analyze the behavior of a digital circuit before she or he ever goes through the slow and painful process of decomposing the circuit down to the transistor level and simulating it in greater detail using PSpice.

We then returned to DYMOLA [6.3] and discussed the potential of this powerful modeling tool for circuit modeling. While DYMOLA is not yet capable of dealing with complex electronic circuitry due to the frequently cited problems with algebraic loops and structural singularities, I am convinced that eventually DYMOLA can become a true alternative to SPICE.

We shall return once more to electronic circuit simulation in the companion book to discuss, in more detail, how the numerical integration (or rather differentiation) is performed in SPICE.

The aim of this chapter was to discuss circuit analysis from a modeling and simulation perspective, not from the perspective of transistor circuit design. Dozens of texts are on the market that discuss transistor circuits very elegantly and in great detail. We didn't see a need to duplicate these efforts here. The bibliography of this chapter lists a more or less arbitrary collection of textbooks that deal with electronic circuit design issues.

References

[6.1] Burr Brown Corp. (1987), *BBSPICE — User's Manual*, Tucson, Ariz.

[6.2] Hilding Elmqvist (1975), *SIMNON — An Interactive Simulation Program for Nonlinear Systems — User's Manual*, Report CODEN: LUTFD2/(TFRT–7502), Dept. of Automatic Control, Lund Institute of Technology, Lund, Sweden.

[6.3] Hilding Elmqvist (1978), *A Structured Model Language for Large Continuous Systems*, Ph.D. dissertation, Report CODEN: LUTFD2/(TRFT–1015), Dept. of Automatic Control, Lund Institute of Technology, Lund, Sweden.

[6.4] Hewlett–Packard Company (1988), *HP 94430A/94431A Tecap User's Manual, HP 94431A Tecap Device Modeling and Parameter Extraction Manual*, Santa Clara, Calif.

[6.5] Granino A. Korn (1989), *Interactive Dynamic–System Simulation*, McGraw–Hill, New York.

[6.6] Mentor Corp. (1987), *ACCUSIM User's Manual*, Palo Alto, Calif.

[6.7] Mentor Corp. (1987), *ACCULIB User's Manual*, Palo Alto, Calif.

[6.8] Meta–Software, Inc. (1985), *HSPICE User's Manual*, Campbell, Calif.

[6.9] MicroSim Corp. (1987), *PSpice User's Manual*, Irvine, Calif.

[6.10] Paul Van Halen (1988), "A New Semiconductor Junction Diode Space Charge Layer Capacitance Model," *Proceedings IEEE 1988 Bipolar Circuits & Technology Meeting*, (J. Jopke, ed.), Minneapolis, Minn., IEEE Publishing Services, New York, pp. 168–171.

[6.11] Viewlogic Systems, Inc. (1988), *WORKVIEW Reference Guide*, Release 3.0, *VIEWDRAW Reference Guide*, Version 3.0, Marlboro, Mass.

Bibliography

[B6.1] Alfred W. Barber (1984), *Practical Guide to Digital Electronic Circuits*, second edition, Prentice–Hall, Englewood Cliffs, N.J.

[B6.2] David J. Corner (1981), *Electronic Design with Integrated Circuits*, Addison–Wesley, Reading, Mass.

[B6.3] Sergio Franco (1988), *Design with Operational Amplifiers and Analog Integrated Circuits*, McGraw–Hill, New York.

[B6.4] Mohammed S. Ghausi (1984), *Electronic Devices and Circuits: Discrete and Integrated*, Holt, Rinehart, and Winston, New York.

[B6.5] Jerald G. Graeme (1977), *Designing with Operational Amplifiers: Applications Alternatives*, McGraw–Hill, New York.

[B6.6] Adel S. Sedra and Kenneth C. Smith (1982), *Microelectronic Circuits*, Holt, Rinehart, and Winston, New York.

Homework Problems

[H6.1] Tunnel Diode

A forward–biased tunnel diode can be represented through the static characteristic shown in Table H6.1:

Table H6.1. Tunnel Diode Characteristic.

Voltage u_d [V]	Current i_d [mA]
0.00	0.00
0.05	1.70
0.10	2.90
0.15	4.00
0.20	4.75
0.25	5.00
0.30	4.90
0.35	4.25
0.40	3.20
0.45	2.35
0.50	2.00
0.55	2.20
0.60	2.70
0.65	3.50
0.70	4.35
0.75	5.30
0.90	9.00

Use CTRL–C (or MATLAB) to obtain a graphical representation of this static characteristic.

We wish to analyze the behavior of the circuit shown in Fig.H6.1a.

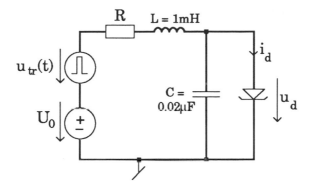

Figure H6.1a. Schematic of a circuit with a tunnel diode.

In the first configuration, we shall use a resistor of $R = 25\ \Omega$, a DC bias of $U_0 = 0.48$ V, and a driver voltage of $u_{tr}(t) = 0.0$ V, i.e., we analyze the behavior of this circuit under "free–running" conditions. Write a simulation program in ACSL that implements this circuit, simulate the circuit during 0.2 msec, and display the diode voltage u_d and the diode current i_d as functions of time. Describe the behavior of this circuit in qualitative terms.

In a second experiment, we wish to analyze the same circuit under modified experimental conditions. This time, we shall use a resistor of $R = 200\ \Omega$ and a DC bias of $U_0 = 1.075$ V. The driver voltage is a pulsed voltage source as depicted in Fig.H6.1b:

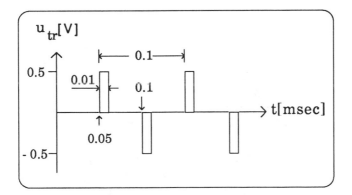

Figure H6.1b. Pulsed driver voltage source.

Implement this modified setup in ACSL using the time–event scheduling facility to describe the pulsed driver voltage. The first two time–events

are scheduled in the INITIAL section of the program, namely, a time–
event by the name of *VON* to occur at $t = 0.05$ msec, and another time–
event by the name of *VOFF* to occur at time $t = 0.1$ msec. As part of
each event description, a new event of the same type is scheduled to occur
$\Delta t = 0.05$ msec later.

[H6.2] Differential Equations

We want to analyze the step response of a system described by the following
transfer function:

$$G(s) = \frac{10}{s^2 + 2s + 10} \tag{H6.2}$$

In a first experiment, use CTRL–C (MATLAB) to transform this frequency–
domain representation into a state–space representation and simulate the
step response of this system during 10 sec directly in CTRL–C (MATLAB).

As an alternative, design a simple passive circuit that could be used to
obtain the same step response. Write a PSpice program that implements
this circuit, perform a transient analysis, and display the obtained step
response with the PROBE option.

Since we don't want SPICE to perform a DC analysis first in this case,
we can prevent this from happening by adding the *UIC* qualifier to our
transient analysis command:

.TRAN 0.1 10.0 *UIC*

[H6.3] Analog Computers

Analog computers are electronic circuits that basically consist of three different element types: analog adders, analog integrators, and potentiome-
ters. We have already met the analog adder (cf. Fig.6.8). Adders used
in commercial analog computers usually provide several inputs with a gain
factor of -1.0 (i.e., 1 V at the input results in -1 V at the output) and
one or several inputs with a gain factor of -10.0. A gain factor of -10.0 is
achieved by making the corresponding input resistor 10 times smaller than
the feedback resistor. The operational amplifier can ideally be represented
through a high–gain amplifier with an amplification of $A = -1.35 \times 10^8$, for
example. An analog integrator can be built in almost the same way, just
by replacing the feedback resistor of $R_f = 1$ MΩ with a feedback capacitor
of $C_f = 1$ μF. In this circuit, 1 V at a unit gain input results in a ramp
at output that, after 1 sec, reaches a value of -1 V (assuming zero initial
conditions). When the 1 V input is applied to an input node with a gain
factor of -10.0, the ramp reaches -1 V already after 0.1 sec.

Build a PSpice subcircuit for an ideal high–gain amplifier. This can be achieved using a voltage–driven voltage source (an *E*–element). Then build subcircuits for the analog adder and the analog integrator using the previously designed ideal high–gain amplifier as a component. Finally, build a subcircuit for the potentiometer. This can again be achieved with a voltage–driven voltage source, but this time, the gain *A* is variable with allowed values anywhere between 0.0 and 1.0. Depending on the SPICE dialect that you use, you may be able to specify the gain *A* as a parameter of the subcircuit, but PSpice does not support this feature. If your SPICE version does not allow you to specify parameters for your subcircuits, it may not be worthwhile to specify the potentiometer as a subcircuit at all.

Draw a block diagram for the state–space model of the differential equation of Hw.[H6.2], and convert it to an analog computer program. Don't forget that both the adder and the integrator contain built–in inverters. The traditional symbols (icons) used to denote analog computer components are shown in Fig.H6.3.

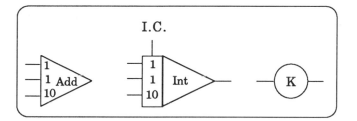

Figure H6.3. Components of analog computers.

Use PSpice to simulate the just–derived analog computer circuit. Since you don't want SPICE to perform a DC analysis first in this case, you can prevent this from happening by adding the *UIC* qualifier to your transient analysis command:

.**TRAN** 0.1 10.0 *UIC*

It would now be possible to replace the ideal high–gain amplifier by the operational amplifier subcircuit presented earlier in this chapter. However, we won't do this since the simulation would take forever. The time step of the transient analysis (the numerical integration) would have to be adjusted to the (very fast) time constants inside the BJTs which are on the order of 100 nsec and a simulation of the overall circuit during 10 sec would thus not be realistic. We ran this model for roughly 1 hour on a VAX–3600 (using BBSPICE) and were able to complete (correctly) just the first 1 msec of the simulation. Consequently, the total simulation would require about 10,000 hours. It was explained previously that modeling must be always goal–driven. A finer granularity of the model does not necessarily make the model any better. Quite the contrary can be true, as this example demonstrates.

[H6.4] Logic Gates

Figure H6.4 shows a circuit for an OR/NOR gate in ECL logic. In this type of logic, -1.8 V corresponds to a logical false (or off), whereas -0.8 V corresponds to a logical true (or on).

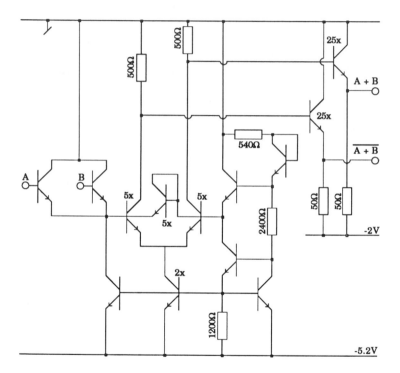

Figure H6.4. Circuit for an OR/NOR gate in ECL logic.

Build a PSpice program that simulates this circuit during 25 μsec. Use our *PROC35.N* model to describe the *NPN* transistors. Connect the substrates of all transistors to the lowest voltage, i.e., to -5.2 V. The *area* parameter to be used for the transistors is indicated on the schematic except for those transistors that use the default value of 1.0.

Keep the input A at -1.8 V during the first 5 μsec, then raise it to -0.8 V and keep it at that level until $t = 15$ μsec. Thereafter, the input A is kept at -1.8 V for the rest of the simulation. The second input, B, behaves similarly, except that it is raised from -1.8 V to -0.8 V at

$t = 10$ μsec and is reset to -1.8 V at $t = 20$ μsec. Use the PULSE function to model the two time–dependent voltage sources. Choose a rise and drop time for the input pulses of 100 nsec each.

Probe several of the voltages and currents in the circuit to come up with an explanation as to how this circuit works.

Reduce the rise and drop times of the two inputs until they are clearly faster than those produced by the circuit. Zoom in on the rising and dropping edges of the two output signals and determine the natural rise and drop times of the the two output signals.

[H6.5] Digital–to–Analog Converter (DAC)

Figure H6.5 shows a four–bit digital–to–analog converter that can be driven by ECL logic.

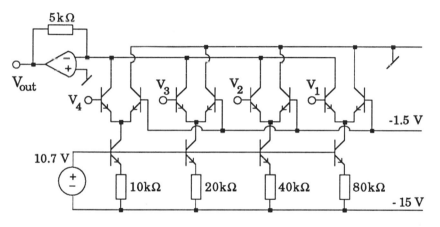

Figure H6.5. Four–bit DAC for ECL logic.

V_1 represents the least significant bit (LSB), while V_4 represents the most significant bit. If your version of SPICE supports subcircuits with formal parameters, design a subcircuit that describes one of the four DAC stages and call it four times with different values of the parameter R. If your SPICE version does not support parameters, you better leave the resistor out of the subcircuit and place it in the main program. Use the ideal opamp of Hw.[H6.3] for the analog output stage. Connect the substrates of all BJTs to the lowest voltage in the circuit, i.e., to -15 V. Apply the following voltages to the four input nodes of the circuit:

```
V1  1  0  PULSE(−1.8 -0.8  24u 100n 100n 23.8u 48u)
V2  2  0  PULSE(−1.8 -0.8  12u 100n 100n 11.8u 24u)
V3  3  0  PULSE(−1.8 -0.8   6u 100n 100n  5.8u 12u)
V4  4  0  PULSE(−1.8 -0.8   3u 100n 100n  2.8u  6u)
```

and simulate the circuit during 60 μsec. Probe several of the voltages and currents in the circuit to understand how the circuit works. Explain the behavior of the output signal $V_{out}(t)$.

This time, we want to replace the ideal opamp by the transistorized opamp described earlier in this chapter. Tie the positive supply of the opamp, V_{CC}, to +15 V and the negative supply, V_{EE}, to −15 V.

Repeat the previous experiment. You may have difficulties with DC convergence. You may have to ramp up all the supplies during 2 μsec and let them settle down during another 2 μsec, in order to obtain a decent initial nodeset. In that case, don't forget to switch all active devices (BJTs) off for the ramping experiment. Use the resulting OP point as a nodeset for the subsequent experiment in which you perform the desired transient analysis.

[H6.6] Circuit Modeling in DYMOLA

In Fig.H6.6, a simple passive circuit is presented. The only new component is a current source for which a DYMOLA model type *csource* must be derived in analogy to the voltage source *vsource*.

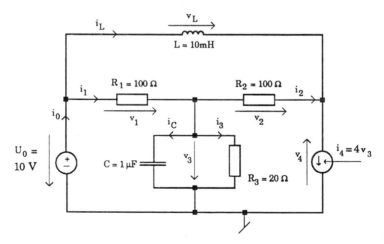

Figure H6.6. Circuit diagram of a simple passive circuit.

Augment DYMOLA's electrical component library by a model type describing the current source. The fact that the current source in our example is a dependent current source does not make any difference here. Unlike SPICE, DYMOLA does not force us to distinguish between independent and dependent sources.

Derive a model for the overall circuit and partition the equations once using the **partition** command and once using the **partition eliminate**

command. In each of these cases, print out the equations, the sorted
equations, and finally the solved equations. Thereafter, generate either
a SIMNON [6.2] or a DESIRE [6.5] program and simulate the system over
50 μsec. Use a step size of 50 nsec. Compare the results with those of
Hw.[H3.4].

[H6.7]* Logic Inverter Modeled in DYMOLA [6.3]

Figure H6.7 shows the schematic of a simple logic inverter model.

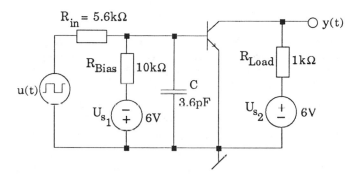

Figure H6.7. Schematic of a logic inverter model [6.3].

For the transistor model, we want to use the default values except for the
following parameters:

$$BF = 190.0$$
$$BR = 0.1$$
$$TF = 50 \text{ psec}$$
$$TR = 1 \text{ μsec}$$
$$IS = 1.1 \text{ fA}$$
$$RB = 1109.9 \ \Omega$$

However, before you can invoke the DYMOLA preprocessor, you will have
to modify the model manually to some extent. Remove the external base–
collector capacitance and replace the base resistance by a regular resistor.
You will also have to modify the junction diode model. Since we know
that we must solve the diode equation for u_d rather than for i_d, we must
invert this nonlinear equation manually. Simplify the equation until it has
an analytical inverse and invert it. You may also want to limit the diode
current i_d to prevent the inverted diode equation from trying to compute
the logarithm of a negative number. With these modifications, DYMOLA
will be able to handle the transistor model.

I suggest that you print out the equations, but then immediately use the
eliminate parameters command followed by the **eliminate equations**

command to get rid of the many remaining defaulted parameters and trivial equations. Observe what happens to the equations on the way. Thereafter, partition your equations, and observe the sorted equations and the solved equations. Then generate a simulation model and simulate the circuit during 150 nsec with a step size of 15 psec using either DESIRE [6.5] or SIMNON [6.2]. The pulsed input is supposed to be $u(t) = 0.0$ V except during the time from 20 nsec to 100 nsec, when the voltage is $u(t) = 7.0$ V.

Now return to your original DYMOLA program and add two more parameters:

$$RE = 13.3 \ \Omega$$
$$RC = 750.0 \ \Omega$$

Notice that you must return to your original DYMOLA program since both parameters had been optimized away with the **eliminate parameters** command. Repeat the same sequence of operations as before. You will observe that, this time, an algebraic loop occurs. After partitioning the equations, apply the various elimination algorithms and observe what happens to the algebraic loop. After the equations have been totally optimized, analyze the algebraic loop using CTRL–C (MATLAB). You will notice that all loop variables appear linearly in all loop equations, i.e., this is a linear algebraic loop. Concatenate all loop variables into a column vector, **x**, and concatenate all previously computed variables into another column vector, **u**. Using this notation, the algebraic loop can be written in a matrix form as:

$$\mathbf{A} \cdot \mathbf{x} = \mathbf{B} \cdot \mathbf{u}$$

A and **B** are constant matrices that depend only on parameter values. Use CTRL–C (or MATLAB) to solve these equations numerically for the loop vector **x**:

$$\mathbf{x} = \mathbf{A}^{-1} \cdot \mathbf{B} \cdot \mathbf{u}$$

Replace the loop equations in your DESIRE [6.5] or SIMNON [6.2] program manually by the set of solved equations obtained from CTRL–C (MATLAB) and repeat the simulation as before. It is foreseen to eventually automate this procedure in DYMOLA by providing an **eliminate loops** command.

Projects

[P6.1] Transistor Models
From the HSPICE manual [6.8] (which, to my knowledge, is currently the most extensive and explicit SPICE manual available), develop DYMOLA

models for the MOSFET and JFET devices similar to the BJT (NPNlat) model presented in this chapter.

[P6.2] Linear Algebraic Loops

Design and implement an algorithm that will automatically solve all sets of linearly coupled algebraic equations using the mechanism advocated in Hw.[H6.7]. Test this algorithm by means of transistor models that contain emitter and collector resistance values.

Research

[R6.1] Junction Capacitance Model

Paul Van Halen [6.10] proposed a new junction capacitance model that avoids problems with a singularity that was inherent in the original SPICE junction capacitance model. As we have seen, this model cannot directly be applied to a DYMOLA circuit simulator since the charge equation must be solved for the voltage v rather than for the charge Q_c. Unfortunately, neither the original SPICE junction capacitance charge equation nor the modified junction capacitance charge equation has an analytical inverse (except in the simplified case that was treated in Hw.[H6.7]). Use curve–fitting techniques to find yet another junction capacitance charge equation that computes the voltage across the junction capacitance v as a function of the charge Q_c stored in the junction capacitance.

Also, develop a better mechanism to represent the distributed base–collector junction capacitance than that of an internal and external capacitance, a mechanism that avoids the structural singularity inherent in this artificial separation.

The goal of this research is to minimize the number of algebraic loops and structural singularities in a transistor model by at least eliminating those that are currently inside one transistor model. However, as we have seen in Chapter 5, new algebraic loops and/or structural singularities may be introduced in the course of coupling different submodels together.

7

Bond Graph Modeling

Preview

When I read a technical paper written by an author from whom I have not read anything before, I usually find that I have the biggest problems not with mastering the intellectual challenges that the paper presents, but with understanding the author's terminology and relating it back to concepts with which I am familiar. In this context, modeling is a notoriously difficult subject since it is so utterly interdisciplinary. It is therefore one of the major goals of this text to introduce a variety of different modeling concepts (terminologies) and relate them to each other. In this chapter, we shall discuss several graphical modeling techniques among which, in more detail, the *bond graph modeling technique* has found widespread acceptance and use among a number of modelers from several application areas, particularly mechanical engineers. We shall see that bond graphs are very easy to relate back to previously introduced concepts, and yet they are somewhat difficult to read on first glance.

7.1 Block Diagrams

We have used block diagrams previously in this text on an *ad hoc* basis without properly introducing them. I felt that this approach was quite adequate since the interpretation of a block diagram is straightforward. But let me now go back and introduce block diagrams formally.

Block diagrams consist of *blocks* that are connected by *paths*. The paths represent signals and the blocks are transducers that transform one (set of) signal(s) into another. While these two types of modeling elements would, in theory, suffice to draw any block diagram, two frequently used types of transducers are usually represented by special symbols: the *take–off point*, and the *summer*. Figure 7.1 shows the four elementary block diagram modeling types.

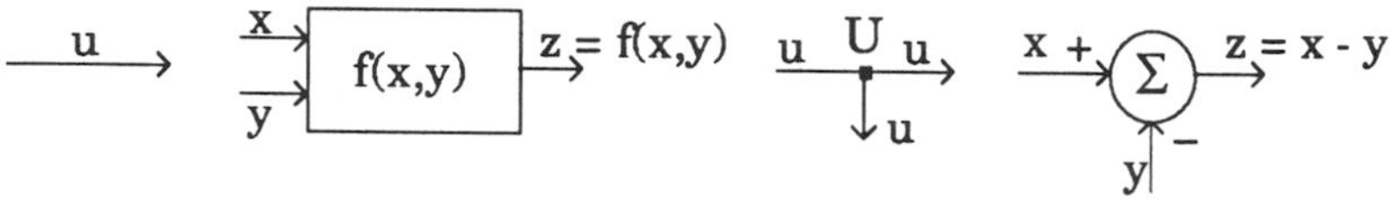

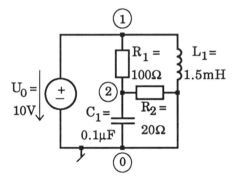

Figure 7.1. Modeling elements of a block diagram.

Let us return once more to our simple electrical circuit.

$$\text{Figure}$$

Figure 7.2. Simple passive circuit.

We wish to construct a block diagram for it. We start with a set of equations that describes the circuit correctly in terms of a *computational structure*. One such set of equations was derived in Chapter 3. Let us write down these equations here once more.

$$i_1 = u_1/R_1 \tag{7.1a}$$

$$i_2 = u_2/R_2 \tag{7.1b}$$

$$\frac{di_L}{dt} = u_L/L_1 \tag{7.1c}$$

$$\frac{du_C}{dt} = i_C/C_1 \tag{7.1d}$$

$$u_1 = U_0 - u_C \tag{7.1e}$$

$$u_2 = u_C \tag{7.1f}$$

$$u_L = u_1 + u_2 \tag{7.1g}$$

$$i_0 = i_1 + i_L \tag{7.1h}$$

$$i_C = i_1 - i_2 \tag{7.1i}$$

When constructing a block diagram, I always start by drawing all integrators as boxes (blocks) below each other. The inputs to these boxes are currently *unresolved variables*. For each unresolved variable, I simply proceed to the equation that defines that variable and draw this equation into the block diagram, until all unresolved variables have been reduced to true inputs of the system. Each equation that has been used in the block diagram is removed from the set of unused equations. In many cases, no unused equations are left at the end. If equations have not been used, these are *output equations* that can simply be added to the block diagram in the end.

This algorithm leads to the block diagram shown in Fig.7.3,

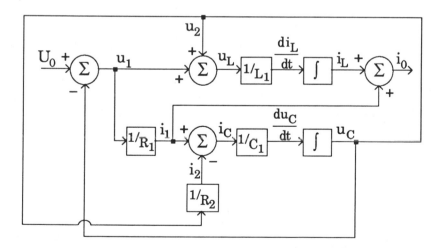

Figure 7.3. Block diagram of the passive circuit.

which can thereafter be simplified as shown in Fig.7.4.

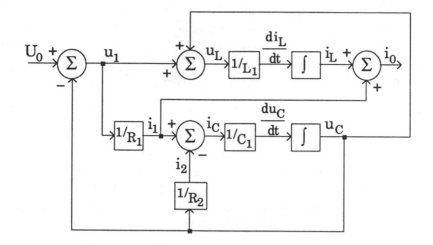

Figure 7.4. Simplified block diagram of the passive circuit.

While the block diagram clearly shows the computational structure of the system, it does not preserve any of the topological properties of the system. As we have seen in the past, a small change in the circuit may force us to rearrange the computational structure entirely, and therefore, its corresponding block diagram may bear little resemblance to the previously used one.

This is a clear disadvantage of block diagrams, and therefore, block diagrams are not commonly used to describe electrical circuits. They are mostly used by control engineers because control engineers *force* their circuits to behave in such a way that the computational structure and topological structure coincide. This can be achieved by placing a voltage follower circuit between any two consecutive blocks of the block diagram, as shown in Fig.7.5:

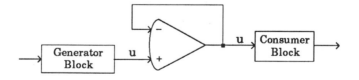

Figure 7.5. Impedance decoupling with a voltage follower.

The voltage follower circuit decouples the two consecutive blocks. The *generator block* produces a control signal u that is independent of the *consumer block*, i.e., the voltage produced by the generator does not depend on the current that is drawn by the consumer. While this technique is common practice among control engineers, the voltage follower circuits are never shown in block diagrams. They are simply assumed to be there.

However, if we connect two wires in an arbitrary electrical circuit, we actually connect *two* variables at the same time, namely, one *across variable*, the potential *v*, and one *through variable*, the current *i*. In the block diagram, these two variables get separated from each other, and it is this fact that destroys the symmetry between the topological structure and the computational structure.

7.2 Signal Flow Graphs

Another frequently used graphical modeling tool is the signal flow graph. A strong relation exists between block diagrams and signal flow graphs. Figure 7.6 shows the correspondence between the elementary modeling tools in a block diagram (top) and the equivalent elementary modeling tools in a signal flow graph (bottom).

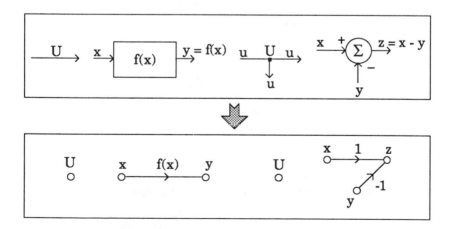

Figure 7.6. Elements of block diagrams versus signal flow graphs.

Each "path" in a block diagram turns into a "node" in the signal flow graph. Each "box" in the block diagram becomes a "path" in the signal flow graph. In this respect, the two diagrams could be called *dual representations*. However, no signal flow graph equivalent exists for a multiport block diagram box. In this respect, signal flow graphs are a little less powerful than block diagrams.

When converting a block diagram into a signal flow graph, I always proceed along the following lines. First, I assign a name to all signals (paths) in the block diagram that have not yet been named. I usually call these signals e_1, e_2, etc. In our example, all signals have been named already. Next, I draw nodes (little circles) wherever the block diagram had a signal path. Finally, I connect these nodes with paths according to the correspondence shown in Fig.7.6.

For our passive circuit, we find the signal flow graph shown in Fig.7.7.

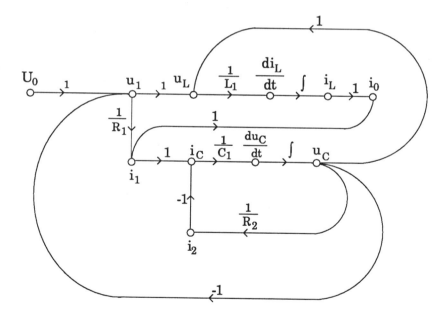

Figure 7.7. Signal flow graph of the passive circuit.

The signal flow graph can be simplified by eliminating all nodes that represent series connections of paths. The corresponding path functions are simply multiplied with each other. The result of this simplification is shown in Fig.7.8.

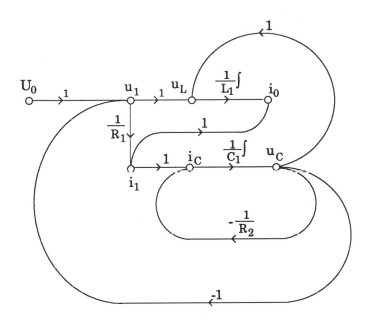

Figure 7.8. Simplified signal flow graph of the passive circuit.

When interpreting a signal flow graph, it is essential to realize that nodes represent both take–off points and summers at the same time. Incoming paths represent summing functions, while outgoing paths represent take–offs. Looking at the node u_1, for example, we find that:

$$u_1 = 1 * U_0 + (-1) * u_C \qquad (7.2)$$

and *not*:

$$u_1 = 1 * U_0 + (-1) * u_C - 1 * u_L - \frac{1}{R_1} * i_1 \qquad (7.3)$$

Once this small detail has been understood, signal flow graphs become as easy to read as block diagrams, and they really are more or less equivalent to each other. In particular, they share the same advantages and disadvantages, namely, they capture the computational structure while they do not preserve the topological structure of the system they represent. They are about equally frequently found in control engineering texts. Some texts operate on signal flow graphs only while others use block diagrams exclusively. The most prominent use of signal flow graphs is for the determination of a transfer function between any two nodes (signals) in a linear circuit using Mason's rule. I don't care too much for signal flow graphs since they are not "computer–friendly," i.e., I have a hard time drawing them neatly on my Macintosh. Therefore, I won't use them much in this text.

7.3 Power Bonds

While block diagrams and signal flow graphs preserve the computational structure of a system only, circuit diagrams reflect the topological structure exclusively. Moreover, they are restricted to use in electrical systems. For these reasons, H.M. Paynter, a professor at MIT, recognized around 1960 the need for yet another graphical representation of systems, which could show simultaneously the topological and computational structure, and which would be general, i.e., could be applied to all kinds of physical systems. He came up with the bond graph [7.15]. A bond, represented by a bold harpoon, is nothing but a connector that simultaneously connects two variables, one across variable, in bond graph terminology usually referred to as the "effort" e, and one through variable, called the "flow" f. The bond is shown in Fig.7.9.

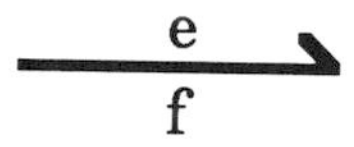

$$\xrightarrow{\quad e \quad}$$
$$f$$

Figure 7.9. The bond.

The bond graph literature is not systematic with respect to the bond graph conventions. The harpoon is sometimes shown to the left and sometimes to the right of the bond, and the effort variable is sometimes indicated on the side of the harpoon and sometimes away from the harpoon. This inconsistency can be explained by the fact that most bond graphers viewed the bond graph methodology as a pure modeling aid to be used with paper and pencil. The fact that a model represents a codified form of knowledge occurred to them at best as an afterthought.

However, if we wish to use bond graphs as a tool to formalize a model and to formulate it as input to a computer program (as we shall do in this chapter), we need to be more rigorous. For this purpose, I decided that the harpoon must sit always on the left of the bond, and the effort variable is always indicated on the side of the harpoon, while the flow variable is indicated on the side away from the harpoon.

The bond is able to preserve the topological structure because the two types of variables are not dislocated from each other. Bonds connect either to system elements, such as a resistor **R**, which in bond graph terminology is a *single port element* (since both variables u_R and i_R are connected simultaneously), or to other bonds in a *junction*. Two different types of junctions exist, the so–called 0–junction, and the so–called 1–junction [7.12]. In a 0–junction, all effort variables are equal while all flow variables add up to zero. A 0–junction is thus equivalent to a node in an electric circuit diagram or in a DYMOLA [7.7] program. In a 1–junction, all flow variables are equal while all effort variables add up to zero. The two junction types are shown in Fig.7.10.

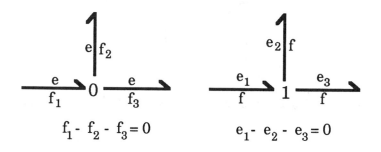

$$f_1 - f_2 - f_3 = 0 \qquad e_1 - e_2 - e_3 = 0$$

Figure 7.10. The two junction types.

The 0–junction thus represents Kirchhoff's current law, while the 1–junction represents Kirchhoff's voltage law. If a bond connects two junctions, one will always be of the 0–junction type, while the other is of the 1–junction type, i.e., in a bond graph, 0–junctions and 1–junctions toggle among each other. Neighboring junctions of the same gender can be combined into one.

7.4 Bond Graphs for Electrical Circuits

Let me explain by means of my passive circuit how a bond graph can be constructed.

Electric circuit designers are in the habit of choosing one node as their *reference node*, usually the ground. Since the ground appears at many places in a complex circuit, they usually do not bother to connect all the grounds together in a circuit diagram.

The circuit diagram shown in Fig.7.11 is equivalent to the diagram of Fig.7.2.

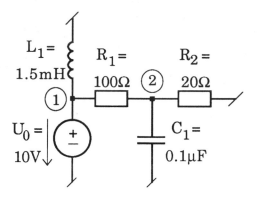

Figure 7.11. Electric circuit diagram of the passive circuit.

In the bond graph, we shall do the same thing. We start by representing each circuit node by a 0–junction except for the reference node, which is drawn like in a circuit diagram. We then represent each branch of the circuit diagram by a pair of bonds connecting two 0–junctions with a 1–junction between them. We let the harpoons point in the same direction that we picked for the branch currents. Finally, we attach the circuit elements to the 1–junctions with the harpoons directed away from the junction for passive circuit elements and toward the junction for sources [7.1].

This algorithm leads to the bond graph shown in Fig.7.12.

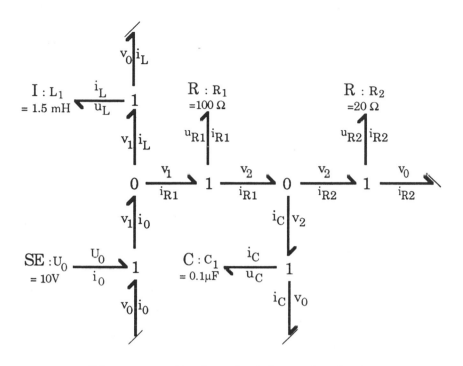

Figure 7.12. Bond graph of the passive circuit.

This bond graph can still be simplified quite a bit. Remember that when writing down the circuit equations we always skipped Kirchhoff's current law for the reference node. In DYMOLA [7.7] notation, we placed a "." in the current's position of the *Common* submodel's cut, which instructs the DYMOLA compiler to skip the current equation. We did this because we knew that this equation would be redundant. Accordingly, since the reference node in the bond graph actually represents a degenerated 0–junction, which represents Kirchhoff's current law for the reference node, we can leave out the reference node together with all the bonds connecting to it. This rule makes physical sense. The power that flows through bonds connecting to the reference node is the product of the current flowing through the bond and the potential of the reference node. Since this potential can, without loss of generality, be assumed to be zero, we can say that no power flows into or out of the reference node. Thus, it makes physical sense to eliminate such bonds from the bond graph.

Also, if a junction has only two bonds attached to it and if their harpoons both point in the same direction, we can eliminate this junction and amalgamate the two bonds into one [7.1]. This procedure leads to the simplified bond graph of Fig.7.13:

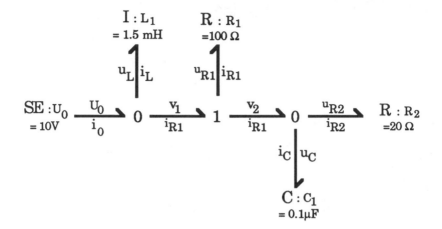

Figure 7.13. Simplified bond graph of the passive circuit.

Notice that the bold printed element type denotes the type of equation that describes the circuit element. **R** stands for resistance, **C** for capacitance (or compliance), **I** for inductance (or inertia), and **SE** for effort source. Correspondingly, **SF** stands for flow source.

It is quite evident that the bond graph preserves the topological structure. However, we have not seen yet how it represents the computational structure at the same time. For this purpose, we introduce the notion of *bond graph causality* [7.12].

Each bond is involved in two equations, one to determine its effort e, the other to determine its flow f. Each of these two equations is formulated at one of the two ends of the bond. The causality is indicated by a short stroke perpendicular to the bond that is placed at one of the two ends of the bond. It marks the side of the bond at which the flow variable is being determined.

Let me explain this concept by means of a simple example. If a bond connected to a resistance **R** has its stroke at the end at which **R** is attached to the bond, as shown in Fig.7.14a,

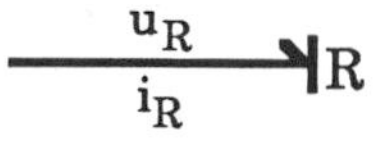

Figure 7.14a. Causal bond connected to a resistance.

then we need to compute the flow variable at the resistance that leads to the equation:

$$i_R = u_R/R \qquad (7.4)$$

This means that u_R must be computed elsewhere, namely, at the other end of the bond. However, if the causality is assigned the other way around, such as shown in Fig.7.14b,

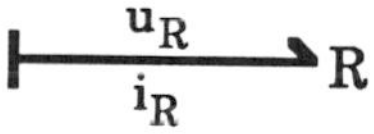

Figure 7.14b. Causal bond connected to a resistance.

then the effort needs to be computed at the resistance and we obtain the equation,

$$u_R = R \cdot i_R \qquad (7.5)$$

whereas the equation used to compute the flow will be formulated at the other end of the bond.

For a resistance, both causalities are physically and computationally meaningful. However, for effort sources and flow sources the causality is physically determined as shown in Fig.7.15,

$$\xrightarrow{u_0(t)} \text{SE} \xrightarrow[\;i_0\;]{\;u_0\;} \qquad\qquad \xrightarrow{i_0(t)} \text{SF} \xrightarrow[\;i_0\;]{\;u_0\;}$$

Figure 7.15. Necessary causalities for effort and flow sources.

whereas for capacitances and inductances the desired causality is dictated by computational requirements since we wish to numerically *integrate* over all state variables rather than to *differentiate* them. The desired causalities for these elements are shown in Fig.7.16.

$$\vdash \underset{i_C}{\overset{u_C}{\longrightarrow}} C \qquad \qquad \underset{i_L}{\overset{u_L}{\longrightarrow}} I$$

Figure 7.16. Desired causalities for capacitances and inductances.

Finally, rules can be specified for the two junction types. Since only one flow equation can be specified for every 0–junction, only one of the bonds can compute the flow at the junction, i.e., exactly one stroke must be located at every 0–junction. Since only one effort equation can be formulated for every 1–junction, only one of the bonds can compute the effort, i.e., only one stroke can be located away from every 1–junction. Figure 7.17 shows the causal bond graph for our passive circuit.

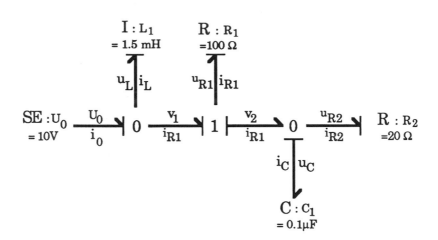

Figure 7.17. Causal bond graph of the passive circuit.

In this example, all conditions can be satisfied and the solution is unique. This is the preferred situation. If not all necessary conditions can be satisfied, we are confronted with a *noncausal system*. This case occurs, for example, if we try to parallel connect two voltage sources with different voltage values. If we cannot satisfy all desired conditions, i.e., if we run into wrong causalities at either **C** or **I** elements, we are confronted with a *degenerate system*, i.e., the true system order is lower than the number of integrators would indicate. We are thus confronted with a *structural singularity*. If we have a choice in assigning the causalities without offending any of the rules, the model contains an *algebraic loop*.

Please notice that the specification of the voltages as "efforts" and the currents as "flows" is somewhat arbitrary. The bond graph is completely symmetrical in this respect. Had we decided to call the currents "efforts" and the voltages "flows," then all junctions would change their gender, and the causality laws for capacitances, inductances, and the two types of sources would be reversed. We call this the *dual bond graph*. It will be discussed later in this chapter in more detail.

7.5 Bond Graphs for Mechanical Systems

If all we could do with bond graphs were to have yet another tool to describe electrical circuits, this would not be very exciting since we already had a topological description mechanism (the circuit diagram) that comes more natural to electrical engineers than bond graphs. However, the concepts of effort and flow are much more general than that.

If we take three levers and join them in one point, we notice that the velocities of the three levers at that point must be equal, while the forces add up to zero. In simple analogy, we are therefore inclined to formulate the following equivalencing scheme:

$$
\begin{array}{ccccc}
\text{effort} & \Longleftrightarrow & \text{potential} & \Longleftrightarrow & \text{velocity} \\
\text{flow} & \Longleftrightarrow & \text{current} & \Longleftrightarrow & \text{force}
\end{array}
$$

Once we made this decision, all other quantities in the bond graph are determined. Let us look at Newton's law:

$$
m \cdot \frac{dv}{dt} = \sum_{\forall i} f_i \quad \Rightarrow \quad \frac{dv}{dt} = \frac{1}{m} \cdot \sum_{\forall i} f_i \quad \Rightarrow \quad v = \frac{1}{m} \cdot \int_0^t \sum_{\forall i} f_i d\tau + v_0 \quad (7.6)
$$

We immediately notice the similarity with the voltage/current relationship for a capacitor:

$$C \cdot \frac{du_C}{dt} = i_C \quad \Rightarrow \quad \frac{du_C}{dt} = \frac{1}{C} \cdot i_C \quad \Rightarrow \quad u_C = \frac{1}{C} \cdot \int_0^t i_C d\tau + u_{C_0} \quad (7.7)$$

Thus, the mechanical mass corresponds to an electrical capacitor. Let us now look at a spring. For the spring, we have the relationship:

$$f_{Sp} = k \cdot x \quad \Rightarrow \quad \frac{df_{Sp}}{dt} = k \cdot v \quad\quad\quad (7.8)$$

which can immediately be compared to the voltage/current relationship for an inductor:

$$\frac{di_L}{dt} = \frac{1}{L} \cdot u_L \quad\quad\quad (7.9)$$

Thus, the inductance can be equivalenced to the inverse of the spring constant, which is sometimes called the *compliance* of the spring. Finally, we can look at friction phenomena:

$$f_{Fr} = b \cdot v \quad \Rightarrow \quad v = \frac{1}{b} \cdot f_{Fr} \qu\quad\quad (7.10)$$

which can be compared to Ohm's law for resistors, i.e., the electric conductance can be compared to the friction constant b.

Unfortunately, "bond graphers" around the world did it the other way around. They once decided to let the forces be called efforts and the velocities be called flows. As I explained before, due to symmetry, both assumptions are equally acceptable. I assume that the original reason for this decision was related to a mixup of the terms "effort" and "flow" with the terms "cause" and "effect." In many mechanical systems, the forces are considered the "causes" and the velocities are considered their "effects" (except if I drive my car against a tree). In any event, the decision is an arbitrary one, and, in order to be in agreement with the existing literature on bond graphs, I shall bow to the customary convention.

In this case, of course, we now need to compare Newton's law to the inductor rather than the capacitor:

$$L \cdot \frac{di_L}{dt} = u_L \quad \Rightarrow \quad \frac{di_L}{dt} = \frac{1}{L} \cdot u_L \quad \Rightarrow \quad i_L = \frac{1}{L} \cdot \int_0^t u_L d\tau + i_{L_0} \quad (7.11)$$

Thus, the mechanical mass corresponds now to an electrical inductor. In accordance with this convention, the spring compares to the capacitor:

$$\frac{du_C}{dt} = \frac{1}{C} \cdot i_C \tag{7.12}$$

i.e., the capacitor corresponds to the compliance of the spring. Finally, the friction constant b now corresponds to the electrical resistance rather than to the electrical conductance.

Notice that the terms effort and flow were defined by researchers dealing with bond graphs. On the other hand, the terms across variable and through variable were defined by other researchers who probably weren't aware of the ongoing research on bond graphs. These researchers defined the velocities as across variables and the forces as through variables (which makes sense). For this reason, the correspondence of effort with across variable, and flow with through variable is not consistent throughout the literature. However, in this text, I shall use effort and across variable interchangeably. To this end, I simply redefined the term across variable to mean effort. Thus, *my* mechanical across variables are the forces and torques, whereas *my* mechanical through variables are the velocities and angular velocities. As I mentioned earlier, this assignment is purely arbitrary. It simply made sense to use a consistent terminology, at least within this text, bearing the risk of potential confusion for readers who are familiar with the more conventional definitions of mechanical across and through variables.

Let us go ahead and derive a bond graph for the simple mechanical system that was presented in Fig.4.2. To refresh our memory, it is shown again in Fig.7.18.

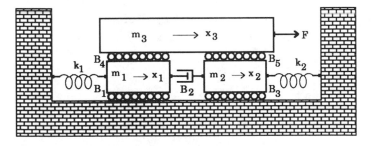

Figure 7.18. Simple translational problem.[†]

[†] Adapted figure from CONTROL SYSTEMS ENGINEERING by Norman Nise, ©1991 by Benjamin/Cummings, Inc., printed by permission of the publisher.

We start by identifying all free–moving bodies. We place 1–junctions for each of their velocities. Wherever two bodies interact with each other, we connect their junctions with branches consisting of two bonds and one 0–junction in between, and attach all interacting elements to that 0–junction. Newton's law (or rather the d'Alembert principle) is formulated at the 1–junctions themselves [7.1]. Figure 7.19 shows the bond graph for the simple translational system of Fig.7.18.

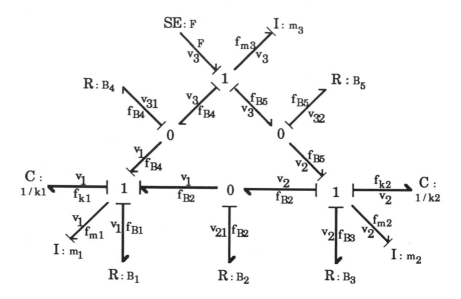

Figure 7.19. Bond graph for the translational problem.

Rotational systems work the same way. Here, the torques are taken for the effort variables and the angular velocities are taken for the flows.

7.6 Generalization to Other Types of Systems

Besides the two basic quantities, effort e and flow f, we often make use of two additional derived quantities, namely, the *generalized momentum*:

$$p = \int_0^t e \, d\tau \qquad (7.13)$$

and the *generalized displacement*:

$$q = \int_0^t f \, d\tau \qquad (7.14)$$

In electrical systems, the generalized momentum is the flux through a coil and the generalized displacement is the charge in a capacitor. In translational mechanical systems, these are the momentum and displacement (bond graphs were invented by a mechanical engineer) and in rotational mechanical systems, they are the angular momentum and angular position.

All these quantities are common to a large variety of other physical systems as well, as are the two Kirchhoff laws. Hydraulic, pneumatic, and acoustic systems operate similarly to the electrical and mechanical ones. In all these systems, the *pressure* is defined as the effort variable, while the *volume flow rate* is defined as the flow variable. The derived quantities are the *pressure momentum* and the *volume*.

The element laws, however, may look different for different types of systems. In particular, it may be noted that the equivalent to Ohm's law for these types of systems is often *nonlinear*. For example, the relation between the pressure (effort) p and the flow q in a (turbulent) hydraulic valve is quadratic:

$$\Delta p \propto q^2 \qquad (7.15)$$

Notice the confusing nomenclature. The symbols p and q are the most commonly used symbols in the hydraulic and pneumatic literature to denote pressures and flows. However, these are *effort* and *flow* variables, not *generalized momenta* and *generalized displacements*.

Table 7.1 presents a summary of the four generic variables for the most commonly used physical system types.

Table 7.1. Power (e,f) and Energy (p,q) Variables [7.18].

	Effort	Flow	Generalized momentum	Generalized displacement
	e	f	p	q
Electrical	Voltage u [V]	Current i [A]	Flux Φ [V sec]	Charge q [A sec]
Translational	Force F [N]	Velocity v [m sec^{-1}]	Momentum $\mathcal{I}$ [N sec]	Displacement x [m]
Rotational	Torque T [N m]	Angular velocity ω [rad sec^{-1}]	Twist T [N m sec]	Angle ϕ [rad]
Hydraulic	Pressure p [N m^{-2}]	Volume flow q [m^3 sec^{-1}]	Pressure momentum Γ [N m^{-2} sec]	Volume V [m^3]
Chemical	Chemical potential μ [J mole^{-1}]	Molar flow ν [mole sec^{-1}]	—	Number of moles n [mole]
Thermo-dynamical	Temperature T [K]	Entropy flow $\frac{dS}{dt}$ [W K^{-1}]	—	Entropy S [J K^{-1}]

7.7 Energy Transducers

Until now, we have looked at different types of systems in isolation. However, one of the true strengths of the bond graph approach is the ease with which transitions from one form of system to another can be made, while ensuring that the energy (or power) conservation rules are satisfied.

The power in an electrical system can be expressed as the product of voltage and current or in terms of the bond graph terminology:

$$P = e \cdot f \tag{7.16}$$

The power is measured in watts [W] = [V · A]. The energy is the integral of the power over time

$$E = \int_0^t e \cdot f \, d\tau \qquad (7.17)$$

which is measured in joules $[J] = [W \cdot sec] = [V \cdot A \cdot sec]$.

Until now, we assumed that we could select the effort and flow variables more or less freely. However, this assumption is incorrect. In all system types, the variables are chosen such that their product results in a variable of type *power* [7.18].

In an energy transducer (such as a transformer or a DC motor), the energy (or power) that is fed into the transducer is converted from one energy form to another, but it is never lost. Consequently, the energy that enters the transducer at one end must come out in one or more different form(s) at the other. A "lossless" energy transducer may, for example, transform electrical energy into mechanical energy. In reality, every energy transducer "loses" some energy, but the energy does not really disappear — it is simply transformed into heat.

This energy conservation law can be satisfied in exactly two ways in an "ideal" (i.e., lossless) energy transducer. One such transducer is the *ideal transformer*. It is governed by the following set of relationships:

$$e_1 = m \cdot e_2 \qquad (7.18a)$$
$$f_2 = m \cdot f_1 \qquad (7.18b)$$

The m value has been (arbitrarily) defined as the flow gain from the primary to the secondary side, or the effort gain from the secondary to the primary side.

The ideal transformer is placed between two junctions. Two types of causalities are possible, as shown in Fig.7.20.

Figure 7.20. Causality bond graphs for the ideal transformer.

Examples of transformers are the electrical transformer (shown in Fig.7.21a), the mechanical gear (shown in Fig.7.21b), and a mechanohydraulic pump (shown in Fig.7.21c).

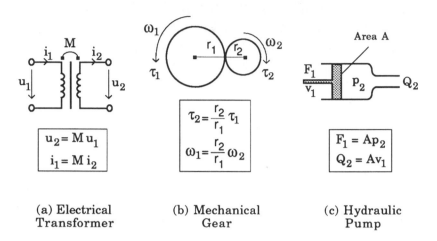

$u_2 = M u_1$	$\tau_2 = \dfrac{r_2}{r_1}\,\tau_1$	$F_1 = A p_2$
$i_1 = M i_2$	$\omega_1 = \dfrac{r_2}{r_1}\,\omega_2$	$Q_2 = A v_1$

(a) Electrical Transformer	(b) Mechanical Gear	(c) Hydraulic Pump

Figure 7.21a–c. Examples of ideal transformers.

The other type of energy transducer is the *ideal gyrator*. Its behavior is governed by the equations:

$$e_1 = r \cdot f_2 \tag{7.19a}$$
$$e_2 = r \cdot f_1 \tag{7.19b}$$

The r value has been (arbitrarily) defined as the gain from the primary flow to the secondary effort.

Also, the ideal gyrator exhibits two forms of causalities, as shown in Fig.7.22.

Figure 7.22. Causality bond graphs for the ideal gyrator.

Examples of gyrators are most electromechanical converters, for example, the DC motor shown in Fig.7.23.

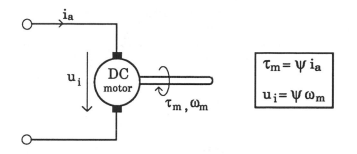

Figure 7.23. Example of an ideal gyrator $(r = \psi)$.

Notice that no real difference exists between the two transducer types [7.2, 7.3]. If the effort and flow variables in the mechanical system were exchanged (as earlier suggested), the DC motor would in fact become a transformer.

Let us go through an example. We want to model the mechanical system of Fig.4.5 driving it with an armature controlled DC motor with constant field, such as the one depicted in Fig.4.13. The resulting bond graph is shown in Fig.7.24.

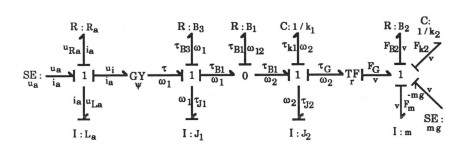

Figure 7.24. Bond graph of a DC motor controlled mechanical system.

Let us look at how the causalities were assigned. We start from the left with the mandatory causality assignment for the effort source. On the 1–junction to its right, only one branch must be without a stroke. We proceed by satisfying the desired causality constraint on the "inertia" L_a, which fixes all causalities for the 1–junction. At this point in time, the gyrator is also fixed. At the next 1–junction, we proceed in the same way, by satisfying the desired causality constraint for the inertia J_1. This fixes all causalities for that junction. At the 0–junction, we still have a choice. Since we cannot decide at once, we proceed to the next 1–junction, where we satisfy the desired causality constraint for the inertia J_2. This fixes all causalities for that junction, and thereby fixes the causalities for the 0–junction in between. At this point, the following transformer is also fixed. Unfortunately, now we are in trouble. All causalities at the final 1–junction are already determined and we are unable to satisfy the causality constraints for the "inertia" m. Consequently, we have detected a structural singularity. The system contains four "inertias" and two "compliances," and therefore, we would expect this system to be of sixth order. However, it is in fact only a fourth–order system. We determined, analyzed, and solved this problem in Chapter 4 by reducing the forces that are attached to the secondary side of the transformer to its primary side. We shall not repeat the analysis at this point.

7.8 Bond Graph Modeling in DYMOLA

Now that we have seen how bond graphs can be constructed, let us discuss how we can use these bond graphs to perform actual simulation runs.

The first bond graph simulation language written in the early 1970s was ENPORT [7.16,7.17]. This software used an approach similar to SPICE, i.e., it did not request causalities to be specified and it transformed the topological input description into a branch–admittance matrix that could then be solved employing similar techniques to those used in SPICE. Consequently, ENPORT is able to handle structurally singular problems. The current version of the code, ENPORT–7 [7.17], offers an alphanumerical topological input language similar to SPICE and a menu–driven graphical input language which, however, is not yet very user–friendly. A full-fledged graphical window system is currently under development. ENPORT–7 runs on various mainframe computers, but a slightly reduced version, ENPORT/PC, exists for IBM PCs and compatibles. ENPORT offers also a macro capability (somewhat comparable to the subcircuits in SPICE) that, however, does not provide for full hierarchical decomposition capabilities.

In the late 1970s, another bond graph simulation language was developed at Twente University in the Netherlands, called THT-SIM in Europe and TUTSIM in the United States [7.18]. TUTSIM translates bond graphs into a state–space representation. The user is required to specify the causalities. Structurally singular systems cannot be handled. TUTSIM's simulation engine is somewhat poor in comparison to other state–space solvers such as ACSL. The same research group is currently prototyping a new bond graph modeling system, CAMAS [7.4], which runs on Suns, has nice graphics capabilities, and is able to handle algebraic loops. CAMAS employs an object–oriented language (SIDOPS) for the model description, which has properties similar to DYMOLA. Once available, this might become a good product.

The third product on the market is CAMP [7.8,7.9], a preprocessor to ACSL [7.14] that translates bond graphs into ACSL programs. CAMP has the same limitations as TUTSIM, i.e., it does not handle algebraic loops or structural singularities, but it has the better simulation engine (ACSL). The input format is topological (as for the two other products). It is not truly flexible with respect to handling nonstandard circuit elements. Nonlinear elements need to be edited manually into the generated ACSL program, which is inconvenient. A graphical front end exists also for CAMP [7.10]. However, as in the case of ENPORT–7, the graphics editor is menu–driven rather than window–operated.

With the exception of the unfinished CAMAS system, none of these products is able to handle hierarchically structured models in a general fashion. However, this feature is essential for the analysis of complex systems. For this reason, I prefer not to discuss any of these programs in greater detail, but to explain instead how DYMOLA [7.7] can be used as a bond graph modeling engine. The approach is actually straightforward. DYMOLA's "nodes" are equivalent to the 0–junctions in bond graph terminology. DYMOLA has no equivalent for 1–junctions, but as explained before, 1–junctions are the same as 0–junctions with the effort and flow variables interchanged [7.5]. Therefore, we created a model type "bond" that simply exchanges the effort and flow variables:

```
model type bond
  cut A(x/y), B(y/-x)
  main cut C[A, B]
  main path P < A - B >
end
```

The bond acts just like a null modem for a computer. Since neighboring junctions are always of opposite sexes, they can both be described by regular DYMOLA "nodes" if they are connected with a "bond."

Since we don't want to maintain different types of **R, C, I, TF,** and **GY** elements, we add one additional rule: in DYMOLA, all elements (except for the bonds) can be attached to 0–junctions only. If they need to be attached to a 1–junction, we simply must place a bond in between.

The following DYMOLA models suffice to describe the basic bond graphs.

```
model type SE
   main cut A(e/.)
   terminal E0
   E0 = e
end

model type SF
   main cut A(./-f)
   terminal F0
   F0 = f
end

model type R
   main cut A(e/f)
   parameter R = 1.0
   R * f = e
end

model type C
   main cut A(e/f)
   parameter C = 1.0
   C*der(e) = f
end

model type I
   main cut A(e/f)
   parameter I = 1.0
   I*der(f) = e
end
```

```
model type TF
    cut A(e1/f1), B(e2/-f2)
    main cut C[A, B]
    main path P < A - B >
    parameter m = 1.0
    e1 = m * e2
    f2 = m * f1
end

model type GY
    cut A(e1/f1), B(e2/-f2)
    main cut C[A, B]
    main path P < A - B >
    parameter r = 1.0
    e1 = r * f2
    e2 = r * f1
end
```

With these modeling elements, we can formulate a bond graph description of our simple passive circuit. Figure 7.25 shows the DYMOLA expanded bond graph with all elements being attached to 0–junctions only.

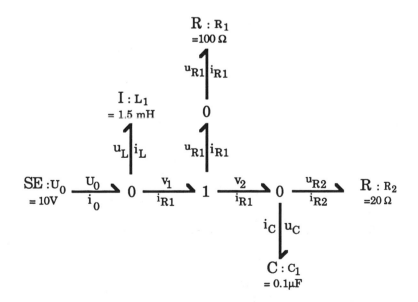

Figure 7.25. DYMOLA expanded bond graph of the passive circuit.

I did not mark down the causalities here since DYMOLA is perfectly able to handle the causality assignment by itself.

Notice that my "bond" model type is actually a gyrator with $r = 1.0$. This special gyrator has sometimes been called a *symplectic gyrator* in the bond graph literature [7.2,7.3].

This bond graph model can directly be coded in DYMOLA as shown in the following code:

```
@se.bnd
@r.bnd
@i.bnd
@c.bnd
@bond.bnd

model RLC

    submodel (SE) U0
    submodel (R) R1(R = 100.0),  R2(R = 20.0)
    submodel (I) L1(I = 1.5E-3)
    submodel (C) C1(C = 0.1E-6)
    submodel (bond) B1, B2, B3
    node v1, ir1, vr1, v2
    output y1, y2

    connect U0 at      v1
    connect L1 at      v1
    connect R1 at      vr1
    connect R2 at      v2
    connect C1 at      v2
    connect B1 from v1    to ir1
    connect B2 from ir1   to v2
    connect B3 from ir1   to vr1

    U0.E0 = 10.0
    y1 = C1.e
    y2 = R2.f

end
```

The interpretation of this code is straightforward.

Let us see how the DYMOLA compiler preprocesses this code. We enter DYMOLA and specify the model to be compiled as follows:

```
$ dymola
 > enter model
 − @rlc.dym
 > outfile rlc.eq
 > output equations
```

DYMOLA's answer is shown in the next code segment.

$$
\begin{array}{ll}
\textbf{U0} & E0 = e \\
\textbf{R1} & R * f = e \\
\textbf{C1} & C * dere = f \\
\textbf{L1} & L * der f = e \\
\textbf{R2} & R * f = e \\
\textbf{RLC} & U0.E0 = 10.0 \\
& y1 = C1.e \\
& y2 = R2.f \\
& L1.e = B1.x \\
& U0.e = L1.e \\
& C1.e = B2.y \\
& R2.e = C1.e \\
& C1.f + R2.f = B2.x \\
& B2.x = B3.x \\
& B1.y = B2.x \\
& B3.y + B2.y = B1.x \\
& R1.e = B3.y \\
& R1.f = B3.x
\end{array}
$$

We can now execute the algorithm that assigns the causalities, i.e., determines what variable to compute from each of the equations. In DYMOLA, this is achieved with the following set of instructions:

> **partition**
> **outfile** *rlc.sor*
> **output sorted equations**

which results in the following answer:

$$
\begin{array}{ll}
\textbf{RLC} & [U0.E0] = 10.0 \\
\textbf{U0} & E0 = [e] \\
\textbf{RLC} & U0.e = [L1.e] \\
& L1.e = [B1.x] \\
& C1.e = [B2.y] \\
& [B3.y] + B2.y = B1.x \\
& [R1.e] = B3.y \\
\textbf{R1} & R * [f] = e \\
\textbf{RLC} & R1.f = [B3.x] \\
& [B2.x] = B3.x \\
& [B1.y] = B2.x \\
& [R2.e] = C1.e \\
\textbf{R2} & R * [f] = e \\
\textbf{RLC} & [C1.f] + R2.f = B2.x \\
\textbf{C1} & C * [dere] = f \\
\textbf{L1} & L * [der f] = e \\
\textbf{RLC} & [y1] = C1.e \\
& [y2] = R2.f
\end{array}
$$

The variables enclosed in "[ ]" are the variables for which each equation must be solved. This set of equations contains many trivial equations of the type $a = b$. DYMOLA is capable of throwing those out. This is accomplished through the following set of instructions:

> partition eliminate
> outfile *rlc.sr2*
> output sorted equations

which results in the following answer:

R2	$R * [y2] = y1$
RLC	$[B3.y] + y1 = 10.0$
R1	$R * [B3.x] = B3.y$
RLC	$[C1.f] + y2 = B3.x$
C1	$C * [dere] = f$
L1	$L * [der f] = 10.0$

which is a much reduced set of equivalent equations. The next step will be to actually perform the symbolic manipulation on the equations. In DYMOLA, this is done in the following way:

> outfile *rlc.sov*
> output solved equations

which results in the following answer:

R2	$y2 = y1/R$
RLC	$B3.y = 10.0 - y1$
R1	$B3.x = B3.y/R$
RLC	$C1.f = B3.x - y2$
C1	$dere = f/C$
L1	$der f = 10.0/L$

We are now ready to add the experiment description to the model. We can, for instance, use the one presented in Chapter 6. The set of DYMOLA instructions:

> enter experiment
  − @*circuit.ctl*
> outfile *rlc.des*
> output desire program

tells DYMOLA to generate the following DESIRE program:

```
- - - - - - - - - - - - - - - - - - - - - - - - - - - - - -
-- CONTINUOUS SYSTEM RLC
- - - - - - - - - - - - - - - - - - - - - - - - - - - - - -
-- STATE y1 L1$f
-- DER dC1$e dL1$f
-- PARAMETERS and CONSTANTS:
R1$R = 100.0
C = 0.1E-6
L = 1.5E-3
R2$R = 20.0
-- INITIAL VALUES OF STATES:
y1 = 0
L1$f = 0
- - - - - - - - - - - - - - - - - - - - - - - - - - - - - -
TMAX = 2E-5 | DT = 2E-7 | NN = 101
scale = 1
XCCC = 1
label TRY
drunr
if XCCC < 0 then XCCC = -XCCC | scale = 2 * scale | go to TRY
          else proceed
- - - - - - - - - - - - - - - - - - - - - - - - - - - - - -
DYNAMIC
- - - - - - - - - - - - - - - - - - - - - - - - - - - - - -
-- Submodel: R2
y2 = y1/R2$R
-- Submodel: RLC
B3$y = 10.0 - y1
-- Submodel: R1
B3$x = B3$y/R1$R
-- Submodel: RLC
C1$f = B3$x - y2
-- Submodel: C1
d/dt y1 = C1$f/C
-- Submodel: L1
d/dt L1$f = 10.0/L
- - - - - - - - - - - - - - - - - - - - - - - - - - - - - -
OUT
dispt y1, y2
- - - - - - - - - - - - - - - - - - - - - - - - - - - - - -
/ - -
/PIC 'rlc.PRC'
/ - -
```

which can be executed at once using the following instructions:

> **stop**
$ **desire** 0
> **load** *'rlc.des'*
> **run**

which will immediately (within less than a second) produce the desired output variables, u_C and i_{R2}, on the screen. Both DY-MOLA [7.7] and DESIRE [7.13] are currently running alternatively on VAX/VMS or PC/MS–DOS. The code will run fine on a PC/XT, but minimum requirements are a 10–MByte hard disk and an 8087 coprocessor. Faster versions exist for the PC/AT and 386–based machines. DESIRE [7.13] supports CGA, EGA, and VGA graphics.

7.9 The Dual Bond Graph

In some cases, bond graphs may result that have many more elements attached to 1–junctions than to 0–junctions. By using the previously introduced methodology, this would force us to introduce many additional 0–junctions and bonds in order to be able to attach all elements to 0–junctions only.

It is possible to circumvent this problem by introducing the concept of the dual bond graph [7.5]. For any bond graph, an equivalent dual bond graph exists in which the role of all effort and flow variables is interchanged. Table 7.2 illustrates what happens to the various bond graph elements under the transition from the regular to the dual bond graph.

Table 7.2. Relation Between Regular and Dual Bond Graph.

Regular Bond Graph	Dual Bond Graph
e	f
R	G
C	L
SE	SF
TF	TF
GY	GY
0–junction	1–junction

In Fig.7.26, this duality transformation has been applied to the bond graph of Fig.7.24.

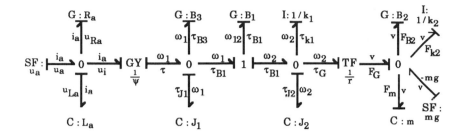

Figure 7.26. Dual bond graph of the electromechanical system.

It can be easily verified that the equations generated from the dual bond graph are the same as those that were generated from the regular bond graph. However, in the regular bond graph, all elements except for the friction element B_1 were attached to 1–junctions. This would have forced us to create 10 additional junctions with 10 additional bonds (one for each element attached to a 1–junction). In the dual bond graph, all elements except for the friction B_1 are attached to 0–junctions. Consequently, we need to expand the bond graph only with one additional 0–junction and bond instead of the former 10.

We just introduced a "new" bond graph element: the *conductance* G. Its governing equation is: $f = G \cdot e$, which DYMOLA can, of course, transform into $e = f/G$. This element is nonessential. Instead of replacing the resistors with conductances in the dual bond graph and writing, for instance, **G**:R_a in Fig.7.26, we could equally well have kept the resistors and written **R**:$1/R_a$. It just seemed more convenient this way. Notice that this decision has absolutely nothing to do with the assumed causality. Both resistors and conductances can assume either causality.

Notice that it is feasible to restrict the application of the duality transformation to subsystems. Natural places where the bond graph can be cut into subsystems are the transformers and gyrators where one sort of energy is transformed into another. Figure 7.27 shows yet another version of the same system. This time, only the electrical subsystem has been transformed to its dual equivalent.

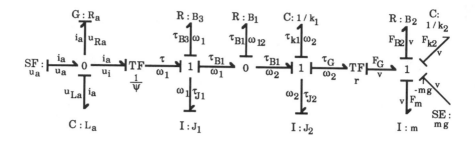

Figure 7.27. Partially transformed bond graph of the mechanical system.

In this duality transformation, gyrators become transformers and vice versa. The reason why the bond graph of the DC motor exhibits a *gyrator* at the interface between its electrical and mechanical subsystems is because of the peculiar way in which Paynter defined his efforts and flows for mechanical and electrical systems [7.15].

However, it is possible to cut systems at other places. Let me illustrate the concept by means of a series of snapshots of a portion of the bond graph of Fig.7.26. Figure 7.28a shows the portion that we want to concentrate on.

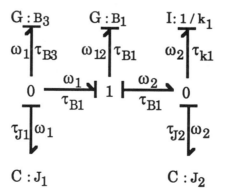

Figure 7.28a. Portion of the DC motor bond graph.

In Fig.7.28b, the two 0–junctions have been stretched out into two junctions each. The arrowless line between the two 0–junctions indicates that this is, in fact, the same junction. The solid line symbolizes a "wire" that exists between the two junctions.

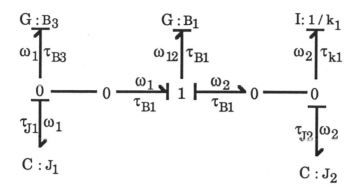

Figure 7.28b. Expanded bond graph.

Obviously, any such wire can be replaced by a "symplectic transformer," i.e., a transformer with $m = 1$. Such a transformer is obviously equivalent to a wire. This replacement is shown in Fig.7.28c.

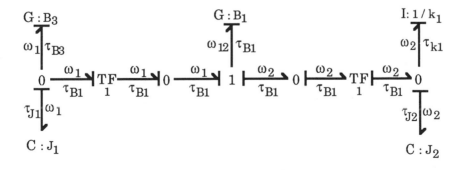

Figure 7.28c. Further expanded bond graph.

At this point, we see that the bond graph contains two 0–junctions with two connections only. These junctions are unimportant and can be eliminated. This is shown in Fig. 7.28d.

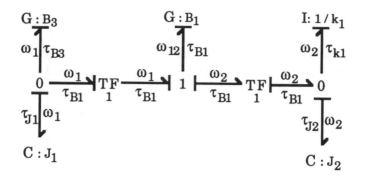

Figure 7.28d. Reduced bond graph.

Now, we have two transformers in the circuit that isolate the portion of the circuit between them. We can, thus, apply a duality transformation to that portion of the circuit. This is shown in Fig.7.28e.

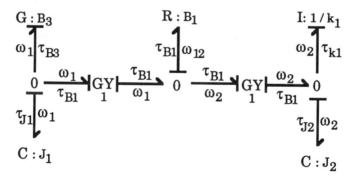

Figure 7.28e. Dually transformed bond graph.

The two symplectic transformers have been transformed into two symplectic gyrators which, of course, are the same as our Dymola bonds. I still prefer to leave these gyrators explicitly in the circuit and use the *GY.dym* model rather than the bond.dym model since a reduction to a normal bond might be graphically confusing. However, we just learned that a bond graph can be cut into subsystems at an arbitrary bond. In the duality transformation, the cutting bond (an implicit symplectic gyrator) is transformed into an explicitly shown symplectic gyrator.

7.10 Summary

In this chapter, we looked at a number of different graphical modeling techniques and analyzed one among them, the bond graph modeling technique, in greater detail. It was the aim of this chapter to relate this seemingly quite different approach to modeling back to the previously introduced methodologies and terminologies.

Future chapters will make more references to bond graphs, and, in particular, Chapter 8 will discuss the application of bond graphs to nonequilibrium state thermodynamics. This application will provide us with even more motivation for the bond graph methodology as a whole.

References

[7.1] Alan Blundell (1982), *Bond Graphs for Modelling Engineering Systems*, Ellis Horwood Publishers, Chichester, United Kingdom, and Halsted Press, New York.

[7.2] Peter C. Breedveld (1982), "Thermodynamic Bond Graphs and the Problem of Thermal Inertance," *J. Franklin Institute*, 314(1), pp. 15–40.

[7.3] Peter C. Breedveld (1984), *Physical Systems Theory in Terms of Bond Graphs*, Ph.D. dissertation, University of Twente, Enschede, The Netherlands.

[7.4] Jan F. Broenink (1990), *Computer–Aided Physical–Systems Modeling and Simulation: A Bond–Graph Approach*, Ph.D. dissertation, University of Twente, Enschede, The Netherlands.

[7.5] François E. Cellier (1990), "Hierarchical Nonlinear Bond Graphs — A Unified Methodology for Modeling Complex Physical Systems," *Proceedings European Simulation MultiConference*, Nürnberg, F.R.G., pp. 1–13.

[7.6] Hilding Elmqvist (1975), *SIMNON — An Interactive Simulation Program for Nonlinear Systems — User's Manual*, Report CODEN: LUTFD2/(TFRT–7502), Dept. of Automatic Control, Lund Institute of Technology, Lund, Sweden.

[7.7] Hilding Elmqvist (1978), *A Structured Model Language for Large Continuous Systems*, Ph.D. dissertation, Report CODEN: LUTFD2/(TRFT–1015), Dept. of Automatic Control, Lund Institute of Technology, Lund, Sweden.

[7.8] Jose J. Granda (1982), *Computer Aided Modeling Program (CAMP): A Bond Graph Preprocessor for Computer Aided Design and Simulation of Physical Systems Using Digital Simulation Languages*, Ph.D. dissertation, Dept. of Mechanical Engineering, University of California, Davis.

[7.9] Jose J. Granda (1985), "Computer Generation of Physical System Differential Equations Using Bond Graphs," *J. Franklin Institute*, **319**(1/2), pp. 243–255.

[7.10] Jose J. Granda and F. Pourrahimi (1985), "Computer Graphic Techniques for the Generation and Analysis of Physical System Models," in: *Artificial Intelligence, Graphics, and Simulation*, Proceedings of the Western Simulation MultiConference (G. Birtwistle, ed.), SCS Publishing, La Jolla, Calif., pp. 70–75.

[7.11] Gene H. Hostetter, Clement J. Savant, Jr., and Raymond T. Stefani (1989), *Design of Feedback Control Systems*, second edition, Saunders College Publishing, New York.

[7.12] Dean C. Karnopp and Ronald C. Rosenberg (1974), *System Dynamics; A Unified Approach*, John Wiley, New York.

[7.13] Granino A. Korn (1989), *Interactive Dynamic–System Simulation*, McGraw–Hill, New York.

[7.14] Edward E. L. Mitchell and Joseph S. Gauthier (1986), *ACSL: Advanced Continuous Simulation Language — User Guide and Reference Manual*, Mitchell & Gauthier Assoc., Concord, Mass.

[7.15] Henry M. Paynter (1961), *Analysis and Design of Engineering Systems*, MIT Press, Cambridge, Mass.

[7.16] Ronald C. Rosenberg (1974), *A User's Guide to ENPORT-4*, John Wiley, New York.

[7.17] RosenCode Associates, Inc. (1989), *The ENPORT Reference Manual*, Lansing, Mich.

[7.18] Jan J. van Dixhoorn (1982), "Bond Graphs and the Challenge of a Unified Modelling Theory of Physical Systems," in: *Progress in Modelling and Simulation* (F.E. Cellier, ed.), Academic Press, London, pp. 207–245.

Bibliography

[B7.1] Albert M. Bos and Peter C. Breedveld (1985), "Update of the Bond Graph Bibliography," *J. Franklin Institute*, **319**(1/2), pp. 269–286.

[B7.2] Peter C. Breedveld, Ronald C. Rosenberg, and T. Zhou (1991), "Bibliography of Bond Graph Theory and Application," *J. Franklin Institute*, to appear.

[**B7.3**] Vernon D. Gebben (1979), "Bond Graph Bibliography," *J. Franklin Institute*, **308**(3), pp. 361–369.

[**B7.4**] Louis P. A. Robichaud (1962), *Signal Flow Graphs and Applications*, Prentice–Hall, Englewood Cliffs, N.J.

[**B7.5**] Ronald C. Rosenberg and Dean C. Karnopp (1983), *Introduction to Physical System Dynamics*, McGraw–Hill, New York.

[**B7.6**] Jean U. Thoma (1989), *Simulation by Bondgraphs — Introduction to a Graphical Method*, Springer–Verlag, New York.

Homework Problems

[H7.1] Algebraic Loop

Draw a bond graph for the simple resistive circuit of Fig.6.11. To refresh your memory, the circuit is shown once more in Fig.H7.1.

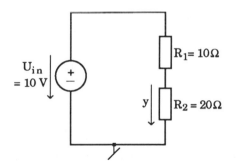

Figure H7.1. Schematic of a trivial resistive circuit.

Prove that this representation contains an algebraic loop by showing that you have a choice in assigning the causalities. Create a DYMOLA program that implements this bond graph and show that during the partitioning process the set of equations that initially looks quite different from the one obtained in Chapter 6 is reduced to the same set of three algebraically coupled equations that we came across in Chapter 6.

[H7.2] Electrical Circuit

Solve Hw.[H6.6] using the bond graph approach. The circuit is presented again in Fig.H7.2.

Since, also in the bond graph models, the sources have been declared as terminals, the dependent current source can be treated in the DYMOLA

program in the same manner as in Chapter 6. Since the cause/effect relationship between the driving voltage and the driven current contains only
one rather than two variables, this connection is not a bond. It is a regular
signal path (as in a block diagram), and in the bond graph, it is represented
through a *thin full arrow* that emanates from the 0–junction at which all
bonds have the driving voltage as their effort variable and ends at the **SF**
element that is being driven by this signal.

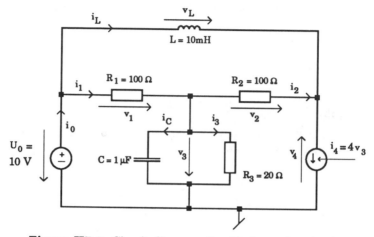

Figure H7.2. Circuit diagram of a simple passive circuit.

[H7.3] Cervical Syndrome [7.11]

Solve Hw.[H4.3] using bond graphs. Figure H7.3 shows the system again.
Create a bond graph for this system, transform the bond graph into a
DYMOLA program, and generate either a SIMNON [7.6] or DESIRE [7.13]
program from it. Contrary to the simulation of Chapter 4, we shall simulate
the system in the time domain for a sinusoidal input of varying frequency.
In general, a sinusoidal input of any frequency ω_0 can be written as:

$$u = \sin(\omega_0 t) \qquad\qquad (H7.3a)$$

In our experiment, we wish to vary ω_0 using a slow ramp of time, i.e.:

$$\omega_0 = k \cdot t \qquad\qquad (H7.3b)$$

By plugging Eq.(H7.3b) into Eq.(H7.3a), we find:

$$u = \sin(k \cdot t^2) \qquad\qquad (H7.3c)$$

Set $k = 0.01$ and simulate the system during 100 sec. Use a step size of
0.01 sec. Observe the input and output over time.

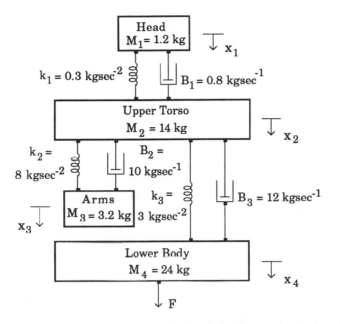

Figure H7.3. Mechanical model of a sitting human body [7.11].[†]

[H7.4] Electromechanical System

For the system shown in Fig.H7.4, generate a bond graph, assign the proper causalities, and discuss the energy flow through the system.

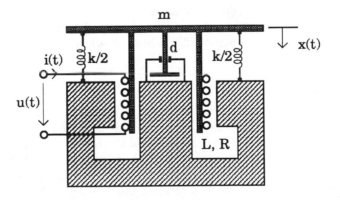

Figure H7.4. Electromechanical system.

[†] Adapted figure from DESIGN OF FEEDBACK CONTROL SYSTEMS, Second Edition by Gene Hostetter *et al.*, ©1989 by Saunders College Publishing, a division of Holt, Rinehart and Winston, Inc., reprinted by permission of the publisher.

This system had once before been discussed in Hw.[H4.4]. Use a duality transformation prior to coding the bond graph in DYMOLA and generate a minimal state–space model using the DYMOLA preprocessor. Compare the resulting model with the one that you found in Hw.[H4.4].

[H7.5] Nonideal Transformer

Figure H7.5a shows a nonideal transformer and Fig.H7.5b shows an equivalent circuit that reduces the transformer to elements we already know.

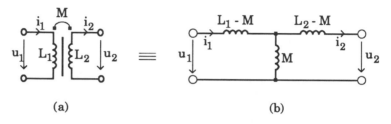

(a)   (b)

Figure H7.5. Nonideal transformer.

Create a bond graph for the transformer using the equivalent circuit approach. Introduce the causalities and determine that the system contains a structural singularity. Extract the equations from the bond graph and verify the structural singularity by means of these equations. Manually reduce the structural singularity. This will unfortunately introduce an algebraic loop. Reduce the algebraic loop and determine a state–space model that describes the nonideal transformer. Create a DYMOLA model (not a bond graph model) that describes this executable set of equations, but assign cuts to this model such that the model can be used as a component model type (named RTF) anywhere within a bond graph model.

[H7.6] Hydraulic System

Figure H7.6a shows a schematic diagram of a hydraulic motor with a four–way servo valve. The input to this system is the position of the piston, x. If the piston of the servo valve is moved far to the right, then the pressure p_1 in the first chamber is the same as the high–pressure P_S, and the pressure p_2 of the second chamber is the same as the low–pressure P_0. Consequently, the hydraulic motor will wish to increase the volume of the first chamber and decrease the volume of the second by moving the motor block to the right. In the given setup, the axis of the hydraulic motor is a screw that therefore starts rotating. If the piston of the servo valve is moved far to the left, then the pressure p_1 is equal to the low–pressure P_0, the pressure p_2 is equal to the high–pressure P_S, and the hydraulic motor rotates in the opposite direction. In between, one position exists where $p_1 = p_2$. In that position, the hydraulic motor does not move. We call this position $x = 0.0$. Hydraulic motors are able to move large masses with little control power, and yet, the inertia of the motor is much smaller than for an equivalent electrical motor.

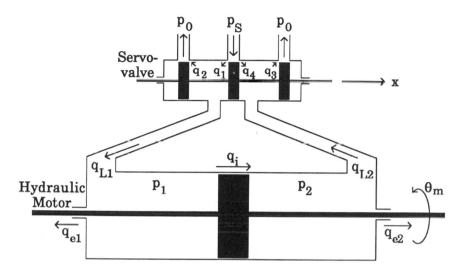

Figure H7.6a. Hydraulic motor with a four–way servo valve.

The flows from the high–pressure line into the servo valve and from the servo valve back into the low–pressure line are turbulent. Consequently, the relation between flow and pressure is quadratic:

$$q_1 = k(x_0 + x)\sqrt{P_S - p_1} \qquad (H7.6a)$$

$$q_2 = k(x_0 - x)\sqrt{p_1 - P_0} \qquad (H7.6b)$$

$$q_3 = k(x_0 + x)\sqrt{p_2 - P_0} \qquad (H7.6c)$$

$$q_4 = k(x_0 - x)\sqrt{P_S - p_2} \qquad (H7.6d)$$

For our servo valve, we want to use the following parameter values: $P_S = 0.137 \times 10^8$ N m^{-2}, $P_0 = 0.0$ N m^{-2}, $x_0 = 0.05$ m, and $k = 0.248 \times 10^{-6}$ kg$^{-1/2}$ m$^{5/2}$. You have to be a little careful with these equations since the expressions $x_0 \pm x$ must never become negative and the same is true for the expressions under the square roots. A negative pressure difference is even physically possible due to cavitation, but our model does not represent cavitation phenomena properly. In your computer programs, you'd better limit these expressions explicitly.

The flow q_i is an internal leakage flow that is laminar and therefore linear in the load pressure p_L, i.e.

$$q_i = c_i \cdot p_L = c_i(p_1 - p_2) \qquad (H7.6e)$$

where $c_i = 0.737 \times 10^{-13}$ kg^{-1} m^4 sec. The flows q_{e1} and q_{e2} are external leakage flows that are also laminar:

$$q_{e1} = c_e \cdot p_1 \qquad (H7.6f)$$
$$q_{e2} = c_e \cdot p_2 \qquad (H7.6g)$$

where $c_e = 0.737 \times 10^{-12}$ kg^{-1} m^4 sec.

The change in the chamber pressures is proportional to the effective flows in the two chambers:

$$\dot{p}_1 = c_1(q_{L1} - q_i - q_{e1} - q_{ind}) \qquad (H7.6h)$$
$$\dot{p}_2 = c_1(q_{ind} + q_i - q_{e2} - q_{L2}) \qquad (H7.6i)$$

where $c_1 = 5.857 \times 10^{13}$ kg m^{-4} sec^{-2}. q_{ind} is the *induced flow*, which is a result of the moving motor block:

$$q_{ind} = \psi \cdot \dot{\theta}_m \qquad (H7.6j)$$

and it is proportional to the angular velocity of the hydraulic motor; in our case, $\psi = 0.575 \times 10^{-5}$ m^3.

The torque produced on the hydraulic motor is proportional to the load pressure p_L:

$$T_m = \psi \cdot p_L = \psi(p_1 - p_2) \qquad (H7.6k)$$

On the mechanical side of the motor, we have an inertia of $J_m = 0.08$ kg m^2 and a viscous friction of $\rho = 1.5$ kg m^2 sec^{-1}.

Figure H7.6b shows the overall position control circuit.

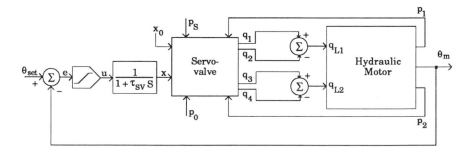

Figure H7.6b. Hydraulic motor position control circuit.

The first–order controller represents the translation from the (electrical) control signal u to the (mechanical) position of the piston of the servo valve. $\tau_{sv} = 0.005$ sec is the mechanical time constant of the servo valve. The error signal e is limited between -1.0 and $+1.0$.

Generate a block diagram that describes the overall system. Code this position control problem in any CSSL and simulate the step response of this system during 0.2 sec.

[H7.7] Hydraulic System

For the system of Hw.[H7.6], generate a bond graph description. The control signals are represented by signal paths but the hydraulic and mechanical parts can be easily described in terms of a bond graph. Discuss the energy flow in this system. Code the bond graph model in DYMOLA, generate either a DESIRE [7.13] or a SIMNON [7.6] program, and simulate the step response of this system during 0.2 sec. Compare the results with those found in Hw.[H7.6].

[H7.8]* Surge Tank Simulation

In Hw.[H5.4], we analyzed the behavior of a water turbine with a surge tank. To refresh your memory, the schematic of that system is shown again in Fig.H7.8.

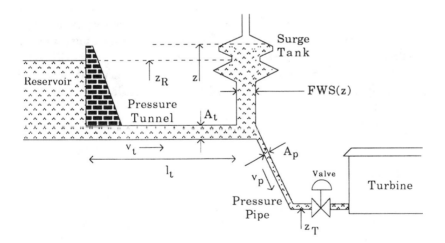

Figure H7.8. Power generation using a water turbine.

This time we wish to represent this system using a bond graph notation. This will ensure that we have modeled the energy flow through the system correctly. For this purpose, we need to know one more piece of information that was not required for the previous model: the pressure tunnel is at an altitude of 3100 m.

Code your bond graph model in DYMOLA, generate either a DESIRE [7.13] or a SIMNON [7.6] program, and simulate the system. Compare your results with those found in Hw.[H5.4].

Projects

[P7.1] Code Generation

Develop a new code generator for DYMOLA for the generation of ACSL [7.14] programs in addition to the currently implemented DESIRE [7.13] and SIMNON [7.6] interfaces.

Research

[R7.1] Graphical Input

Develop a graphical front end for bond graph models. Using this software, we should be able to draw bond graphs on the screen using a zoom and pan capability. We should also be able to assign new bond graph symbols (icons) to subnetworks for hierarchical modeling of bond graphs. Finally, we should be able to describe new atomic bond graph models as DYMOLA model types and assign these model types to new bond graph icons. A graphical preprocessor is to be developed that can translate the schematic into a DYMOLA program.

8

Modeling in Nonequilibrium Thermodynamics

Preview

Until now, we have dealt with applications from either classical mechanics or electrical circuits exclusively. In this chapter, we shall discuss nonequilibrium state thermodynamics. Most engineering students consider thermodynamics a rather difficult topic. The reason for this seeming difficulty lies in the fact that basically all available treatises of thermodynamics have been written by physicists rather than by engineers. Physicists are, by education, phenomenologically rather than systemically oriented. They do not wish to change the world, only to understand it. Therefore, their approach to dealing with problems is quite different from ours. Rather than looking at a system as a whole and trying to analyze the couplings of its subsystems (as we engineers do), they always try to single out individual phenomena and discuss those in isolation. As a consequence, most physics texts present the topic through a collection of various formulae, which are all individually correct and meaningful but hard to relate to each other. It is the aim of this chapter to bridge the gap between those individually well–known equations that govern the behavior of nonequilibrium state thermodynamic systems. According to Jean Thoma, another reason why most thermodynamics textbooks are obscure is the fact that they avoid to work with *entropy flow* as a physical variable. He remarks rightly that textbooks for electrical circuit theory would be equally obscure if they were to describe all electrical phenomena in terms of the electrical potential and the stored energy alone while avoiding the concept of current flow [private communication]. In this text, thermodynamic phenomena will be described through a set of adjugate variables comparable to those used in electrical circuits and in mechanical motion.

8.1 Power Flow

Traditionally, two separate approaches have been used for dealing with thermodynamics, a *macroscopic* approach that describes thermodynamics through variables such as temperature, heat, and energy and is deterministic in nature [8.8,8.19], and a *microscopic* approach that analyzes the movement of particles and is stochastic in nature [8.9,8.10,8.17]. In this text, we deal with the macroscopic, and therefore deterministic, aspects of thermodynamics exclusively. For this purpose, we shall employ the bond graph methodology that was introduced in the previous chapter. It turns out that this methodology can present us with a good insight into the macroscopic processes that govern thermodynamic systems.

Let us start by revealing one more property of the bond graph approach to modeling that is related to the flow of power.

Let us perform the following experiment. We take the simple electrical circuit drawn in Fig.8.1

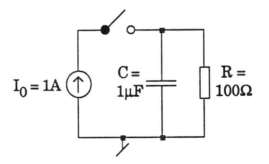

Figure 8.1. Simple RC circuit.

and simulate this circuit during 0.6 msec. During the first 0.3 msec, the switch is closed; thereafter it is open. We coded this problem in ACSL [8.12] (a trivial exercise). The results of the simulation are shown in Fig.8.2. From Fig.8.2, we see that the capacitor is being *charged* whenever the current and voltage (proportional to the charge) have the same sign, and it is being *discharged* otherwise.

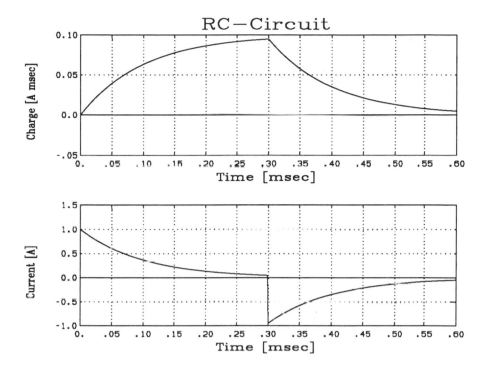

Figure 8.2. Trajectory behavior of the RC circuit.

Figure 8.3 shows the relation between current and voltage in the $[i_c, u_C]$ plane, which is sometimes referred to as the *phase plane* of this first–order system.

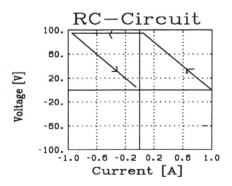

Figure 8.3. Trajectory in the phase plane.

It turns out that as long as we operate the capacitor in the first or third quadrant of its phase plane, power flows *into* the capacitor, otherwise it flows *out of* the capacitor.

This property actually holds not only for capacitors, but for *all* circuit elements: power flows into the element if the effort and flow variables have the same sign, and it flows out of the element otherwise.

Let us now look again at the bond graph describing our simple passive circuit that was shown in Fig.7.13. This bond graph is repeated in Fig.8.4.

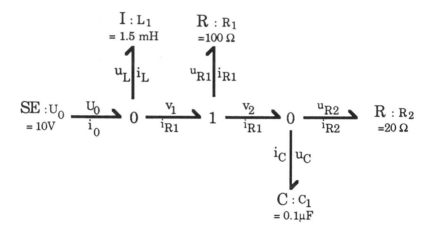

Figure 8.4. Bond graph of the passive circuit.

The directions of the harpoons were purposefully chosen such that they indicate the *direction of power flow*. The bond graph shows clearly how the power is generated in the voltage source and then spreads through the circuit and gets absorbed by the passive components. Of course, in case of an oscillation, the power flow in the capacitor and inductor can temporarily be reversed, i.e., the capacitor or inductor can temporarily be operated in the second and fourth quadrants of their phase planes.

What happens with the power flow in a resistor? Obviously, since the voltage and the current in a resistor are proportional to each other, they both change their sign simultaneously, i.e., the phase plane plot of the resistor occupies the first and third quadrants only. Consequently, power can flow *into* the resistor, but never back out.

This property actually holds for all types of resistors, not only the linear ones. All resistors are represented in the $[f, e]$ plane (the phase plane) through (possibly nonlinear) functions that are located in the first and third quadrants of the phase plane exclusively. Figure 8.5 shows the nonlinear relationship between flow and effort variables in a turbulent hydraulic "resistor."

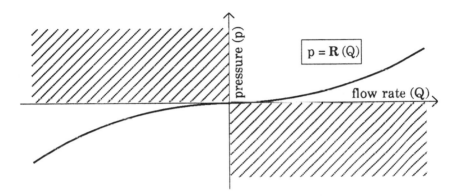

Figure 8.5. Power dissipation in a hydraulic system.

The electrical diode shown in Fig.6.5 is another example of a non-linear resistor.

How does this agree with the energy conservation law (the first law of thermodynamics), which states that, in every closed system, the total amount of energy must be preserved? This discrepancy cannot be solved within the concepts of electrical systems alone. However, we have no difficulty solving this problem if we consider the thermic behavior of the resistor. The resistor simply gets heated. As we see, our previously advocated resistor bond graph representation was actually incomplete. We should replace it by the enhanced bond graph representation shown in Fig.8.6 [8.18].

$$\frac{u_R}{i_R} \searrow RS \frac{T}{\frac{ds}{dt}} \searrow$$

Figure 8.6. Enhanced bond graph of a resistor.

Consequently, resistors are actually *two–port elements*. They have one electrical port and one thermal port. Power can flow from the electrical side to the thermal side, but never the other way around. On the thermal side, we use *temperature* as the effort variable and *entropy flow* as the flow variable. In many textbooks, the entropy flow is replaced by the heat flow, but this is not such a good idea since the product of temperature and heat flow is not of type power. Consequently, temperature and heat flow cannot be considered adjugate variables. The equations governing this enhanced resistor model are as follows:

$$u_R = R(\Delta T) \cdot i_R, \quad R(\Delta T) \approx R_0 + R_1 \cdot \Delta T + R_2 \cdot \Delta T^2 \quad (8.1a)$$

$$\dot{S} = \frac{1}{T} \cdot P_{elect}, \quad P_{elect} = u_R \cdot i_R \quad\quad\quad\quad (8.1b)$$

The relationship between the resistance R and temperature T is empirical. ΔT denotes the difference between the temperature of the resistor and the temperature of its environment, R_0 is the resistive value at room temperature, and R_1 and R_2 are the first and second temperature coefficients. Equation (8.1b) simply denotes the continuity of power flow through the resistor. The causalities on either side of the RS element is arbitrary, i.e., four different causalities exist that an RS element can assume.

Since the equations for the resistor are now different from those we used before, it has become customary among bond graphers to denote this enhanced resistor with the symbol RS rather than R. The S symbolizes the source character of this element's thermic side [8.18].

Obviously, the RS model can also be used to symbolize the heat produced in mechanical friction or any other related dissipative phenomenon.

What happens with the heat once it has been produced? Three separate physical phenomena provide mechanisms for heat transport or heat flow, namely, *conduction*, *convection*, and *radiation*. Let us discuss these three phenomena one at a time.

8.2 Thermal Conduction

Heat conduction occurs naturally whenever there is a gradient in temperature. Heat flows from the warmer to the colder spot in order to approach the thermal equilibrium. The reasons for this behavior require insight into the microscopic aspects of thermodynamics and will not be discussed here.

Let us analyze how heat is dissipated along a rod, one end of which is hot while the other is cold. The physical law that governs the thermal conduction along the rod is the heat equation:

$$\frac{\partial T}{\partial t} = \sigma \frac{\partial^2 T}{\partial x^2} \tag{8.2}$$

which is a *partial differential equation* (PDE) in the two independent variables t (time) and x (space). One way to approximately solve this PDE is by discretizing the space axis x while leaving the time axis t continuous. We can approximate the second derivative in space through:

$$\frac{\partial^2 T(t, x_k)}{\partial x^2} \approx \frac{T(t, x_{k+1}) - 2T(t, x_k) + T(t, x_{k-1})}{\Delta x^2} \tag{8.3}$$

where x_k denotes any particular value x and $x_{k\pm1}$ are abbreviations for $x \pm \Delta x$. By applying this transformation, the PDE is reduced to a set of ordinary differential equations (ODEs) of the type:

$$\frac{dT_k(t)}{dt} = \frac{\sigma}{\Delta x^2}[T_{k+1}(t) - 2T_k(t) + T_{k-1}(t)], \quad k \in \{1\}, \dots, \{n\} \tag{8.4}$$

where $T_k(t)$ denotes the temperature T at $x = x_k$ as a function of time. Now we are back in business since we already know how to solve a set of ODEs. This technique is usually referred to as the *method of lines*.

Let us now look at one such equation, namely, that for $k = i$.

$$\left(\frac{\Delta x^2}{\sigma}\right) \cdot \frac{dT_i}{dt} = T_{i+1} - 2T_i + T_{i-1} \tag{8.5}$$

Since T_i is an effort variable, this equation looks exactly like the electrical circuit equation that would result from modeling the circuit shown in Fig.8.7

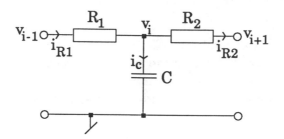

Figure 8.7. Electrical circuit diagram of an RC 'T' link.

since:

$$C \cdot \frac{dv_i}{dt} = i_C, \quad i_C = i_{R1} - i_{R2} \tag{8.6a}$$

$$i_{R1} = \frac{v_{i-1} - v_i}{R_1}, \quad i_{R2} = \frac{v_i - v_{i+1}}{R_2} \tag{8.6b}$$

Comparison of coefficients suggests that:

$$C = \frac{\Delta x^2}{\sigma}, \quad R_1 = R_2 = 1.0 \tag{8.7}$$

However, we could also multiply the constant $\Delta x^2/\sigma$ to the other side of the equal sign and then we find that:

$$C = 1.0, \quad R_1 = R_2 = \frac{\Delta x^2}{\sigma} \tag{8.8}$$

Obviously, the analogy is not completely determined. Any combination of Rs and Cs is possible as long as:

$$R_1 \cdot C = R_2 \cdot C = \frac{\Delta x^2}{\sigma} \tag{8.9}$$

It is obvious that we can model the entire chain of ODEs through the electrical circuit analogon shown in Fig.8.8.

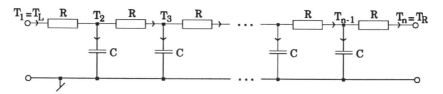

Figure 8.8. Electrical circuit analogon of a diffusion chain.

A bond graph representation of this RC chain is shown in Fig.8.9

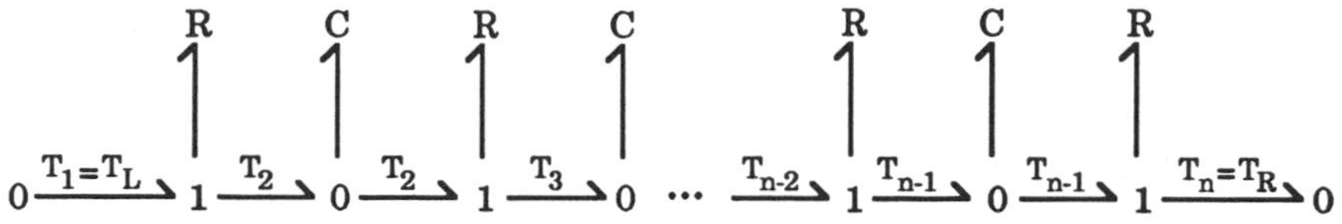

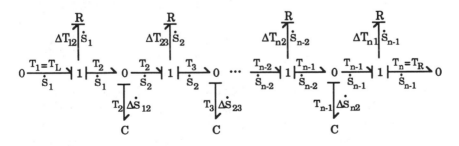

Figure 8.9. Bond graph representation of the diffusion chain.

This representation would be perfectly correct if all we wanted was to model the temperature distribution along the rod as a function of time. However, our rod lives in an environment and we must somehow model the interaction of our rod with its environment by applying appropriate boundary conditions. Here, we run into difficulties with our approach. How do we attach an entropy source to the hot end of the rod since we threw out entropy as a variable in our model altogether? Remember that in order to represent a physical system correctly, it is insufficient to model it through a set of individual signals. The most important property of any physical system is the fact that it conserves energy, but energy was thrown out from our "model" altogether. Figure 8.10 shows the same bond graph, but this time the effort and flow variables have been appropriately named, and the causalities have been introduced. We shall find that the energy conservation requirement will present us with the missing condition to find a unique formula for the correct distribution of the $\Delta x^2/\sigma$ factor between the R and C elements.

Figure 8.10. Causal bond graph of the diffusion chain.

According to the first law of thermodynamics, the total energy E_t of a closed system is constant. The total energy is defined as the sum of the free energy E_f and the thermal energy or heat Q:

$$E_t = E_f + Q \tag{8.10}$$

The free energy is the sum of all types of energy except for the thermal one.

Equation (8.10) can be reformulated in terms of differentials. If the total energy of a closed system is constant, then the power flow into and out of the system must be zero:

$$\dot{E}_f + \dot{Q} = 0.0 \tag{8.11}$$

We shall define the *entropy flow* $\dot{S}$ of a body as its heat flow $\dot{Q}$ divided by its temperature:

$$\dot{S} = \frac{\dot{Q}}{T} \tag{8.12}$$

Notice that our definition of entropy flow is not totally in accordance with the usual thermodynamic definition of entropy. However, it is commonly used among bond graphers, and it is useful for our purpose.

The capacity of a body to transport heat in a dissipative manner can be described by the equation:

$$\Delta T = \theta \cdot \dot{Q} = (\theta \cdot T) \cdot \dot{S} \tag{8.13}$$

where θ is the thermal resistance of the body. This looks very much like Ohm's law; thus we can also write:

$$\Delta T = R \cdot \dot{S}, \quad R = \theta \cdot T \tag{8.14}$$

Here, the reason why many textbooks prefer to use the heat flow $\dot{Q}$ rather than the entropy flow $\dot{S}$ as the bond graph's flow variable becomes evident: using heat flow, the resistor assumes a constant value, while using entropy flow, the resistor is modulated (multiplied) with the effort variable T. However, using DYMOLA [8.6] as a modeling tool, this does not cause any problem and is therefore a small price to pay for the convenience of dealing with true power flows.

For a rod of length ℓ and cross section A, we find that the thermal resistance can be written as:

$$\theta = \left(\frac{1}{\lambda}\right) \cdot \left(\frac{\ell}{A}\right) \tag{8.15}$$

where λ denotes the specific thermal conductance of the material. Again, this looks the same as the corresponding equation for a rod–shaped electrical resistor. Since we have cut our rod into small segments of length Δx, we replace Eq.(8.15) by:

$$R = \theta \cdot T = \frac{\Delta x \cdot T}{\lambda \cdot A} \qquad (8.16)$$

The capacity of a body to store heat is expressed through the equation:

$$\Delta \dot{Q} = \gamma \frac{dT}{dt} \qquad (8.17)$$

where γ denotes the thermal capacitance of the body. This equation can also be written as:

$$\Delta \dot{S} = C \frac{dT}{dt}, \quad C = \frac{\gamma}{T} \qquad (8.18)$$

Notice that the terms thermal resistance and thermal capacitance were traditionally introduced for the relationship between temperature and heat, not for the relationship between temperature and entropy, which is truly regrettable. The thermal capacitance of a body can be written as:

$$\gamma = c \cdot m \qquad (8.19)$$

where m is the mass of the body and c is the specific thermal capacitance of the material. The mass can further be written as the product of density ρ and volume V:

$$m = \rho \cdot V \qquad (8.20)$$

and, for our rod segment:

$$V = A \cdot \Delta x \qquad (8.21)$$

and thus:

$$C = \frac{\gamma}{T} = \frac{c \cdot \rho \cdot A \cdot \Delta x}{T} \qquad (8.22)$$

Consequently, we can determine the time constant of our diffusion equation to be:

$$R \cdot C = \theta \cdot \gamma = \frac{c \cdot \rho}{\lambda} \Delta x^2 \qquad (8.23)$$

or

$$\sigma = \frac{\lambda}{c \cdot \rho} \tag{8.24}$$

At this point, we know how to dimension both the resistive and capacitive elements in our bond graph in order to reflect the physicality of our equations.

However, the bond graph shown in Fig.8.10 still exhibits a small problem. As in the electrical case, we seem to have resistances that dissipate heat and thereby lose energy. This is a rather dubious concept. Since we are dealing with thermic variables throughout, we have a problem. Where is the dissipated heat flowing to? Since the dissipated power cannot vanish, we simply reintroduce it right away at the next node, as shown in Fig.8.11.

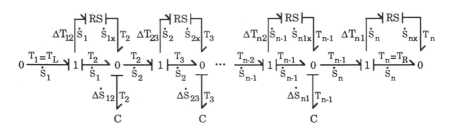

Figure 8.11. Corrected causal bond graph for the diffusion chain.

Therefore, the ith computational cell of our diffusion chain can be described through the following set of equations:

$$\frac{dT_i}{dt} = \frac{1}{C} \Delta \dot{S}_i \tag{8.25a}$$

$$\Delta T_i = T_{i-1} - T_i \tag{8.25b}$$

$$\dot{S}_{i-1} = \frac{1}{R} \Delta T_i \tag{8.25c}$$

$$\dot{S}_{ix} = \dot{S}_{i-1} \frac{\Delta T_i}{T_i} \tag{8.25d}$$

$$\Delta \dot{S}_i = \dot{S}_{i-1} + \dot{S}_{ix} - \dot{S}_i \tag{8.25e}$$

where T_{i-1} is being determined by the computational cell to the left, while $\dot{S}_i$ is being determined by the computational cell to the right.

Notice that our modified bond graph is no longer exactly equivalent to the electrical circuit analogon. While the electrical circuit is able to represent the temperature distribution correctly, it fails to represent the power flow adequately.

The bond graph shown in Fig.8.11 is not symmetrical, i.e., it favors heat flow from the left to the right. This is, of course just an approximation to what is really going on in the distributed parameter system. We could have decided to reintroduce the lost heat one element further left instead of one element further right or we could have split the RS element into two equal parts, one turning left and the other turning right. However, the last alternative is not such a desirable choice since it introduces algebraic loops, as can be easily verified since we have now some freedom in assigning the causalities at the RS elements, as shown in Figs.8.12a–b.

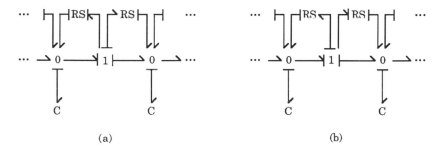

(a)   (b)

Figure 8.12a–b. Causality choices in symmetric causal bond graph.

It is usually a good idea to bias the RS elements away from the hot and toward the cold end of the rod.

How about the boundary conditions? Let me first assume that the rod being modeled is the electrical resistor itself. In that case, we must cut our electrical resistor into small resistors of length Δx, which are connected in series. Each one of these resistors contains a small entropy source that introduces entropy into the thermal network as shown in Fig.8.13.

A series connection of resistors is represented in the bond graph as a set of resistors all attached to the same 1–junction which, for topological reasons, has been split into several 1–junctions. However, the harpoons between those junctions were left out to symbolize that these represent, in reality, one and the same junction. Since the capacitances determine the temperatures at each 0–junction, the thermal input at each of these junctions must assume the causality of an entropy source rather than that of a temperature source. Had we decided to introduce the external power at the 1–junctions instead of at the 0–junctions, we would have introduced algebraic loops, as can be easily verified. Also, it is important to introduce the electrical power in the form of a current source rather than in the form of a voltage source in order to avoid the creation of another algebraic loop.

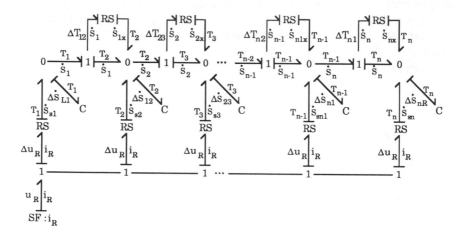

Figure 8.13. Causal bond graph for the heated rod resistor.

If the resistor represents a source of heat or temperature at one end of a narrow and radially well–insulated thin air channel (say, my meerschaum pipe), then the entropy source is simply introduced at the hot end of the rod, as shown in Fig.8.14.

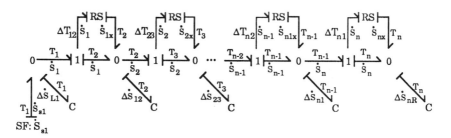

Figure 8.14. Causal bond graph for the heated air channel.

If we decide to introduce the external power at the nearest 0–junction, we must use an entropy source in order to avoid a structural singularity. Had we decided to introduce the external power at the nearest 1–junction instead, we could have used either type of power source. Distributed parameter systems often provide us with some flexibility, which can be exploited in avoiding algebraic loops and/or structural singularities.

The concepts shown so far can easily be extended to multidimensional heat flow problems. In this case, the heat equation is modified as follows:

$$\frac{\partial T}{\partial t} = \sigma \cdot \nabla^2 T \qquad (8.2^{alt})$$

where ∇^2, for the three–dimensional case, is the Laplacian operator:

$$\nabla^2 = \frac{\partial^2}{\partial x^2} + \frac{\partial^2}{\partial y^2} + \frac{\partial^2}{\partial z^2} \qquad (8.26)$$

Let us discuss the two–dimensional case. In order to apply the method of lines approach to an n–dimensional problem, we always discretize $n-1$ independent variables and leave one variable (usually the time t) continuous. Consequently:

$$\nabla^2 T(t, x_k, y_k) \approx \frac{T(t, x_{k+1}, y_k) - 2T(t, x_k, y_k) + T(t, x_{k-1}, y_k)}{\Delta x^2} +$$
$$\frac{T(t, x_k, y_{k+1}) - 2T(t, x_k, y_k) + T(t, x_k, y_{k-1})}{\Delta y^2} \qquad (8.27)$$

which leads to the electrical circuit analogon shown in Fig.8.15, as can be easily verified.

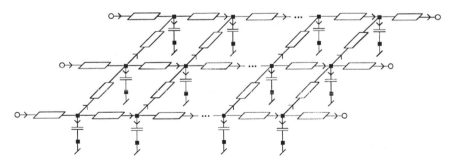

Figure 8.15. Electrical analogon for a two–dimensional diffusion.

Such electrical circuit analogies for the representation of distributed parameter systems are sometimes referred to as *Beuken models* [8.7].

I refrain from presenting here the correct bond graph for the two–dimensional case, since the graph would be rather busy. However, generating such a bond graph is not conceptually difficult, just a little more work.

At this point, one may jump to the conclusion that the bond graph representation is necessarily more complicated than the "equivalent" electrical analogon, but this is not really the case. Remember that the bond graph is not simply another representation scheme for the same thing. It provides more information than the electrical circuit diagram since it represents the topological as well as computational structure of the system, and moreover, since the electrical circuit diagram is even incorrect with respect to the representation of the power flow in the system.

Let us model a computational cell of a three–dimensional heat flow model in DYMOLA. [8.6] The DYMOLA equivalent of the RS element is presented here.

```
model type RS
    cut A(e1/f1), B(e2/−f2)
    main cut C[A, B]
    main path P < A − B >
    parameter R = 1.0
    R ∗ f1 = e1
    e1 ∗ f1 = e2 ∗ f2
end
```

However, RS elements with a thermal primary side must be modeled using a modulated RS element, since the resistance is multiplied with the temperature (i.e., the effort variable).

```
model type mRS
    cut A(e1/f1), B(e2/−f2)
    main cut C[A, B]
    main path P < A − B >
    parameter theta = 1.0
    R = theta ∗ e2
    R ∗ f1 = e1
    e1 ∗ f1 = e2 ∗ f2
end
```

Notice that the modulation uses the secondary effort e_2 and not the primary effort e_1, since e_2 denotes an absolute temperature whereas e_1 is a temperature difference.

Similarly, thermal capacitances must also be "modulated," since the capacitor must be divided by the temperature.

```
model type mC
    main cut A(e/f)
    parameter gamma = 1.0
    C = gamma/e
    C∗der(e) = f
end
```

Notice that the "modulation" of a capacitance is a rather dubious undertaking. How do we ensure that the modulated capacitance is still an energy storage element and does not suddenly start to dissipate energy? This problem requires further contemplation.

The energy stored in a capacitor (or inductor) is the integrated power that flows into that capacitor (or inductor); thus:

$$E(t) = \int_0^t P(\tau)d\tau = \int_0^t e(\tau) \cdot f(\tau)d\tau \qquad (8.28)$$

whereby the energy for $t = 0.0$ has arbitrarily been normalized to zero. Using the formula for the general displacement (the charge) of the capacitor:

$$q(t) = \int_0^t f(\tau)d\tau \qquad (8.29)$$

we can write:

$$E(t) = \int_0^t e(\tau) \cdot \dot{q}(\tau)d\tau = \int_0^q e(q)dq \qquad (8.30)$$

Thus, in order for an element to behave like a capacitor, the effort e must be expressible as a (possibly nonlinear) function of q:

$$e_C = \Phi_C(q_C) \qquad (8.31)$$

Similarly, we can use the formula for the generalized momentum (the flux) of an inductor:

$$p(t) = \int_0^t e(\tau)d\tau \qquad (8.32)$$

Therefore, in order for an element to behave like an inductor, the flow f must be expressible as a (possibly nonlinear) function of p:

$$f_I = \Phi_I(p_I) \qquad (8.33)$$

Let us check whether our modulated capacitance satisfies Eq.(8.31). We know that:

$$f_C(t) = C \cdot \dot{e}_C(t) = \frac{\gamma}{e_C(t)} \cdot \dot{e}_C(t) \qquad (8.34)$$

and therefore:

$$q_C(t) = \int_0^t f_C(\tau)d\tau = \gamma \int_0^t \frac{\dot{e}_C(\tau)}{e_C(\tau)}d\tau = \gamma \cdot \log(e_C) \qquad (8.35)$$

The capacitive charge q_C is indeed a nonlinear function of the effort e_C, and the capacitive nature of our modulated capacitance has thus been verified.

Let us return to our heat flow problem now. We want to assume that each cell consists of one modulated capacitor, and three modulated resistors, namely, the one to its left, the one to its front, and the one below, as shown in Fig.8.16. Remember that, for DYMOLA modeling, bond graphs must be enhanced to avoid the necessity of attaching any elements to 1–junctions.

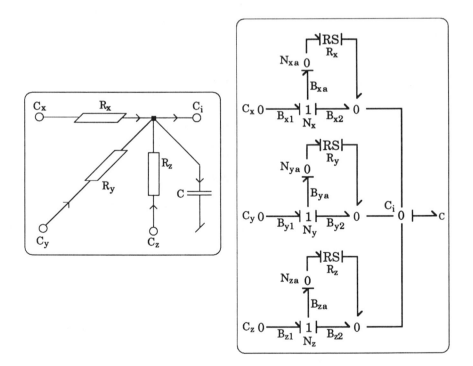

Figure 8.16. Three–dimensional diffusion cell.

The following DYMOLA model describes a three–dimensional diffusion cell:

model type *c3d*

> **submodel** (*mRS*) Rx, Ry, Rz
> **submodel** (*mC*) C(**ic** $e = 298.0$)
> **submodel** (*bond*) $Bx1$, $Bx2$, Bxa, $By1$, $By2$, Bya, $Bz1$, $Bz2$, Bza
> **node** Nx, Nxa, Ny, Nya, Nz, Nza
>
> **cut** $Cx(ex/fx)$, $Cy(ey/fy)$, $Cz(ez/fz)$, $Ci(ei/-fi)$
> **path** $Px < Cx - Ci >$, $Py < Cy - Ci >$, $Pz < Cz - Ci >$
>
> **connect** $Bx1$ **from** Cx **to** Nx
> **connect** $By1$ **from** Cy **to** Ny
> **connect** $Bz1$ **from** Cz **to** Nz
> **connect** $Bx2$ **from** Nx **to** Ci
> **connect** $By2$ **from** Ny **to** Ci
> **connect** $Bz2$ **from** Nz **to** Ci
> **connect** Bxa **from** Nx **to** Nxa
> **connect** Bya **from** Ny **to** Nya
> **connect** Bza **from** Nz **to** Nza
> **connect** Rx **from** Nxa **to** Ci
> **connect** Ry **from** Nya **to** Ci
> **connect** Rz **from** Nza **to** Ci
> **connect** C **at** Ci

end

Notice one additional new concept: While invoking the submodel C of model type mC, we assigned an initial condition of $T = 298.0$ K to the integrator inside the mC model.

Now, we can create a diffusion model that contains as many computational cells as we need, say the 27 *c3d* cells $C111, \ldots C333$, three in each direction. These cells are then best connected through their paths with statements such as:

> **connect** (Px) $C111 - C211 - C311$, $C112 - C212 - C312$
> **connect** (Py) $C111 - C121 - C131$, $C112 - C122 - C132$
> **connect** (Pz) $C111 - C112 - C113$, $C121 - C122 - C123$

Altogether, 27 such connections will be needed. Then, we must attach additional capacitors to each of the left–most, front–most, and bottom–most *c3d* cells, and set the entropy flow equal to zero at each of the right–most, back–most, and top–most cells. Finally, we may attach entropy sources to any of the cells as desired for the purpose of modeling thermal input.

8.3 Thermal Convection

So far, we have discussed only one mechanism of heat transfer, namely, *diffusion*, i.e., heat transfer that is caused by the microscopic motion of individual particles. However, heat is attached to matter, and a second form of heat transfer is by means of moving matter around on a macroscopic scale. Obviously, the heat stored in any moving piece of matter is transferred together with that matter. This physical phenomenon is called *convection*. It describes the transfer of heat as a result of macroscopic rather than microscopic movement.

Convection can occur autonomously. If we heat the floor of a room, the surrounding air expands and thereby its density is being reduced. Consequently, the hot air moves upward toward the ceiling, and is replaced by colder air that moves down toward the floor. This is a typical example of a convective phenomenon.

Convection can also be artificially induced. In a room that is heated through a warm water radiator, the convection can be increased by installing a fan that blows at the radiator. This fan will increase the air circulation in the vicinity of the radiator and thereby increase the convection.

Let us model convection mechanisms in a solar–heating system. Water is contained in a pipe that connects the water heater with the solar collector. At the solar collector, the water is heated by the sun (mostly through thermal radiation). The sun also feeds a solar battery, which drives a small pump that circulates the water from the water heater to the collector and back, i.e., the mechanism of transferring the heat from the solar collector to the water heater is primarily convective.

Let us assume that the pipe is free of air and that the water in the pipe is totally incompressible. Under this assumption, water will flow through the entire pipe with a constant velocity v_w as soon as the solar battery turns on the pump. Heat will thereby be transferred from one computational cell to the next.

Let q denote the hydraulic flow rate expressed in m^3 sec^{-1}. The volume of water in one computational cell is $V = A \cdot \Delta x$. Therefore, the amount of entropy that leaves the ith computational cell per unit time to the right is:

$$\dot{S}_{i\ out} = \Delta S_i \cdot \frac{q}{V} \tag{8.36}$$

which can also be written as:

$$\dot{S}_{i\ out} = (C \cdot \frac{q}{V}) \cdot T_i \tag{8.37}$$

In the same time unit, a similar amount of heat is transferred into the cell from its left neighbor:

$$\dot{S}_{i\ in} = (C \cdot \frac{q}{V}) \cdot T_{i-1} \tag{8.38}$$

These two equations can be combined into:

$$\dot{S}_{i\ conv} = G_{conv} \cdot \Delta T_i, \quad G_{conv} = C \cdot \frac{q}{V} \tag{8.39}$$

Consequently, the effect of the convection is simply a second convective resistance which is connected in parallel with the conductive resistance, i.e., convection simply increases the thermal conductivity.

Obviously, these equations contain a number of implicit simplifications. In reality, we ought to consider the friction between the liquid and the wall and the friction within the liquid. By doing so, we would see that the liquid flows faster at the center of the pipe and slower in the vicinity of the wall. The hydraulic friction is again a dissipative process that produces more heat and therefore should result in additional small entropy sources applied to the thermal model.

If we let go of the assumption of incompressibility, for example, if we model a gas flowing through the pipe rather than a liquid, the situation becomes much more complicated. Now we need to model the pneumatic process in addition to the thermal process. The pneumatic process will generate a time– and space–dependent flow rate $q(t, x)$ that can be used to modulate the convective resistance of the thermal model. The fluid dynamics model would also have to provide equations for the pneumatic dissipation, which could then be used to drive additional RS elements that bridge over from the pneumatic to the thermal subsystem. However, we shall not pursue this avenue.

8.4 Thermal Radiation

The third mechanism of heat transport is through thermal radiation, i.e., the emission/absorption of light. In order to fully understand the rationale behind thermal radiation, we would again need to look into the microscopic aspects of thermodynamics. However, we can analyze the macroscopic effects of thermal radiation using the law of Stefan–Boltzmann, which states that the emitted/absorbed radiation of a body is proportional to the fourth power of its temperature:

$$\mathcal{R} = \sigma \cdot T^4 \tag{8.40}$$

where σ is different from the σ that we met in thermal conduction. Here σ denotes the capability to emit and/or absorb light, which depends on the body's color. Black bodies emit/absorb much more strongly than white bodies — that is why there aren't so many dark painted cars here in Arizona. $\mathcal{R}$ is the emitted/absorbed power per unit surface. Consequently, this equation can be rewritten as:

$$\dot{Q} = \sigma \cdot A \cdot T^4 \tag{8.41}$$

where A denotes the emitting surface or in terms of the emitted/absorbed entropy:

$$\dot{S} = \sigma \cdot A \cdot T^3 \tag{8.42}$$

which is the version of the Stefan–Boltzmann law that I prefer. Also the radiation phenomenon is clearly dissipative. It can be described by yet another (nonlinear) R or RS element where:

$$R = \frac{T}{\dot{S}} = \frac{1}{\sigma \cdot A \cdot T^2} \tag{8.43}$$

which is again a modulated resistor. As with all other dissipative elements, the causality of this element is not predetermined. Emitted radiation usually uses temperature as a cause while absorbed radiation usually assumes the causality of an entropy source. Therefore, RS elements describing radiation are usually located between two 0–junctions, as shown in Fig.8.17:

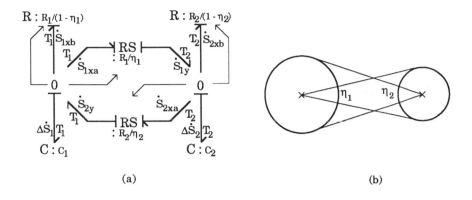

(a)　　　　　　　　　　　　　　　　　　　　(b)

Figure 8.17. Power exchange between two bodies through radiation.

η_1 denotes the percentage of the surface of the first body that radiates toward the second body and η_2 denotes the percentage of the surface of the second body that radiates toward the first. This is shown in Fig.8.17b. The thin full arrows are *signal paths*. They symbolize the temperature modulation of the dissipative elements. This bond graph leads to the following set of equations:

$$\dot{S}_{1xa} = \frac{1}{R_1/\eta_1} \cdot T_1 = \eta_1 \cdot \frac{1}{R_1} \cdot T_1 \tag{8.44a}$$

$$\dot{S}_{1xb} = (1 - \eta_1) \cdot \frac{1}{R_1} \cdot T_1 \tag{8.44b}$$

$$\dot{S}_{1y} = \frac{T_1}{T_2} \cdot \dot{S}_{1xa} \tag{8.44c}$$

$$\dot{S}_{2xa} = \eta_2 \cdot \frac{1}{R_2} \cdot T_2 \tag{8.44d}$$

$$\dot{S}_{2xb} = (1 - \eta_2) \cdot \frac{1}{R_2} \cdot T_2 \tag{8.44e}$$

$$\dot{S}_{2y} = \frac{T_2}{T_1} \cdot \dot{S}_{2xa} \tag{8.44f}$$

$$R_1 = \frac{1}{\sigma_1 \cdot A_1 \cdot T_1^2} \tag{8.44g}$$

$$R_2 = \frac{1}{\sigma_2 \cdot A_2 \cdot T_2^2} \tag{8.44h}$$

8.5 Thermal Inertance: The Missing Link

A strange discrepancy may be noticed in Table 7.1. The table suggests that thermodynamic systems don't possess a generalized momentum. Since the generalized momentum can always be expressed as the product of inertance and flow, this is equivalent to saying that thermodynamic systems don't possess inertance. Indeed, none of the bond graphs shown in this chapter exhibits any inertances.

It has been shown that the existence of a thermic inertance would be in contradiction with the second law of thermodynamics, which states that, in a closed system, the total entropy can never decrease [8.1]:

$$\dot{S}_{total} \geq 0 \qquad (8.45)$$

In reversible thermodynamics, according to Eq.(8.17), the total entropy is always kept in balance, while in irreversible thermodynamics [8.15,8.20], according to Eq.(8.13), the total entropy always grows, unless the RS element has two thermic ports, in which case the entropy stays in balance. One of the consequences of the second law is the fact that spontaneous heat flow can never occur from a point of lower temperature to a point of higher temperature. (A cold body can radiate heat to a hot body, but the radiation in reverse direction will always overcompensate, i.e., in the balance, heat still flows from the hot to the cold body.)

Now, let us assume that we found a thermic inertance. Its state equation would be:

$$\Delta T = I \cdot \frac{d\dot{S}}{dt} \qquad (8.46)$$

i.e., the thermic inertance can be used as a storage element for entropy flow, which indicates that a constant entropy flow may exist even for ΔT being equal to zero. This is clearly in contradiction with the second law, and thus, thermic inertances indeed do not exist.

Is this statement in contradiction with the observed heat wave (or "second sound" wave) in superfluid Helium (Helium–II) [8.11,8.13, 8.14]? It is correct that a linear system with *constant* R and C coefficients can never oscillate. However, this statement is no longer true when we allow the R and C parameters to be modulated by other variables. The thermal capacitance and resistance are strongly nonlinear in the vicinity of the so–called λ point, i.e., the point of transition between liquid Helium–I and superfluid Helium–II. One possible explanation for the second sound wave is thus through parameter modulation. Time–variant linear systems with constant negative real eigenvalues have been discovered which are unstable. It should be investigated whether the temperature dependencies of the R and C elements alone might explain the observed phenomenon. Another possible explanation for the second sound wave is through convection. If the hydraulic subsystem produces two oscillating flow rates q_i with opposite phases, then these flow rates could modulate the convective resistance in such a way that a heat wave can indeed occur which is not detectable in the hydraulic subsystem. A nice research topic would be to formulate the so–called two–fluid theory in terms of the bond graph methodology and thereby develop a simulation model that reproduces the observed second sound wave while conserving both energy and mass in the system.

8.6 Irreversible Thermodynamics

One of the properties of state–space models is the fact that they allow us to "reverse time" (at least in a mathematical sense) in a trivial manner. Let me explain this concept. We want to discuss the general nonlinear state–space model:

$$\frac{d\mathbf{x}(t)}{dt} = \mathbf{f}(\mathbf{x}(t), \mathbf{u}(t), t) \qquad (8.47)$$

which we can simulate forward in time:

$$t : t_0 \longrightarrow t_f, \quad t_f > t_0$$

Now, we wish to "reverse time," and simulate the system backward in time:

$$\tau : t_f \longrightarrow t_0, \quad t_0 < t_f$$

such that the final state of the reversed model is the same as the initial state of the original model. Obviously, this can be achieved with the simple transformation:

$$\tau = t_0 + t_f - t \qquad (8.48)$$

and therefore:

$$\frac{dt}{d\tau} = -1 \tag{8.49}$$

Consequently, we can rewrite our state–space model in the new independent variable τ as:

$$\frac{d\mathbf{x}(\tau)}{d\tau} = \frac{d\mathbf{x}(t)}{dt} \cdot \frac{dt}{d\tau} = -\mathbf{f}(\mathbf{x}(\tau), \mathbf{u}(\tau), \tau) \tag{8.50}$$

In other words, by simply placing an inverter at the input of every integrator in the state–space model, we can reverse the time behavior of any state–space model.

Let us demonstrate this concept by means of a second–order autonomous time–invariant system, the famous Van–der–Pol oscillator, in which the time reversal can be easily visualized. This example will, in addition, teach us some basic properties of trajectory behavior of state–space models.

The Van–der–Pol oscillator is described by the following second–order differential equation:

$$\ddot{x} - \mu(1 - x^2)\dot{x} + x = 0 \tag{8.51}$$

By choosing the outputs of the two integrators as our two state variables:

$$x_1 = x, \quad x_2 = \dot{x}$$

we obtain the state–space model:

$$\dot{x}_1 = x_2 \tag{8.52a}$$
$$\dot{x}_2 = \mu(1 - x_1^2)x_2 - x_1 \tag{8.52b}$$

We coded this model in ACSL [8.12] (another trivial exercise) and ran six different simulations with various sets of initial conditions for the two state variables x_1 and x_2. Figure 8.18 shows the results of this simulation.

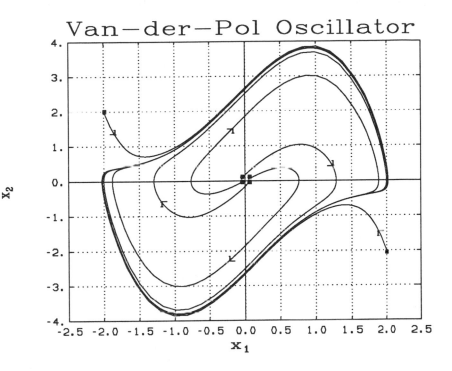

Figure 8.18. Trajectory behavior of the Van–der–Pol oscillator.

Figure 8.18 shows a phase portrait of x_2 plotted as a function of x_1. Four of the initial conditions were chosen very close to the origin, which is an unstable singularity. It becomes obvious that the solution is extremely sensitive to the initial condition when the initial condition is chosen close to the origin. The other two initial conditions were chosen far away from the origin. It turns out that a stable limit cycle exists that attracts all trajectories emanating from any point in the phase plane except for the origin itself.

We then applied the time reversal algorithm, which leads us to the following set of equations:

$$\dot{x}_1 = -x_2 \qquad\qquad (8.53a)$$
$$\dot{x}_2 = -\mu(1 - x_1^2)x_2 + x_1 \qquad\qquad (8.53b)$$

which we again simulated using four different initial conditions as shown in Fig.8.19. The former limit cycle is here not part of any trajectory. It was simply copied from Fig.8.18 for an easier comparison between the two figures.

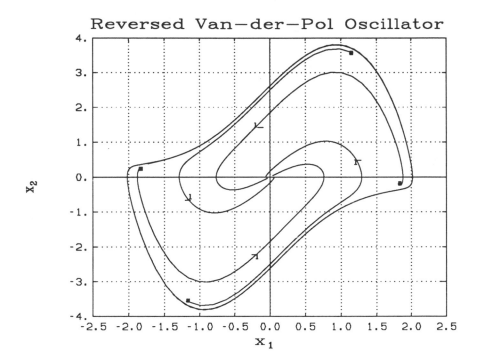

Figure 8.19. Trajectory behavior of the reversed Van–der–Pol equation.

The formerly unstable singularity at the origin has now become a stable singularity, a so–called *attractor*. The formerly stable limit cycle has turned into an unstable limit cycle. Trajectories emanating from anywhere inside the limit cycle are attracted by the singularity at the origin, whereas trajectories emanating from anywhere outside the limit cycle escape to infinity. The limit cycle has thus become a border line between two domains, the region of stability (the domain of attraction) of the singularity at the origin and the unstable domain encompassing it.

It can be observed that, in the phase plane, the two sets of time trajectories look the same, only the direction of the arrows has been reversed.

We also ran the following experiment. Starting from the initial condition $[x_1, x_2] = [0.1, 0.1]$, we simulated the original Van–der–Pol equation over 2 sec. Thereafter we reversed the sign of the inputs to the two integrators and continued to simulate the system for another 2 sec. This ACSL [8.12] program is a little more interesting, therefore, let me write down the code:

```
PROGRAM oscillator
INITIAL
    constant tmx = 4.0, x10 = 0.1, x20 = 0.1, c = 1.0
    cinterval cint = 0.05
    schedule revers .at. 2.0
END $ "of INITIAL"
DYNAMIC
    DERIVATIVE
        x1d = x2
        x2d = 2.0 * (1.0 - x1 * x1) * x2 - x1
        x1  = integ(c * x1d, x10)
        x2  = integ(c * x2d, x20)
    END $ "of DERIVATIVE"
    DISCRETE revers
        c = -1.0
    END $ "of DISCRETE revers"
    termt(t.ge.tmx)
END $ "of DYNAMIC"
END $ "of PROGRAM"
```

The time–event *revers* was used to model the discrete event of switching between the two modes of the simulation. Figure 8.20 shows the results of this simulation. The final values of the two state variables were again $[x_1, x_2] = [0.1, 0.1]$, correct up to eight digits, i.e., we have successfully reversed the model to the original initial conditions.

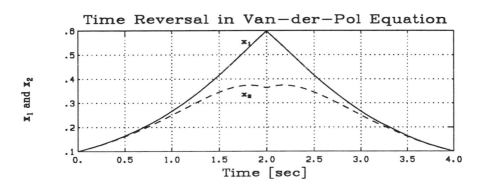

Figure 8.20. Time reversal for the Van–der–Pol equation.

However, let us see what this seeming time reversal really means.

Figure 8.21 shows the plot of the two state derivatives $\dot{x}_1$ and $\dot{x}_2$ over time.

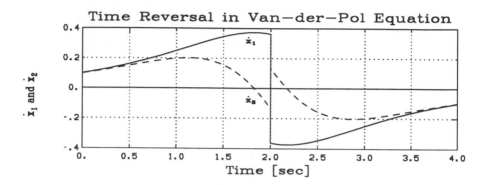

Figure 8.21. Time reversal for the Van–der–Pol equation.

Obviously, these two variables were not properly reversed. Time reversal, in a mathematical sense, means the reversal of all *state variables* in the model. However, the *state derivatives* are not reversed; they change their sign. This can be easily seen from the reversed state–space model.

In an electrical circuit, the most commonly used state variables are the currents through the inductances and the voltages across capacitors. Consequently, in a time reversal, those will remain the same. However, the voltages across inductances and the currents through capacitors will change their sign. Therefore, time reversal is an illusion.

Let us now repeat the experiment, but this time we shall simulate the original system forward in time over 20 sec before we apply time reversal.

The final state after 40 sec is now $[x_1, x_2] = [1.59 \times 10^{-5}, 1.59 \times 10^{-5}]$, i.e., completely wrong. Figure 8.22 shows the time trajectories of the two state variables. Shortly after time 20 sec, the time reversal seems to work fine, but thereafter, around time 28 sec, the trajectories suddenly deviate from their expected paths and skip one whole oscillatory cycle. What is the reason for this unexpected behavior?

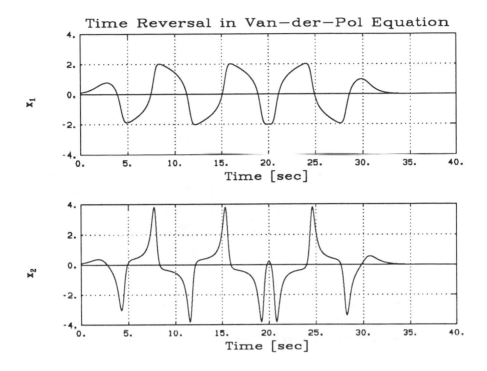

Figure 8.22. Time reversal for the Van–der–Pol equation.

The Van–der–Pol oscillator exhibits two types of singularities. The origin, i.e., the point $[x_1, x_2] = [0, 0]$, is an unstable equilibrium point. If we start at the origin, x_1 and x_2 will stay at the origin forever, i.e., the trajectory consists of a single point. However, the smallest disturbance away from the origin in any direction will make the trajectory move away from the origin, and approach the second singularity, which is a stable limit cycle.

It becomes immediately evident that the sensitivity of the trajectory to the initial condition grows larger and larger, the closer to the unstable singularity the initial condition is chosen. At the singularity itself, the sensitivity is infinite. This can be easily seen from Fig.8.18, where four of the initial conditions were chosen very close to the origin and the resulting trajectories were rather different. For all practical purposes, we can say that the trajectory behavior is *nondeterministic* if the initial condition is chosen in the vicinity of the unstable singularity.

If we reverse the time, the previously unstable equilibrium point turns into a stable equilibrium point (an attractor), whereas the previously stable limit cycle turns into an unstable limit cycle, i.e., the region inside the unstable limit cycle is now the *domain of attraction* of the stable equilibrium point. This provides us with an excellent technique to determine the domain of attraction of any stable equilibrium point of a second–order system. We simply reverse time and simulate the system until the trajectory becomes periodic, i.e., traverses its (meanwhile stable) limit cycle. If we start the reversed simulation from somewhere in the vicinity of the unstable limit cycle, the trajectory behavior is extremely sensitive to the precise choice of the initial condition, although, as Fig.8.22 demonstrates, the sensitivity does not have to become evident immediately.

In our example, since both types of simulations tend toward a singular solution, time becomes irreversible for all practical purposes if we wait long enough, i.e., if we perform one type of simulation for sufficiently long and then "forget" the trajectory that we just generated, we cannot hope to retrieve the trajectory by executing the reversed simulation starting from the final value of the original one.

It is true for all (arbitrarily nonlinear) systems, that the trajectories are always repelled by unstable singularities and attracted by stable singularities. For a long time, it was therefore believed that, if we wait long enough, all trajectories of autonomous systems do either approach stable singularity points, stable limit cycles, or escape to infinity. This is unfortunately not so. Other types of system behavior exist that are stable, nonstationary, and nonperiodic. These are called *chaotic motions*, and we shall analyze examples of such behavior in Chapter 10 of this text.

According to a number of highly reputed physicists, it is this type of time irreversibility that is at the origin of the seemingly stochastic microscopic behavior of thermic systems (the Brown movement), and that is ultimately responsible for the irreversibility condition that is expressed in the second law of thermodynamics [8.16]. This rationale was thought to explain the seeming discrepancy between the time reversibility of any state–space model and the obvious time irreversibility of thermic systems due to the second law. However, a much simpler explanation can be found for the time irreversibility of thermal systems.

Let us look a little more closely at the physics of time reversal. What does the mathematical operation of inverting the inputs of all integrators mean physically? For this purpose, let us analyze time reversal in the context of a simple electrical circuit. What does it mean in this context to invert the inputs of all integrators? Let us look at the simple state equation:

$$\frac{du_C}{dt} = \frac{1}{C} \cdot i_C \qquad (8.54)$$

Time reversal transforms this equation into:

$$\frac{du_C}{dt} = -\frac{1}{C} \cdot i_C \qquad (8.55)$$

and since we wish to retain the capacitive current i_C, we need to accept the necessity of a "negative capacitor," which is not exactly a physically sound concept. But let us accept this answer for the time being. Time reversal in a simple passive circuit is accomplished by replacing every capacitor in the circuit by a negative capacitor, and every inductor by a negative inductor. If we then let all sources run backward through time, we have achieved the desired time reversal in terms of the electrical variables, i.e., all voltages and currents in the circuit run backward in time.

However, since the currents and voltages over the resistors have not changed after the time reversal (according to our premises), power is still being dissipated by them and the resistors continue to heat up rather than cool down (in accordance with the second law). How can we get the thermal variables to reverse as well? Mathematically, the answer is straightforward. We simply must include dissipated power into our model as an additional state variable, i.e.:

$$\frac{dQ_{irr}}{dt} = P_{elect} = u_R \cdot i_R \qquad (8.56)$$

Time reversal turns this equation into:

$$\frac{dQ_{irr}}{dt} = -P_{elect} = -u_R \cdot i_R \qquad (8.57)$$

Unfortunately, this equation is physically simply wrong. The dissipated energy is the integral of the electrical power over the resistor, not the integral of the negative electrical power. We could defer the negative sign to either the resistive voltage or current (by introducing a negative resistance), but this would not be in accordance with the time reversal of the electrical variables.

What we learned is the fact that the operation of "simply placing an inverter in front of every integrator in the system" is not a physically meaningful proposition. This is what I meant when I wrote in Chapter 1 of this book about the danger of "falling in love with our model." Many mathematically correct manipulations can be applied to a model, which could never be applied to the real system because they violate its physicality conditions. Therefore, the mathematically feasible time reversal of state–space models is not in contradiction with the second law of thermodynamics. The facts are much simpler than that. The mathematical operation of time reversal simply violates the physicality of the model.

8.7 Summary

In this chapter, we discussed thermodynamics from a systemic rather than a phenomenological viewpoint. We have seen that bond graphs present us with a tool to ensure adherence to physicality in modeling thermodynamic systems, and we have seen by means of an extended (drastic) example what can happen if physicality is ignored in the process of model manipulations. Bond graphs are not just another tool for mathematical modeling. In fact, bond graphs are quite meaningless when applied to the description of mathematical equations bare of their physical interpretation. It is therefore not currently feasible to apply bond graphs to the description of a macroeconomic model, for example, since we don't know what energy conservation means in such a model. What does economic power mean in a system theoretical rather than in a political sense? We don't know. Consequently, we cannot define a set of adjugate variables that describe the behavior of a macroeconomy.

However, I would like to go one step further. While I cannot prove this to be correct, I am personally convinced that any real system that can meaningfully be described by a differential equation model (and macroeconomic systems are among those without any question), possesses some sort of "energy" that obeys the law of energy conservation. It is just that, to my knowledge, nobody has ever looked into systems, such as macroeconomies, from quite that perspective and tried to formulate a meaningful and consistent definition of the terms "energy" and "power," and from there derived a set of adjugate variables, the product of which is "power." This would be a very worthwhile topic for a Ph.D. dissertation.

Bond graphs have been successfully applied to most areas of physical systems that result in ordinary differential equation models. In this chapter, we have seen a meaningful application of bond graphs to one case of a partial differential equation as well, namely, the diffusion equation Eq.(8.2). However, bond graphs have not yet been successfully applied to most other types of PDE models, such as those occurring in fluid dynamics [8.3,8.4]. The reason is again a simple one. Most systems governed by PDEs require more than energy conservation. Liquids can change their shape while they still preserve their volume. Gases don't even preserve their volume. Yet all these systems conserve mass. Consequently, we must ensure that our models not only conserve energy but also mass. In a bond graph, mass appears as a parameter. Distributed pneumatic systems would therefore have to be described through modulated parameters, a technique that does not guarantee that the mass is properly conserved. It is currently unclear how we can design a systematic methodology, a generalized bond graph maybe, that ensures conservation of both energy and mass simultaneously. Such an investigation might therefore be a fruitful topic for another Ph.D. dissertation — not exactly an easy task either. I shall pursue this avenue quite a bit further in Chapter 9, but the final answer to this question has certainly not yet been given.

Notice that the theoretical foundations from which the bond graph methodology was derived are much deeper than I was willing to reveal in this book. In particular, notice the similarity between the concept of adjugate variables as used in the bond graph approach to the modeling of physical systems and the adjugate variables that were briefly introduced in Chapter 4 relating to the Hamiltonian of a system, and yet, the Hamiltonian was only defined for conservative (i.e., nondissipative) systems, while bond graphs extend naturally to cover dissipative processes as well. This similarity is not accidental and needs to be explored further. The question of finding a generalized bond graph may be just another formulation of the desire to find a generalized Hamiltonian, a hot topic among applied mathematicians interested in the study of dynamical systems.

Finally, I would like to acknowledge the contributions of Peter Breedveld of the University of Twente (Enschede, The Netherlands). His insight into the principles of thermodynamics and into the bond graph methodology, as expressed in his Ph.D. dissertation [8.1], were essential to my understanding of the material presented in this chapter.

References

[8.1] Peter C. Breedveld (1984), *Physical Systems Theory in Terms of Bond Graphs*, Ph.D. dissertation, University of Twente, Enschede, The Netherlands.

[8.2] François E. Cellier (1990), "Hierarchical Nonlinear Bond Graphs — A Unified Methodology for Modeling Complex Physical Systems," *Proceedings European Simulation MultiConference*, Nürnberg, F.R.G., pp. 1–13.

[8.3] N. Curle and Hubert J. Davies (1968,1971), *Modern Fluid Dynamics*, two volumes, Van Nostrand Reinhold, London, U.K.

[8.4] Iain G. Currie (1974), *Fundamental Mechanics of Fluids*, McGraw–Hill, New York.

[8.5] John A. Duffie and William A. Beckman (1980), *Solar Engineering of Thermal Processes*, John Wiley, New York.

[8.6] Hilding Elmqvist (1978), *A Structured Model Language for Large Continuous Systems*, Ph.D. dissertation, Report CODEN: LUTFD2/(TRFT–1015), Dept. of Automatic Control, Lund Institute of Technology, Lund, Sweden.

[8.7] International Colloquium on Field Simulation (1976), Proceedings of the Fourth International Colloquium on the Beuken Model, September 1974, Polytechnic of Central London, London, U.K.

[8.8] Aharon Katzir–Katchalsky and Peter F. Curran (1965), *Nonequilibrium Thermodynamics in Biophysics*, Harvard University Press, Cambridge, Mass.

[8.9] Bernard H. Lavenda (1985), *Nonequilibrium Statistical Thermodynamics*, John Wiley, New York.

[8.10] Derek F. Lawden (1987), *Principles of Thermodynamics and Statistical Mechanics*, John Wiley, New York.

[8.11] Fritz London (1954), *Superfluids — Volume II: Macroscopic Theory of Superfluid Helium*, John Wiley, New York.

[8.12] Edward E. L. Mitchell and Joseph S. Gauthier (1986), *ACSL: Advanced Continuous Simulation Language — User Guide and Reference Manual*, Mitchell & Gauthier Assoc., Concord, Mass.

[8.13] V. Peshkov (1944), "'Second Sound' in Helium II," *J. Phys. USSR*, **8**, p.381.

[8.14] V. Peshkov (1946), "Determination of the Velocity of Propagation of the Second Sound in Helium II," *J. Phys. USSR*, **10**, pp. 389–398.

[8.15] Ilya Prigogine (1967), *Thermodynamics of Irreversible Processes*, third edition, John Wiley Interscience, New York.

[8.16] Ilya Prigogine (1980), *From Being to Becoming: Time and Complexity in the Physical Sciences*, Freeman, San Francisco, Calif.

[8.17] Keith S. Stowe (1984), *Introduction to Statistical Mechanics and Thermodynamics*, John Wiley, New York.

[8.18] Jean U. Thoma (1975), "Entropy and Mass Flow for Energy Conversion," *J. Franklin Institute*, **299**(2), pp. 89–96.

[8.19] Clifford Truesdell (1984), *Rational Thermodynamics*, second edition, Springer–Verlag, New York.

[8.20] Y. L. Yao (1981), *Irreversible Thermodynamics*, Science Press, Beijing, distributed by Van Nostrand Reinhold, New York.

Bibliography

[B8.1] Albert M. Bos and Peter C. Breedveld (1985), "Update of the Bond Graph Bibliography," *J. Franklin Institute*, **319**(1/2), pp. 269–286.

[B8.2] Peter C. Breedveld, Ronald C. Rosenberg, and T. Zhou (1991), "Bibliography of Bond Graph Theory and Application," *J. Franklin Institute*, to appear.

[B8.3] Vernon D. Gebben (1979), "Bond Graph Bibliography," *J. Franklin Institute*, **308**(3), pp. 361–369.

Homework Problems

[H8.1] Heat Flow Along a Copper Rod

A copper rod of length $\ell = 1$ m and radius $r = 0.01$ m is originally in an equilibrium state at room temperature $T = 298.0$ K. At time $t = 0.0$ sec, the left end of the rod is brought in contact with a body that is kept at a temperature of $T_L = 390.0$ K.

Model the rod through a set of 10 one–dimensional cells. Model the boundary conditions through an effort source attached to the input port of the first segment and specify that the heat flow out of the output port of the last segment is zero. We want to assume that the heat transport in the rod occurs purely by means of diffusion and that the rod is so well insulated that no heat escapes through its surface.

The density of copper is $\rho = 8960.0$ kg m^{-3}, the specific thermal conductance is $\lambda = 401.0$ J m^{-1} sec^{-1} K^{-1}, and the specific thermal capacitance is $c = 386.0$ J kg^{-1} K^{-1}.

Simulate the system during $20,000.0$ sec and display the temperature in the middle and at the end of the rod.

[H8.2] Lightning Rod

A copper lightning rod of length $\ell = 5$ m and radius $r = 0.01$ m is originally in an equilibrium state at room temperature $T = 298.0$ K. At time $t = 0.0$ sec, the left end of the rod is hit by lightning, which results in a current flow of $I_0 = 200$ kA which lasts for a duration of $t_{pulse} = 75$ μsec.

Model the rod through a set of 10 one–dimensional cells with RS elements attached to each cell that represent the heat input through electrical dissipation. Model the boundary conditions through a flow source attached to the single electrical port and specify that the heat flow out of the output port of the last cell is zero. We want to assume that the heat transport in the rod occurs purely by means of diffusion and that the rod is so well insulated that no heat escapes through its surface.

The density of copper is $\rho = 8960.0$ kg m^{-3}, the specific thermal conductance is $\lambda = 401.0$ J m^{-1} sec^{-1} K^{-1}, the specific thermal capacitance is $c = 386.0$ J kg^{-1} K^{-1}, and the specific electrical resistance is $\rho_{el} = 1.7 \cdot 10^{-8}$ Ω m.

Simulate the system during $5 \cdot 10^{-4}$ sec, and display the temperature in the middle and at the end of the rod. Which is the maximum temperature increase that the lightning rod experiences?

Projects

[P8.1] Solar–Heated House

Figure P8.1a depicts a solar–heated house. One or several collectors act as black bodies which absorb incoming solar radiation. Consequently, the temperature inside the collectors rises. The collectors can be filled with any material with a large heat capacity. Usually, it is simply air. Inside the collectors, a water pipe meanders back and forth between the two ends of the collector, thereby maximizing the exposed pipe surface. We shall call this a "water spiral" [8.2]. A (mostly conductive) heat exchange takes place between the collector chamber and the water pipe, thereby heating the water in the pipe. A pump circulates the water from the collectors to the storage tank, thereby transporting the heat convectively from the collectors to the tank. We call this the "collector water loop" [8.2]. The water spirals in the various collectors can be either series–connected or connected in parallel. The pump is usually driven by a solar panel. In the panel, the solar light is converted to electricity, which drives the pump. Thereby, the pump circulates the water only while the sun is shining, which is what we want. In addition, a freeze protection device is often installed that also switches the pump on whenever the outside temperature falls below 5°C.

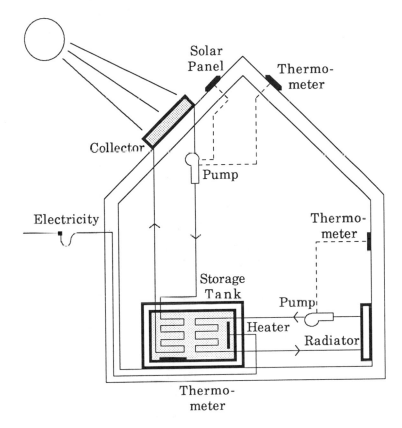

Figure P8.1a. A solar–heated house.

The storage tank is often realized simply as a large well–insulated water container (a water heater). However, such a solution would get us into mixing thermodynamics and may be a little difficult to model at this point. Therefore, we shall assume that a solid body storage tank is used together with another water spiral that deposits the heat in the storage tank the same way as it was picked up in the collectors. Consequently, the water from the collector loop and from the heater loop never mix.

A second water spiral inside the storage tank belongs to the heater water loop. It picks up the heat from the storage tank. An additional electrical heater is also installed that heats the storage tank electrically whenever the storage tank temperature falls below a critical value, but does so only during night hours when electricity is cheap.

The heater water loop is driven by another pump that is switched on whenever the room temperature falls below 20°C during the day or 18°C during the night, and which is switched off whenever the room temperature rises beyond 22°C during the day or 20°C during the night.

In the house, we use one or several radiators (more water spirals) which, contrary to what their name suggests, exchange heat with the room in a partly conductive and partly convective manner.

Figure P8.1b depicts the collector in more detail.

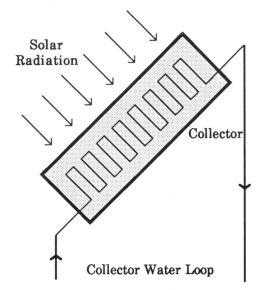

Figure P8.1b. The solar collector.

The water spiral is modeled through a series of one–dimensional cells as introduced in this chapter. We want to model each such cell as shown in Fig.P8.1c.

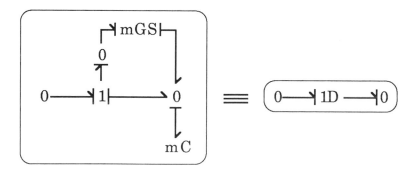

Figure P8.1c. Bond graph of a one–dimensional cell.

Each cell is described by a DYMOLA model type called *c1d.dym* which, from now on, can be used as an additional bond graph element. The correct causalities have been marked on the graph. The *mGS* element is a "modulated conductive source." It is modulated with temperature (as always in thermal systems), but it is also modulated with the water velocity in the pipe, as shown in Fig.P8.1d. Since the conductance changes linearly with the water velocity v_w, it was preferred to model this element through its conductance rather than through its resistance.

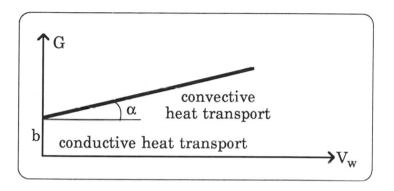

Figure P8.1d. Modulated conductive source.

The *c1d* model references three submodels, a temperature–modulated capacitance *mC*, a temperature and water velocity–modulated conductive source *mGS*, and finally the regular *bond* submodel. Remember to augment the bond graph by additional 0–junctions to ensure that all elements are attached to 0–junctions only.

Figure P8.1e shows the heat exchanger model that is used to describe the exchange of heat across the border of two media.

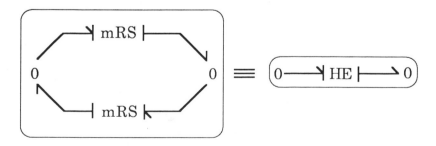

Figure P8.1e. Bond graph of a heat exchanger.

The heat exchanger is used here to model the transfer of heat from the collector chamber to the water spiral.

The water spiral is modeled through a series connection of several *c1d* elements with heat exchangers attached in between. Figure P8.1f shows the water spiral. We decided to cut the spiral into three discrete links. Obviously, this is an approximation of a process with distributed parameters.

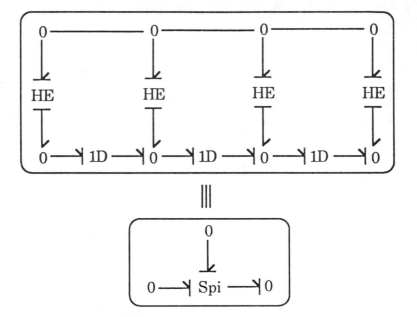

Figure P8.1f. Bond graph of a water spiral.

Notice that the newly introduced bond graph symbol representing the water spiral is a three–port element.

We need to model also the loss from the collector chamber to the environment. This loss is partly conductive and partly convective. Figure P8.1g depicts the loss element (a one–port element).

Figure P8.1g. Bond graph of thermic loss.

The effort source denotes the outside temperature. The mG element denotes the heat dissipation to the environment. The dissipated heat is proportional to the difference in temperatures between the inside and outside. mG is a modulated conductance similar to the mGS element found earlier, but this time, the secondary port (the environment) is not modeled, and the modulation is now with respect to the wind velocity v_{wind} rather than with respect to the water velocity v_w.

We are now ready to model the overall collector. Figure P8.1h shows the overall collector.

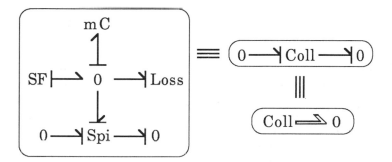

Figure P8.1h. Bond graph of the collector.

The mC element is the (temperature–modulated) heat capacitance of the collector chamber. The SF element denotes the heat input from solar radiation.

We use the hierarchical cut concept of DYMOLA to combine the two cuts (i.e., bonds), *inwater* and *outwater*, into one hierarchical cut, *water*. This can be pictorially represented by a double bond. This *aggregated bond graph* representation has, of course, the disadvantage that causalities can no longer be depicted.

Let us now model the heat input from the solar radiation. Figure P8.1i shows a typical heat flow curve over a period of three days.

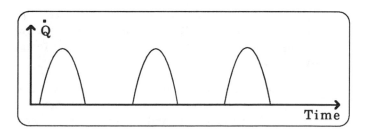

Figure P8.1i. Heat flow from solar radiation.

Of course, the curve changes somewhat with the time of the year and the location on the globe (the latitude). We wish to model the heat flow correctly for an arbitrary time of the year and an arbitrary latitude. For this purpose, create a model that describes the motion of planet Earth around the sun. Neglect the influence of the moon and the other planets. The output of that model will be the celestial coordinates (declination and right ascension) of planet Earth as a function of the time of the year expressed in sun–centered coordinates. Create a second model that converts these coordinates to the celestial position of the sun expressed in Earth–centered coordinates. Create a third model that, for any position on our globe, converts the celestial coordinates into surface–bound coordinates, i.e., use the sidereal time to convert the right ascension of the sun to its hour angle equivalent.

By now, we know the apparent position of the sun for any latitude, for any day of the year, and for any time of the day. We need to convert this to the available solar heat flow. This depends on the angle of the sun above the horizon, or more precisely, on the thickness of the atmospheric layer that the solar rays must travel through before reaching the surface. This function has been tabulated and can be found in the literature. Create a model that converts the solar position to the available heat flow (assuming optimal visibility and a spotless blue sky).

At this point, we know how much solar heat is available per unit time and per visible unit surface. The visible surface of the collector depends on its physical surface and its position. Let us assume that we operate with one flat collector. We shall certainly position the collector exactly toward the south when located anywhere on the northern hemisphere, and exactly toward the north when located on the southern hemisphere. The optimal slanting angle with the horizontal depends on the latitude. At the equator, the optimal angle is $0°$, at the pole it is $90°$. It turns out that a good choice for the slanting angle is the latitude itself. Standard collectors come in sizes of 1 m × 2 m. Create a model that converts the available solar heat to effectively used solar heat per optimally positioned but fixed flat collector at any given latitude, on any given day of the year, and at any given time of the day.

Since heat flow is the product of entropy flow and temperature, we can divide the effectively used heat by the collector temperature and model the resulting entropy flow into the collector as a (time– and temperature–modulated) heat source.

We are now ready to model the transport of heat from the collector to the storage tank, i.e., the collector water loop. We model each of the pipes through a series of one–dimensional cells and we shall assume that the pipes are thermally well insulated, i.e., that no heat is lost to the environment on the way. Figure P8.1j depicts the water loop.

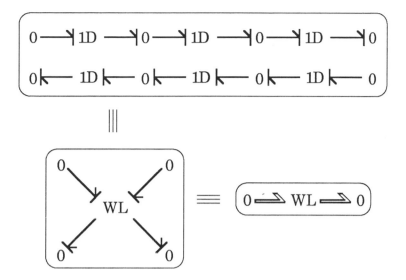

Figure P8.1j. Bond graph of the water loop.

This bond graph element is a four–port element. We shall combine the cut *inwater1* with the cut *outwater2* to the hierarchical cut *inwater*, and the cut *outwater1* with the cut *inwater2* to the hierarchical cut *outwater*. We shall furthermore declare a main path *water*, which creates a logical bridge from the hierarchical cut *inwater* to the hierarchical cut *outwater*.

The storage tank contains two water spirals, one that belongs to the collector water loop and one that belongs to the heater water loop. In addition, an electrical resistance heater has been installed as a backup device. The electrical heater is turned on only if the temperature in the storage tank falls below a critical value. Furthermore, the backup device is never used during daytime hours when the electricity is expensive; instead, we shall wait with electrically heating the storage tank until the evening hours when the price for electricity is lower. Figure P8.1k shows the storage tank.

The mC element denotes the heat capacity of the storage tank. The flow source together with the mRS element denote the electrical backup heater. The primary side of the resistive source is electrical while the secondary side is thermic.

This is another four–port element. This time, we shall combine the cut *inwater1* with the cut *outwater1* to the hierarchical cut *inwater*, and the cut *outwater2* with the cut *inwater2* to the hierarchical cut *outwater*. We shall again declare a main path *water*, which creates a logical bridge from the hierarchical cut *inwater* to the hierarchical cut *outwater*.

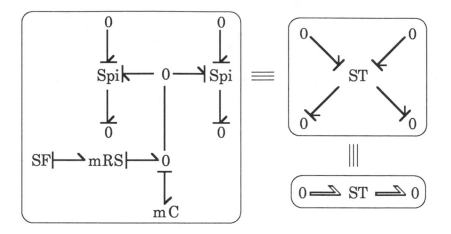

Figure P8.1k. Bond graph of the storage tank.

The heater water loop is modeled in exactly the same manner as the collector water loop.

Finally, let us discuss the house itself. For simplicity, we shall assume that the house is a cube of 10 m × 10 m × 10 m. We shall model the house with 27 three-dimensional cells, three in each direction. Let us assume that the entire house consists of one room only, and that a single (large) radiator is used to heat the house. The radiator is attached to the left wall of the house somewhere close to the floor, i.e., heat input occurs at the left low outside center three–dimensional cell. We shall not model the radiator explicitly since it is much smaller in dimensions than the house itself. Therefore, we shall simply connect the *outwater1* of the heater water loop with the *inwater2* of the heater water loop. At this node, we attach another heat exchanger responsible for the exchange of heat between the heater water loop and the house.

We shall also assume that the house loses some heat through the four walls and the roof, but not through the floor. Attach *Loss* elements to each of the 0–junctions as appropriate. If a cell is adjacent to two or three outside walls, attach one combined loss element to the corresponding node only since otherwise an algebraic loop will occur.

This concludes the description of the system. Figure P8.1l depicts the overall system as a series connection of the previously presented aggregated bond graph elements.

Figure P8.11. Aggregated bond graph of the overall system.

Since in a solar–heating system, we pay for the installed energy rather than for the used energy, such a heating system is only economical in a climate with extended heating periods combined with lots of sunshine. Let us assume that our solar house is situated in Denver, Colorado. We wish to analyze the effectiveness of our heating system. For this purpose, we study the behavior of our heating system for December 21, the beginning of winter. The sun shines from a spotless blue sky. The outside temperature is time–dependent. The time dependence can be modeled with a sine function. The low temperature is $-10°C$. It is reached at 3:00 a.m. The high temperature is $+5°C$. It is reached at 3:00 p.m.

We first want to determine the critical temperature of the storage tank. For ths purpose, we simulate the heating of the house with a constant storage tank temperature. Let us assume that our initial temperature in the house is 15°C. We should be able to heat the house to 20°C within four hours. Determine the size of the radiator (i.e., the conductance of the heat exchanger between the heater water loop and the house), and an appropriate storage tank temperature that will allow you to attain this goal. Of course, the smaller you choose the radiator, the higher must be the storage tank temperature. You can find decent values in Duffie and Beckman [8.5]. This will thus determine our critical temperature.

Next we want to dimension the electrical backup system. The electrical heater should be able to raise the storage tank temperature from room temperature, i.e., 20°C, to the critical temperature within one hour.

Next we want to dimension the collector system. Determine how many of the (series connected) standard collectors are required to keep the storage tank temperature at a periodic steady–state for the December 21 situation without activating the backup heating device.

Many parameter values must be selected. Use recommended values from the literature [8.5] where available, otherwise use physical intuition and common sense to determine appropriate values for these parameters. Simulate the overall system for various climatic conditions. Consult published weather data to determine the frequency of use of the electric backup system. What is the reduction in the utility bill achievable with this system in comparison to an electric only solution? Simulate the electric only solution by permanently disabling the collector water loop pump.

Research

[R8.1] Second Sound Wave in Superfluid Helium–II

Formulate the two–fluid theory, which explains the behavior of liquid Helium in the vicinity of the so–called λ point, in terms of a bond graph. Develop a simulation model that reproduces the observed second sound wave while conserving both energy and mass in the system.

[R8.2] Bond Graphs for Maxwell's Equations, Nonlinear Optics, and Fluid Dynamics

In this chapter, we showed one successful application of bond graphs for modeling distributed parameter systems. For the reasons explained in the summary section, other types of PDE problems don't lend themselves as easily to a bond graph formulation. The next simplest type of practical PDE problems is the wave equation:

$$\frac{\partial^2 u}{\partial t^2} = c^2 \cdot \frac{\partial^2 u}{\partial x^2} \tag{R8.2a}$$

$$\frac{\partial^2 i}{\partial t^2} = c^2 \cdot \frac{\partial^2 i}{\partial x^2} \tag{R8.2b}$$

which can describe the transport of voltage and current along a lossless transmission line, the pressure and flow rate of a compressible liquid or gas in a pipe, the longitudinal oscillation of an elastic rod, sound waves in gases or liquids, and optical waves.

Also the wave equation can be modeled easily in terms of an equivalent electrical circuit, and therefore, in terms of a bond graph. Figure R8.2a shows the equivalent electrical circuit,

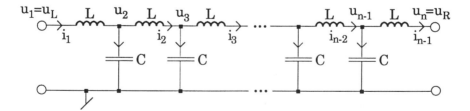

Figure R8.2a. Equivalent electrical circuit modeling the wave equation.

and Fig.R8.2b shows a corresponding bond graph:

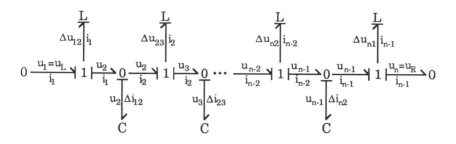

Figure R8.2b. Bond graph for the wave equation.

The variables are specified for an electromagnetic wave.

Unfortunately, this brings us to the end of the simple analogies. Problems in electromagnetics and nonlinear optics are probably a little easier to tackle than those in fluid dynamics since we don't need to concern ourselves with an additional equation (the mass conservation equation), and yet, already these problems are quite tough.

Let us look at a fairly simple problem, the electrical transmission line with dissipative losses. This system can be modeled by the telegraph equation:

$$\frac{\partial^2 u}{\partial x^2} = RGu + (RC + LG)\frac{\partial u}{\partial t} + LC\frac{\partial^2 u}{\partial t^2} \qquad (R8.2c)$$

$$\frac{\partial^2 i}{\partial x^2} = RGi + (RC + LG)\frac{\partial i}{\partial t} + LC\frac{\partial^2 i}{\partial t^2} \qquad (R8.2d)$$

where

$$Z_{sh} = R + sL \qquad (R8.2e)$$

is the impedance of a unit–length segment of the short–circuited transmission line and

$$Y_{op} = G + sC \qquad (R8.2f)$$

is the admittance of a unit–length segment of the open transmission line. An approximate solution can be found by combining the bond graph of Fig.R8.2b with that of Fig.8.10, i.e., by adding RS elements to the bond graph of Fig.R8.2b. However, this bond graph does not model the telegraph equation exactly. It is possible to add RS elements emanating from the 1–junctions and GS elements emanating from the 0–junctions without introducing algebraic loops, but it is not clear that this will bring much of an improvement. On the other hand, if we solve a PDE problem numerically, we discretize anyway, and therefore approximate the true distributed parameter system by a lumped parameter system. It is not clear that this proposed circuit analogy is worse than any other approximation that we may choose.

Fluid dynamics problems are governed by the Navier–Stokes equation (for the power balance) combined with an additional equation (for the mass balance). These are tough problems to solve. A bond graph model that shows how the energies and masses flow through the system would indeed be very useful. I am convinced that any good numerical algorithm ought to take power and mass flows into consideration and should try to map into a numerical scheme what the physics dictate in the real system, not only in the overall solution, but also in the implementation of individual steps.

[R8.3] Bond Graphs for Macroeconomies

Define a set of adjugate variables for models of macroeconomic systems together with a formulation of the energy conservation law as applied to these adjugate variables. Show, by means of examples, how this concept can be used to model macroeconomic systems.

9

Modeling Chemical Reaction Kinetics

Preview

In previous chapters of this text, we have considered various forms of energy, however, we treated them mostly in isolation. Transitions between different forms of energy were accomplished by so–called energy transducers, but, until now, no more than two forms of energy were involved in any such transition. Chapter 8 shed some light on distributed processes and showed how one form of energy, namely, heat, is transported by and transmitted through matter. In this chapter, we shall discuss what happens to the total energy balance when one type of matter interacts with another, i.e., we shall look at the thermodynamics of chemical reaction systems. Notice that, in this analysis, we are not so much interested in the chemical properties of matter (such as color, taste, smell, or toxicity), as we are interested in the physical properties of matter (such as temperature, pressure, volume, flow rate, or concentration). In this discussion, it will be unavoidable to consider a multitude of simultaneously occurring energy transitions.

9.1 Introduction

We have seen that all physical systems require two different types of variables for a complete description of their dynamics. In an electrical circuit, we cannot usually write state equations in terms of voltages or currents alone, but, in any circuit that contains both capacitors and inductors, we always need a combination of the two. This is due to the fact that in a capacitor charge is accumulated which leads to the selection of voltage as a natural state variable, whereas in an inductor flux is accumulated which leads to the selection of current as a natural state variable.

In Chapter 8, we saw that thermodynamic systems can be described by either temperature or heat alone, i.e., the state equations can be described either in terms of temperatures or in terms of heat only. This is due to the fact that no thermal inertance exists. Any system that contains only one type of storage element can be described in terms of one type of state variables. The thermodynamic literature exploits this fact frequently. It was shown, however, that this simplification has a very unwelcome side effect. If we concentrate on one form of state variable (such as the temperature), we lose the capability of balancing the energy (power) correctly, and therefore, it becomes very difficult to connect the thermal subsystem to other subsystems. In other words, as long as we are happy to analyze a thermal system in isolation, we can concentrate on temperatures alone, but if we wish to connect the thermal system to other systems, we must carry along both types of adjugate variables, i.e., temperatures and entropy flows.

Table 7.1 shows that chemical systems do not possess inertia either. The two adjugate variables, the chemical potential μ and the molar flow ν, can therefore be decoupled. It is possible to describe the dynamics of a chemical reaction in terms of number of moles and molar flow rates alone, without consideration of the chemical potential. Since chemical reactions are highly nonlinear, we can't express the number of moles in terms of molar flow rates, and the resulting differential equations will thus use the number of moles as state variables rather than the molar flow rates. Frequently, the number of moles is furthermore normalized with the reactor volume, resulting in a set of differential equations using molar densities/concentrations as state variables.

The literature on *chemical reaction kinetics* operates traditionally on molar concentrations and their time derivatives exclusively, and it ignores the energy and its conservation entirely [9.8]. Many of these references do not introduce the chemical potential as a system property at all.

On the other hand, the literature on *chemical thermodynamics* concentrates on the overall energy balance in a chemical reaction using terms such as the enthalpy, the entropy (differently defined than in Chapter 8), and the Gibbs free energy [9.15]. Some of these references introduce the chemical potential as an auxiliary quantity that can be derived from the Gibbs free energy; others don't [9.10]. However, all of these references discuss the equilibrium state thermodynamics only, i.e., they ignore the dynamic behavior of the system. Therefore, flow rates are of no concern to them.

As in the case of the thermal systems, the disadvantage of this dichotomy is the fact that it prevents us from connecting a chemical reaction system dynamically with other subsystems. As long as we are happy to analyze chemical reaction systems in isolation, the decomposition into the dynamic reaction kinetics and the static thermodynamics may be quite appropriate. However, if we wish to analyze dynamically the interactions among chemical, thermic, pneumatic, mechanical, and electric subsystems, this approach becomes inadequate. It is the goal of this chapter to derive a methodology that will enable us to describe the dynamic behavior of chemical reaction systems through a set of adjugate variables, and thereby enable us to study the dynamic interactions of chemical reaction systems with their environment.

9.2 Chemical Reaction Kinetics

The physics behind chemical reactions among gases or aqueous solutions are very similar to those discussed in Chapter 8. As in thermodynamics, we can distinguish between *macroscopic* and *microscopic* aspects of such reactions. As in the previous chapter, we shall mostly concentrate on the macroscopic aspects, but we need some understanding of the microscopic aspects in order to derive the correct macroscopic equations.

Let us look at the chemical reaction:

$$C_3H_8 + 5O_2 \rightleftharpoons 3CO_2 + 4H_2O \qquad (9.1)$$

By convention, the chemical species on the left–hand side of the arrow are called the *reactants*, while the species to the right of the arrow are called *products*. However, the two–sided arrow indicates that, in reality, two different reactions take place. Equation (9.1) is an abbreviation for the reaction system:

$$C_3H_8 + 5O_2 \rightarrow 3CO_2 + 4H_2O \qquad (9.2a)$$
$$3CO_2 + 4H_2O \rightarrow C_3H_8 + 5O_2 \qquad (9.2b)$$

The first reaction states that when one molecule of propane C_3H_8 meets with five molecules of oxygen O_2, the atoms may regroup and form seven new molecules, namely, three molecules of carbon dioxide (CO_2) and four molecules of water vapor (H_2O). The second reaction states that the reverse may also happen.

Any chemical reaction describes a trade of atoms between a number of different molecules. In this trade, the mass balance must be preserved, i.e., the number of atoms of each type on both sides of the arrow must be the same, and an integer number of molecules is involved in the trade. The act of balancing the masses in a chemical reaction is called *stoichiometry*, and the integer constants that multiply each species (i.e., the number of molecules involved in the trade) are called the *stoichiometric coefficients*. The mass balance is achieved by ensuring that the total number of atoms of each type is equal on both sides of the arrow.

It is interesting to discuss how frequently such a trade occurs. For this purpose, we need to look at the microscopic aspects of the chemical reaction. In order for the trade to take place, all the involved partner molecules must meet at the same time at one place. Let us assume that the molecules are mixed in a completely homogeneous manner. Consequently, the probability of any one molecule to be at any one place is the same. In other words, the probability of a molecule to be in any unit volume element of the mixture is the same and is proportional to the concentration of the molecule in the overall mixture (solution). We can also assume that the probability of any one molecule to be in a considered unit volume element is statistically independent from the probability of any other molecule to be in the same unit volume element. Thus, the joint probability of all partner molecules to be in the same unit volume element at the same time is proportional to the product of their concentrations.

Let us denote the concentration of a molecule M by c_M. For example, we shall denote the concentration of water vapor as c_{H_2O}. Therefore, the reaction according to Eq.(9.2a) should occur with a reaction rate of:

$$k_1 \cdot c_{C_3H_8} \cdot c_{O_2}^5 \qquad (9.3a)$$

and similarly, for the second reaction:

$$k_2 \cdot c_{CO_2}^3 \cdot c_{H_2O}^4 \qquad (9.3b)$$

We denote this by placing the reaction rate constants on the arrows:

$$C_3H_8 + 5O_2 \xrightarrow{k_1} 3CO_2 + 4H_2O \qquad (9.4a)$$

$$3CO_2 + 4H_2O \xrightarrow{k_2} C_3H_8 + 5O_2 \qquad (9.4b)$$

and thus, we are made to believe that we can write the following set of differential equations for the reaction system:

$$\frac{d}{dt}c_{C_3H_8} = k_2 \cdot c_{CO_2}^3 \cdot c_{H_2O}^4 - k_1 \cdot c_{C_3H_8} \cdot c_{O_2}^5 \qquad (9.5a)$$

$$\frac{d}{dt}c_{O_2} = 5k_2 \cdot c_{CO_2}^3 \cdot c_{H_2O}^4 - 5k_1 \cdot c_{C_3H_8} \cdot c_{O_2}^5 \qquad (9.5b)$$

$$\frac{d}{dt}c_{CO_2} = 3k_1 \cdot c_{C_3H_8} \cdot c_{O_2}^5 - 3k_2 \cdot c_{CO_2}^3 \cdot c_{H_2O}^4 \qquad (9.5c)$$

$$\frac{d}{dt}c_{H_2O} = 4k_1 \cdot c_{C_3H_8} \cdot c_{O_2}^5 - 4k_2 \cdot c_{CO_2}^3 \cdot c_{H_2O}^4 \qquad (9.5d)$$

Notice that each reaction is driven by the concentrations of the reactants only, and not by the concentrations of the products. Notice furthermore that the stoichiometric coefficients multiply the reaction equations. This is easily understandable since, in each trade of the first type, only one molecule of C_3H_8 is consumed for every five molecules of O_2. Consequently, the concentration of O_2 decreases five times faster than that of C_3H_8. Similarly, in the second reaction, five molecules of O_2 are produced for every one molecule of C_3H_8.

And yet, the set of differential equations, Eqs.(9.5a–d), is certainly incorrect. The reason is that the chances for six or seven different molecules to meet at any one place at any one time are minuscule, and therefore, such a reaction is very unlikely to occur. Equation (9.1) is not meant to be an adequate description of what really happens in the chemical reaction, but rather, it describes an *equilibrium of flow*, i.e., it describes the balance of various atomic trades between molecules near steady–state conditions. Whether a particular stoichiometrically feasible reaction actually can take place or not, and if so at what reaction rate, are difficult questions to answer. We cannot answer them without looking in detail at the energy balance in the trade in addition to the mass balance governed by the stoichiometry. In general, if more than three or (maximum) four reactant molecules are involved in a proposed chemical reaction, that reaction will never occur in one step.

Let us look at some simpler reactions to explain this concept. Hydrogen gas and iodine vapor can react to form gaseous hydrogen iodide:

$$H_2 + I_2 \rightleftharpoons 2HI \qquad (9.6)$$

As before, the reaction can take place in both directions. We decompose the equilibrium reaction into two separate reactions:

$$H_2 + I_2 \xrightarrow{k_1} 2HI \tag{9.7a}$$

$$2HI \xrightarrow{k_2} H_2 + I_2 \tag{9.7b}$$

which can be described by the following (correct) set of differential equations:

$$\frac{d}{dt}c_{H_2} = k_2 \cdot c_{HI}^2 - k_1 \cdot c_{H_2} \cdot c_{I_2} \tag{9.8a}$$

$$\frac{d}{dt}c_{I_2} = k_2 \cdot c_{HI}^2 - k_1 \cdot c_{H_2} \cdot c_{I_2} \tag{9.8b}$$

$$\frac{d}{dt}c_{HI} = 2k_1 \cdot c_{H_2} \cdot c_{I_2} - 2k_2 \cdot c_{HI}^2 \tag{9.8c}$$

We can investigate what happens in the equilibrium, i.e., after a long time. We assume that an equilibrium of flow is reached in the form of a steady–state solution of our set of differential equations. In the steady–state, all derivatives are zero, thus:

$$k_2 \cdot c_{HI}^2 - k_1 \cdot c_{H_2} \cdot c_{I_2} = 0.0 \tag{9.9a}$$

$$2k_1 \cdot c_{H_2} \cdot c_{I_2} - 2k_2 \cdot c_{HI}^2 = 0.0 \tag{9.9b}$$

and therefore:

$$\frac{c_{HI}^2}{c_{H_2} \cdot c_{I_2}} = \frac{k_1}{k_2} = \text{constant} \tag{9.10}$$

In the equilibrium, a constant ratio exists between the concentrations of the different species irrespective of the chosen initial conditions. (This constant may still depend on temperature since the reaction rate constants k_1 and k_2 are temperature–dependent.)

Let us now look at another similar reaction:

$$H_2 + Br_2 \rightleftharpoons 2HBr \tag{9.11}$$

From the previous discussion, we would expect that, in the steady–state:

$$\frac{c_{HBr}^2}{c_{H_2} \cdot c_{Br_2}} = \text{constant} \tag{9.12}$$

However, experimental laboratory results prove Eq.(9.12) to be incorrect. This indicates that, in reality, a somewhat different reaction takes place. By experiment, it was verified that the production rate of HBr is proportional to the concentration of H_2, but proportional to the square root of the concentration of Br_2:

$$\frac{d}{dt}c_{HBr} \propto c_{H_2} \cdot c_{Br_2}^{1/2} \qquad (9.13)$$

This experimental result can be explained by the following set of reactions:

$$Br_2 \rightleftharpoons 2Br^\bullet \qquad (9.14a)$$

$$Br^\bullet + H_2 \rightleftharpoons HBr + H^\bullet \qquad (9.14b)$$

$$Br_2 + H^\bullet \rightarrow HBr + Br^\bullet \qquad (9.14c)$$

It turns out that H_2 and Br_2 are much too inert to engage in any "extramarital" relationships. However, if either a $Br^\bullet$ atom or an $H^\bullet$ atom is "single," it would rather steal the "partner" of an existing "couple" than stay single. Equation (9.14a) denotes a *fast equilibrium* reaction, i.e., Br_2 "couples" sometimes split for a while and recombine with other "single" $Br^\bullet$ atoms. Equation (9.14c) denotes a *fast* reaction, i.e., $H^\bullet$ atoms are very unhappy individuals who will reengage at once. Consequently, very few single $H^\bullet$ atoms are around at any one time, and the chances for two such $H^\bullet$ atoms to meet are negligible. "Marriages" between single $Br^\bullet$ and single $H^\bullet$ atoms are more likely to happen, but even they hardly occur because the chances to meet are so small. Equation (9.14b) denotes a *slow equilibrium* reaction, i.e., while most single $Br^\bullet$ atoms will recombine again with other single $Br^\bullet$ atoms, some get "frustrated" before they can find an appropriate new "partner" and attract a partner of an existing H_2 "couple" instead. Single $H^\bullet$ atoms are so eager to reengage that they are sometimes able to "steal" the $H^\bullet$ "partner" of an existing HBr "couple."

The three reactions in Eqs.(9.14a–c) can be expanded into:

$$Br_2 \xrightarrow{k_1} 2Br^\bullet \qquad (9.15a)$$

$$2Br^\bullet \xrightarrow{k_2} Br_2 \qquad (9.15b)$$

$$Br^\bullet + H_2 \xrightarrow{k_3} HBr + H^\bullet \qquad (9.15c)$$

$$HBr + H^\bullet \xrightarrow{k_4} Br^\bullet + H_2 \qquad (9.15d)$$

$$Br_2 + H^\bullet \xrightarrow{k_5} HBr + Br^\bullet \qquad (9.15e)$$

with typical reaction rate constants of $k_1 = 25.13$ sec^{-1}, $k_2 = 1.057 \cdot 10^7$ m^3 sec^{-1} mole^{-1}, $k_3 = 4.44 \cdot 10^6$ m^3 sec^{-1} mole^{-1}, $k_4 = 188.6$ m^3 sec^{-1} mole^{-1}, and $k_5 = 1886.0$ m^3 sec^{-1} mole^{-1}, at a temperature of 900 K. The molar concentrations are measured in mole m^{-3}. Equations (9.15a–e) lead to the following set of differential equations:

$$\frac{d}{dt}c_{Br_2} = -k_1 \cdot c_{Br_2} + k_2 \cdot c_{Br\bullet}^2 - k_5 \cdot c_{H\bullet} \cdot c_{Br_2} \tag{9.16a}$$

$$\frac{d}{dt}c_{Br\bullet} = 2k_1 \cdot c_{Br_2} - 2k_2 \cdot c_{Br\bullet}^2 - k_3 \cdot c_{H_2} \cdot c_{Br\bullet}$$
$$+ k_4 \cdot c_{HBr} \cdot c_{H\bullet} + k_5 \cdot c_{H\bullet} \cdot c_{Br_2} \tag{9.16b}$$

$$\frac{d}{dt}c_{H_2} = -k_3 \cdot c_{H_2} \cdot c_{Br\bullet} + k_4 \cdot c_{HBr} \cdot c_{H\bullet} \tag{9.16c}$$

$$\frac{d}{dt}c_{H\bullet} = k_3 \cdot c_{H_2} \cdot c_{Br\bullet} - k_4 \cdot c_{HBr} \cdot c_{H\bullet} - k_5 \cdot c_{H\bullet} \cdot c_{Br_2} \tag{9.16d}$$

$$\frac{d}{dt}c_{HBr} = k_3 \cdot c_{H_2} \cdot c_{Br\bullet} - k_4 \cdot c_{HBr} \cdot c_{H\bullet} + k_5 \cdot c_{H\bullet} \cdot c_{Br_2} \tag{9.16e}$$

The overall reaction rate is dominated by the slowest reaction, i.e., by Eq.(9.16e). Since the concentration of H$^\bullet$ atoms is very small at all times, Eq.(9.16e) is dominated by the k_3 term. Therefore, we can write:

$$\frac{d}{dt}c_{HBr} \approx k_3 \cdot c_{H_2} \cdot c_{Br\bullet} \tag{9.17}$$

Since the equilibrium reaction Eq.(9.14a) is much faster, we can analyze this equation under steady–state conditions:

$$k_1 \cdot c_{Br_2} - k_2 \cdot c_{Br\bullet}^2 = 0.0 \tag{9.18}$$

and thus:

$$c_{Br\bullet} = \sqrt{\frac{k_1}{k_2} \cdot c_{Br_2}} = k_{eq} \cdot c_{Br_2}^{1/2} \tag{9.19}$$

Plugging Eq.(9.19) into Eq.(9.17), we find:

$$\frac{d}{dt}c_{HBr} \approx k_3 k_{eq} \cdot c_{H_2} \cdot c_{Br_2}^{1/2} \tag{9.20}$$

as shown in the laboratory experiments.

While this formulation is the traditionally used form of reaction kinetics equations, we wish to reformulate the reaction equations in terms of number of moles and molar flow rates.

We first need to introduce the concept of a "mole." Traditionally, the amount of a substance was either measured in terms of its weight (for solid materials) or in terms of its volume (for liquids). In physics, it became more practical to express the amount of any substance through its mass, since this simplifies the formulation of Newton's law. In chemistry, this approach is not convenient, since chemical reactions trade substances in a fixed ratio among the involved types of molecules (or atoms). Unfortunately, 1 kg of one chemical substance contains a different number of molecules (atoms) than 1 kg of another chemical substance. Thus, it is more convenient to express the amount of a chemical substance by counting the number of molecules (or atoms) contained in the measured quantity.

A *pure substance* is a substance that contains only one chemical species, i.e., one type of molecule. The *molecular mass* of a chemical species can be roughly expressed as the number of heavy particles (protons and neutrons) contained in the species multiplied by the mass of a heavy particle. Consequently, the number of molecules contained in any amount of a pure substance can be expressed as the ratio between its mass and its molecular mass.

However, by using the molecular mass, we end up with very large numbers at all times. Therefore, it is customary to normalize masses in a different way. We count the number L of atoms contained in 12 g of carbon C^{12}. In some references, L is called *Avogadro's number*, while other texts refer to L as *Loschmidt's number*. Similarly, the number of molecules contained in 2 g of hydrogen gas H_2 is also L. In general, L is a constant that measures the number of molecules (atoms) contained in k g of any pure substance, where k denotes the number of heavy particles (protons and neutrons) contained in one molecule (atom) of the substance. L has a value of $6.025 \cdot 10^{23}$.

One *mole* of any pure substance is the amount that contains L molecules (atoms). The *molar mass* of any substance is defined as the mass of one mole of the substance, i.e., the molar mass of any substance is the product of L and its molecular mass. The molar mass of carbon C^{12} is 12 g, while the molar mass of hydrogen gas H_2 is 2 g. We can, thus, express the amount of a pure substance as the ratio between its mass and its molar mass, i.e., by counting the number of moles contained in the measured quantity. The *mole* is the measurement unit introduced to measure the number of moles of any pure substance.

The *molar concentration* of any pure substance is defined as the number of moles contained in the measured quantity divided by the volume that this quantity occupies. Consequently, molar concentrations are measured in mole m^{-3}.

The *molar flow rate* describes the change of the number of moles over time, i.e., it is the derivative of the number of moles with respect to time. Consequently, molar flow rates are measured in mole sec^{-1}.

Sometimes, it is useful to measure the amount of a mixture of pure substances. The number of moles of any substance is the sum of the number of moles of the pure substances contained in the mixture.

It is our goal to reformulate the reaction kinetics equations in terms of number of moles and molar flow rates. Let me explain this concept by means of the previously used H_2–Br_2 reaction system. We can rewrite the set of differential equations (9.16a–e) in terms of the number of moles. Let the quantity n_{H_2O} denote the number of moles of water vapor. The molar concentration is the number of moles divided by the total volume, for example:

$$c_{H_2O} = \frac{n_{H_2O}}{V} \tag{9.21}$$

Therefore, Eqs.(9.16a–e) become:

$$\frac{d}{dt}\left(\frac{n_{Br_2}}{V}\right) = -k_1 \cdot \frac{n_{Br_2}}{V} + k_2 \cdot \left(\frac{n_{Br\bullet}}{V}\right)^2 - k_5 \cdot \frac{n_{H\bullet}}{V} \cdot \frac{n_{Br_2}}{V} \tag{9.22a}$$

$$\frac{d}{dt}\left(\frac{n_{Br\bullet}}{V}\right) = 2k_1 \cdot \frac{n_{Br_2}}{V} - 2k_2 \cdot \left(\frac{n_{Br\bullet}}{V}\right)^2 - k_3 \cdot \frac{n_{H_2}}{V} \cdot \frac{n_{Br\bullet}}{V}$$
$$+ k_4 \cdot \frac{n_{HBr}}{V} \cdot \frac{n_{H\bullet}}{V} + k_5 \cdot \frac{n_{H\bullet}}{V} \cdot \frac{n_{Br_2}}{V} \tag{9.22b}$$

$$\frac{d}{dt}\left(\frac{n_{H_2}}{V}\right) = -k_3 \cdot \frac{n_{H_2}}{V} \cdot \frac{n_{Br\bullet}}{V} + k_4 \cdot \frac{n_{HBr}}{V} \cdot \frac{n_{H\bullet}}{V} \tag{9.22c}$$

$$\frac{d}{dt}\left(\frac{n_{H\bullet}}{V}\right) = k_3 \cdot \frac{n_{H_2}}{V} \cdot \frac{n_{Br\bullet}}{V} - k_4 \cdot \frac{n_{HBr}}{V} \cdot \frac{n_{H\bullet}}{V} - k_5 \cdot \frac{n_{H\bullet}}{V} \cdot \frac{n_{Br_2}}{V} \tag{9.22d}$$

$$\frac{d}{dt}\left(\frac{n_{HBr}}{V}\right) = k_3 \cdot \frac{n_{H_2}}{V} \cdot \frac{n_{Br\bullet}}{V} - k_4 \cdot \frac{n_{HBr}}{V} \cdot \frac{n_{H\bullet}}{V} + k_5 \cdot \frac{n_{H\bullet}}{V} \cdot \frac{n_{Br_2}}{V} \tag{9.22e}$$

Let ν_{H_2O} denote the molar flow rate of water vapor, i.e., the change of the number of moles of water vapor with respect to time:

$$\nu_{H_2O} = \frac{d}{dt} n_{H_2O} \tag{9.23}$$

and let q be the volume flow rate, i.e., the derivative of the total volume V with respect to time:

$$q = \frac{dV}{dt} \tag{9.24}$$

With this notation, Eqs.(9.22a–e) can be rewritten as follows:

$$\nu_{Br_2} = -k_1 \cdot n_{Br_2} + k_2 \cdot \left(\frac{n_{Br\bullet}^2}{V}\right) - k_5 \cdot \left(\frac{n_{H\bullet} \cdot n_{Br_2}}{V}\right) + q \cdot \left(\frac{n_{Br_2}}{V}\right) \quad (9.25a)$$

$$\nu_{Br\bullet} = 2k_1 \cdot n_{Br_2} - 2k_2 \cdot \left(\frac{n_{Br\bullet}^2}{V}\right) - k_3 \cdot \left(\frac{n_{H_2} \cdot n_{Br\bullet}}{V}\right)$$

$$+ k_4 \cdot \left(\frac{n_{HBr} \cdot n_{H\bullet}}{V}\right) + k_5 \cdot \left(\frac{n_{H\bullet} \cdot n_{Br_2}}{V}\right) + q \cdot \left(\frac{n_{Br\bullet}}{V}\right) \quad (9.25b)$$

$$\nu_{H_2} = -k_0 \cdot \left(\frac{n_{H_2} \cdot n_{Br\bullet}}{V}\right) + k_4 \cdot \left(\frac{n_{HBr} \cdot n_{H\bullet}}{V}\right) + q \cdot \left(\frac{n_{H_2}}{V}\right) \quad (9.25c)$$

$$\nu_{H\bullet} = k_3 \cdot \left(\frac{n_{H_2} \cdot n_{Br\bullet}}{V}\right) - k_4 \cdot \left(\frac{n_{HBr} \cdot n_{H\bullet}}{V}\right)$$

$$- k_5 \cdot \left(\frac{n_{H\bullet} \cdot n_{Br_2}}{V}\right) + q \cdot \left(\frac{n_{H\bullet}}{V}\right) \quad (9.25d)$$

$$\nu_{HBr} = k_3 \cdot \left(\frac{n_{H_2} \cdot n_{Br\bullet}}{V}\right) - k_4 \cdot \left(\frac{n_{HBr} \cdot n_{H\bullet}}{V}\right)$$

$$+ k_5 \cdot \left(\frac{n_{H\bullet} \cdot n_{Br_2}}{V}\right) + q \cdot \left(\frac{n_{HBr}}{V}\right) \quad (9.25e)$$

We can introduce the *reaction flow rates*:

$$\nu_{k1} = k_1 \cdot n_{Br_2} \quad (9.26a)$$

$$\nu_{k2} = k_2 \cdot \left(\frac{n_{Br\bullet}^2}{V}\right) \quad (9.26b)$$

$$\nu_{k3} = k_3 \cdot \left(\frac{n_{H_2} \cdot n_{Br\bullet}}{V}\right) \quad (9.26c)$$

$$\nu_{k4} = k_4 \cdot \left(\frac{n_{HBr} \cdot n_{H\bullet}}{V}\right) \quad (9.26d)$$

$$\nu_{k5} = k_5 \cdot \left(\frac{n_{H\bullet} \cdot n_{Br_2}}{V}\right) \quad (9.26e)$$

and therefore:

$$\frac{d}{dt} n_{Br_2} = -\nu_{k1} + \nu_{k2} - \nu_{k5} + q \cdot \left(\frac{n_{Br_2}}{V}\right) \quad (9.27a)$$

$$\frac{d}{dt} n_{Br\bullet} = 2\nu_{k1} - 2\nu_{k2} - \nu_{k3} + \nu_{k4} + \nu_{k5} + q \cdot \left(\frac{n_{Br\bullet}}{V}\right) \quad (9.27b)$$

$$\frac{d}{dt} n_{H_2} = -\nu_{k3} + \nu_{k4} + q \cdot \left(\frac{n_{H_2}}{V}\right) \quad (9.27c)$$

$$\frac{d}{dt} n_{H\bullet} = \nu_{k3} - \nu_{k4} - \nu_{k5} + q \cdot \left(\frac{n_{H\bullet}}{V}\right) \quad (9.27d)$$

$$\frac{d}{dt} n_{HBr} = \nu_{k3} - \nu_{k4} + \nu_{k5} + q \cdot \left(\frac{n_{HBr}}{V}\right) \quad (9.27e)$$

Equations (9.27a–e) look funny. Let me propose the following experiment: We store a certain amount of reactants in a closed container with a movable piston. We wait until an equilibrium has been reached, i.e., until all time derivatives have died out. At such time, we pull the piston further out, i.e., we apply an external force to artificially increase the volume. Equations (9.27a–e) claim that the number of moles will start to grow, but we know that this can't be true. The total number of moles can grow only if we add more substance to the container.

Let us look a little more closely at one of these equations, for example, the one for HBr. The flow rate balance of the substance HBr from all the reactions k_3, k_4, and k_5 in which the substance HBr is involved can be computed from the equation:

$$\nu_{\mathrm{HBr}} = \nu_{k3} - \nu_{k4} + \nu_{k5} \tag{9.28}$$

Obviously, the change in the number of moles of HBr must be equal to the balance among the reaction flow rates:

$$\frac{d}{dt} n_{\mathrm{HBr}} = \nu_{\mathrm{HBr}} \tag{9.29}$$

Comparing Eq.(9.29) with Eq.(9.27e), we see that the q term in Eq.(9.27e) is surplus. In order to correct Eqs.(9.27a–e), we must modify the original set of state equations (9.16a–e) in the following way:

$$\frac{d}{dt} c_{\mathrm{Br_2}} = -k_1 \cdot c_{\mathrm{Br_2}} + k_2 \cdot c_{\mathrm{Br\bullet}}^2 - k_5 \cdot c_{\mathrm{H\bullet}} \cdot c_{\mathrm{Br_2}} - \frac{q}{V} \cdot c_{\mathrm{Br_2}} \tag{9.30a}$$

$$\frac{d}{dt} c_{\mathrm{Br\bullet}} = 2k_1 \cdot c_{\mathrm{Br_2}} - 2k_2 \cdot c_{\mathrm{Br\bullet}}^2 - k_3 \cdot c_{\mathrm{H_2}} \cdot c_{\mathrm{Br\bullet}}$$
$$+ k_4 \cdot c_{\mathrm{HBr}} \cdot c_{\mathrm{H\bullet}} + k_5 \cdot c_{\mathrm{H\bullet}} \cdot c_{\mathrm{Br_2}} - \frac{q}{V} \cdot c_{\mathrm{Br\bullet}} \tag{9.30b}$$

$$\frac{d}{dt} c_{\mathrm{H_2}} = -k_3 \cdot c_{\mathrm{H_2}} \cdot c_{\mathrm{Br\bullet}} + k_4 \cdot c_{\mathrm{HBr}} \cdot c_{\mathrm{H\bullet}} - \frac{q}{V} \cdot c_{\mathrm{H_2}} \tag{9.30c}$$

$$\frac{d}{dt} c_{\mathrm{H\bullet}} = k_3 \cdot c_{\mathrm{H_2}} \cdot c_{\mathrm{Br\bullet}} - k_4 \cdot c_{\mathrm{HBr}} \cdot c_{\mathrm{H\bullet}} - k_5 \cdot c_{\mathrm{H\bullet}} \cdot c_{\mathrm{Br_2}}$$
$$- \frac{q}{V} \cdot c_{\mathrm{H\bullet}} \tag{9.30d}$$

$$\frac{d}{dt} c_{\mathrm{HBr}} = k_3 \cdot c_{\mathrm{H_2}} \cdot c_{\mathrm{Br\bullet}} - k_4 \cdot c_{\mathrm{HBr}} \cdot c_{\mathrm{H\bullet}} + k_5 \cdot c_{\mathrm{H\bullet}} \cdot c_{\mathrm{Br_2}}$$
$$- \frac{q}{V} \cdot c_{\mathrm{HBr}} \tag{9.30e}$$

This correction term makes a lot of sense. Let us apply the same experiment as before. If we artificially increase the volume of the reactor, we know that the concentrations will start decreasing. The correction term takes care of this. The uncorrected equations (9.16a–e) did not reflect the necessary change in the concentrations. Obviously, the uncorrected equations (9.16a–e) are only valid for reactions that occur under constant volume conditions where $q = 0.0$ at all times.

9.3 Chemical Thermodynamics

Until now, we have looked at the mass transfer (the stoichiometry) of a chemical reaction only. If we wish to understand why chemical reactions take place, we need to look at the energy transfer as well.

Most chemical reactions are either *exothermic* or *endothermic*, i.e., they either generate or absorb heat. Quite frequently, other things happen as well. In a closed container (i.e., under constant volume), the pressure of a gas or liquid may change during the reaction, and in an open container (i.e., under constant pressure), the volume of the gas or liquid may change. This volume/pressure change can sometimes assume violent dimensions, for instance, in an explosion.

This indicates that, in a chemical reaction, two different goods are actually traded, namely, mass and energy. The energy traded in a chemical reaction must come from somewhere. Consequently, we must assume that a chemical substance can store energy similar to a capacitor in an electrical circuit. Let us try to define this mechanism.

According to the first and second laws of thermodynamics, the change of the *internal energy U* of a chemical substance in any process (such as a change of temperature or a chemical reaction) can be described as follows:

$$\dot{U} = T \cdot \dot{S} - p \cdot \dot{V} + \sum_{\forall i} \mu_i \cdot \dot{n}_i \qquad (9.31)$$

where T denotes the temperature, S stands for the entropy, p is the pressure, V denotes the volume, and n_i is the number of moles of substance i. Since, in any chemical reaction, energy is traded as well as mass, the flow of mass across the arrow must carry chemical power along with it. Therefore, we must assume that an adjugate variable μ_i to the molar flow rate ν_i exists such that the product of the molar flow rate and its adjugate variable is chemical power. The adjugate variable has been coined the *chemical potential* of a substance, and is measured in J mole^{-1}. The chemical potential of a substance describes the amount of chemical energy stored in a mole of that substance. When the substance is traded across the arrow in a chemical reaction, its chemical energy is traded along with its mass. Equation (9.31) is frequently referred to as the *Gibbs equation*.

If a substance is separated into several components during a chemical reaction, the sum of the chemical potentials of the products can be different from the chemical potential of the reactant. It can be either smaller or larger. If this is the case, the energy must be balanced elsewhere. For instance, if the sum of the chemical potentials of the products is smaller than the chemical potential of the reactant, the excess energy must be converted to another form, by producing heat or increasing the pressure or increasing the volume. The same argument holds for several reactants being combined into one product, or for several reactants being rearranged into several products.

We realize that matter can store energy simultaneously using three different mechanisms, a *thermic storage* (expressed through the heat capacity), a *hydraulic/pneumatic storage*, and finally a *chemical storage* (a structural storage).

Notice that the variables U, S, V, and n_i depend on the quantity of substance that acts as a storage. These are called *extensive variables*. On the other hand, T, p, and μ_i are independent of the quantity. They are called *intensive variables*. It is sometimes useful to normalize the extensive variables (i.e., make them intensive) by dividing through the total number of moles:

$$\mathcal{U} = \frac{U}{n} \quad , \quad \mathcal{S} = \frac{S}{n} \quad , \quad \mathcal{V} = \frac{V}{n} \quad , \quad x_i = \frac{n_i}{n} \tag{9.32}$$

$\mathcal{U}$ is called the *molar internal energy*, $\mathcal{S}$ is called the *molar entropy*, $\mathcal{V}$ is called the *molar volume*, and x_i is called the *mole fraction*. Equation (9.31) can thus be rewritten as follows:

$$\frac{d}{dt}(n\mathcal{U}) - T \cdot \frac{d}{dt}(n\mathcal{S}) + p \cdot \frac{d}{dt}(n\mathcal{V}) - \sum_{\forall i} \mu_i \cdot \frac{d}{dt}(n\,x_i) = 0.0 \tag{9.33}$$

which can be evaluated as follows:

$$n\Big(\frac{d\mathcal{U}}{dt} - T \cdot \frac{d\mathcal{S}}{dt} + p \cdot \frac{d\mathcal{V}}{dt} - \sum_{\forall i} \mu_i \cdot \frac{dx_i}{dt}\Big)$$
$$+ \frac{dn}{dt}(\mathcal{U} - T \cdot \mathcal{S} + p \cdot \mathcal{V} - \sum_{\forall i} \mu_i \cdot x_i) = 0.0 \tag{9.34}$$

Since dn/dt is independent of n, the two parentheses must both be equal to zero:

$$\mathcal{U} = T \cdot \mathcal{S} - p \cdot \mathcal{V} + \sum_{\forall i} \mu_i \cdot x_i \tag{9.35a}$$

$$\frac{d\mathcal{U}}{dt} = T \cdot \frac{d\mathcal{S}}{dt} - p \cdot \frac{d\mathcal{V}}{dt} + \sum_{\forall i} \mu_i \cdot \frac{dx_i}{dt} \tag{9.35b}$$

Equation (9.35a) provides us with an explicit formula for the internal energy stored in a mole of any substance.

It is now time to revisit the previously discussed electrical and mechanical energy storages. An electrical circuit has two separate means to store energy. The energy stored in a capacitor is:

$$E_C = \frac{1}{2} C \cdot u_C^2 \tag{9.36}$$

and thus the power flow into and out of a capacitor is:

$$P_C = \frac{dE_C}{dt} = C \cdot u_C \cdot \frac{du_C}{dt} = u_C \cdot i_C \tag{9.37}$$

Similarly, for an inductor:

$$E_L = \frac{1}{2} L \cdot i_L^2 \quad , \quad P_L = \frac{dE_L}{dt} = L \cdot i_L \cdot \frac{di_L}{dt} = u_L \cdot i_L \tag{9.38}$$

i.e., while the energy formulae look different, the power flow is always expressed as $u \cdot i$.

The situation is similar for the potential and kinetic energy stored in physical bodies. While the energy formulae look different, the power is always $F \cdot v$ for translational motions and $\tau \cdot \omega$ for rotational motions.

This is a general truth. While different types of systems may contain various different mechanisms to store energy, each type of system offers exactly one mechanism for power flow. This is what makes the bond graph so much more versatile than the Hamiltonian. The Hamiltonian is an *energy storage* description technique in which the energy is described in terms of generalized momenta p and generalized displacements q, whereas the bond graph is a *power flow* description technique in which the power is described in terms of efforts e and flows f. Since thermodynamic systems don't have a generalized momentum, they obviously don't possess a Hamiltonian.

However, we shall demonstrate that the bond graph approach still works.

Equation (9.35a) describes the energy that is stored in the matter itself. It is an energy storage which has its origin in the *microscopic* aspects of chemical substances. It describes the cumulative energy contained in the molecular bonds of the chemical substance. This energy storage is a function of six different variable types. Its time derivative, i.e., the power flowing into and out of that energy storage, is described in Eq.(9.35b). It contains three components: a thermal component expressed in the variables T and dS/dt, a hydraulic/pneumatic component expressed in the variables p and dV/dt, and finally a chemical component expressed in the variables μ and dn/dt. Notice that Eq.(9.35a) does not describe three different energy storages, only three facets of one and the same energy storage, whereas Eq.(9.35b) describes three physically different power flows.

However, from a mathematical point of view it is perfectly legitimate to treat the internal energy as three different energy storages: a thermal, a hydraulic/pneumatic, and a chemical energy storage. Let us analyze the internal hydraulic/pneumatic energy storage of a liquid or gaseous substance. The energy is stored as:

$$U_{h/p} = p \cdot V \tag{9.39}$$

Therefore, the power flow can be described as:

$$\dot{U}_{h/p} = \frac{d}{dt}(p \cdot V) = \dot{p} \cdot V + p \cdot \dot{V} \tag{9.40}$$

Equation (9.40) introduces a funny power flow variable. The price that we pay for treating the internal energy as three distinct energy storages is the introduction of three fictitious power flows: $\dot{T} \cdot S$, $\dot{p} \cdot V$, and $\dot{\mu} \cdot n$. The real power flows $T \cdot \dot{S}$, $p \cdot \dot{V}$, and $\mu \cdot \dot{n}$ describe physical power flows into and out of the internal energy storage, whereas the fictitious power flows describe the power flow among the three components of the internal energy storage. Yet, it turns out that this separation is a fruitful concept.

Hydraulic/pneumatic power can flow under conditions of constant pressure (i.e., the so–called isobaric conditions) using the expression $p \cdot \dot{V}$ or under conditions of constant volume (i.e., the so–called isochoric conditions) using the expression $\dot{p} \cdot V$. If both variables are time–dependent, both power flow mechanisms occur simultaneously. Under isobaric conditions, the internal energy storage can accept or release hydraulic/pneumatic power but no hydraulic/pneumatic energy is converted internally to other forms of energy, whereas under isochoric conditions, no hydraulic/pneumatic power flows into or out of the internal energy storage, but the internally stored hydraulic/pneumatic energy can be converted to other energy forms inside the internal energy storage.

The same is true for the thermal storage. Thermal power can flow under conditions of constant temperature (i.e., the so–called isothermic conditions) using the expression $T \cdot \dot{S}$ or under conditions of constant entropy (i.e., the so–called isentropic conditions) using the expression $\dot{T} \cdot S$. If both variables are time–dependent, both power flow mechanisms occur simultaneously.

Similarly, two power flow mechanisms exist for the chemical power. However, by specifying the free thermal and hydraulic/pneumatic conditions, the system is completely determined, i.e., we cannot choose to execute a chemical reaction under conditions of constant chemical potential or constant molar flow rate.

Notice that all three internal energy storages describe properties of the matter itself. They exist irrespective of whether the matter is moved around or kept in one place. If we move a liquid or gas around macroscopically, we must also consider its potential and kinetic energies, which are also expressed in terms of the pressure p, mass flow rate q, and volume V. These energy storages are different from the internal hydraulic/pneumatic energy storage of the matter itself and cannot be written as $p \cdot V$. However, also the macroscopic hydraulic/pneumatic power flows are expressed as $p \cdot \dot{V}$.

Notice that the sign of the internal hydraulic/pneumatic energy term in Eq.(9.35a) is opposite to the sign of the internal thermic and chemical energy terms. We shall explain this fact later.

Two other potential functions are commonly used in thermodynamics:

$$H = U + p \cdot V = T \cdot S + \sum_{\forall i} \mu_i \, n_i \qquad (9.41a)$$

$$G = H - T \cdot S = \sum_{\forall i} \mu_i \, n_i \qquad (9.41b)$$

H is called the *enthalpy* of the substance and G is called the *Gibbs free energy* of the substance. G measures the total amount of chemical (structural) energy stored in the substance. Of course, H and G are also extensive variables that can be made intensive by normalization:

$$\mathcal{H} = \frac{H}{n} = \mathcal{U} + p \cdot \mathcal{V} = T \cdot \mathcal{S} + \sum_{\forall i} \mu_i \, x_i \qquad (9.42a)$$

$$\mathcal{G} = \frac{G}{n} = \mathcal{H} - T \cdot \mathcal{S} = \sum_{\forall i} \mu_i \, x_i \qquad (9.42b)$$

$\mathcal{H}$ is called the *molar enthalpy* of the substance and $\mathcal{G}$ is called the *molar Gibbs free energy* of the substance. $\mathcal{G}$ measures the total amount of chemical (structural) energy stored in 1 mole of the substance.

Let me analyze this relation a little further:

$$G = H - T \cdot S = U + p \cdot V - T \cdot S \qquad (9.43)$$

and therefore:

$$\dot{G} = \dot{U} + \dot{p} \cdot V + p \cdot \dot{V} - \dot{T} \cdot S - T \cdot \dot{S} \qquad (9.44)$$

We plug $\dot{U}$ into Eq.(9.44) from Eq.(9.31):

$$\dot{G} = -\dot{T} \cdot S + \dot{p} \cdot V + \sum_{\forall i} \mu_i \cdot \nu_i \qquad (9.45)$$

However, since:

$$G = \sum_{\forall i} \mu_i \cdot n_i \qquad (9.46)$$

we can also write:

$$\dot{G} = \sum_{\forall i} \dot{\mu}_i \cdot n_i + \sum_{\forall i} \mu_i \cdot \nu_i \qquad (9.47)$$

A comparison to Eq.(9.45) shows that:

$$\sum_{\forall i} \dot{\mu}_i \cdot n_i = \dot{p} \cdot V - \dot{T} \cdot S \qquad (9.48)$$

Notice that the time derivatives of the enthalpy H and Gibbs free energy G contain fictitious power flows. H and G are potential functions from which constitutive laws can be derived, similarly to the Lagrangian in a mechanical system. They are not truly energy functions in a physical sense. Yet if we apply the time derivatives mathematically and allow fictitious power flows to be present in these derivatives, the analysis is mathematically correct and we *can* use H and G as energy functions.

Finally, we know that the total energy in a closed system is constant. If we can neglect other forms of energy (such as mechanical or electrical energy), the internal energy of the closed system is constant:

$$U = \text{constant} \tag{9.49}$$

and therefore, from Eq.(9.31):

$$\dot{U} = T \cdot \dot{S} - p \cdot \dot{V} + \sum_{\forall i} \mu_i \cdot \nu_i = 0.0 \tag{9.50}$$

Equations (9.48) and (9.50) together provide us with two separate power balance equations:

$$p \cdot \dot{V} = T \cdot \dot{S} + \sum_{\forall i} \mu_i \cdot \nu_i \tag{9.51a}$$

$$\dot{p} \cdot V = \dot{T} \cdot S + \sum_{\forall i} \dot{\mu}_i \cdot n_i \tag{9.51b}$$

Equation (9.51a) is a special form of the Gibbs equation, whereas Eq.(9.51b) is one form of the Gibbs–Duhem equation. Equation (9.51a) will have to be modified whenever additional energy storages contribute to the total power flow. We shall show this in Section 9.8 where we describe the mechanisms of photochemistry and electrochemistry. Equation (9.51b) is more general. Since it describes the power balance among the three fictitious energy storages within the true internal energy storage U, Eq.(9.51b) is always correct.

Notice that our discussion evolves around *power* rather than *energy*. Traditionally, the chemical thermodynamics literature discusses the equilibrium (i.e., the steady–state behavior) of the chemical reactions rather than analyzing their dynamics. Consequently, these treatises always concentrate on the conservation of energy rather than the power balance. However, for our purposes, it is more natural to concentrate on the power balance.

In a chemical reaction system, chemical energy is traded among the various substances involved in the various reactions. However, the trade is not always balanced. Either a surplus or lack of chemical power may exist. If the chemical power balance of a reaction system results in surplus chemical power (i.e., $G < 0.0$), the excess power is converted into either thermic power, hydraulic/pneumatic power, or both. If the chemical power balance results in a lack of chemical power (i.e., $G > 0.0$), power must be imported from either the thermic or pneumatic/hydraulic side. A chemical reaction system strives to maintain the smallest amount of chemical energy possible. Consequently, reactions with $G < 0.0$ will take place spontaneously, while reactions with $G > 0.0$ can only occur forcefully, i.e., by forcing power into the chemical system, for example, through external heating.

In a chemical reaction, the total power must be balanced. The chemical power stored in all products minus the chemical power stored in all reactants is the power available for conversion into other forms of power. This power (which can assume either positive or negative values) will be split among thermic power (through either production or consumption of heat) and hydraulic/pneumatic power (through either volume or pressure increase or reduction).

9.4 The Equation of State

We have seen that chemical systems that are completely described through their internal energy contain three internal energy storages: a chemical storage, a thermic storage, and a hydraulic/pneumatic storage. Thus, such systems are defined through six different types of variables, namely, T, S, p, V, μ_i, and n_i. We thus need six types of equations to completely describe the dynamics of the overall system. Until now, we met the *reaction equations*, which are used to compute the ν_i, and the two *power balance equations*, of which Eq.(9.51a) is usually employed to compute the entropy flow $\dot{S}$, and Eq.(9.51b) is used to determine the changes in the chemical potentials $\dot{\mu}_i$.

Since the system has two ports to the environment, a thermal port and a hydraulic/pneumatic port, one relation exists between the two thermal variables and another exists between the two hydraulic/pneumatic variables that are determined by the environment. We can decide to control the temperature of the reaction system such that T stays constant during the entire reaction. In that case, T is modeled as a temperature source, while Eq.(9.51a) determines the resulting entropy flow. This is a very common scenario. A chemical reaction that takes place under conditions of constant temperature is called an *isothermic reaction*. Independently, we can choose one variable between the two hydraulic/pneumatic variables.

If we let the reaction take place in an open container, we can assume that the pressure p stays constant throughout the entire reaction. This is the normal assumption for batch reactors. A chemical reaction that takes place under conditions of constant pressure is called an *isobaric reaction*. Equation (9.51b) shows that, under isothermic and isobaric conditions:

$$\sum_{\forall i} \dot{\mu}_i \cdot n_i = 0.0 \qquad (9.52)$$

We shall see that, in this case:

$$\mu_i = \text{constant} \qquad (9.53)$$

i.e., the chemical potentials of all substances i involved in the reaction are constant.

Alternatively, if we let the reaction take place in a closed container that is constantly kept full, we can assume that the volume V stays constant throughout the entire reaction. This is the normal assumption for tubular reactors. A chemical reaction that takes place under conditions of constant volume is called an *isochoric reaction*.

We are still missing the sixth and last equation. Yet another additional equation exists that relates the various state variables to each other. This is commonly referred to as the *equation of state*.

For an ideal gas, the equation of state is:

$$p \cdot V = n \cdot R \cdot T \qquad (9.54)$$

where p denotes the pneumatic pressure of the gas, V represents its volume, R is the gas constant ($R = 8.314$ J K^{-1} mole^{-1}), T is the absolute temperature of the gas (measured in K), and n denotes the total number of moles.

The equation of state relates the pressure p, volume V, and temperature T to each other. A real gas does not obey Eq.(9.54) exactly. Yet a static relation still exists between the three variables p, V, and T. Several approximate equations for various types of gases have been developed and can be found in the literature. Systems characterized by a static relation among p, V, and T are often called PVT *systems*.

Liquids, on the other hand can be assumed incompressible. Under this assumption, a direct proportionality relation exists between the number of moles of a liquid and its volume:

$$V \propto n \qquad (9.54^{alt})$$

For example, 1 mole of H_2O has a mass of 18 g which, in the liquid phase, occupies a volume of 18 cm^3. The equation of state is used to compute one of the hydraulic/pneumatic variables.

Unfortunately, there is a problem with the equation of state. This equation is actually a steady–state equation, i.e., it is only valid in a strict sense under equilibrium conditions. It should be replaced by a more general equation that is true far from equilibrium. Unfortunately, we haven't been able to find such a generalized equation yet.

Let us now discuss ideal gas phase reactions in more detail. According to Eq.(9.54), if we increase the temperature of a fixed amount of gas, either its pressure or its volume must grow. The same is true if we increase the amount of gas under constant temperature conditions. The power balance equations must reflect this fact. Consequently, Eq.(9.50) must exhibit the variables p and V on one side of the equal sign, and n and T must show up on the other side. This explains the opposite sign of the hydraulic/pneumatic energy term in Eq.(9.35a).

The product $p \cdot V$ is the internal pneumatic energy content of the gas. It would be useful if we could assign a value to the pneumatic energy content of each gaseous component in a gas mixture. Traditionally, this has been accomplished through the introduction of the *partial gas pressure* of a component gas. Let us perform the following experiment. We shall decompose the gas mixture into its component gases and store each component gas in a container of the same size (volume) that the original container had. We measure the pressure of the gases in each of these containers. This is called the partial gas pressure. According to *Dalton's law*, the sum of all partial gas pressures equals the pressure of the gas mixture. We can thus rewrite Eq.(9.54) as follows:

$$\left(\sum_{\forall i} p_i\right) \cdot V = \left(\sum_{\forall i} n_i\right) \cdot R \cdot T \qquad (9.55)$$

where n_i denotes the number of moles of the component gas i. We can then decompose this equation into its individual components:

$$p_i \cdot V = n_i \cdot R \cdot T \qquad (9.56)$$

However, if we operate under conditions of constant temperature T and constant pressure p, it may be more appealing to perform a different experiment. This time, we keep all the gas molecules together, but we sort them such that all Br_2 molecules are in one corner and all H_2 molecules are in another. We can then "measure" the *partial volume* that each of the component gases occupies, and using the same argument as before, we can write:

$$p \cdot V_i = n_i \cdot R \cdot T \tag{9.57}$$

where V_i is the partial volume of component gas i and p is the pressure of the gas mixture. This experiment may be hard to perform in practice, but according to *Avogadro's law*, one mole of an ideal gas under constant pressure and temperature occupies exactly the same volume as one mole of any other ideal gas (a direct consequence of the equation of state). At a pressure of $p = 1$ atm $= 760$ Torr $= 1.0132 \cdot 10^5$ N m^{-2} and a temperature of $T = 0°C = 273.15$ K, a mole of any ideal gas occupies a volume of $V = R \cdot T/p = 22.4$ liter $= 0.0224$ m^3. Therefore, we can compute the partial pressures (Dalton's law) and the partial volumes (Avogadro's law) by use of the mole fractions:

$$p_i = \frac{n_i}{n} \cdot p = x_i \cdot p \tag{9.58a}$$

$$V_i = \frac{n_i}{n} \cdot V = x_i \cdot V \tag{9.58b}$$

We can use the equation of state to determine the volume:

$$V = \frac{n \cdot R \cdot T}{p} \tag{9.59}$$

and if we assume the pressure p and temperature T to be constant, we can compute the derivative of Eq.(9.59) as follows:

$$q = \frac{\nu \cdot R \cdot T}{p} \tag{9.60}$$

or for any component gas:

$$q_i = \frac{\nu_i \cdot R \cdot T}{p} \tag{9.61}$$

Let us plug Eq.(9.61) into the reaction equations:

$$q_{Br_2} = \nu_{Br_2} \cdot (\frac{R \cdot T}{p}) = (-\nu_{k1} + \nu_{k2} - \nu_{k5}) \cdot (\frac{R \cdot T}{p}) \tag{9.62a}$$

$$q_{Br^\bullet} = \nu_{Br^\bullet} \cdot \left(\frac{R \cdot T}{p}\right) = (2\nu_{k1} - 2\nu_{k2} - \nu_{k3} + \nu_{k4} + \nu_{k5}) \cdot \left(\frac{R \cdot T}{p}\right) \quad (9.62b)$$

$$q_{H_2} = \nu_{H_2} \cdot \left(\frac{R \cdot T}{p}\right) = (-\nu_{k3} + \nu_{k4}) \cdot \left(\frac{R \cdot T}{p}\right) \quad (9.62c)$$

$$q_{H^\bullet} = \nu_{H^\bullet} \cdot \left(\frac{R \cdot T}{p}\right) = (\nu_{k3} - \nu_{k4} - \nu_{k5}) \cdot \left(\frac{R \cdot T}{p}\right) \quad (9.62d)$$

$$q_{HBr} = \nu_{HBr} \cdot \left(\frac{R \cdot T}{p}\right) = (\nu_{k3} - \nu_{k4} + \nu_{k5}) \cdot \left(\frac{R \cdot T}{p}\right) \quad (9.62e)$$

The total volume flow is the sum of all component volume flows:

$$q = \sum_{\forall i} q_i = \nu \cdot \left(\frac{R \cdot T}{p}\right) = (\nu_{k1} - \nu_{k2}) \cdot \left(\frac{R \cdot T}{p}\right) \quad (9.63)$$

We can split this equation differently into reaction terms rather than component terms:

$$q_{k1} = (2 - 1) \cdot \nu_{k1} \cdot \frac{R \cdot T}{p} \quad (9.64a)$$

$$q_{k2} = (1 - 2) \cdot \nu_{k2} \cdot \frac{R \cdot T}{p} \quad (9.64b)$$

$$q_{k3} = (2 - 2) \cdot \nu_{k3} \cdot \frac{R \cdot T}{p} \quad (9.64c)$$

$$q_{k4} = (2 - 2) \cdot \nu_{k4} \cdot \frac{R \cdot T}{p} \quad (9.64d)$$

$$q_{k5} = (2 - 2) \cdot \nu_{k5} \cdot \frac{R \cdot T}{p} \quad (9.64e)$$

where the numbers between the parentheses denote the difference of the number of gaseous product particles minus the number of gaseous reactant particles. In the reactions k_3, k_4, and k_5, two gaseous molecules are traded for two other gaseous molecules. Consequently, in accordance with Avogadro's law, the volume does not change, and thus, pneumatic power is neither generated nor absorbed. In reaction k_1, two $Br^\bullet$ atoms are produced for each Br_2 molecule. Consequently, the partial volume grows. In reaction k_2, two $Br^\bullet$ atoms are consumed for each Br_2 molecule that is being generated. Consequently, the partial volume decreases.

Let us call the balance factor between the parentheses multiplied by the reaction flow rate ν_{k_i} the effective reaction flow rate, and denote it with the symbol νe_{k_i}.

Molecules in their liquid and/or solid phases don't contribute to the effective reaction flow rate since the volume occupied by liquids and solid bodies is negligible in comparison to the volume occupied by gases and since they can be assumed noncompressible.

9.5 Chemical Reaction Bond Graphs

The six types of dynamic equations relating the six types of variables to each other can be represented in a bond graph. Figure 9.1 shows the bond graph for the (slightly simplified) hydrogen–bromine reaction system:

$$\text{Br}_2 \xrightarrow{k_1} 2\text{Br}^\bullet \tag{9.65a}$$

$$2\text{Br}^\bullet \xrightarrow{k_2} \text{Br}_2 \tag{9.65b}$$

$$\text{Br}^\bullet + \text{H}_2 \xrightarrow{k_3} \text{HBr} + \text{H}^\bullet \tag{9.65c}$$

$$\text{Br}_2 + \text{H}^\bullet \xrightarrow{k_5} \text{HBr} + \text{Br}^\bullet \tag{9.65d}$$

where the least important reaction k_4 was left out in order to keep the bond graph a planar graph. The bond graph has been drawn under the assumptions of isothermic and isobaric conditions.

In the bond graph, each of the five species Br_2, $\text{Br}^\bullet$, H_2, $\text{H}^\bullet$, and HBr is represented by a 0–junction. Attached to each of these 0–junctions is a new element, the *CS* element. The *CS* element represents a *capacitive source*. In the capacitive source, mass is accumulated. The molar flow into (out of) the capacitive source is the balance of the flows into (out of) the various reactions in which the species is involved, for example:

$$\nu_{\text{Br}_2} = \nu_{k2} - \nu_{k1} - \nu_{k5} \tag{9.66}$$

The species Br_2 is involved in the reactions k_1, k_2, and k_5. The chemical power $\mu_{\text{Br}_2} \cdot \nu_{k1}$ flows into the reaction k_1 and the chemical power $\mu_{\text{Br}_2} \cdot \nu_{k5}$ flows into the reaction k_5. The reaction k_2 delivers the chemical power $\mu_{\text{Br}_2} \cdot \nu_{k2}$ back to the species Br_2. These powers are balanced by the chemical power $\mu_{\text{Br}_2} \cdot \nu_{\text{Br}_2}$, which flows into (out of) the capacitive source. In the capacitive source, the (constant) chemical potential μ_{Br_2} is computed using a formula we haven't discussed yet.

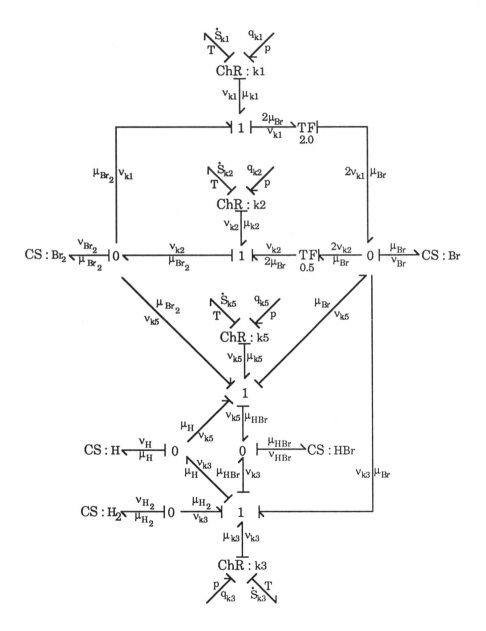

Figure 9.1. Bond graph of the isothermic and isobaric H_2–Br_2 reaction.

Each chemical reaction is represented by a 1–junction. The 1–junction balances the chemical powers into (out of) the reaction with the other types of power, namely, thermic power and hydraulic/pneumatic power. For example, the reaction k_3 receives chemical power from the species Br$^\bullet$ and H_2, and delivers chemical power to the species H$^\bullet$ and HBr. The balance among these chemical powers $\mu_{k3} \cdot \nu_{k3}$ is the power which, in the chemical reaction, will be converted into other types of power. μ_{k3} can thus be computed from the formula:

$$\mu_{k3} = \mu_{H^\bullet} + \mu_{HBr} - \mu_{Br^\bullet} - \mu_{H_2} \qquad (9.67)$$

Attached to each of the 1–junctions is another new bond graph element, the *ChR* element, which describes the chemical reaction. The *chemical reactor ChR* converts power among its chemical, thermic, and hydraulic/pneumatic forms. Assuming that the temperature T is constant (i.e., we operate under isothermic conditions) and the hydraulic/pneumatic pressure p is constant (i.e., we operate under isobaric conditions), the chemical reactor computes the three flow rates, namely, the partial entropy flow $\dot{S}_{k_i}$ of the reaction (from Eq.(9.51a)), the partial volume flow q_{k_i} of the reaction (from Eq.(9.64)), and the partial molar flow ν_{k_i} of the reaction (from Eq.(9.26)).

The *ChR* element is very similar to the previously encountered *RS* element. As the *RS* element, the *ChR* element is basically an ideal energy transducer that balances the power of its ports. No energy is stored inside the *ChR* element.

The equation for the molar flow rate is modulated by the number of moles of all the reactant species. The modulating signal paths have been omitted from the bond graph of Fig.9.1 to avoid overcrowding.

The stoichiometric coefficients are represented by transformers. Reaction k_1 delivers twice the flow rate ν_{k1} to the species Br$^\bullet$. Since the chemical potential of this species is $\mu_{Br^\bullet}$ and since the flow rate out of the 1–junction must be ν_{k1}, we have to assume a chemical potential of $2\mu_{Br^\bullet}$ at the 1–junction to balance the power along this power path.

Let us look a little more closely at the relations between the flow rates and chemical potentials of the component gases on the one hand versus those of the chemical reactions on the other. We can write these relations in a matrix form as follows:

$$
\begin{pmatrix} \nu_{Br_2} \\ \nu_{Br^\bullet} \\ \nu_{H_2} \\ \nu_{H^\bullet} \\ \nu_{HBr} \end{pmatrix}
=
\begin{pmatrix}
-1 & 1 & 0 & 0 & -1 \\
2 & -2 & -1 & 1 & 1 \\
0 & 0 & -1 & 1 & 0 \\
0 & 0 & 1 & -1 & -1 \\
0 & 0 & 1 & -1 & 1
\end{pmatrix}
\cdot
\begin{pmatrix} \nu_{k_1} \\ \nu_{k_2} \\ \nu_{k_3} \\ \nu_{k_4} \\ \nu_{k_5} \end{pmatrix}
\tag{9.68a}
$$

$$
\begin{pmatrix} \mu_{k_1} \\ \mu_{k_2} \\ \mu_{k_3} \\ \mu_{k_4} \\ \mu_{k_5} \end{pmatrix}
=
\begin{pmatrix}
-1 & 2 & 0 & 0 & 0 \\
1 & -2 & 0 & 0 & 0 \\
0 & -1 & -1 & 1 & 1 \\
0 & 1 & 1 & -1 & -1 \\
-1 & 1 & 0 & -1 & 1
\end{pmatrix}
\cdot
\begin{pmatrix} \mu_{Br_2} \\ \mu_{Br^\bullet} \\ \mu_{H_2} \\ \mu_{H^\bullet} \\ \mu_{HBr} \end{pmatrix}
\tag{9.68b}
$$

which can be written in a more compact way:

$$
\vec{\nu}_s = \mathbf{N} \cdot \vec{\nu}_r \quad , \quad \vec{\mu}_r = \mathbf{M} \cdot \vec{\mu}_s
\tag{9.69}
$$

where:

$$
\mathbf{M} = \mathbf{N}'
\tag{9.70}
$$

Notice that, in this example, the matrices $\mathbf{N}$ and $\mathbf{M}$ are singular. In fact, these two matrices don't even have to be square. Consequently, we can compute the vector of the substance flow rates $\vec{\nu}_s$ from the vector of the reaction flow rates $\vec{\nu}_r$, but not vice versa, and we can compute the vector of the reaction potentials $\vec{\mu}_r$ from the vector of the substance potentials $\vec{\mu}_s$, but not vice versa.

Let us analyze the chemical power balance for the hydrogen–bromine reaction under isothermic and isobaric conditions.

$$
\begin{aligned}
\dot{G} &= \sum_{\forall i} (\mu_i \cdot \nu_i) \\
&= \mu_{Br_2} \nu_{Br_2} + \mu_{Br^\bullet} \nu_{Br^\bullet} + \mu_{H_2} \nu_{H_2} + \mu_{H^\bullet} \nu_{H^\bullet} + \mu_{HBr} \nu_{HBr} \\
&= \mu_{Br_2} (-\nu_{k1} + \nu_{k2} - \nu_{k5}) \\
&\quad + \mu_{Br^\bullet} (2\nu_{k1} - 2\nu_{k2} - \nu_{k3} + \nu_{k4} + \nu_{k5}) \\
&\quad + \mu_{H_2} (-\nu_{k3} + \nu_{k4}) \\
&\quad + \mu_{H^\bullet} (\nu_{k3} - \nu_{k4} - \nu_{k5}) \\
&\quad + \mu_{HBr} (\nu_{k3} - \nu_{k4} + \nu_{k5}) \\
&= (-\mu_{Br_2} + 2\mu_{Br^\bullet}) \nu_{k1} \\
&\quad + (\mu_{Br_2} - 2\mu_{Br^\bullet}) \nu_{k2} \\
&\quad + (-\mu_{Br^\bullet} - \mu_{H_2} + \mu_{H^\bullet} + \mu_{HBr}) \nu_{k3} \\
&\quad + (\mu_{Br^\bullet} + \mu_{H_2} - \mu_{H^\bullet} - \mu_{HBr}) \nu_{k4} \\
&\quad + (-\mu_{Br_2} + \mu_{Br^\bullet} - \mu_{H^\bullet} + \mu_{HBr}) \nu_{k5}
\end{aligned}
$$

$$= \mu_{k1}\nu_{k1} + \mu_{k2}\nu_{k2} + \mu_{k3}\nu_{k3} + \mu_{k4}\nu_{k4} + \mu_{k5}\nu_{k5}$$
$$= \sum_{\forall i}(\mu_{k_i} \cdot \nu_{k_i}) \tag{9.71}$$

or using the more compact matrix notation:

$$\dot{G} = \vec{\mu}_s' \cdot \vec{\nu}_s = \vec{\mu}_s' \cdot (\mathbf{N} \cdot \vec{\nu}_r) = \vec{\mu}_s' \cdot (\mathbf{M}' \cdot \vec{\nu}_r)$$
$$= (\vec{\mu}_s' \cdot \mathbf{M}') \cdot \vec{\nu}_r = (\mathbf{M} \cdot \vec{\mu}_s)' \cdot \vec{\nu}_r = \vec{\mu}_r' \cdot \vec{\nu}_r \tag{9.72}$$

Equation (9.72) shows that we can decompose the total chemical power either in terms of component substances or component reactions.

Let us look once more at the power balance equation Eq.(9.51a). We can rewrite this equation in terms of partial volume flows and partial entropy flows as follows:

$$p \cdot \sum_{\forall i} q_i = T \cdot \sum_{\forall i} \dot{S}_i + \sum_{\forall i} \mu_i \cdot \nu_i \tag{9.73}$$

which can also be written in terms of individual reactions rather than individual component gases:

$$p \cdot \sum_{\forall k_i} q_{k_i} = T \cdot \sum_{\forall k_i} \dot{S}_{k_i} + \sum_{\forall k_i} \mu_{k_i} \cdot \nu_{k_i} \tag{9.74}$$

which can be written separately for reaction k_i as:

$$p \cdot q_{k_i} = T \cdot \dot{S}_{k_i} + \mu_{k_i} \cdot \nu_{k_i} \tag{9.75}$$

Equation (9.75) can be used to compute the contribution of the reaction k_i to the overall entropy flow.

The preceding discussion allows us to create a DYMOLA model for each chemical reactor [9.5]. The code for the chemical reactor k_3 is given here.

```
model type ChRk3
    main cut chem(muk3/−nuk3)
    cut thermk3(Temp/−Sdotk3), pneumk3(p/qk3)
    terminal nH2, nBr, V
    parameter R = 8.314
    local k3, nuek3
    k3 = (10 * *11.43) * exp(−82400/(R * Temp))
    nuek3 = 0.0
    p * qk3 = Temp * Sdotk3 + muk3 * nuk3
    p * qk3 = nuek3 * R * Temp
    nuk3 = k3 * nH2 * nBr/V
end
```

The reaction rate constant k_3 depends explicitly on the temperature T. The experimentally determined k_3 characteristic was taken from Moore *et al.* [9.11]. The effective reaction flow rate νe_{k3} is zero. In addition, the model contains three equations describing (i) the power balance across the reaction, (ii) the equation of state for the reaction, and (iii) the reaction rate equation.

Due to the rather complex modulation (the number of terms in the reaction rate equation is variable), we need to create one such model for each reaction in the system.

The *CS* model type can be coded in DYMOLA as follows:

```
model type CS
    main cut chem(mu/nu)
    terminal n, mu0
    mu = mu0
    der(n) = nu
end
```

The *CS* model is sufficiently simple to allow the creation of a generic model type that can be reused for all *CS* elements. The chemical potential μ_i of any substance is constant under isothermic and isobaric conditions. We still need to discuss how to determine the chemical potential μ_{i_0}.

Finally, we need to sum up all the partial volumes (pressures) and entropies in a pneumatic and a thermic model.

```
model type Pneumatic
    main cut pneutot(p/q)
    cut pneuk1(p/-qk1), pneuk2(p/-qk2)
    cut pneuk3(p/-qk3), pneuk4(p/-qk4)
    cut pneuk5(p/-qk5)
    terminal V
    q = qk1 + qk2 + qk3 + qk4 + qk5
    der(V) = q
end

model type Thermic
    main cut thermtot(Temp/-Sdot)
    cut thermk1(Temp/Sdotk1), thermk2(Temp/Sdotk2)
    cut thermk3(Temp/Sdotk3), thermk4(Temp/Sdotk4)
    cut thermk5(Temp/Sdotk5)
    terminal S
    Sdot = Sdotk1 + Sdotk2 + Sdotk3 + Sdotk4 + Sdotk5
    der(S) = Sdot
end
```

We also still need to discuss how we can determine the initial condition for the entropy S.

If we operate under conditions of constant volume rather than constant pressure, i.e., we operate under *isochoric* conditions instead of the previously assumed *isobaric* conditions, the partial pressure approach is more adequate.

Now we need both power balance equations:

$$0.0 = T \cdot \dot{S} + \sum_{\forall i} \mu_i \cdot \nu_i \qquad (9.76a)$$

$$\dot{p} \cdot V = \sum_{\forall i} \dot{\mu}_i \cdot n_i \qquad (9.76b)$$

We know that we can decompose Eq.(9.76a) into either component substances or component reactions. Let us check whether we can do the same with Eq.(9.76b). As a consequence of Eq.(9.69), we can write:

$$\mathbf{n}_s = \mathbf{N} \cdot \mathbf{n}_r + \mathbf{n}_0 \quad , \quad \vec{\mu}_r = \mathbf{M} \cdot \vec{\mu}_s \qquad (9.77)$$

where $\mathbf{n}_0$ is the integration constant. Unfortunately, due to the singularity of $\mathbf{N}$, we cannot determine initial conditions for $\mathbf{n}_r$, which are consistent with the given initial conditions of $\mathbf{n}_s$, and consequently, $\mathbf{n}_0$ cannot be normalized to zero. One possible assignment would be to normalize all initial conditions of $\mathbf{n}_r$ to zero and let the constant vector $\mathbf{n}_0$ equal the vector of initial conditions of $\mathbf{n}_s$:

$$\mathbf{n}_r(t = 0.0) = \mathbf{0.0} \quad , \quad \mathbf{n}_0 = \mathbf{n}_s(t = 0.0) \qquad (9.78)$$

As a consequence, Eq.(9.72) now becomes:

$$\dot{G} = \vec{\mu}'_s \cdot \vec{\nu}_s + \vec{\mu}'_s{}' \cdot \mathbf{n}_s = \vec{\mu}'_r \cdot \vec{\nu}_r + \vec{\mu}'_r{}' \cdot \mathbf{n}_r + \vec{\mu}'_s{}' \cdot \mathbf{n}_0 \qquad (9.79)$$

and therefore:

$$\vec{\mu}'_s{}' \cdot \mathbf{n}_s \neq \vec{\mu}'_r{}' \cdot \mathbf{n}_r \qquad (9.80)$$

Unfortunately, we cannot decompose the second power balance equation into component reactions.

$$0.0 = T \cdot \dot{S}_{k_i} + \mu_{k_i} \cdot \nu_{k_i} \qquad (9.81a)$$
$$\dot{p}_i \cdot V = \dot{\mu}_i \cdot n_i \qquad (9.81b)$$

We use Eq.(9.81a) to determine the partial entropy flow $\dot{S}_{k_i}$ of the reaction k_i, and we use Eq.(9.81b) to determine the pressure change $\dot{p}_i$ induced by the chemical substance i.

We replace Eq.(9.59) with:

$$p = \frac{n \cdot R \cdot T}{V} \tag{9.59^{alt}}$$

and therefore, Eq.(9.61) becomes:

$$\dot{p}_i = \frac{\nu_i \cdot R \cdot T}{V} \tag{9.61^{alt}}$$

We can plug Eq.(9.81b) into Eq.(9.61^{alt}), and obtain:

$$\dot{\mu}_i = \frac{\nu_i}{n_i} \cdot R \cdot T = \frac{\dot{x}_i}{x_i} \cdot R \cdot T \tag{9.82}$$

We shall use Eq.(9.82) to compute the chemical potentials. However, a small problem remains to be solved. Notice that Eq.(9.82) exhibits a singularity at $n_i = 0.0$, i.e., if we start with an unexisting product that is to be generated during the chemical reaction, we run into numerical problems. Physically, this singularity signifies a discontinuity in the chemical potential. No matter has no chemical potential, but already the first molecule that is being generated comes with its own free energy of formation, thus, the chemical potential jumps from an arbitrarily set value of zero to its value of formation. Since we shall assign the free energy of formation as an initial condition to the chemical potential of the product species even before the first molecule of that species has been generated, we can ignore this discontinuity. Numerically, this is best done by adding a small number ϵ to the denominator. ϵ should be at least three to four orders of magnitude smaller than the expected number of moles of the product species.

In the isochoric case, the chemical potentials μ_i are no longer constant. Therefore, the chemical power will now be computed as follows:

$$P_{i_{chem}} = \dot{G}_i = \mu_i \cdot \nu_i + \dot{\mu}_i \cdot n_i \tag{9.83}$$

$P_{i_{chem}}$ is the power that flows into the CS element of the component gas i. The CS element is therefore now a two–port element.

Figure 9.2 shows the bond graph of the isothermic and isochoric hydrogen–bromine reaction:

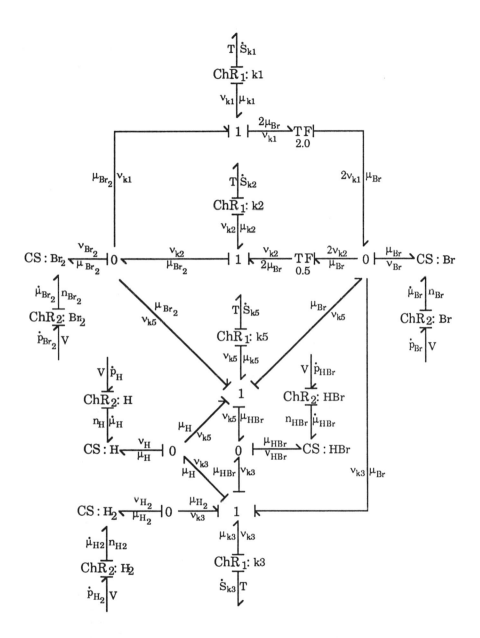

Figure 9.2. Bond graph of the isothermic and isochoric H_2–Br_2 reaction.

Notice that the isochoric bond graph contains two separate *ChR* elements per reaction. The *ChR1* element represents the first power balance equation Eq.(9.51*a*) and the *ChR2* element represents the second power balance equation Eq.(9.51*b*). The isothermic and isobaric reaction was obviously a special case. Since all $\dot{\mu}_i = 0.0$, the second power bond into the *CS* element was eliminated and the n_i were computed directly inside the *CS* element.

The chemical reactor model *ChR1* for the isothermic and isochoric hydrogen–bromine reaction k_3 can be coded as follows:

```
model type ChR1k3
    main cut chem1(muk3/−nuk3),
    cut thermk3(Temp/−Sdotk3)
    terminal nH2, nBr, V
    parameter R = 8.314
    local k3
    k3 = (10 ∗ ∗11.43) ∗ exp(−82400/(R ∗ Temp))
    0.0 = T ∗ Sdotk3 + muk3 ∗ nuk3
    nuk3 = k3 ∗ nH2 ∗ nBr/V
end
```

The chemical reactor model *ChR2* for substance Br_2 can be coded as follows:

```
model type ChR2Br2
    main cut chem2(mudotBr2/−nBr2)
    cut pneuBr2(pdotBr2/V)
    terminal nuBr2
    pdotBr2 ∗ V = mudotBr2 ∗ nBr2
    der(nBr2) = nuBr2
end
```

The second chemical reactor imports the flow rate ν_{Br_2} as a modulation signal. The corresponding *CS* element can be coded as follows:

```
model type CS
    cut chem1(mu/nu), chem2(mudot/n)
    terminal Temp
    parameter R = 8.314, eps = 1.0E-15
    mudot ∗ (n + eps) = nu ∗ R ∗ Temp
    der(mu) = mudot
end
```

Again, the model is sufficiently simple so that we can create a generic model type that will work for all *CS* elements.

The pneumatic model is replaced by:

```
model type Pneumatic
  main cut pneutot(pdot/V)
  cut pneuBr2(pdotBr2/-V), pneuBr(pdotBr/-V)
  cut pneuH2(pdotH2/-V), pneuH(pdotH/-V)
  cut pneuHBr(pdotHBr/-V)
  terminal p
  pdot = pdotBr2 + pdotBr + pdotH2 + pdotH + pdotHBr
  der(p) = pdot
end
```

The thermic model remains the same as before.

Let us discuss next what happens if we operate under the assumption of constant total entropy rather than constant temperature, i.e., if we operate under isentropic conditions rather than isothermic conditions. Let us discuss the combination of isentropic and isobaric conditions. This time, it is more practical to operate on *partial temperature flows* rather than the previously used partial entropy flows. "Partial temperatures" do not exactly represent a physical concept, but mathematically they work well. The power balance equations now turn into:

$$p \cdot q_{k_i} = \mu_{k_i} \cdot \nu_{k_i} \tag{9.84a}$$

$$0.0 = \dot{T}_i \cdot S + \dot{\mu}_i \cdot n_i \tag{9.84b}$$

In this case, the equation of state turns into:

$$p \cdot q_i = \nu \cdot R \cdot T_i + n \cdot R \cdot \dot{T}_i \tag{9.85}$$

where n denotes the total number of moles and μ denotes the total molar flow rate, or plugging in from Eq.(9.84a) this time expressed in terms of component substances:

$$\mu_i \cdot \nu_i = \nu \cdot R \cdot T_i + n \cdot R \cdot \dot{T}_i \tag{9.86}$$

Expanding Eq.(9.86) with S and plugging in Eq.(9.84b), we find:

$$\dot{\mu}_i \cdot n_i \cdot n \cdot R = \nu \cdot R \cdot T_i \cdot S - \mu_i \cdot \nu_i \cdot S \tag{9.87}$$

Figure 9.3 illustrates such reactions by means of the bond graph of the isentropic and isobaric hydrogen–bromine reaction.

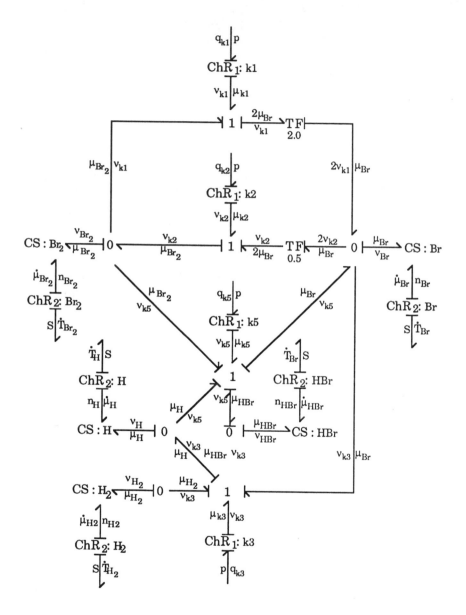

Figure 9.3. Bond graph of the isentropic and isobaric H_2–Br_2 reaction.

The bond graph looks very similar to the previous one. Only the thermic and hydraulic/pneumatic ports have swapped positions. Let me write down the appropriate DYMOLA model types. The isobaric and isentropic *ChR1* reactor for k_3 is now coded as follows:

```
model type ChR1k3
    main cut chem1(muk3/−nuk3)
    cut pneuk3(p/qk3)
    terminal nH2, nBr, V, Temp
    parameter R = 8.314
    local k3
    k3 = (10 ∗ ∗11.43) ∗ exp(−82400/(R ∗ Temp))
    p ∗ qk3 = muk3 ∗ nuk3
    nuk3 = k3 ∗ nH2 ∗ nBr/V
end
```

and the *ChR2* reactor for Br_2 is coded as follows:

```
model type ChR2Br2
    main cut chem2(mudotBr2/−nBr2)
    cut thermBr2(TdotBr2/−S)
    terminal nuBr2
    parameter eps = 1.0E-15
    0.0 = TdotBr2 ∗ (S + eps) + mudotBr2 ∗ nBr2
    der(nBr2) = nuBr2
end
```

The *CS* model type computes $\dot{\mu}_i$ from Eq.(9.87):

```
model type CS
    cut chem1(mu/nu), chem2(mudot/n)
    terminal Ttot, S, ntot, nutot
    local Temp
    parameter R = 8.314, eps = 1.0E-15
    Temp = Ttot ∗ n/ntot
    mudot ∗ (n + eps) ∗ ntot ∗ R = nutot ∗ R ∗ Temp ∗ S − mu ∗ nu ∗ S
    der(mu) = mudot
end
```

We notice the same singularity problem as in the isothermic and isochoric case. The pneumatic model type is the same as for the isothermic and isobaric case. The thermic model looks now as follows:

```
model type Thermic
  main cut thermtot(Tdot/−S)
  cut thermBr2(TdotBr2/S), thermBr(TdotBr/S)
  cut thermH2(TdotH2/S), thermH(TdotH/S)
  cut thermHBr(TdotHBr/S)
  terminal Temp
  Tdot = TdotBr2 + TdotBr + TdotH2 + TdotH + TdotHBr
  der(Temp) = Tdot
end
```

This time we need an additional global model that computes the total number of moles and the total molar flow:

```
model type Chemical
  terminal nuBr2, nuBr, nuH2, nuH, nuHBr
  terminal ntot, nutot
  nutot = nuBr2 + nuBr + nuH2 + nuH + nuHBr
  der(ntot) = nutot
end
```

The *chemical* model is not connected to the bond graph by means of power bonds, only by means of modulating signal paths.

The isentropic and isochoric case is more difficult. The first power balance equation turns into:

$$\sum_{\forall i} \mu_i \cdot \nu_i = 0.0 \qquad (9.88a)$$

which cannot be decomposed into individual components. Since the ν_i are computed from the reaction rate equations, at least one of the μ_i must be computed from Eq.(9.88a). However, since the other power balance equation now turns into:

$$\dot{p}_i \cdot V = \dot{T}_i \cdot S + \dot{\mu}_i \cdot n_i \qquad (9.88b)$$

which references all $\dot{\mu}_i$, the set of equations is structurally singular.

Finally, if we assume that the reactions operate under complete thermic insulation, i.e., that the total thermic power adds up to zero, we call this an *adiabatic* reaction system. In an adiabatic reaction system, neither the temperature nor the entropy are constant. Instead, the external condition can be written as:

$$\sum_{\forall i} (T \cdot \dot{S}_i + \dot{T} \cdot S_i) = 0.0 \qquad (9.89)$$

Let us look at the adiabatic and isobaric reaction system. We decide to operate on partial entropy flows again. The power balance equations are thus written as:

$$p \cdot q_i = T \cdot \dot{S}_i + \mu_i \cdot \nu_i \qquad (9.90a)$$
$$0.0 = \dot{T} \cdot S_i + \dot{\mu}_i \cdot n_i \qquad (9.90b)$$

and the equation of state is written as:

$$p \cdot q_i = n_i \cdot R \cdot \dot{T} + \nu_i \cdot R \cdot T \qquad (9.91)$$

Equation (9.89) can be used to compute $\dot{T}$, Eq.(9.90a) is used to compute $\dot{S}_i$, Eq.(9.90b) is used to determine $\dot{\mu}_i$, and Eq.(9.91) is used to find q_i. Unfortunately, this equation system contains a nasty algebraic loop that goes right across all the equations.

Let us discuss some additional properties of chemical reaction networks as they are exposed by the bond graph. Looking at the definitions for the reaction flow rates according to Eq.(9.26), it is obvious that the reaction flow rates can never be negative and that they are zero only if one of the reactant populations has been exhausted. It is also clear that the Gibbs free energy (and therefore the chemical potential) of any chemical species is always positive except for $T = 0$ K. Consequently, the chemical bonds in the chemical reaction bond graph describe correctly the direction of power flow through the network.

According to Eq.(9.67), the chemical potential of a reaction is defined as the balance between the sum of the chemical potentials of all products minus the sum of the chemical potentials of all reactants. Consequently, the chemical potential of a reaction can be either positive or negative. The chemical bonds that connect the chemical network with the two reactors do not necessarily represent the direction of power flow. A *spontaneous reaction* requires that $G_{k_i} < 0.0$, and therefore, $\mu_{k_i} < 0.0$. Consequently, spontaneous reactions occur "downhill," i.e., they take place if the sum of the chemical potentials of all products is smaller than the sum of the chemical potentials of all reactants.

According to Eq.(9.66), the flow rate of a chemical species is defined as the balance between the sum of the flow rates of all reactions that produce that species minus the sum of the flow rates of all reactions that consume the species. Consequently, the chemical bonds that connect the chemical network with the capacitive source do not

necessarily represent the direction of power flow. Chemical energy
is being stored in the CS element of a given species whenever more
of that species is produced than consumed and it is being depleted
otherwise.

9.6 Energies of Formation

Before we can solve our set of differential equations, we need to
discuss what the correct initial conditions are. The n_i, V_i, p_i, and T
are easily measurable. The T_i from the isentropic and isobaric case
can be easily determined from the equation:

$$p \cdot V_i = n \cdot R \cdot T_i \qquad (9.92)$$

The only difficulties that we have are related to the S_i and μ_i. In
the isothermic and isobaric case, they are obviously related to each
other through the equation:

$$U_i = -p \cdot V_i + T \cdot S_i + \mu_i \cdot n_i \qquad (9.93)$$

which can be rewritten as:

$$H_i = U_i + p \cdot V_i = T \cdot S_i + \mu_i \cdot n_i \qquad (9.94)$$

Thus, it will suffice to determine two initial energy values, such as
H_i and S_i, and then we can determine U_i and μ_i from Eq.(9.94).

Let me explain how these energies are determined in practice.
The reaction enthalpy H_{k_i} was introduced because it is a quantity
that can be conveniently measured. Since the enthalpy is a state
function, the total reaction enthalpy generated/consumed during the
transformation of a set of reactants into a set of products does not
depend on how this transformation was accomplished. For example,
we can compute the reaction enthalpy ΔH_{k1} of the reaction:

$$C_3H_8 + 5O_2 \overset{k_1}{\to} 3CO_2 + 4H_2O \qquad (9.95)$$

in the following way: We decompose the reaction into *primitive reactions*, i.e., into reactions that produce individual molecules out of
their elementary atoms:

$$3C + 4H_2 \rightarrow C_3H_8 \Rightarrow H^f_{C_3H_8} \qquad (9.96a)$$

$$C + O_2 \rightarrow CO_2 \Rightarrow H^f_{CO_2} \qquad (9.96b)$$

$$H_2 + \frac{1}{2}O_2 \rightarrow H_2O \Rightarrow H^f_{H_2O} \qquad (9.96c)$$

The enthalpies of chemical elements have been (arbitrarily) normalized to zero except for elements that under normal conditions (i.e., room temperature and atmospheric pressure) are in their gas phase. In those cases, the enthalpy of the stable gas form has been normalized to zero. Therefore, $H^f_C = 0.0$, and $H^f_{H_2} = 0.0$. Consequently, if we know the so-called *enthalpies of formation* H^f of the individual components of the chemical reaction, it is easy to compute the reaction enthalpy:

$$\Delta H_{k1} = 3 \cdot H^f_{CO_2} + 4 \cdot H^f_{H_2O} - H^f_{C_3H_8} \qquad (9.97)$$

It is common to work with the normalized quantities, i.e., with *molar enthalpies of formation*:

$$\Delta \mathcal{H}_{k1} = 3 \cdot \mathcal{H}^f_{CO_2} + 4 \cdot \mathcal{H}^f_{H_2O} - \mathcal{H}^f_{C_3H_8} \qquad (9.98)$$

The molar enthalpy of formation of any pure substance depends on the temperature. The molar enthalpy of formation of a nonideal gas depends furthermore on the gas pressure:

$$\mathcal{H}^f_i = \mathcal{H}^f_i(T, p) \qquad (9.99)$$

Since $\mathcal{H}^f_i(T, p)$ is a state function, i.e., it is independent of how the compound is being produced, we can "manufacture" the compound under "normal" conditions (such as $T = T_0 = 25°C$, and $p = p_0 = 760$ Torr) and then modify the temperature and pressure of the already "manufactured" compound separately:

$$\mathcal{H}^f_i(T_1, p_1) = \mathcal{H}^f_i(T_0, p_0) + \int_{T_0}^{T_1} \frac{\partial \mathcal{H}^f_i(T, p_0)}{\partial T} dT + \int_{p_0}^{p_1} \frac{\partial \mathcal{H}^f_i(T_1, p)}{\partial p} dp \qquad (9.100)$$

The constant term $\mathcal{H}^f_i(T_0, p_0)$ is called the *standard molar enthalpy of formation* of the pure substance i. For most pure chemical substances, the standard molar enthalpies of formation have been tabulated and can be found in the literature, for example, in Perry *et al.* [9.14] (Table 3–206).

The temperature correction term can be computed from the heat capacity of the substance since:

$$\frac{\partial \mathcal{H}_i^f(T, p_0)}{\partial T} = c_{i_p}(T) \qquad (9.101)$$

i.e., the partial derivative of the molar enthalpy with respect to temperature is the heat capacity c_{i_p} of the pure substance i measured under the assumption of constant pressure. The heat capacity of pure substances depends on the temperature. In the vicinity of atmospheric pressure conditions, we can approximate the heat capacity through a quadratic polynomial:

$$c_{i_p}(T) = \alpha_i + \beta_i \cdot T + \gamma_i \cdot T^2 \qquad (9.102)$$

The α_i, β_i, and γ_i coefficients have also been tabulated, at least for most pure gases. $c_{i_p}(T)$ changes abruptly at the transition from one phase to another, for example, when going from the liquid phase to the gas phase. Therefore, Eq.(9.102) should not be used to integrate across a phase transition.

The pressure correction is only necessary for nonideal gases. Correction terms have also been tabulated.

We are now able to compute the enthalpy of formation $H_i^f(T, p)$ for most pure substances at arbitrary values of temperature and pressure and therefore we can compute the reaction enthalpy for an arbitrary reaction. Notice that the reaction enthalpy does not appear explicitly in the reaction kinetics equations. However, it enters the reaction kinetics equations implicitly through the temperature dependence of the reaction rate constants. The enthalpy of formation will prove useful for the computation of the chemical potential. The chemical potential of the involved substances is needed to determine the chemical potential of the reaction. This in turn enables us to compute the excess/lacking chemical power needed to balance the total power of the reaction. This power, together with the equation of state, will determine the change in the thermic and hydraulic/pneumatic variables that finally influence the reaction rate constants.

I mentioned before that since chemical systems contain only one sort of chemical energy storage, the molar flow rate equations (the reaction kinetics equations) are decoupled from the equations that determine the chemical potential (the thermodynamics equations). However, this is only true as long as the chemical energy is considered in isolation. As soon as we couple the chemical model with the thermic and hydraulic/pneumatic models, additional energy storage elements enter the system. Thereby we create an indirect coupling between the equations that determine the flow rate and those that determine the potential.

A second property that we need to consider is the entropy of the system. We shall again use the normalized molar entropy. The entropy is another state function, thus we can use the same approach as for the enthalpy:

$$S_i^f(T_1, p_1) = S_i^f(T_0, p_0) + \int_{T_0}^{T_1} \frac{\partial S_i^f(T, p_0)}{\partial T} dT + \int_{p_0}^{p_1} \frac{\partial S_i^f(T_1, p)}{\partial p} dp \quad (9.103)$$

The entropy of any pure substance has been arbitrarily normalized to zero for $T = 0$ K. Therefore, if we take T_0 to be zero, we can cancel the constant term $S_i^f(T_0, p_0)$ from Eq.(9.103). The second term can again be computed from the heat capacity since:

$$\frac{\partial S_i^f(T, p_0)}{\partial T} = \frac{c_{i_p}}{T} \quad (9.104)$$

Again, we have to be a little careful with phase transitions. Let us denote the melting temperature of a substance i with T_{m_i} and the boiling temperature of that substance with T_{b_i}. In that case, we can compute the *molar entropy of formation* (sometimes called the *absolute entropy*) of gas i as follows:

$$S_i^f(T) = \int_0^{T_{m_i}} \frac{c_{i_p}}{T} dT + \frac{\Delta \mathcal{H}_{m_i}^f}{T} + \int_{T_{m_i}}^{T_{b_i}} \frac{c_{i_p}}{T} dT + \frac{\Delta \mathcal{H}_{b_i}^f}{T} + \int_{T_{b_i}}^{T} \frac{c_{i_p}}{T} dT \quad (9.105)$$

where $\Delta \mathcal{H}_{m_i}^f$ denotes the abrupt change of the molar enthalpy of formation from the solid to the liquid phase and $\Delta \mathcal{H}_{b_i}^f$ denotes the abrupt change of the molar enthalpy of formation from the liquid to the gas phase.

We still need to discuss the third term. One of the so-called Maxwell relations shows that:

$$\frac{\partial S}{\partial p} = -\frac{\partial V}{\partial T} \quad (9.106)$$

Let me prove the correctness of this relation. We have seen before in Eq.(9.45) that:

$$\dot{G} = -\dot{T} \cdot S + \dot{p} \cdot V + \sum_{\forall i} \mu_i \cdot \nu_i$$

However, since we realize that $G = G(T, p, n_1, \ldots, n_c)$, we can also evaluate $\dot{G}$ from the complete differential:

$$dG = \frac{\partial G}{\partial T} \cdot dT + \frac{\partial G}{\partial p} \cdot dp + \sum_{\forall i} \left(\frac{\partial G}{\partial n_i} \cdot dn_i \right)$$

$$= -S \cdot dT + V \cdot dp + \sum_{\forall i} (\mu_i \cdot dn_i) \tag{9.107}$$

A comparison of coefficients shows that:

$$\frac{\partial G}{\partial T} = -S \quad , \quad \frac{\partial G}{\partial p} = V \tag{9.108}$$

and therefore:

$$\frac{\partial}{\partial p} \left(\frac{\partial G}{\partial T} \right) = -\frac{\partial S}{\partial p} = \frac{\partial}{\partial T} \left(\frac{\partial G}{\partial p} \right) = \frac{\partial V}{\partial T} \tag{9.109}$$

as claimed earlier.

For an ideal gas, we can compute V from the equation of state Eq.(9.54), and thus:

$$\int_{p_0}^{p_1} \frac{\partial S_i^f(p)}{\partial p} dp = -\int_{p_0}^{p_1} \frac{\partial V_i(p)}{\partial T} dp = -n_i \cdot R \cdot \int_{p_0}^{p_1} \frac{1}{p} dp = -n_i \cdot R \cdot \log\left(\frac{p_1}{p_0}\right) \tag{9.110}$$

The normalized version of Eq.(9.110) looks as follows:

$$\Delta S_i^f(p) = -x_i \cdot R \cdot \log\left(\frac{p_1}{p_0}\right) \tag{9.111}$$

As in the case of the enthalpy, molar entropies of formation have been tabulated and can be found in the literature, usually for room temperature ($T = 25°C$). The tabulated values contain only the temperature term, not the pressure term.

Finally, let us look at the third energy function:

$$G_i^f = H_i^f - T \cdot S_i^f \tag{9.112}$$

G_i^f is the Gibbs free energy of formation of the pure substance i, which is a measure of the chemical energy stored in the substance.

The normalized version of Eq.(9.112) can be written as:

$$\mathcal{G}_i^f = \mathcal{H}_i^f - T \cdot \mathcal{S}_i^f \tag{9.113}$$

Of course, there is no need to tabulate $\mathcal{G}_i^f$ if we have tables for $\mathcal{H}_i^f$ and $\mathcal{S}_i^f$. Most references therefore tabulate only two of the three variables. Usually, the "enthalpy of formation" $\mathcal{H}_i^f$ is tabulated, but while some references tabulate the "absolute entropy" $\mathcal{S}_i^f$, others tabulate the "free energy" $\mathcal{G}_i^f$.

The chemical potential of the pure substance i is just another name for its *molar Gibbs free energy of formation*. Using the definitions given earlier, we find that:

$$
\begin{aligned}
\mu_i^f(T,p) &= \mathcal{G}_i^f(T) + x_i \cdot R \cdot T \cdot \log(\frac{p}{p_0}) \\
&= \mathcal{H}_i^f(T) - T \cdot \mathcal{S}_i^f(T) + x_i \cdot R \cdot T \cdot \log(\frac{p}{p_0}) \tag{9.114}
\end{aligned}
$$

where $\mathcal{G}_i^f(T)$ contains the tabulated constant values plus the temperature correction, which is a function of the heat capacity. The logarithmic term stems from the pressure correction of the entropy. Notice that Eq.(9.114) is not in conflict with the second power balance equation for isothermic and isobaric reactions. Although μ_i^f is time–varying (with x_i), the sum over all μ_i^f is still constant. However, if we choose the reference pressure p_0 to be identical to the operating pressure p, the time–dependent term drops out of Eq.(9.114), and μ_i^f is indeed constant.

At this point, we have learned how to determine the chemical potential of any substance for arbitrary values of temperature and pressure. Under isothermic and isobaric conditions, the chemical potential μ_i is a constant.

The literature on chemical thermodynamics is unfortunately full of myths. The majority of references include in their formula for the chemical potential a term that is logarithmic in the mole fraction. This is justified for ideal gas reactions under isothermic and isochoric conditions, where it is possible to integrate the differential equations for the μ_i explicitly. According to Eq.(9.82):

$$\dot{\mu}_i = \frac{\dot{x}_i}{x_i} \cdot R \cdot T$$

which can be integrated into:

$$\mu_i = \mu_i^f + (R \cdot T) \cdot \log(x_i) \tag{9.115}$$

However, this is a special case. Most references suppress the information that Eq.(9.115) is only valid for isothermic and isochoric reactions among ideal gases. In our approach, we don't need to use Eq.(9.115), since we treat the thermal system dynamically and can simply integrate the differential equations for μ_i. Our approach is thus much more general.

Most references derive Eq.(9.115) from one of the Maxwell relations, and therefore claim that it holds true for all ideal gases under arbitrary operation conditions. However, they use a partial derivative with respect to the pressure p in the derivation. This doesn't make sense under isobaric conditions.

Notice that all the equations shown in this section are strictly steady–state equations. They should not be used as part of a dynamic simulation study, and indeed, we won't use them for such a purpose. However, we use them to determine initial conditions for the chemical potentials of the component species and sometimes to determine initial conditions for their entropies. This is justified if we assume that the reaction to be simulated starts out from steady–state conditions. Once we have established the initial state, we don't use these equations any longer.

9.7 Continuous Reactors

Until now, we have discussed closed (autonomous) reaction systems. However, many chemical reactors are operated under conditions of constant inflow of reactants and constant outflow of products, whereby the total volume inside the reactor remains constant.

The simplest continuous reactor is the *continuously stirred tank reactor*, which is frequently abbreviated CSTR. The CSTR can be modeled easily by adding flow sources to all 0–junctions in the reaction system. The flow sources represent the balance between inflow and outflow. Let us look at the simple reaction:

$$A \xrightarrow{k_1} B + C \tag{9.116}$$

which is to describe an isothermic and isobaric reaction among three liquids. Since neither the pressure nor the volume will change, the hydraulic/pneumatic port of the reactor can be eliminated. We assume that a constant flow rate of liquid A is added to the system:

$$\nu_{A_{in}} = c = \text{constant} \tag{9.117}$$

Under the assumption of an ideal mixture of the three liquids, we can model the outflow through modulated flow sources. The amount of liquid of type i is proportional to its own mole fraction:

$$\nu_{i_{out}} = c \cdot \dot{x}_i = \frac{c \cdot \dot{n}_i}{n} \tag{9.118}$$

Therefore, the three flow sources can be modeled as follows:

$$\nu_{A_0} = c \cdot \left(\frac{n - n_A}{n}\right) \tag{9.119a}$$

$$\nu_{B_0} = -c \cdot \left(\frac{n_B}{n}\right) \tag{9.119b}$$

$$\nu_{C_0} = -c \cdot \left(\frac{n_C}{n}\right) \tag{9.119c}$$

Figure 9.4 shows the bond graph of the isothermic and isobaric CSTR.

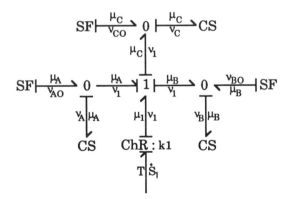

Figure 9.4. Bond graph of a simple CSTR reaction.

The most commonly used reactors are fixed bed reactors. Fluids are transported through a tube that is filled with gravel. The fluid reacts with the surface of the grains. This reactor type is much more complicated to model. The common approach is to compartmentalize the tubular reactor, i.e., cut it into pieces of length Δz. We want to assume that each compartment is homogeneous, i.e., the same temperature, entropy, pressure, volume, and chemical composition can be used throughout the compartment. For each compartment, we can set up a chemical reaction model similar to the one shown earlier. In addition, we need to model the heat flow through the tube using a diffusion model, as presented in Chapter 8. The hydraulic/pneumatic system is modeled through wave equations with an additional friction term as shown in the research portion of Chapter 8. It is important to model each component fluid separately, since the compressibility and friction are different for each of them. Consequently, each chemical species moves through the tube with a different speed. This fact is being exploited by chemical engineers in some of the common separation techniques, such as thin layer chromatography or column chromatography. These separation techniques sort the molecules according to either their affinity or their molecular sizes. Both are dissipative phenomena. The thermic and hydraulic/pneumatic models are coupled through the ChR elements within each compartment. The volume flow can be used to compute the molar flow between the compartments. These molar flows will modulate flow sources that are attached to the 0–junctions, just as in the case of the CSTR. The modulated flow sources are now two–port elements since the molar flow out of one compartment flows directly into the next, i.e., modulated flow sources are placed between each 0–junction of one compartment and the corresponding 0–junction of the next compartment.

Let us assume that a reactant A in an aqueous solution travels through the tubular reactor. The reactant A is decomposed in the reactor into the components B and C. Thus, further down the tube, A, B, and C are mixed in the solution. Each of these components travels with a different speed. We have seen in Chapter 8 that the flow of a fluid through a tube can be described by the wave equation:

$$\frac{\partial^2 p}{\partial t^2} = c^2 \cdot \frac{\partial^2 p}{\partial z^2} \tag{9.120}$$

The wave equation (9.120) assumes that the friction of the fluid is negligible, while its compressibility is not negligible. For our purposes, we don't want to make this assumption, since it is precisely the friction that is responsible for the varying speed. The R elements in the bond graph model the friction (affinity) effects. We notice that this model contains two different energy storages. This seems to be in contradiction with our previous statement that the hydraulic/pneumatic energy is stored in a single (capacitive) container of the type $p \cdot V$. In the past, we assumed that the fluid was stationary. $p \cdot V$ is a structural energy storage that is part of the internal energy of the matter. However, now our fluid moves around. Therefore, the mechanical properties of the matter become important, i.e., the fluid carries both potential energy and kinetic energy.

Figure 9.5 shows a bond graph of the transport of a liquid through a tube.

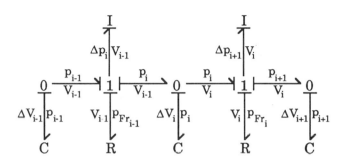

Figure 9.5. Bond graph of liquid transport through a tube.

The C elements in the bond graph in Fig. 9.5 store the potential energy $C \cdot p^2/2$ of the fluid, whereas the I elements store the kinetic energy $I \cdot q^2/2$ of the fluid.

How does the fluid transport equation interact with the chemical reaction model? Figure 9.6 shows an excerpt of the isothermic and isochoric bond graph of the previously described reaction. Only two compartments and only the A species are depicted.

The RS element produces friction heat, which is combined with the reaction heat in the thermic model. Also, we need a diffusion equation (not shown in the graph) that models the flow of heat through the system. The total heat produced in one compartment is split into a portion that is lost from the system (diffusion through the tubular walls) and another portion that is transported down to the next compartment.

The C element is now a CS element since the change in the partial pressure of component A in the compartment i is now simultaneously caused by the potential energy of the fluid and the chemical reaction in the compartment:

$$\dot{p}_{A_i} = \frac{1}{C} \cdot \Delta V_{A_i} + \text{reaction term} \qquad (9.121)$$

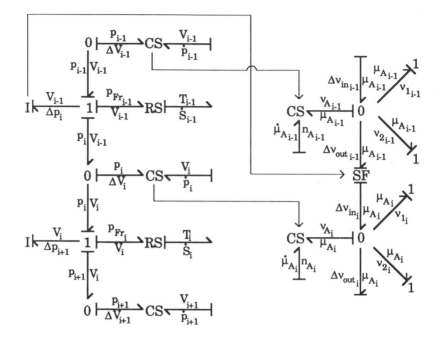

Figure 9.6. Excerpt of a bond graph of a tubular reactor.

The partial pressure that is computed in the CS element of the transport model modulates the equation for the computation of $\dot{\mu}_{A_i}$ in the CS element of the reaction model.

The I element computes the volume flow rate q_{A_i} from which the molar flow rate $\Delta\nu_{A_i}$ can be derived. Thus, the I element of the transport model modulates the SF element of the reaction model.

If we model the flow of a pure liquid such as H_2O through a tube under isothermic conditions, we find that a fixed ratio exists between the volume of the liquid and the number of moles, or between the volume flow rate and the molar flow rate:

$$n = r \cdot V \quad , \quad \nu = r \cdot q \tag{9.122}$$

However, from Eq.(9.81b), we find that:

$$\frac{n}{V} = \frac{\dot{p}}{\dot{\mu}} = r \tag{9.123}$$

and therefore:

$$p = r \cdot \mu + \text{constant} \tag{9.124}$$

where the integration constant can be normalized to zero. Consequently, the chemical variables μ and ν are related to the hydraulic variables p and V through a transformer. Therefore, we can write the wave equation in either hydraulic or chemical variables, i.e.:

$$\frac{\partial^2 p}{\partial t^2} = c^2 \cdot \frac{\partial^2 p}{\partial z^2} \quad , \quad \frac{\partial^2 q}{\partial t^2} = c^2 \cdot \frac{\partial^2 q}{\partial z^2} \tag{9.125a}$$

$$\frac{\partial^2 \mu}{\partial t^2} = c^2 \cdot \frac{\partial^2 \mu}{\partial z^2} \quad , \quad \frac{\partial^2 \nu}{\partial t^2} = c^2 \cdot \frac{\partial^2 \nu}{\partial z^2} \tag{9.125b}$$

Since water flows from a point of high pressure to a point of low pressure, and since, in the equilibrium state, all pressures are identical, the same holds true for the chemical variables. Molar flows within one species occur from a point of high chemical potential to a point of low chemical potential, and in the equilibrium state, the chemical potentials of all molecules of the same species are equal.

Notice a possible point of confusion here. The term "molar flow" has now been used in two quite different contexts. On the one hand, it denotes the *physical transport* of matter from one point in space and time to another, while on the other hand, it describes the *transformation* of one chemical species into another during a chemical reaction. For illustration purposes, the change per unit time in the number of plumbers in a community can be computed as the difference between the number of plumbers who move into the community minus those who move away (transport), plus the difference between the number of young adults who finish their education as plumbers minus the number of plumbers who die or decide to go for a career change (transformation).

The concept of matter flowing from a point of high chemical potential to a point of low chemical potential applies to both phenomena, i.e., matter "flows" (in both senses) "downhill" from points of high chemical potential to points of low chemical potential. In a true equilibrium, all chemical potentials would thus have to be constant and equal. However, the term "equilibrium" has been slightly generalized in the context of chemical reaction systems. We talk about a *chemical equilibrium state* when all molar flow rates ν_i into and out of the CS elements have died out and when all chemical reaction potentials μ_{k_i} are zero, but this does not imply that the molar flow rates inside the chemical network (i.e., the reaction flow rates ν_{k_i})

have died out as well, nor does it imply that all chemical potentials of the species μ_i are equal. We call this generalized type of equilibrium the *equilibrium of flow*.

9.8 Photochemistry and Electrochemistry

We have seen that we can determine one thermal variable and one hydraulic/pneumatic variable externally, for example, by externally controlling the reaction temperature and by operating under atmospheric pressure conditions. In this way, we can manipulate the chemical potential and thereby the chemical energy, but we seem not to influence the flow rate equations at all. The flow rate equations were described as autonomous except for the volume influence mentioned earlier in this chapter. However, this is a simplification. It turns out that the reaction rate constants are strongly temperature-dependent. Figure 9.7 shows the k_1 parameter of the hydrogen–bromine reaction plotted over temperature.

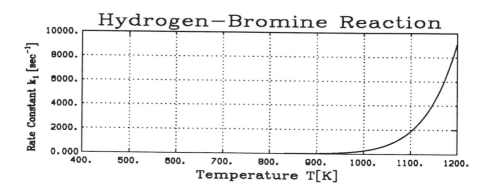

Figure 9.7. Temperature dependence of reaction rate constants.

How can this be explained? Most reactions require an *activation energy* to occur, i.e., they need to pick up some energy before they can take place, even if they release more energy during the reaction than what they needed to pick up in the first place. Thus a reaction of the type shown in Eq.(9.116):

$$A \overset{k_1}{\rightarrow} B + C$$

will usually require activation energy since otherwise A would be a highly unstable species (unless it is being regenerated by another simultaneously occurring reaction). Where does this activation energy come from? It stems from a collision with another molecule, i.e., the reaction of Eq.(9.116) should probably be rewritten as:

$$A + M \xrightarrow{k_1} B + C + M \qquad (9.116^{alt})$$

where M is a catalyst that is not further involved in the reaction except by shedding some of its own energy. Consequently, the k_1 reaction of the hydrogen bromine system should be rewritten as:

$$Br_2 + M \xrightarrow{k_1} 2Br^{\bullet} + M \qquad (9.126)$$

where M can be anything, even another Br_2 molecule. The energy that can be shed most easily is the microscopic thermally induced kinetic energy of the molecules, which increases with temperature. During a collision, the colliding molecules change their direction and speed. It is then easy for a molecule to steal some of the kinetic energy of the colliding partner and transform it into the required activation energy. Thus, reaction rate constants always grow with increasing temperature.

In Fig.9.7^{alt}, we replotted Fig.9.7 using a double logarithmic scale.

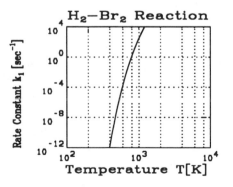

Figure 9.7^{alt}. Temperature dependence of reaction rate constants.

We notice the almost linear relation between $\log(k_1)$ and $\log(T)$. On the basis of this experimental observation, Arrhenius formulated the following law for reaction rate constants:

$$k = A \cdot \exp(-\frac{E_a}{R \cdot T}) \tag{9.127}$$

where A is the frequency factor (denoting the frequency of collisions between molecules) and E_a is the activation energy. The collision frequency is also temperature–dependent. More modern treatises have refined Arrhenius' law in the following way:

$$k = k_0 \cdot T^m \cdot \exp(-\frac{E_a}{R \cdot T}) \tag{9.127^{alt}}$$

where m is a real number between 0.0 and 4.0, often $m \approx 0.5$.

By changing the reaction temperature, we can influence the reaction kinetics. At higher temperatures, all reactions are accelerated; reactions that were not possible at low temperatures can suddenly take place if the collision among molecules has become sufficiently violent so that it can deliver the required activation energy. Since the temperature dependence of the reaction rate constants is different for the various k_i coefficients in a reaction system, even the equilibrium state can be influenced by a change in the reaction temperature. Detailed information on reaction rate coefficients and their temperature dependence can be found in Kerr and Moss [9.7].

Is temperature the only mechanism that we have at our disposal for influencing the reaction dynamics? It is logical that the collision frequency among gas molecules also depends on the gas pressure. However, since the exponential factor in the Arrhenius equation is the dominant factor, the pressure dependence is not usually explored.

However, other mechanisms can also provide the required activation energy. One such mechanism is light. From Fig.9.7, we see that Br_2 does not spontaneously decompose into $Br^{\bullet}$ at room temperature, but without any $Br^{\bullet}$ radicals, the other reactions cannot take place. This is what we observe when we keep the hydrogen bromine mixture at room temperature in a dark closet. Yet if we expose the gas mixture to the influence of light, a reaction takes place.

The energy content of one photon is $h \cdot \nu$, where h is the *Planck constant* $h = 6.625 \cdot 10^{-34}$ J sec, and ν is the frequency (color) of the light. Thus, optical power can be expressed as:

$$P_{opt} = (h\nu) \cdot I \tag{9.128}$$

where $h\nu$ is the optical across variable and I is the optical through variable. I denotes the number of photons per unit time, i.e., the light intensity. Now, what happens physically when photons arrive at the hydrogen bromine mixture? Two things happen: (i) the light is absorbed and eventually transformed into heat (radiation), and (ii) the photons collide with the Br_2 molecules and provide the (small) activation energy necessary to separate the Br_2 molecule into two $Br^\bullet$ radicals. The energy balance occurs inside the *ChR1* reactor, i.e., Eq.(9.51a) is now modified to:

$$p \cdot q + (h\nu) \cdot I = T \cdot \dot{S} + \sum_{\forall i} \mu_i \cdot \nu_i \qquad (9.129)$$

and the k_1 reaction is modified to:

$$Br_2 + M \overset{k_1}{\to} 2Br^\bullet + M \qquad (9.130a)$$

$$Br_2 + h\nu \overset{k_1'}{\to} 2Br^\bullet \qquad (9.130b)$$

and therefore:

$$\nu_{Br^\bullet} = 2k_1 \cdot \left(\frac{n_{Br_2} \cdot n_M}{V}\right) + 2k_1' \cdot I \cdot n_{Br_2} + \dots \qquad (9.131)$$

At room temperature, k_1 is almost zero, but k_1' is not zero.

Another mechanism through which chemical reactions can sometimes be influenced is an electrical field. Salts dissolve in aqueous solutions into individual ions, for example:

$$H_2O + HCl \to H_3O^+ + Cl^- \qquad (9.132)$$

Ions are electrically charged and they can therefore be separated through the application of an electrical field. If two metal plates are dipped into an ionized solution and if an electrical field is created by connecting a voltage source to the two plates, we can observe that the positively charged ions migrate toward the cathode, while the negatively charged ions migrate toward the anode. This works even with pure water (but better with a drop of sulphoric acid H_2SO_4 or potash lye KOH added to it). In the water, an equilibrium exists between H_2O molecules on the one hand, and HO^- and H_3O^+ ions on the other. The acid or alkali will drastically enhance the number of ions in the solution. In the *electrolysis*, the following reactions take place:

$$4H_3O^+ + 4e^- \overset{k_1}{\to} 2H_2 + 4H_2O \qquad (9.133a)$$

$$4OH^- \overset{k_1}{\to} 2H_2O + O_2 + 4e^- \qquad (9.133b)$$

$$2H_2O \overset{k_3}{\to} H_3O^+ + OH^- \qquad (9.133c)$$

$$H_3O^+ + OH^- \overset{k_3}{\to} 2H_2O \qquad (9.133d)$$

Reactions k_2 and k_3 are fast equilibrium reactions. The surplus electrons of Eq.(9.133b) wander from the anode through the voltage source back to the cathode where they are available for Eq.(9.133a). The acid or alkali acts as a catalyst, i.e., it does not participate in the reaction.

The energy balance occurs again inside the *ChR1* reactor, i.e., Eq.(9.51a) is now modified to:

$$p \cdot q = T \cdot \dot{S} + \sum_{\forall i} \mu_i \cdot \nu_i + u \cdot i \qquad (9.134)$$

The difference in the electrical potential between the two electrodes creates an electrical field. Since the ions are charged components, the field generates an electromotorical force, which pulls the ions toward the electrodes. However, since the ions are dipoles, they themselves create a small electrical field that must first be overcome by the external field. Therefore, up to a given built–in voltage, the applied voltage will only be able to polarize the ions. Only if a voltage larger than the built–in voltage is applied will an actual transport of ions start to occur. Due to various friction phenomena, an opposing force builds up during the transport, which is proportional to the speed of the ions. In the steady–state, the two forces must balance each other out. Consequently, the speed (and therefore the reaction rate) is proportional to the applied voltage minus the built–in voltage. The number of ions that arrive at the two electrodes is proportional to their traveling speed and the number of negatively charged ions. At the so–called anode, an *oxidation reaction* takes place in accordance with Eq.(9.133b), i.e., the surplus electrons are absorbed by the anode and can travel through the voltage source to the so–called cathode where they are available for a *reduction reaction* in accordance with Eq.(9.133a). (Notice that the cathode is not necessarily the negative pole. By definition, the cathode is the electrode at which the reduction takes place, while the anode is the electrode at which the oxidation takes place.)

The electrically induced k_1 reaction flow rate can thus be computed as follows:

$$\nu_{el} = k_1 \cdot (u - u_0) \cdot n_{OH^-} \qquad (9.135)$$

Equation (9.135) assumes a model of homogeneously distributed ions. In the electrolysis of water, this assumption is justified since new ion pairs are constantly produced in accordance with Eq.(9.133c). In other types of electrolysis, this assumption may be incorrect. Since the electrical field is constant everywhere, all ions travel simultaneously with the same speed toward their attractor and the region far away from the attracting electrode is soon depleted of one type of ion. Consequently, we may want to replace the current number of moles of donor ions in Eq.(9.135) by the initial number of moles of donor ions and build in a discontinuity that resets ν_{el} to zero as soon as the donor ions have been exhausted.

Notice that, both in the case of photochemistry and electrochemistry, Eq.(9.51a) had to be modified to accommodate the additional types of dissipated energy. In reality, we should have done this already in Section 9.7 where we discussed the CSTRs. The mechanical process of stirring dissipates mechanical energy which should have been taken into account in the first power balance equation.

9.9 Summary

In this chapter, we demonstrated how the previously introduced modeling concepts can be employed to describe chemical reaction systems. It became clear that the bond graph approach to modeling supports the process of understanding the underlying physical phenomena significantly. DYMOLA [9.5] proved to be a valuable tool for hierarchical and modular modeling of chemical reaction systems, far superior in flexibility and generality to previously used tools such as DYNSYL [9.13], LARKIN [9.4], or MACKSIM [9.2]. For the first time, we can hope to create a chemical model library from which modules can be picked more or less arbitrarily that can then be combined to make up complex systems, but a lot of research still needs to be done.

Many quite common chemical reactions are still poorly understood. Measurement techniques to experimentally determine the required model parameters are often inaccurate, and even more frequently, nonexistent. Impurities are the cause of large fluctuations in the behavior of real chemical reaction systems. These are just some of the problems that the chemical engineer is faced with in his or her everyday work. Yet we are confident that this treatise presents a significant step forward in the mathematical description of chemical reaction kinetics.

Notice that our methodology avoided the distinction between reversible and irreversible thermodynamics. This dichotomy had to be introduced in earlier treatises because of the fact that these discussions all centered around the thermodynamic steady–state exclusively. In a steady–state analysis, it is important to distinguish between conservative (reversible) and dissipative (irreversible) phenomena, and the steady–state values do not preserve this information. Therefore, it became important to distinguish between them. However, this dichotomy is an artifact. If we describe the nonequilibrium thermodynamics through a bunch of differential equations, as we did in this chapter, the problem vanishes.

It can be noticed that our thermodynamic equations are quite a bit simpler and more compact than those found in most texts on thermodynamics. Again, this is a consequence of the dynamic versus the static approach. On first glance, this seems counterintuitive. Why should an approach that provides more information about the system result in simpler equations? This puzzle cannot be resolved without an understanding of the mechanisms of *complexity*. Let me, therefore, postpone an answer to this puzzle until the next chapter, in which complexity will play a pertinent role.

What has been previously accomplished in the context of modeling chemical reaction systems? Important research results have been reported by Oster *et al.* [9.12]. They already used a bond graph representation to describe chemical reaction systems. They already discovered that substance accumulation is a capacitive phenomenon and attached C elements to 0–junctions for this purpose. They also discovered that the reactions themselves are of a dissipative nature and attached R elements to 1–junctions to describe those. However, they were not interested in simulation. They wanted to apply system–theoretic techniques to the analytical treatment of chemical reaction systems. For this reason, they made use of a result by Onsager according to whom all thermodynamical equations can be linearized in the vicinity of the flow equilibrium. They did not formalize the connection to the thermal and hydraulic/pneumatic subsystems (although they were aware of the Gibbs equation, Eq.(9.51a)) and did not realize the importance of the influence of the experimental conditions (isothermic versus isentropic and isobaric versus isochoric reactions). It is somewhat surprising to me that the important discoveries made by Oster and his colleagues were not further exploited by either them or other researchers for over 15 years.

Finally, I wish to acknowledge my gratitude to Peter Breedveld, Dean Karnopp, and Jean Thoma for their constructive criticisms of Chapters 7 to 9 of this text. Their comments were highly appreciated.

References

[9.1] Ihsan Barin and Ottmar Knacke (1973), *Thermochemical Properties of Inorganic Substances*, Springer–Verlag, Berlin.

[9.2] Mike B. Carver, D. H. Hanley, and K. R. Chaplin (1978), *MACK-SIM: Mass Action Chemical Kinetics Simulation User's Manual*, Report: AECL–6413, Atomic Energy of Canada, Ltd., Chalk River Nuclear Laboratories, Chalk River, Ontario, Canada.

[9.3] John T. Davies (1950), "The Mechanism of Diffusion of Ions Across a Phase Boundary and Through Cell Walls," *J. Physical and Colloid Chemistry*, **54**, pp. 185–204.

[9.4] Peter Deuflhard, G. Bader, and U. Nowak (1980), *LARKIN — A Software Package for the Numerical Simulation of LARge Systems Arising in Chemical Reaction KINetics*, Report 100, University of Heidelberg, Institut für Angewandte Mathematik, Heidelberg, F.R.G.

[9.5] Hilding Elmqvist (1978), *A Structured Model Language for Large Continuous Systems*, Ph.D. dissertation, Report CODEN: LUTFD2/(TRFT–1015), Dept. of Automatic Control, Lund Institute of Technology, Lund, Sweden.

[9.6] Frank C. Hoppensteadt, P. Alfeld, and Richard Aiken (1981), "Numerical Treatment of Rapid Chemical Kinetics by Perturbation and Projection Methods," in: *Modelling of Chemical Reaction Systems* (K.H. Ebert, P. Deuflhard, and W. Jäger, eds.), Springer–Verlag, Berlin, Series in Chemical Physics, **18**, pp. 31–37.

[9.7] J. Alistair Kerr and Stephen J. Moss (1981), *Handbook of Bimolecular and Termolecular Gas Reactions*, CRC Press, Boca Raton, Fla.

[9.8] Joseph L. Latham (1962), *Elementary Reaction Kinetics*, Butterworth & Co., London, U.K.

[9.9] Mark L. Levine and Milan Bier (1990), "Electrophoretic Transport of Solutes in Aqueous Two–Phase Systems," submitted to *Electrophoresis*.

[9.10] Bruce H. Mahan (1965), *University Chemistry*, Addison–Wesley, Reading, Mass.

[9.11] John W. Moore, Ralph G. Pearson, and Arthur A. Frost (1981), *Kinetics and Mechanism*, John Wiley & Sons, New York.

[9.12] George F. Oster, Alan S. Perelson, and Aharon Katzir–Katchalsky (1973), "Network Thermodynamics: Dynamic Modelling of Biophysical Systems," *Quarterly Reviews of Biophysics*, **6**(1), pp. 1–134.

[9.13] Gary K. Patterson and R. B. Rozsa (1978), *DYNSYL: A General-Purpose Dynamic Simulator for Chemical Processes*, Report: UCRL–52561, Lawrence Livermore National Laboratory, Livermore, Calif.

[9.14] Robert H. Perry, Don W. Green, and James O. Maloney, Eds. (1984), *Perry's Chemical Engineers' Handbook*, sixth edition, McGraw–Hill, New York.

[9.15] Peter A. Rock (1983), *Chemical Thermodynamics*, University Science Books, Mill Valley, Calif.

[9.16] Terry Triffet, Ed. (1989) *Automation of Extraterrestrial Systems for Oxygen Production*, Proceedings of the Workshop AESOP'89, La Jolla, Calif., July 13–15, published by University of Arizona Space Engineering Research Center, UA/NASA SERC, Tucson, Ariz.

Bibliography

[B9.1] Guillermo J. Ferraudi (1988), *Elements of Inorganic Photochemistry*, John Wiley, New York.

[B9.2] Jerry Goodisman (1987), *Electrochemistry: Theoretical Foundations, Quantum and Statistical Mechanics, Thermodynamics, the Solid State*, John Wiley, New York.

[B9.3] Laszlo Kiss (1988), *Kinetics of Electrochemical Metal Dissolution*, Elsevier, Amsterdam, The Netherlands.

[B9.4] Irving M. Klotz and Robert M. Rosenberg (1986), *Chemical Thermodynamics: Basic Theory and Methods*, fourth edition, Benjamin/Cummings Publishing, Menlo Park, Calif.

[B9.5] Keith J. Laidler (1987), *Chemical Kinetics*, third edition, Harper & Row, New York.

[B9.6] Harold Nicholson, Ed. (1980), *Modelling of Dynamical Systems*, two volumes, Peter Peregrinus Ltd., Stevenage, U.K.

[B9.7] Joseph M. Smith (1981), *Chemical Engineering Kinetics*, third edition, McGraw–Hill, New York.

[B9.8] Jeffrey I. Steinfeld, Joseph S. Francisco, and William L. Hase (1989), *Chemical Kinetics and Dynamics*, Prentice–Hall, Englewood Cliffs, N.J.

[B9.9] Jean Thoma (1976), "Simulation, Entropy Flux and Chemical Potential," *BioSystems*, **8**, pp. 1–9.

[B9.10] Jean Thoma and Henri Atlan (1985), "Osmosis and Hydraulics by Network Thermodynamics and Bond Graphs," *J. Franklin Institute*, **319**(1/2), pp. 217–226.

Homework Problems

[H9.1] Hydrogen–Bromine Reaction

For the hydrogen–bromine reaction discussed in this chapter, derive a set of differential equations that describes the dynamic behavior of the five species Br_2, $Br^\bullet$, H_2, HBr, and $H^\bullet$ over time. Assume isothermic and isochoric conditions.

Moore *et al.* [9.11] gave the following experimental temperature characteristics for the five reaction rate constants:

$$ak_1 = 1.39 \cdot 10^8 \cdot \sqrt{T} \cdot (\frac{189243.0}{R \cdot T})^{1.97} \qquad (H9.1a)$$

$$k_1 = ak_1 \cdot \exp(\frac{-189243.0}{R \cdot T}) \qquad (H9.1b)$$

$$k_2 = \frac{k_1}{K(T)} \qquad (H9.1c)$$

$$k_3 = 10^{11.43} \cdot \exp(\frac{-82400.0}{R \cdot T}) \qquad (H9.1d)$$

$$k_5 = 10^{11.97} \cdot \exp(\frac{-149800.0}{R \cdot T}) \qquad (H9.1e)$$

$$k_4 = 0.1 \cdot k_5 \qquad (H9.1f)$$

where $R = 8.314$ J K^{-1} mole^{-1} is the gas constant and $T = 800$ K is the absolute temperature measured in Kelvin. The temperature dependence of the k_i coefficients of the hydrogen–bromine reaction were taken from Barin and Knacke [9.1]. The initial numbers of moles of Br_2 and H_2 are both 0.0075, and the total reaction volume is $V = 10^{-3}$ m^3.

The static characteristic $K(T)$ denotes the equilibrium constant of the third reaction. It is strongly temperature–dependent. Values have been experimentally determined. They are given in Table H9.1.

Program your model in ACSL. Since the problem is very stiff, use Gear's stiff integration algorithm ($ialg = 2$) with $nstp = 2001$. During the first five seconds of the simulation use a communication interval of $cint = 0.01$ sec, thereafter increase the communication interval to $cint = 10.0$ sec. Simulate the reaction system over 5000 sec.

Since the concentrations are very small, you must help ACSL to determine appropriate absolute error bounds for the numerical integration. Specify the absolute error (XERROR) for the five state variables as follows: $Br_2 = 10^{-9}$, $Br^\bullet = 10^{-8}$, $H_2 = 10^{-10}$, HBr $= 10^{-6}$, and $H^\bullet = 10^{-8}$.

Plot the numbers of moles of Br_2, $Br^\bullet$, H_2, and $H^\bullet$ over time during the first five seconds of the simulation and plot the number of moles of HBr over the entire simulation period.

Table H9.1. Equilibrium Constant as a Function of Temperature.

Abs. Temperature T [K]	Equilibrium Const. K [mole m^{-3}]
300.0	7.7446×10^{-29}
400.0	1.9543×10^{-20}
500.0	2.2182×10^{-15}
600.0	5.2844×10^{-12}
700.0	1.3867×10^{-9}
800.0	9.0782×10^{-8}
900.0	2.3768×10^{-6}
1000.0	3.2509×10^{-5}
1100.0	2.7861×10^{-4}
1200.0	1.6788×10^{-3}
1300.0	7.6913×10^{-3}
1400.0	2.8510×10^{-2}
1500.0	8.8716×10^{-2}
1600.0	2.4044×10^{-1}
1700.0	5.8344×10^{-1}
1800.0	1.7947
1900.0	2.6061
2000.0	4.9431

[H9.2] Oxyhydrogen Gas Reaction

When oxygen and hydrogen gases are mixed in similar proportions, a spark can bring the mixture to explosion. We wish to describe this process. From the literature [9.5], it can be found that this process can be described by the following set of individual reactions:

$$H_2 + O_2 \overset{k_0}{\rightarrow} H^\bullet + HO_2^\bullet \qquad (H9.2a)$$

$$H_2 + OH^\bullet \overset{k_1}{\rightarrow} H^\bullet + H_2O \qquad (H9.2b)$$

$$O_2 + H^\bullet \overset{k_2}{\rightarrow} OH^\bullet + O^\bullet \qquad (H9.2c)$$

$$H_2 + O^\bullet \overset{k_3}{\rightarrow} H^\bullet + OH^\bullet \qquad (H9.2d)$$

$$OH^\bullet + W \overset{a_1}{\rightarrow} \qquad (H9.2e)$$

$$H^\bullet + W \overset{a_2}{\rightarrow} \qquad (H9.2f)$$

$$O^\bullet + W \overset{a_3}{\rightarrow} \qquad (H9.2g)$$

where W stands for the wall. At the wall, the unstable atoms $H^\bullet$ and $O^\bullet$ and the unstable radical $OH^\bullet$ can be absorbed. The absorption rates at the wall are strictly proportional to the concentrations of the absorbed species themselves, for example:

$$\frac{d}{dt} c_{OH^\bullet\,wall} = -a_1 \cdot c_{OH^\bullet} \qquad (H9.2h)$$

The reaction rate constants given were as follows: $k_0 = 60.0$, $k_1 = 2.3 \cdot 10^{11}$, $k_2 = 4.02 \cdot 10^9$, $k_3 = 2.82 \cdot 10^{12}$, $a_1 = 920.0$, $a_2 = 80.0$, and $a_3 = 920.0$. The initial conditions are: $H_2 = 10^{-7}$ and $O_2 = 0.5 \cdot 10^{-7}$. The reference fails to specify for which temperature the given reaction rate constants are valid.

Generate a set of differential equations describing this system. Program your model in ACSL. Simulate the system over 0.1 sec. During the first 0.2 msec, use a communication interval of 10 μsec, thereafter use a communication interval of 1 msec. Because of the inherent stiffness of the equations, use Gear's integration algorithm ($ialg = 2$) with $nstp = 1001$. Because of the small numerical values of the concentrations, specify the absolute error (XERROR) for all state variables as 10^{-10}.

[H9.3] Isothermic and Isobaric Hydrogen–Bromine Reaction

Create a DYMOLA [9.5] model that describes the total dynamics of the hydrogen–bromine reaction under the assumption of isothermic ($T = 800$ K) and isobaric ($p = 1$ atm) conditions. The initial reaction volume can be determined from the equation of state. The initial values of the total entropy and the total free energy are arbitrarily set to zero (since we are only interested in the relative values ΔS and ΔG of the reaction). The constant μ_i values can be found from the literature. For this example, we find: $\mu_{Br_2} = -204,493.0$, $\mu_{Br\bullet} = -53,772.8$, $\mu_{H_2} = -129,023.0$, $\mu_{H\bullet} = +106,772.0$, and $\mu_{HBr} = -226,828.0$.

Use the model types described in this chapter and connect the overall model according to the bond graph of Fig.9.1.

Simulate the system and compare your results with those of Hw.[H9.1].

[H9.4] Isothermic and Isochoric Hydrogen–Bromine Reaction

Create a DYMOLA [9.5] model that describes the total dynamics of the hydrogen–bromine reaction under the assumption of isothermic ($T = 800$ K) and isochoric ($V = 10^{-3}$ m^3) conditions. The initial pressure can be determined from the equation of state. The initial values of the total entropy and the total free energy are arbitrarily set to zero (since we are only interested in the relative values ΔS and ΔG of the reaction). The constant μ values of Hw.[H9.3] are now used as initial conditions.

Replace the model types according to the description given in this chapter and enhance the bond graph by adding the missing chemical bonds for the $\dot{\mu}_i \cdot n_i$ terms.

Simulate the system and compare your results with those of Hw.[H9.3].

[H9.5] Model Validation

We wish to validate the models of Hw.[H9.3] and Hw.[H9.4]. For this purpose, we shall perform a number of experiments.

(1) *Mass flow balance*: Since we operate under closed conditions, the total number of Br atoms and H atoms must remain constant at all times. Add equations to the models of Hw.[H9.3] and Hw.[H9.4] that compute (stoichiometrically) the total numbers of moles of the two types: na_H and na_{Br}. Verify that these numbers remain constant during the simulation. This test helps you gain confidence in the accuracy of the numerical integration.

(2) *Energy flow balance*: Since we operate under closed conditions, the total internal energy U of the reaction system must remain constant at all times. Add equations to the models of Hw.[H9.3] and Hw.[H9.4] that compute the internal energy U of the reaction systems and verify that the internal energy remains constant over time. This test is even better than the previous one, since the numerical value of U is more sensitive to numerical integration errors than the mass balance values na_H and na_{Br}.

(3) *Steady–state conditions*: The steady–state numbers of moles of the five species can be analytically determined. We start out with the set of equations:

$$\frac{d}{dt}n_{Br_2} = -\nu_{k1} + \nu_{k2} - \nu_{k5} \qquad (H9.5a)$$

$$\frac{d}{dt}n_{Br\bullet} = 2\nu_{k1} - 2\nu_{k2} - \nu_{k3} + \nu_{k4} + \nu_{k5} \qquad (H9.5b)$$

$$\frac{d}{dt}n_{H_2} = -\nu_{k3} + \nu_{k4} \qquad (H9.5c)$$

$$\frac{d}{dt}n_{H\bullet} = \nu_{k3} - \nu_{k4} - \nu_{k5} \qquad (H9.5d)$$

$$\frac{d}{dt}n_{HBr} = \nu_{k3} - \nu_{k4} + \nu_{k5} \qquad (H9.5e)$$

and set all derivatives equal to zero. Since the **N** matrix of this reaction is singular, we cannot determine the steady–state (equilibrium of flow) values of all reaction flow rates. We obtain three equations for five unknowns. We replace the flow rates in these equations by their definition equations, Eq.(9.26), and obtain three highly nonlinear equations in the five variables n_{Br_2}, $n_{Br\bullet}$, n_{H_2}, $n_{H\bullet}$, and n_{HBr}. However, we can obtain the missing two equations from the knowledge that the final numbers of atoms of both types must be equal to the initial numbers of atoms of the two types. By solving this set of equations, we obtain a unique solution determining the steady–state numbers of moles of the five species. Verify that the steady–state values obtained from the simulations of Hw.[H9.3] and Hw.[H9.4] are in accordance with the theoretically found values. This test will quickly unravel potential bugs in your reaction rate equations.

(4) *Total reaction energy:* Since we know the steady–state mass distribution, we can compute the enthalpies and free energies of formation before and after the reaction. We can then compute the total reaction energy ΔG produced or consumed during the reaction. Add equations to the two models of Hw.[H9.3] and Hw.[H9.4] that compute the reaction energy G as a function of time. Subtract the initial value of G (which was arbitrarily set equal to zero in the model) from the final value of G. This number should be the same as the theoretically found value of ΔG. This test detects potential bugs in your thermodynamics equations.

[H9.6] Boiling Water

Take 1 liter of water and put it on your stove, which you have previously heated up using the highest stove position. Start from room temperature and measure the time that it takes to reach the boiling point. Continue to boil your water and measure the time that it takes until no water is left in the pot.

Create a model that reproduces this experiment. Assume isentropic and isobaric conditions. Adjust the heat source in your model such that the time constants match those of the experiment. Notice that you have to take into account that you don't operate in a closed system. The evaporating water carries heat away that will keep the temperature from rising beyond the boiling point. What is the highest temperature that you will observe?

Repeat your modeling effort for a pressure cooker. This time, we assume isochoric conditions up to a pressure of $p = 2.5$ atm. At this point, a pressure release valve will release some of the water vapor until the pressure is below a value of $p = 2$ atm. Thereafter, the pressure starts building up again.

Solve manually the resulting structural singularity using one of the techniques proposed in Chapter 5. Simulate your isentropic and piecewise isochoric system. What is the highest water temperature that you will observe in this case?

[H9.7]* Rechargeable Battery

Rechargeable car batteries operate on the following principle: Two lead plates are dipped into a solution of sulphoric acid. During the electrolysis (i.e., the recharging process), the following reactions occur:

$$PbSO_4 + 2e^- \rightarrow Pb + SO_4^{2-} \quad \text{(cathode)} \qquad (H9.7a)$$
$$PbSO_4 + 6H_2O \rightarrow PbO_2 + SO_4^{2-} + 4H_3O^+ + 2e^- \quad \text{(anode)} \qquad (H9.7b)$$

During the utilization phase, the same reactions occur in opposite direction. When the battery is charged, metallic lead Pb builds up at the cathode and brown lead dioxide PbO_2 builds up at the anode. When the battery is used, $PbSO_4$ goes into solution. If no lead dioxide is left on the anode, the battery is empty.

Unfortunately, this type of battery does not hold its charge very well over an extended period of time. Even without drawing current, the lead dioxide PbO_2 gets dissolved through a reaction with the lead of the plate itself:

$$Pb + PbO_2 + 2H_2SO_4 \rightarrow 2PbSO_4 + 2H_2O \qquad (H9.7c)$$

Another problem with this sort of battery is the following. A certain percentage of the lead sulphate $PbSO_4$ crystallizes out and can no longer be used for the oxidation/reduction process. If all sulphoric acid H_2SO_4 has been converted to crystalline lead sulphate $PbSO_4$, the battery is dead.

The built–in voltage of the rechargeable battery is $u_0 = 2.02$ V. Place six such elements in series to get a 12 V battery. Create a DYMOLA [9.5] model that represents the dynamics of the rechargeable battery. Simulate a complete recharging/utilization cycle. If the battery is initially empty, how long does it take to recharge it? Assume that you forgot to switch off your headlights. How long does it take until the battery is empty?

Projects

[P9.1] Mining the Asteroids

The following description is extracted from the AESOP report [9.16]. It is due to Andy Cutler.

In future decades, it may be advantageous to extract oxygen from extraterrestrial sources such as lunar ilmenite or asteroidal chondrite. In this way, we may be able to produce propellant for interplanetary missions much more cheaply than by lifting it up from the gravity well of planet Earth.

Candidates for oxygen production in a low–gravity environment are lunar or asteroidal rock. On the moon, a high concentration of ilmenite $FeTiO_3$ exists, in particular in the mares. The Mare Tranquilitatis consists of almost 50% of surface ilmenite that could be mined. The following reaction describes an oxygen reduction process in which ilmenite is reduced to iron Fe, to rutile TiO_2, and to water vapor:

$$FeTiO_3 + H_2 \rightarrow Fe + TiO_2 + H_2O \qquad (P9.1a)$$

The hydrogen gas must be imported from Earth since the moon is very poor in hydrogen. However, in a second electrolysis stage, the water vapor can be reduced to oxygen gas and to hydrogen gas:

$$2H_2O \rightarrow 2H_2 + O_2 \qquad\qquad (P9.1b)$$

Thus, the hydrogen gas can be recovered and recycled for reuse in the first stage. Therefore, the hydrogen import may not be as expensive as we might believe on first sight.

Similarly, many near–Earth asteroids are likely to be rich in carbonaceous chondritic material. Water vapor can be extracted from such material in a pyrolysis stage. The water content of asteroidal chondrites is believed to be between 5% and 10%. Figure P9.1 shows a chemical process that could be used for oxygen production from asteroidal chondritic material.

While such a flow diagram is fairly easy to generate, it does not provide us with precise quantitative information as to the rendition of oxygen and the total amount of energy that is being used up by the process. However, these two parameters will ultimately decide the economic attractiveness of the operation. Moreover, energy is a scarce resource in Space, and we need to plan ahead in order to determine what type of energy generation plant we must have if we wish to ensure the availability of the required amount of energy at all times. It is therefore not sufficient to be able to analyze the energy requirements under steady–state conditions since we must install energy at the peak value. Consequently, we should be able to quantitatively model and simulate such a mining operation before it is built and deployed. The model should simultaneously explore the mass and energy flows under dynamic conditions. Thus, the methodology that was introduced in this chapter seems to be the most suitable approach to tackle this problem.

Let me briefly explain how the process depicted in Fig.P9.1 works. The *sizer* consists of a crusher that decomposes the rock mechanically and a sieve that lets sufficiently small material pass and reroutes larger pieces to the crusher for further decomposition. The material is then forwarded to an inlet *lock hopper*, which accepts the presized material under vacuum and forwards it to the preheater under pressure. The *preheater* brings the entire charge to a fixed temperature slightly below that at which pyrolysis begins to occur. The preheater is also responsible for controlling the gas pressure, which may rise due to beginning evaporation of the charge. The *pyrolyzer* devolatilizes the preheated feed by the addition of microwave heat. Rapid pressure control is important in order to keep the gas pressure inside the pyrolyzer within acceptable bounds at all times. The *heat recovery* stage is a heat exchanger that recovers a high percentage of the heat of the spent feed after pyrolysis. The temperature and pressure must be controlled carefully in order to recover as much heat as possible without condensing water vapor, which would be thrown out with the spent feed and would thereby be lost. The outlet *lock hopper* accepts the spent feed and discharges it to the vacuum with little loss of the pressurized gas inside the system. This completes the description of the top row of Fig.P9.1.

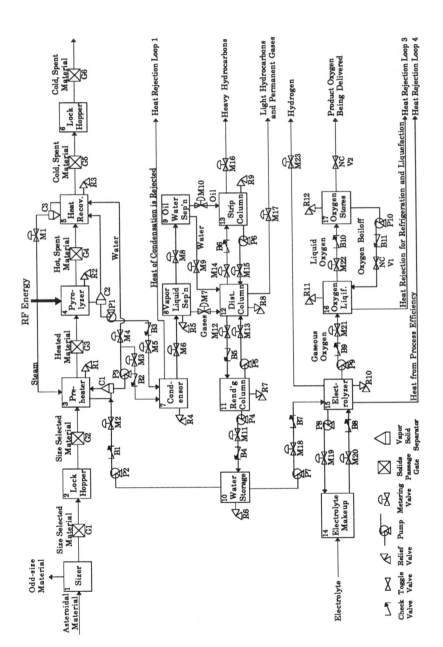

Figure P9.1. Process flow diagram for chondritic material pyrolysis.

The second row contains a *condenser*, which precools the pyrolyzate vapor. A *vapor liquid separator* sends the remaining gases to the third row for further processing and lets the liquid proceed to an *oil–water separator*. Both products of the oil–water separator are processed further in the third row. This completes the description of the second row.

The third row contains three columns (tubular reactors). The *distillation column* accepts the remaining gases and the impure water from the second row. It separates the material further. The heavy material ends up at the bottom, while the lightest material is at the top. The bottom will contain hydrocarbons, which are fed to the *stripping column* for further processing. The top will contain gases that can either be disposed of or recycled for production of hydrogen. The medium portion of the distillation column contains prepurified water, which is passed on to the *rendering column* for further purification. The stripping column is used to dewater the heavy hydrocarbons. The rendering column removes the remaining heavy organic material from the prepurified water. Its top product is pure water, which is passed on to the *water storage area* to await further processing. This completes the description of the third row.

The fourth and final row mixes the pure water with an electrolyte (such as potash lye KOH) for processing by the *electrolyzer*. The electrolyte will invariably get contaminated by impurities in the water and must therefore be periodically recycled to an *electrolyte makeup* chamber where the quality of the electrolyte is controlled and improved when needed. The electrolyzer produces hydrogen, which can either be burned off or recycled, and oxygen. In order to prevent losses of the oxygen product due to leakage, an *oxygen liquefier* is used, which will convert the oxygen to a form that can more easily be stored in the *oxygen storage area*. Oxygen that boils off is captured and rerouted to the oxygen liquefier. The heat that is produced in both the electrolysis and the oxygen refrigeration stage is also rerouted to the top column where it can be reused.

Create separate DYMOLA [9.5] models for each of the boxes of Fig.P9.1. Connect the entire model and simulate the mining operation. What is the peak value of energy that you must make available during the operation? What is the rendition of oxygen?

Research

[R9.1] Hierarchical Modular Model Library for Chemical Process Simulation

Build a library of DYMOLA [9.5] model types for chemical reaction kinetics. The library should contain modules for all common reactor types and it should be possible to solve problems such as Pr.[P9.1] by simply connecting the modules of the library with each other.

[R9.2] Electrophoretic Transport of Solutes in Aqueous Two–Phase Systems

Levine and Bier [9.9] recently reported a phenomenon relating to the transport of proteins across a phase barrier under the influence of electrical fields in an electrophoretic setting. Proteins partition in aqueous two–phase systems as a result of differential affinities for the polymers. It turns out that proteins are readily transported across the phase barrier from low to high affinity, but they accumulate below the interface when electrically driven in the opposite direction. Surprisingly, the electrical field is not a sufficiently strong force to overcome the thermodynamics of solute partitioning. The fact that electrophoretic transport as well as diffusion [9.3] can be influenced by the partitioning behavior of solute is of significance for the understanding of affinity and it has direct relevance to biological systems with their many semipermeable and electrically charged membranes.

Levine and Bier believe that a full explanation of their experimental results will require an integration of kinetic terms governing electrophoretic transport with thermodynamic terms governing partitioning. It is hoped that the methodology derived in this chapter may provide us with such a tool.

Model the transport of proteins within individual phases and compare the simulation results with experimentally found results. Thereafter, model the phase barrier in an aqueous two–phase system and simulate the transport of proteins across this barrier. Include chemical, thermic, hydraulic, and electrical terms in the model. Compare your simulation results with the experimental results reported in Levine and Bier [9.9].

10

Population Dynamics Modeling

Preview

Until now, we have dealt with *deductive modeling* exclusively, i.e., all our models were created on the basis of a physical understanding of the processes that we wished to capture. Therefore, this type of modeling is also frequently referred to as *physical modeling*. As we proceed to more and more complex systems, less and less meta–knowledge is available that would support physical modeling. Furthermore, the larger uncertainties inherent in most physical parameters of such systems make physical models less and less accurate. Consequently, researchers in fields such as biology or economics often prefer an entirely different approach to modeling. They make observations about the system under study and then try to fit a model to the observed data. This modeling approach is called *inductive modeling*. The structural and parametric assumptions behind inductive models are not based on physical intuition, but on factual observation. This chapter illustrates some of the virtues and vices associated with inductive modeling and lists the conditions under which either of the two approaches to modeling is more adequate than the other. This chapter documents how population dynamics models are created, discusses the difference between structural and behavior complexity of models in general, introduces the concept of chaotic motion, and addresses a rather difficult issue, namely, the question of self–organization within systems.

10.1 Growth and Decay

In any population dynamics study, the change of the population per unit time can be described through the difference between the birth rate and the death rate:

$$\dot{P} = BR - DR \tag{10.1}$$

where BR denotes the birth rate, which includes both physical birth and migration into the system, and DR stands for the death rate, which again includes both physical death and migration out of the system.

It is natural to assume that both the birth rate and the death rate are proportional to the current population:

$$BR = k_{BR} \cdot P \quad , \quad DR = k_{DR} \cdot P \tag{10.2}$$

and therefore:

$$\dot{P} = (k_{BR} - k_{DR}) \cdot P \tag{10.3}$$

If we assume that k_{BR} and k_{DR} are two constants, the model exhibits either an exponential growth of the population (if $k_{BR} > k_{DR}$) or an exponential decay (if $k_{BR} < k_{DR}$).

However, is the assumption of constant birth and death factors k_{BR} and k_{DR} indeed justified? In order for a population to grow, it must consume some form of energy. Since in any closed system, the energy is limited, a population cannot grow exponentially forever. If we assume that the available energy is equally distributed among all members of the population, then we can conclude that the *per capita* energy $E_{p.c.}$ is inverse proportional to the population count:

$$E_{p.c.} = \frac{E_{total}}{P} \tag{10.4}$$

Both the birth factor and the death factor will somehow depend on the available energy. If the *per capita* energy becomes small, the birth factor declines (since malnutrition leads to impotence and increased infant mortality) and the death factor increases (due to famine and an increased vulnerability to epidemic diseases). Eventually, an equilibrium will be reached in which the death factor balances the birth factor out, thus the population reaches a steady–state; therefore, the *per capita* energy does not decrease any further, and therefore, the birth and death factors remain constant as well. The population dependence of the birth and death factors is usually referred to as the *crowding effect*.

Unfortunately, a physical law that would allow us to specify an explicit and accurate relationship between the *per capita* energy and the birth and death factors is unknown, and we must therefore base our model on inductive rather than deductive evidence.

The most commonly made assumption is the following: A one–species ecosystem can support a fixed number of animals (plants) over an extended period of time. Let us call this number P_{max}. We can model the smooth growth and saturation of the population in the following way:

$$\dot{P} = k \cdot (1.0 - \frac{P}{P_{max}}) \cdot P \tag{10.5}$$

where k denotes the difference between k_{BR} and k_{DR}. If the population P approaches its maximum allowed value P_{max}, the factor in the parentheses becomes smaller and smaller and finally reaches zero for $P = P_{max}$. Thus, the population stops growing. Figure 10.1 shows the behavior of this model for the same initial population P_0, the same maximum population P_{max}, but three different values of k.

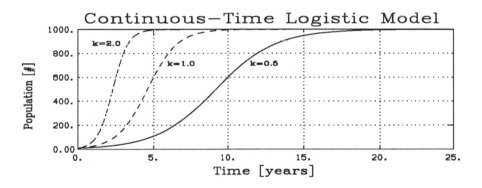

Figure 10.1. Trajectory behavior of the continuous–time logistic model.

Equation (10.5) is called the *continuous–time logistic equation*. Notice that the crowding effect is modeled through a quadratic term in the population. For instance, the two–species model of Hw.[H10.1] uses the logistic approach to model the crowding effects within each of the two species.

Notice that absolutely no physical evidence is available that supports the correctness of the logistic model. However, it is true that many biological species exhibit a population dynamics behavior that can be fitted rather accurately with a logistic model.

However, the logistic model has several obvious disadvantages:

(1) In order to identify the P_{max} value of a logistic model with reasonable accuracy, we must wait until a steady–state has actually occurred in the real system. However, this may be undesirable if the purpose of the model was to predict where the real system might settle to its steady–state value, and possibly, to find a means to influence the steady–state value of the real system before it occurs. For example, in world modeling, it is clearly undesirable if we have to wait until the human population has reached its steady–state value before we can construct a decent model for this fact.

(2) If an ecosystem contains several species, many different stable steady–state points may exist. It is the total energy of the system that is limited, but we have ample reason to assume that the ecosystem can support either more animals of one species and less of another or vice versa if both populations forage on the same food. Thus, it becomes difficult to formulate a meaningfully founded value for P_{max} under any circumstance, even if a steady–state has been observed in the real system.

Since inductive models are based on observation, such models are difficult to validate beyond the observed facts. Consequently, we may be forced to observe a disaster before we can devise a strategy that might have prevented it in the first place had the strategy been known earlier. This is a clear disadvantage of inductive modeling.

Let me illustrate this problem a little further. Figure 10.2 shows the answer to Hw.[H10.4].

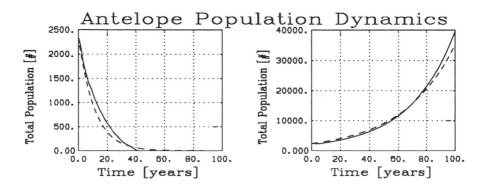

Figure 10.2. Antelope population dynamics with exponential fitting.

Without giving the model away, we can analyze these results a little further. Homework [H10.3] presents a fairly accurate nonlinear physical discrete–time model of an antelope population. Two simulations were performed, which differ in the assumption made about the infant mortality rate. Other than that, the two models were identical. Homework [H10.4] then fitted exponential models to the results of Hw.[H10.3]. Figure 10.2 shows the dramatic effect of the minor variation between the two models on the trajectory behavior. The continuous lines are the simulation results from the physical models and the dashed lines represent the simulation results from the curve–fitted inductive meta–models. Figure 10.2 shows clearly that the exponential approximation is well justified. The available measurement data on which the models were based are certainly not accurate to such an extent that we could claim with confidence that the deductive physical model is more accurate than the inductive exponential model.

Yet a very important difference exists between the two models. While the physical model requires more data items to be measured, we can obtain all the required data within a relatively short time span, say, two or three years. Once we measure the required data, we can predict the future behavior of the antelope population, and if the model predicts the extinction of the entire population, we may be able to propose a countermeasure that could prevent the disaster from happening. As Fig.10.2 shows clearly, a minor modification in only a few physical parameters, such as a minor change in rainfall statistics, an overgrazing by farm animals who compete for the same food, a new viral infection that increases the infant mortality by a few percent, or other seemingly insignificant changes in the environmental conditions, can have drastic long–term effects on the antelope population, which may easily go undetected for quite some time. The fitting parameter of the exponential model is highly sensitive to changes in the physical parameters of the model. Without performing a simulation of the physical model over an extended time span, we have no decent way to estimate an accurate value of the fitting parameter. Consequently, if we rely on the inductive model alone, we may have to wait for a long time before we discover from measurements that a modification of the system parameters has occurred that endangers the population. By that time, it may be too late for any countermeasure to be effective.

We just learned that, while the simpler inductive model explains the data as well as the more complex deductive model does, and while it is normally advisable to operate on the simplest model that explains the available data adequately, other considerations may exist as well. In the preceding example, the fact that the deductive model can be identified accurately within a much shorter time span is clearly a strong argument in favor of the deductive physical model.

Something else is interesting in the preceding example. While the physical system is clearly of high order, it behaves in a manner very similar to that of a low–order system, i.e., inherent *structural complexity* of a system does not necessarily imply that the system must also exhibit *behavioral complexity*. This observation is much deeper than what could be expected on first glance. We shall return to this discussion later.

10.2 Predator–Prey Models

What happens if our ecosystem contains more than one species? It is clear that a model of a multispecies ecosystem must contain submodels that describe the behavior of the individual species, following the same arguments as before. However, we need additional terms that describe the interactions among the different species.

Obviously, the interaction terms must be such that no interaction takes place between two species if either of the two populations is extinct. The simplest expression that exhibits this property is the product of the two populations. If either of the two populations is zero, the product is also zero.

The simplest model of this type is the *Lotka–Volterra model* [10.8], which assumes that a population of predators x_{pred} forays on a population of preys x_{prey}. When a predator meets a prey, it gets fed and a certain number of calories are exchanged. The predator population now has more calories and the prey population has less. Usually, the Lotka–Volterra model introduces an efficiency factor, i.e., the prey population loses more calories than the predator population gains. It is also assumed that the predator population would die out when left without prey, while the prey population feeds on another species that is available in abundance and is not itself contained in the model. Thus:

$$\dot{x}_{pred} = -a \cdot x_{pred} + k \cdot b \cdot x_{pred} \cdot x_{prey} \qquad (10.6a)$$

$$\dot{x}_{prey} = c \cdot x_{prey} - b \cdot x_{pred} \cdot x_{prey} \qquad (10.6b)$$

where $a > 0.0$ is the excess death rate of the predator population, $c > 0.0$ is the excess birth rate of the prey population, $b > 0.0$ is the foray (grazing) factor, and $0.0 < k \leq 1.0$ is the efficiency factor.

The Lotka–Volterra model exhibits a number of remarkable properties. In particular, this model does not approach a *continuous steady–state value*. Instead, it approaches a *periodic steady–state value*, i.e., the solution oscillates. The shape of the oscillation is very characteristic.

Figure 10.3 compares the measured data (dashed line) of a population of insects, the larch bud moth, *Zeiraphera diniana* (Guenée), which is endemic in the upper Engadine valley of southeastern Switzerland, to simulation results (continuous line) stemming from an optimized inductive two–species Lotka–Volterra model. The populations are expressed in larvae per kilogram of branches. Approximately once every nine years, we observe a large increase in the insect population. While the adult insects are quite harmless, their larvae feed on the needles of the larch trees, causing a lot of damage to the larch forest. Over several square kilometers, the green trees turn brown [10.6]. Therefore, it can be assumed that the insects are the predators in a predator–prey relationship with the larches being the prey. This assumption provided the rationale for the construction of the inductive model. Only the predator population is shown on the graph.

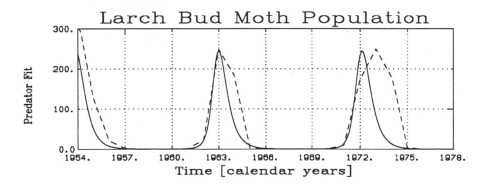

Figure 10.3. Lotka–Volterra model compared to measured insect data.

The attractiveness of the Lotka–Volterra model becomes immediately evident from a comparison of the two curves. Clearly, our mathematical model reflects very well the data that have been observed. However, please remember that we have absolutely no physical evidence that would give credibility to the internal structure of our equations. All we can say is that this model fits well the measured data, and since we operate under periodic steady–state conditions, i.e., no trend exists in either the measured or the simulated data, this model can be used very well to predict the insect population over a long time span, given that no significant change occurs in climatic or other environmental conditions.

Notice that a good curve fit does not prove our equations to be correct. It does not even prove that our initial assumption of the larch bud moth being the predator in a two–species predator–prey relationship is correct.

It was also observed that the adult insects suffer a lot from parasites. Obviously, the danger of epidemic diseases grows with the density of the insect population. Thus, it is equally reasonable to assume that the larch bud moth is in fact the prey in a two–species predator–prey relationship with the parasites being the predator. Figure 10.4 shows another two–species Lotka–Volterra model that was optimized under this new assumption. Again, the dashed line depicts the measured data, while the continuous line represents the simulated data. This time, only the prey population is shown on the graph.

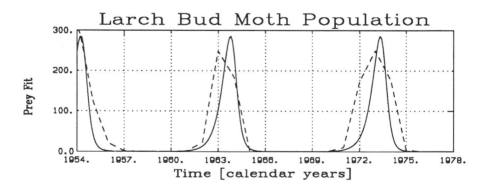

Figure 10.4. Lotka–Volterra model compared to measured insect data.

Just from looking at these two simulations, we cannot say with certainty that either of the two assumptions leads to a significantly better curve fit, and is therefore more likely to be accurate. Consequently, we must be supercautious about concluding causal relationships among variables in a real system on the basis of a good curve fit by an inductively constructed model.

Figure 10.5 shows the simulation results (continuous lines) of the same two models once more, superposed with the measured data (dashed lines), but this time plotted on a semilogarithmic scale.

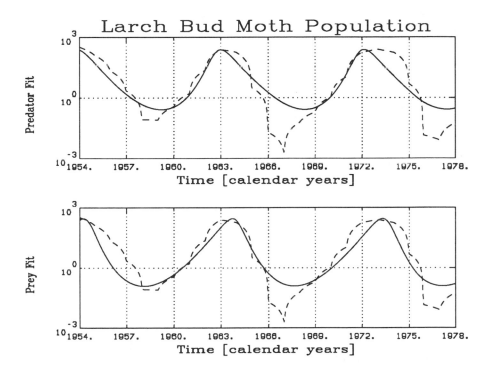

Figure 10.5. Lotka–Volterra models compared to measured insect data.

This presentation shows that the predator fit exhibits a slow decay and a fast rise, whereas the prey fit exhibits a fast decay and a slow rise. The second model gives a slightly better fit, but the difference is certainly not significant.

Maybe some truth lies in both assumptions and we should use a three–species Lotka–Volterra model with the insect being the "sandwich" population between the predator and the prey. Figure 10.6 compares a somewhat optimized three–species Lotka–Volterra model to the measured data, but I didn't try too hard to get the best possible fit. The insect population is shown once using a linear scale and once using a semilogarithmic scale.

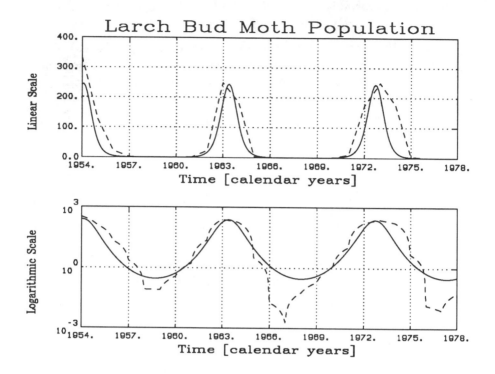

Figure 10.6. Three–species model compared to measured insect data.

This time, we observe a more symmetric rise/decay behavior. I am sure that by trying a little harder I could obtain an even better fit than shown in Fig.10.6. Yet this is not the important message. Clearly, the measured data are insufficient to validate and therefore justify the more complex three–species model.

Notice the difference in argument here. Previously, I argued in favor of a complex physical model over a simple inductive model. That is different. We need much less measurement data to justify a complex physical model than to justify a complex inductive model.

Notice that, just by adding sufficiently many parameters to our model, we can fit virtually any model structure to virtually any data. This is one of the most severe problems with many inductive modeling techniques. Inductive models lure us into love affairs with unhappy endings. In Chapter 13 of this text, I shall present another inductive modeling technique that does not suffer from this disease. It will simply refuse to let us identify models that exhibit a larger complexity than what is justifiable from the amount of available measured data.

Analyzing the preceding example a little further, we notice one of the major problems with modeling biological systems. Since we, the modelers, are a biological system ourselves, it is not surprising that the time constants of the biological systems to be modeled are generally of the same order of magnitude as our own time constants, i.e., in population dynamics, it usually takes years to obtain data that can support modeling. In the case of the larch bud moth ecosystem, the researchers had to spend three weeks every fall in the (undeniably very beautiful) upper Engadine valley during 30 years, climb trees, and count needles and larvae, to obtain the (still–meager) data points used to validate their models. What a difference with electrical circuitry!

Fischlin and Baltensweiler [10.6] carried their model identification even farther. They argued that the insects breed only during a relatively short time period every year. Therefore, a discrete–time model with a sampling period of one year may be a more reasonable assumption than the continuous–time Lotka–Volterra models shown in this chapter. Then they argued that the Engadine valley is not a homogeneous ecosystem. Different parts of the valley exhibit different insect populations at different times. This claim was supported by data. Data had indeed been collected for different spots in the valley separately. Therefore, they compartmentalized the valley into 20 different segments. They used a *semiphysical* modeling technique, which I shall introduce in Chapter 11, to model the dynamics within each segment, and they included the migration of insects between these segments in their model. Their modeling technique allowed them to create a model of this ecosystem that is strictly based on causal relationships among variables. The parameters of the model were identified individually on the basis of measurement statistics. Their model was able to predict the insect population fairly well without applying a global postoptimization to trim the model parameters. In Chapter 11, I shall present a simplified version of their model.

Postoptimization is, of course, a questionable enterprise. This model contains so many parameters that chances are that a good fit with the measured data could be found even if the model structure is incorrect. This problem can be overcome by assigning tolerance bands to all parameters of the model that describe the inaccuracy of the deductive model, i.e., parameters that have been well established are assigned narrow tolerance bands, while parameters for which few measurement data are available are assigned wide tolerance bands.

Then we could restrict the postoptimization in such a way that every one of the parameters must remain within its assigned tolerance band.

The optimizations shown in this section are quite tricky. In the companion book of this text, I shall explain how they were accomplished. In fact, the optimization problems demonstrated in this section will turn into excellent homework problems then. But let's try to swallow one bite at a time.

10.3 Competition and Cooperation

In addition to the previously discussed predator–prey relationships among different species, two other relationships deserve to be mentioned: competition and cooperation.

A *competitive situation* typically occurs when several species compete for the same food source. In a way, this is similar to crowding, i.e., the more densely populated an area is, the more severe the competition will become. Similar to crowding, competition should thus be modeled as a quadratic effect. As in the predator–prey situation, if one of the populations is extinct, the competition stops. Consequently, we again model competition as a cross–product of the competing populations, but this time, it appears with a negative sign in both equations, i.e.:

$$\dot{x}_1 = a \cdot x_1 - b \cdot x_1 \cdot x_2 \qquad (10.7a)$$
$$\dot{x}_2 = c \cdot x_2 - d \cdot x_1 \cdot x_2 \qquad (10.7b)$$

Here, both populations are expected to grow exponentially if they are not inhibited by competition, but due to competition, neither of the two populations will be able to grow forever.

The opposite situation is called *cooperation*. Cooperation occurs naturally in a variety of situations. A typical scenario is the symbiosis among two species. Neither of the two species can survive without the other.

Equation (10.8) describes the typical cooperation model:

$$\dot{x}_1 = -a \cdot x_1 + b \cdot x_1 \cdot x_2 \qquad (10.8a)$$
$$\dot{x}_2 = -c \cdot x_2 + d \cdot x_1 \cdot x_2 \qquad (10.8b)$$

Both populations have a tendency to decay, but in cooperation, a stable equilibrium can be reached that saves both populations from a destiny of extinction.

A similar effect is grouping. *Grouping* is the inverse mechanism of crowding. Many animals travel in herds, because in a herd, they are less vulnerable to being attacked by another species. A typical grouping model is shown below:

$$\dot{x} = -a \cdot x + b \cdot x^2 \tag{10.9}$$

Of course, all these effects can occur together, i.e., the typical population dynamics model involving n species can be written as:

$$\dot{x}_i = \left(a_i + \sum_{j=1}^{n} b_{ij}x_j\right)x_i \quad , \quad \forall i \in [1,n] \tag{10.10}$$

or using a matrix notation:

$$\dot{x} = \bigl(\text{diag}(a) + \text{diag}(x) \cdot \mathbf{B}\bigr) \cdot x \tag{10.11}$$

where the **a** vector captures the balance between birth rate factors and death rate factors, the diagonal elements of the **B** matrix describe the balance between grouping and crowding factors, and the off–diagonal elements of the **B** matrix cover the balance between cooperation and competition factors. They also include the predator–prey situation in which the predator considers the prey to be "cooperative," while the prey considers the predator to be "competitive." The "diag" function turns its argument vector into a diagonal matrix.

This model assumes that all relationships between species are binary, i.e., no cross–products exist between more than two species. This is meaningful. If three species compete for the same food source, the competition terms should be modeled as:

$$-b_{12} \cdot x_1 \cdot x_2 - b_{13} \cdot x_1 \cdot x_3 - b_{23} \cdot x_2 \cdot x_3 \tag{10.12a}$$

and not as:

$$-b \cdot x_1 \cdot x_2 \cdot x_3 \tag{10.12b}$$

since, if one of the three populations is extinct, the other two still compete for the same food. Equation (10.12b) would suggest that competition stops as soon as one of the three populations dies out.

The situation would be quite different if we were to assume a symbiosis among three different species in which neither of the species can survive without the other two, but to my knowledge, such a phenomenon has never been observed on this planet.

What are the lessons learned? In short, we can postulate the following three rules:

(1) If we have a choice between a decent physical model and an inductive model, we always choose the physical model, even if it is more complex than the equivalent inductive model.

(2) If we have to content ourselves with an inductive model, we choose the simplest model that explains the data reasonably well. We start with a linear exponential growth/decay model and add more terms until we obtain a reasonable fit. Among equally reasonable terms, we choose the one with the highest sensitivity, since it is probably the most important among the terms.

(3) We should avoid the love story cliff. We must be extremely self–critical when we try to deduct causal relationships among real system variables from an inductive model. Had this third rule been properly observed, Australia would not be plagued with rabbits and foxes today.

A more detailed discussion of the basic mechanisms of population dynamics can be found in a recent book by Edward Beltrami [10.1].

10.4 Chaos

Until now, every autonomous system that we studied exhibited, in the steady–state, one of three types of behavior:

(1) *continuous steady–state*, i.e., after a long time, every variable in the system assumes a constant value,

(2) *periodic steady–state*, i.e., after all transients have died out, some variables oscillate with a fixed frequency, while others may assume constant values, or

(3) *no steady–state*, i.e., the transients never die out; on the contrary, they grow beyond all bounds, that is, the system is unstable.

An interesting question is whether these are indeed all possible types of behavior that autonomous systems can exhibit. Would it not be feasible that the transients in an autonomous system do not die out and yet stay bounded? Let us discuss this question further by means of a three–species population dynamics model.

The following set of equations describes one predator x_3 foraying on two different preys x_1 and x_2. The preys suffer furthermore from crowding and competition:

$$\dot{x}_1 = x_1 - 0.001 \ x_1^2 - 0.001 \ x_1 x_2 - 0.01 \ x_1 x_3 \qquad (10.13a)$$

$$\dot{x}_2 = x_2 - 0.0015 \ x_1 x_2 - 0.001 \ x_2^2 - 0.001 \ x_2 x_3 \qquad (10.13b)$$

$$\dot{x}_3 = -x_3 + 0.005 \ x_1 x_3 + 0.0005 \ x_2 x_3 \qquad (10.13c)$$

This set of equations has been analyzed by Michael Gilpin [10.7]. Figure 10.7 shows the trajectory behavior of this system simulated over 5000 time units. The initial populations of all three species were arbitrarily set to 100 each.

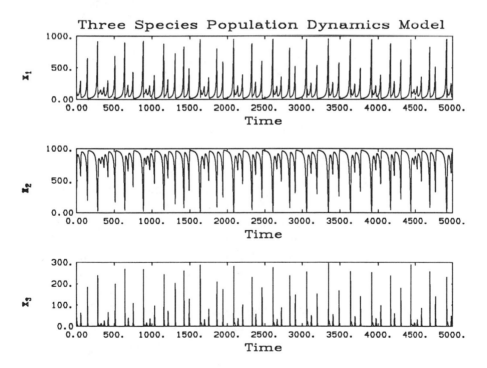

Figure 10.7. Chaotic three–species Lotka–Volterra model.

During most of the time, plenty of preys of the x_2 type are around, but from time to time, the predator population seems to explode.

Then it reduces the x_2 population drastically. Since no more food is around, the predator population decreases again. Now, the x_1 population can grow, which before was hampered by heavy competition from x_2. However, the x_2 population recovers quickly and resumes its dominant position in the ecosystem. And yet each cycle seems to be a little different and no periodic pattern seems to evolve. This type of behavior is called *chaotic motion*. Chaotic motion is characterized by a transient response that does not die out, yet remains stable.

Let us look at the same problem in the phase plane. Figure 10.8 shows two of the three phase portrayals:

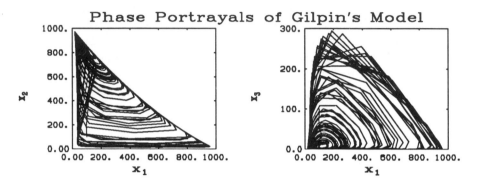

Figure 10.8. Chaotic Lotka–Volterra model in the phase plane.

Most of the time the trajectory lingers around the left upper corner of the $< x_1, x_2 >$ plane and the left lower corner of the $< x_1, x_3 >$ plane. The transitions away from these corners take place very rapidly and last only a very short time. The nonsmooth characteristic of the $< x_1, x_3 >$ graph shows the influence of the numerical integration. In reality, this curve should be smooth. We must therefore be suspicious. Is this behavior for real, or is it simply an artifact of the numerical integration, i.e., if we were to integrate with higher precision on a machine with a large mantissa, would the behavior turn out to be periodic, or is what we observe a real property of a real system?

In a rigorous sense, this question is still unanswered. Let me try to provide a better answer than has been given previously, but before I do so, we want to analyze a much simpler example, an example that is so simple that we can tackle it analytically.

The discrete–time version of the logistic equation can be written as:

$$x_{k+1} = a \cdot x_k(1.0 - x_k) \tag{10.14}$$

Let us analyze the steady–state behavior of this equation as a function of the single parameter a. Figure 10.9 shows the trajectory behavior of this system for a random initial condition between 0.25 and 0.75, and for four different values of a chosen between 0.0 and 1.0. I plotted the left–hand side of Eq.(10.14) as one curve and the right–hand side as another. In this way, we can solve the recursion equation graphically, which will be useful for a full understanding of what is going on.

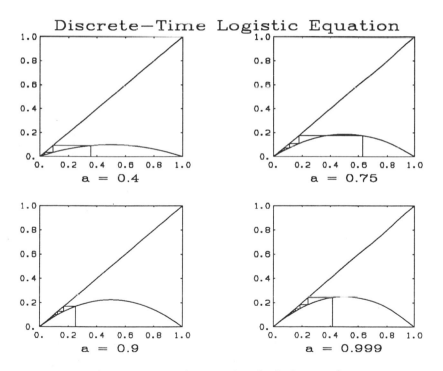

Figure 10.9. Discrete–time logistic equation.

In the given range of the parameter a, the two curves intersect only at a single point, namely, for $x = 0.0$. This turns out to be a stable continuous steady–state point of the discrete–time logistic equation. As we approach $a = 1.0$, the area between the two curves becomes more and more narrow, and consequently, it will take more and more iterations to reach the steady–state point. At $a = 1.0$, this steady–state point becomes unstable. If you are curious about how Fig.10.9 was produced, solve Hw.[H10.6].

Figure 10.10 shows what happens in the range $a \in [1.0, 3.0]$.

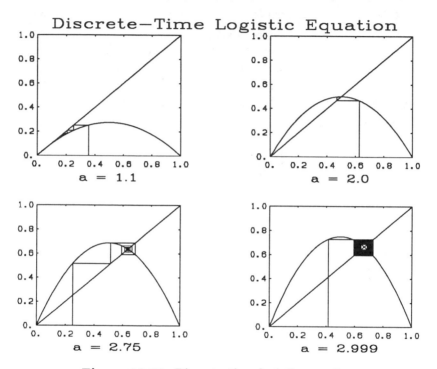

Figure 10.10. Discrete–time logistic equation.

Now the two curves intersect in two points. The second intersection is now a stable continuous steady–state solution of the discrete–time logistic equation. At $a = 1.0$, this solution is identical with the previous one and is thus marginally stable. As we leave the vicinity of $a = 1.0$, the steady–state point becomes more and more stable. For example, at $a = 2.0$, it takes only very few iterations to reach the steady–state point. As we approach $a = 3.0$, this steady–state point again becomes less and less stable. At $a = 3.0$, the system is marginally stable. At $a = 3.0 - \epsilon$, it takes infinitely many iterations to reach the steady–state point.

Figure 10.11 shows what happens in the range $a \in [3.0, 3.5]$.

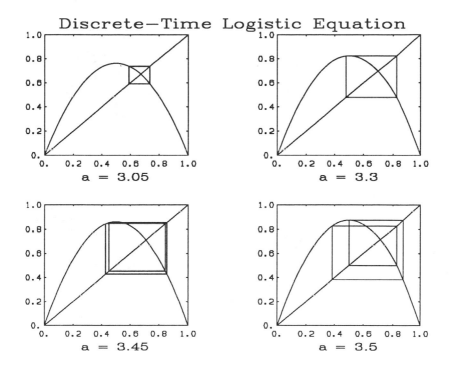

Figure 10.11. Discrete–time logistic equation.

This time, the first 100 iterations were skipped in order to show the steady–state solution directly. At $a = 3.1$, a periodic steady–state exists with a period of 2 around the intersection of the two curves. As we approach $a = 3.0$ from the upper side, the steady-state rectangle shrinks more and more. At $a = 3.0 + \epsilon$, the rectangle is infinitely small, and it takes infinitely many iterations to reach it. As we leave the area of $a = 3.0$ in the positive direction, the periodic steady–state becomes more and more stable, i.e., it takes fewer and fewer iterations to reach the limit cycle. In the vicinity of $a = 3.45$, the next accident happens. Again, the solution has become marginally stable. Now the limit cycle splits and we obtain a new limit cycle with a period of 4.

Figure 10.12 shows what happens in the range $a \in [3.5, 4.0]$.

Discrete—Time Logistic Equation

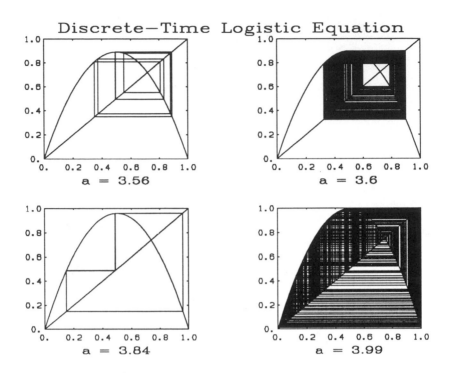

Figure 10.12. Discrete—time logistic equation.

The stable intervals between two new bifurcations become smaller and smaller. Each time, the period of the periodic steady–state doubles. At $a = 3.56$, we observe a period of 8. Then the period becomes *infinite*. At $a = 3.6$, the signal has become nonperiodic, yet stable. We call this a *chaotic steady–state*.

However, even this is not the full truth. At $a = 3.84$, we observe another stable limit cycle, this time with a period of 3. At $a = 3.99$, we obtain a totally aperiodic behavior with random values of x anywhere between 0.0 and 1.0. Finally, at $a = 4.0$, the equation becomes unstable. In the companion book, we shall return to this discussion, and analyze the properties of the logistic equation for $a = 3.99$ as a random–number generator.

Figure 10.13 shows the behavior of the discrete–time logistic equation in a so–called *bifurcation map* for $a \in [2.9, 4.0]$.

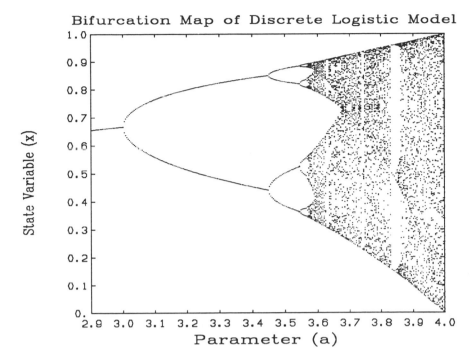

Figure 10.13. Bifurcation map of the discrete–time logistic equation.

On the independent axis, the parameter a is varied. On the dependent axis, possible steady–state values of x are shown. For $a < 3.0$, only one such value exists, but for $a > 3.0$, several such values exist. The two branches of the map show the upper and lower value of x in the periodic steady–state. The dark areas are chaotic. As can be seen, even for larger values of a, chaotic areas are interspersed with nonchaotic areas. If you are curious about how Fig.10.13 was created, solve Hw.[H10.7].

This time, we know for a fact that we have not observed merely an artifact. It is a straightforward (though tedious) exercise to compute the *bifurcation points*, i.e., the points on the bifurcation map where the number of branches doubles. Let me sketch an algorithm with which this can be achieved. We start with the assumption of a continuous steady–state, i.e.:

$$x_{k+1} = a \cdot x_k(1.0 - x_k) \equiv x_k \quad , \quad k \to \infty \qquad (10.15)$$

which has the two solutions $x_k = 0.0$ and $x_k = (a - 1.0)/a$. We move the second solution into the origin with the transformation: $\xi_k = x_k - (a - 1.0)/a$. This generates the difference equation:

$$\xi_{k+1} = -a \cdot \xi_k^2 + (2.0 - a)\xi_k \qquad (10.16)$$

We linearize this difference equation around the origin and find:

$$\xi_{k+1} = (2.0 - a)\xi_k \qquad (10.17)$$

which is marginally stable for $a = 1.0$ and $a = 3.0$. Now we repeat this analysis with the assumption of a periodic steady–state with period 2, thus:

$$x_{k+2} = a \cdot x_{k+1}(1.0 - x_{k+1}) \equiv x_k \quad , \quad k \rightarrow \infty \qquad (10.18)$$

We evaluate Eq.(10.18) recursively, until x_{k+2} has become a function of x_k. This leaves us with a fourth–order polynomial in x_k. Obviously, the previously found two solutions must also be solutions of this new equation. Thus, we can divide through and again obtain a second–order polynomial with two additional solutions. We move those into the origin, linearize, and obtain two new values for a, one of which will again be $a = 3.0$, the other will be the next bifurcation point. We continue in the same manner. Each time, the previous solutions will be solutions of the new equation as well. Of course, it is more meaningful to perform this analysis on a computer rather than manually and software systems exist that allow us to solve such problems elegantly.

Notice that simulation is a very poor tool for determining the logistic bifurcation points accurately, since unfortunately, the logistic equation is marginally stable at every bifurcation point. A better numerical technique might be to exploit the marginal stability explicitly, i.e., compute root loci of the difference equations as a function of a and solve for the points where the root loci intersect with the unit circle. However, this is of little concern to our cause. All we wanted to achieve was to convince ourselves that the bifurcations are for real and are not merely artifacts. A more detailed discussion of analytical and semianalytical techniques for the explicit evaluation of bifurcation points can be found in a paper by Mitchell Feigenbaum [10.5].

Let us now return to the Gilpin equations. These equations are far too complicated to perform a closed analysis, as in the case of the logistic equation. The problem is the following. Even if we assume that we operate on a computer with a mantissa of infinite length, the numerical integration algorithm still converts our differential equations into difference equations, and we don't know for sure whether the observed chaotic behavior is due to the differential equations themselves or whether it was introduced in the process of discretizing the differential equations into difference equations. Consequently, the observed chaotic behavior could, in fact, be a numerical artifact.

Several researchers have argued that an easy way to decide this question is to simply switch the integration algorithm. If the chaotic behavior occurs for the same parameter values, then it can be concluded that the numerical integration is not the culprit. Unfortunately, this argument does not hold. As we see clearly from Fig.10.7, the trajectory behavior of the Gilpin model is highly irregular, almost discontinuous at times. Consequently, a fixed–step integration algorithm will compute garbage, while a variable–step algorithm will reduce the step size to very small values in the vicinity of the spikes. As we shall see in the companion book, all currently advocated numerical integration algorithms are based on polynomial extrapolation. When the step size is reduced, the higher–order terms in the approximation become less and less important. Ultimately, for a sufficiently small step size, every integration algorithm will behave numerically like a forward Euler algorithm. Therefore, by switching the integration algorithm, we haven't really achieved anything. We have just replaced one Euler algorithm with another.

Let me propose another approach. We apply a logarithmic transformation to our populations:

$$y_i = \log(x_i) \tag{10.19}$$

Thereby, the Gilpin equations are transformed into:

$$\dot{y}_1 = 1.0 - 0.001 \ \exp(y_1) - 0.001 \ \exp(y_2) - 0.01 \ \exp(y_3) \tag{10.20a}$$
$$\dot{y}_2 = 1.0 - 0.0015 \ \exp(y_1) - 0.001 \ \exp(y_2) - 0.001 \ \exp(y_3) \tag{10.20b}$$
$$\dot{y}_3 = -1.0 + 0.005 \ \exp(y_1) + 0.0005 \ \exp(y_2) \tag{10.20c}$$

If we simulate Eqs.(10.20a–c) instead of Eqs.(10.13a–c), we are confronted with a numerically different problem. Yet the analytical solution must be the same. Thus, if Eqs.(10.13a–c) and Eqs.(10.20a–c) give rise to the same bifurcation map, we can indeed believe that the map was caused by the differential equations themselves, and had not been introduced in the process of discretization.

Let us apply the following experiment. The parameter that we are going to vary is the competition factor. For this purpose, we rewrite the Gilpin equations in the following way:

$$\dot{x}_1 = x_1 - 0.001\ x_1^2 - k \cdot 0.001\ x_1 x_2 - 0.01\ x_1 x_3 \qquad (10.21a)$$
$$\dot{x}_2 = x_2 - k \cdot 0.0015\ x_1 x_2 - 0.001\ x_2^2 - 0.001\ x_2 x_3 \qquad (10.21b)$$
$$\dot{x}_3 = -x_3 + 0.005\ x_1 x_3 + 0.0005\ x_2 x_3 \qquad (10.21c)$$

For $k = 1.0$, we obtain the same solution as before. $k < 1.0$ reduces competition, while $k > 1.0$ increases competition. Figure 10.14 shows the behavior of the Gilpin model for a slightly reduced competition. Only the first prey x_1 is shown.

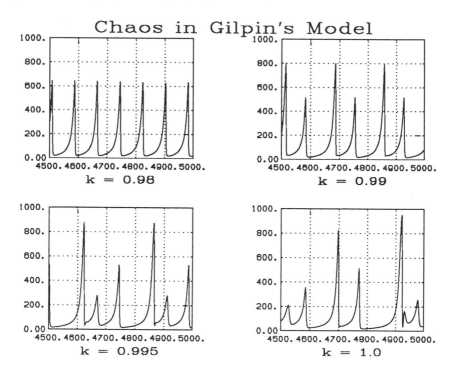

Figure 10.14. The influence of competition in Gilpin's model.

With 98% competition, we observe a stable periodic steady–state with a discrete period of '1,' meaning that each peak reaches the same height. For 99% competition, the discrete period has doubled. For 99.5% competition, the period has tripled. Somewhere just below $k = 1.0$, the model turns chaotic. For lower values of competition, the Gilpin model exhibits a continuous steady–state, i.e., no oscillation occurs at all. Let us see what happens if we increase the competition.

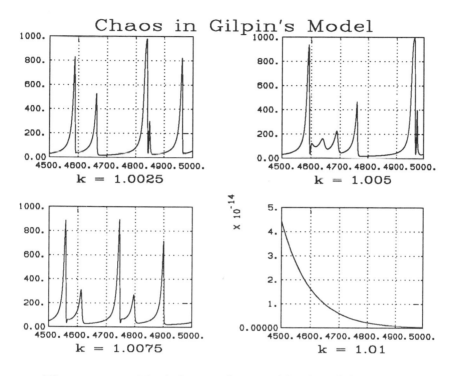

Figure 10.15. The influence of competition in Gilpin's model.

The behavior of the Gilpin model stays chaotic only up to $k = 1.0089$. For higher competition values, the x_1 population dies out altogether. We notice an incredible manifold of different possible behavioral patterns of this structurally simple model.

I repeated the same experiment for the modified Gilpin model, i.e., the model after application of the logarithmic transformation. The results were the same, except that in the chaotic cases the curves looked different, which is easily explained by the high sensitivity of the model to round–off errors. We would need a computer with a mantissa of infinite length in order to obtain the same trajectories.

Figure 10.16 shows the two bifurcation maps of Gilpin's model with and without the logarithmic transformation. For this purpose,

I recorded all extrema of the x_1 population, i.e., the values for which $\dot{x}_1 = 0.0$.

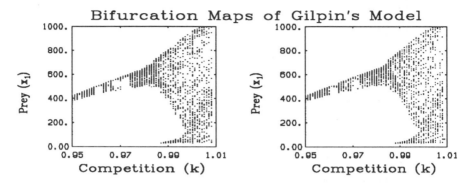

Figure 10.16. Bifurcation map of Gilpin's model.

The bifurcation maps are somewhat inaccurate due to the large gradients in the immediate vicinity of the peaks. A bifurcation map for the Gilpin model was previously given by Schaffer [10.9]. His map looks qualitatively similar, but Schaffer didn't specify which parameter he varied to obtain his map and did not label the axes on his graph. Thus, I cannot really compare the results. Obviously, Schaffer used a better algorithm to determine accurately the maxima since his bifurcation map is much more crisp than mine and resembles more closely the logistic bifurcation map of Fig.10.13. However, Fig.10.16 shows clearly that the two bifurcation maps (with and without logarithmic transformation) are qualitatively the same. Thus, we conclude that the observed chaotic behavior is indeed a property of the physical system, not merely an artifact of the numerical solution technique.

The Gilpin model is of course rather artificial. I don't think that a real system with such a high vulnerability to competition would be able to survive and thrive in a chaotic mode. Yet this is a particularity of the Gilpin model, not a property of chaos *per se*. Chaotic motion has been observed on numerous occasions. A good account of chaotic observations in biological systems can be found in Degn *et al.* [10.3]. Chaotic motion is by no means limited to biological systems alone. I simply found this to be a good place to introduce the concept. A well–written survey of the mechanisms of chaotic motion can be looked up in a recent book by Robert Devaney [10.4].

In the first section of this chapter, we analyzed the antelope population dynamics problem, which exhibited a fairly large structural complexity and yet the behavioral complexity of that model was very low. In the last section, we saw several examples of systems with very limited structural complexity that exhibited a stunning wealth of behavioral complexity. Obviously, structural and behavioral complexity do not have to go hand in hand. This comes as a surprise. In earlier days, observed erratic behavior of physical processes had always been attributed to their structural complexity. Chaos theory has taught us that even structurally simple systems can exhibit an astounding variation of behavioral patterns. However, the reverse is also true. Structural complexity does not necessarily lead to behavioral complexity and in fact it commonly won't. This statement deserves additional consideration. The next section will provide just that.

10.5 The Forces of Creation

The behavioral complexity of a system is usually larger than its structural complexity. We see this already by means of the simplest differential equation system:

$$\dot{x} = -a \cdot x \quad , \quad x(t = 0.0) = x_0 \tag{10.22}$$

has the solution:

$$x(t) = \exp(-at) \cdot x_0 \tag{10.23}$$

While the differential equation itself is linear, its solution is already exponential. The differential equation describes the *structure* of the system, while the solution describes its *behavior*.

What chaos theory taught us is the fact that the behavioral complexity of a system can even be much greater than its structural complexity, more so than we had thought possible before chaos had been discovered.

This is the answer to the question I posed at the end of Chapter 9. In the past, researchers looked at thermodynamics only in the steady–state. To do so, they analyzed a limited subset of the behavioral patterns of thermic systems, namely, those observed under steady–state conditions. Yet even those patterns were already quite complex, and it was necessary to distinguish between different types of behavioral patterns, such as those observed under reversible conditions versus those observed under irreversible conditions. By working with the differential equations directly, we concentrate on the structural patterns of thermic systems and this turns out to be simpler.

The next question is: Which mechanisms exist that limit the behavioral complexity of a system? If we add more and more structural components to a system, why is it that its behavioral complexity does not grow beyond all bounds? I would like to identify three different mechanisms that limit system complexity.

(1) *Physical constraints:* In connecting two subsystems, their combined degrees of freedom are usually lower than the sum of the degrees of freedom of the subsystems.

(2) *Control mechanisms:* Controllers in a system tend to restrict the possible modes of behavior of a system.

(3) *Energy:* The laws of thermodynamics state that each system sheds as much energy as it can, i.e., it maintains the lowest amount of energy feasible. This mechanism also limits the complexity.

Let us discuss these three mechanisms in more detail. If we take a lever and describe all possible motions of that lever we need six differential equations to do so, i.e., we formulate Newton's law three times for the three translational motions and three times for the three rotational motions. When we now take two levers and connect them at one point, we notice that the total number of differential equations needed to describe all possible motions of the combined system is nine rather than twelve. The connection between the two levers has introduced three structural singularities. We noticed this fact before when we analyzed the behavior of a DC motor with a mechanical gear. The gear was responsible for a structural singularity. Consequently, physical constraints restrict the structural complexity of a system.

If we introduce a controller to a system, we reduce the sensitivity of that system to parameter variations. We could also say: A controller reduces the sensitivity of the behavior of a system to structural changes, i.e., the controller restricts the behavioral complexity of the system. A system with a controller will exhibit fewer modes of operation than the same system would if the controller were removed. This is why the device is called a "controller." It controls the behavior of a system. This is equally true for human–made controllers as for controllers in nature. Why is it that all trees grow new leaves in spring and shed their leaves in the fall? The control mechanisms built into the system regulate this behavior. Due to these controls, the trees react uniformly and the overall behavioral complexity is much reduced. Without such control mechanisms, trees could grow leaves arbitrarily and they would have more "freedom" in determining their behavior.

The third mechanism is energy. To return to our previous example, why is it that the motion of a lever can be described by six differential equations, although this metal bar contains billions of atoms? The true structural complexity of this system is phenomenal, and yet its apparent structural complexity is very limited. The reason is simple: it is cheaper that way, i.e., the total energy content of the system can be reduced by forcing the atoms into a crystalline structure.

Obviously, these three mechanisms are not independent of each other. In our population dynamics models, crowding certainly exerts control over a population. Yet the crowding effect is caused by energy considerations. The limited energy content of a closed system is a great regulator in population dynamics. Returning once more to our metal bar: The laws of thermodynamics manifest themselves in an apparent control mechanism that restricts the motion of the individual atoms. From a more macroscopic point of view, if we ignore the dynamics of this control mechanism, we can view the global effects of this control as a constraint. All atoms seem to move in unison, and therefore we experience our metal bar as a rigid body.

So why is there any complexity at all in this universe? Why does our planet exhibit such a wealth of different systems and different behaviors? We don't know the answer to this question. However, we can observe that a second force is at work besides *energy*, which is the *entropy*. Every system tries to maximize its own entropy, i.e., it strives to reach a state of highest disorder. We don't know where this force comes from. We don't know whether the laws of thermodynamics are globally true or local dynamic aberrations. Is entropy a global force or is it simply a reminiscent of the dynamics of energy transfer in the big bang? We are currently unable to answer these questions, and my guess is that we never will. The laws of thermodynamics are *empirical laws*, i.e., they are based on observation, not on deduction. I believe that these laws are so fundamental to the functioning of this universe that we shall never be able to unravel their origin. Yet within the framework of our understanding of physics, these laws have certainly been confirmed over and over again.

We notice that two separate forces are at work. *Entropy* (or rather the law of entropy maximization) is the great innovator. Entropy tries hard to "create a mess." *Energy* (or rather the law of energy minimization) is the great organizer. Energy tries hard to "clean up" the mess that entropy created. Together they manage the *evolution*.

I know that this a very simplistic view of an impossibly complex problem. I know all too well the dangers behind oversimplification. Didn't I warn constantly about the love story cliff? Yet this view raises some interesting questions in the context of modeling. Is it possible to create a computer algorithm that somehow exhibits elements of evolution? First attempts in this direction have been reported. Brown and Vincent [10.2] obtained interesting results in the study of *evolutionary games*. This topic is also closely related to the area of *machine learning*. Several recent advances in machine learning can be viewed in the light of modeling evolution, in particular, some of the reported research efforts in *neural network learning*, and in *genetic algorithms*. We shall return to this discussion in Chapter 14 of this text. Yet this research area is still in its infancy and more research is highly encouraged.

From the preceding discussion, I conclude that any algorithm that attempts to replicate elements of evolution needs two separate mechanisms:

(1) *an innovator*, i.e., a mechanism to generate new behavioral elements; in a computer model, this will probably have to be triggered by some sort of random–number generator, and

(2) *an organizer*, i.e., a mechanism to restrict behavioral complexity and guarantee overall system stability. In a computer model, this will probably have to be some sort of optimization algorithm.

Don't get me wrong. I do not suggest that energy and entropy should be made responsible for every move in our daily lives. I am aware of the fact that modeling (by definition) must be reductionistic. The more highly organized a system is, the more other influencing factors that will eventually dominate the behavior of the system will appear. The shortcomings of "social Darwinism" have long been discovered and the methodology has been discredited as a mechanism to describe, e.g., the social behavior of human organizations. Yet at the bottom of every system, underneath all other competing factors, the two primeval forces are always at work, the tidal forces of our universe, the laws of thermodynamics.

10.6 Summary

In this chapter, we analyzed basic models that describe the behavioral patterns of population dynamics. For this purpose, it was necessary to leave the road of physical modeling and introduce a new concept, the technique of inductive modeling. We then proceeded to more advanced topics, we introduced the concept of chaotic motion, and we dealt with the relationship between structural complexity and behavioral complexity in a model.

Notice that the inductive modeling methodology proposed in this chapter is clearly limited to the discussion of population dynamics. In the next chapter, we shall generalize this idea and introduce a methodology that will allow us to model a much wider variety of systems in a semiphysical and semiinductive manner.

References

[10.1] Edward Beltrami (1987), *Mathematics for Dynamic Modeling*, Academic Press, Boston, Mass.

[10.2] Joel S. Brown and Thomas L. Vincent (1987), "A Theory for the Evolutionary Game," *Theoretical Population Biology*, **31**(1), pp. 140–166.

[10.3] Hans Degn, Arun V. Holden, and Lars F. Olsen, Eds. (1987), *Chaos in Biological Systems*, Plenum Press, New York, NATO ASI Series, Series A: Life Sciences, **138**.

[10.4] Robert L. Devaney (1986), *An Introduction to Chaotic Dynamical Systems*, Benjamin/Cummings, Menlo Park, Calif.

[10.5] Mitchell J. Feigenbaum (1978), "Quantitative Universality for a Chaos of Nonlinear Transformations," *J. Statistical Physics*, **19**(1), pp. 25–52.

[10.6] Andreas Fischlin and Werner Baltensweiler (1979), "Systems Analysis of the Larch Bud Moth System. Part 1: The Larch — Larch Bud Moth Relationship," Mitteilungen der Schweizerischen Entomologischen Gesellschaft, **52**, pp. 273–289.

[10.7] Michael E. Gilpin (1979), "Spiral Chaos in a Predator–Prey Model," *The American Naturalist*, **113**, pp. 306–308.

[10.8] Alfred J. Lotka (1956), *Elements of Mathematical Biology*, Dover Publications, New York.

[10.9] William M. Schaffer (1987), "Chaos in Ecology and Epidemiology," in: *Chaos in Biological Systems*, (H. Degn, A.V. Holden, and L.F. Olsen, eds.), Plenum Press, New York, pp. 233–248.

[10.10] R. B. Williams (1971), "Computer Simulation of Energy Flow in Cedar Bog Lake, Minnesota Based on the Classical Studies of Lindeman," in: *Systems Analysis and Simulation in Ecology, Vol.1*, (B.C. Patten, ed.), Academic Press, New York.

Bibliography

[B10.1]   Alan Berryman, Ed. (1988), *Dynamics of Forest Insect Popula-tions: Patterns, Causes, Implications*, Plenum Press, New York.

[B10.2]   Mitchell J. Feigenbaum (1980), "The Transition to Aperiodic Be-havior in Turbulent Systems," *Commun. Mathematical Physics*, 77, pp. 65–86.

[B10.3]   Tom Fenchel (1987), *Ecology, Potentials and Limitations*, Ecol-ogy Institute, Oldendorf/Luhe, F.R.G.

[B10.4]   Michael Mesterton–Gibbons (1989), *A Concrete Approach to Mathematical Modelling*, Addison–Wesley, Redwood City, Calif.

Homework Problems

[H10.1] Predator–Prey Model

A simple ecosystem consists of a population of seasnails p_s that forages on a population of algae p_a. The populations are measured in number of species per unit surface. The dynamics of this ecosystem can be described by two differential equations:

$$\dot{p}_s = b \cdot c_1 \cdot p_s \cdot p_a - c_2 \cdot p_s - c_3 \cdot p_s^2 \qquad (H10.1a)$$

$$\dot{p}_a = c_4 \cdot p_a - c_5 \cdot p_a^2 - c_1 \cdot p_s \cdot p_a \qquad (H10.1b)$$

where $c_1 = 10^{-3}$ is the grazing factor, $c_2 = 0.9$ is the excess mortality rate factor of the snails, $c_3 = 10^{-4}$ is the crowding ratio of the snails, $c_4 = 1.1$ is the excess reproduction rate of the plants, $c_5 = 10^{-5}$ is the crowding ratio of the plants, and $b = 0.02$ is the grazing efficiency factor. All constants have been converted to a per day basis.

An accident in a nearby chemical plant diminishes the two populations to values of $p_s = 10.0$ and $p_a = 100.0$. Simulate the ecosystem for a period of 30 days to check whether the two populations can recover.

Use simulation to determine the range of initial conditions from which the two populations can recover.

[H10.2] Linear Regression Model

Williams [10.10] developed a linear regression model for the ecosystem of Cedar Bog Lake, Minn. The model includes three biological species: a pop-ulation of carnivores x_c that feeds on a population of herbivores x_h which, in turn, feeds on a population of seaweed x_p. All populations are expressed in terms of their energy content measured in cal cm^{-2}. In addition, the model considers the biological sediment that forms on the bottom of the lake and the loss of biomass to the environment (carried out of the lake with the water). The model is driven by solar energy x_s, which enables the growth of the plants.

The model can be described by the following set of differential equations:

$$x_s = 95.9 \cdot (1.0 + 0.635 \cdot \sin(2\pi t)) \qquad (H10.2a)$$

$$\dot{x}_p = x_s - 4.03 \cdot x_p \qquad (H10.2b)$$

$$\dot{x}_h = 0.48 \cdot x_p - 17.87 \cdot x_h \qquad (H10.2c)$$

$$\dot{x}_c = 4.85 \cdot x_h - 4.65 \cdot x_c \qquad (H10.2d)$$

$$\dot{x}_0 = 2.55 \cdot x_p + 6.12 \cdot x_h + 1.95 \cdot x_c \qquad (H10.2e)$$

$$\dot{x}_e = 1.0 \cdot x_p + 6.9 \cdot x_h + 2.7 \cdot x_c \qquad (H10.2f)$$

All constants have been converted to a per year basis. The equation for x_s models the solar radiation as it changes over the year due to the varying position of the sun in the sky.

Assume the following initial conditions: $x_p = 0.83$, $x_h = 0.003$, $x_c = 0.0001$, $x_0 = x_e = 0.0$, and simulate this ecosystem over a period of 2 years.

Except for the variables x_0 and x_e, a periodic steady–state will occur, i.e., the three variables x_p, x_h, and x_c will become periodic with a period of 1 year. Think of a way how the periodic steady–state could be computed faster than by simply simulating the system over a long time.

[H10.3] Antelope Population Model

We wish to create a detailed model of the population dynamics of a breed of antelopes in Serengeti National Park at the border between Kenya and Tanzania (eastern Africa). The antelopes are monogamous. They choose their partner at age three and stay together for the rest of their lives. When one partner dies, the other stays single except at age three before they had offspring, when the surviving animal would look for a new partner. All antelopes select partners of their own age. If a three–year–old cannot find a partner, it migrates out of the park.

All available adults mate and produce offspring every year between ages four and eight. Each year, every couple produces one calf. The probability of male/female offspring is 55% for male and 45% for female.

The mortality rates are given in Table H10.3a.

Table H10.3a. Mortality Rates.

Age [years]	Male [%]	Female [%]
1	60	40
2	10	10
3	5	5
4	5	8
5	5	8
6	5	8
7	6	9
8	7	11
9	10	10
10	25	25
11	70	70
12	100	100

Young females are a little stronger than young males. However, during the reproduction period, the females have a slightly higher mortality rate than their male partners. The numbers are given as percentages of the population of the same sex one year younger. The chances of dying are assumed to be statistically independent of the animal's societal status, i.e., single animals die equally often as mated animals and the death of a mated animal does not influence the life expectancy of its partner.

We want to model this system through a set of difference equations (i.e., as a discrete–time system). $M_i(k)$ denotes the number of single male animals of age i in year k, $F_i(k)$ denotes the number of single females and $C_i(k)$ denotes the number of couples.

Simulate this discrete–time system over 100 years. Assume the initial conditions as given in Table H10.3b.

Table H10.3b. Initial Conditions.

Age [years]	Male [#]	Female [#]	Couples [#]
1	100	100	0
2	100	100	0
3	100	100	0
4	0	0	100
5	3	3	95
6	7	7	90
7	12	12	85
8	15	15	80
9	30	30	70
10	30	30	50
11	20	20	25
12	5	5	0

Use ACSL to model this system. This application won't require a DERIVA-TIVE block. The model is coded in a single DISCRETE block that is executed once every year (use the INTERVAL statement). Declare all population variables as INTEGER except for the total animal population P_{tot}, which is to be exported to CTRL–C or MATLAB. (CTRL–C does not import INTEGER variables properly from ACSL. I haven't checked whether MATLAB handles this problem any better).

Repeat the simulation with the modified assumption that the first–year mortality rates are half as large as assumed earlier. Plot the total animal population over time on separate plots for the two cases.

Save the results from this simulation on a data file for later reuse (use the SAVE statement of CTRL–C or MATLAB).

[H10.4] Meta–Modeling

We wish to describe the behavior of the antelope population dynamics system of Hw.[H10.3] through a meta–model. The assumed meta–model is of the type:

$$\dot{P} = a \cdot P \qquad (H10.4a)$$

with the solution:

$$P(t) = \exp(at) \cdot P_0 \qquad (H10.4b)$$

where P_0 is known while a is unknown. We want to find the best possible values of a for the two cases of Hw.[H10.3].

For this purpose, we want to apply linear regression analysis. We compute the logarithm of Eq.(H10.4b):

$$\log(P) - \log(P_0) = t \cdot a \qquad (H10.4c)$$

We can read in the results from the Hw.[H10.3] using CTRL–C's (or MAT-LAB's) LOAD command. For each of the two cases, we have a vector of P values and a vector of t values. We can thus interpret Eq.(H10.4c) as an overdetermined set of linear equations with one unknown parameter, a, that can be solved in a least squares sense, i.e.,

$$\log(P) - \log(P_0) - t \cdot a = \text{residua} \qquad (H10.4d)$$

where a should be chosen such that the L_2 norm of the residua vector is minimized. This can be easily achieved in CTRL–C (MATLAB) using the notation:

```
[>   y = log(P) - log(P_0) * ONES(P)
[>   a = t\y
```

Find the best a values for the two cases, compute the meta–model P trajectories by plugging these a values into Eq.(H10.4b), and plot together the "true" populations and meta–model populations on separate plots for the two cases.

In one of the two cases, the population decreases to zero. Make sure to cut the trajectory for the regression analysis before this happens since your computer won't like the request to compute the logarithm of zero.

[H10.5] Lotka–Volterra Models

Whenever you can fit the predator of a two–species Lotka–Volterra model to measurement data exhibiting a periodic steady–state, you can invariably also fit the prey of such a model. Prove this statement by applying the time–reversal algorithm of Chapter 8 to the Lotka–Volterra model.

[H10.6] Logistic Equation

Write a program in either CTRL–C or MATLAB that will reproduce Figs.10.9–12. The concept is actually quite simple. First, you need to create your own scaling. Both languages provide you with such a feature. Next you plot the two curves onto the same plot. Thereafter, you plot the zigzag path onto the plot. All you need to do is to store the x and y values of the path corners into two arrays in the correct sequence. Thereafter, PLOT will produce the path for you.

[H10.7] Logistic Bifurcation Map

Reproduce Fig.10.13. To create Fig.10.13, I wrote an ACSL program. In the initial segment, I set $a = 2.9$, I chose an initial value for x $(x = 0.5)$, and I iterated the logistic equation 1000 times in a DO loop. I used an integer counter, which I set to zero.

The DYNAMIC segment consisted of a single DISCRETE block. In this block, I iterated the logistic equation once and incremented the counter. I then tested whether the counter had reached a value of 50. If this was the case, I reset the counter to zero, and incremented a by 0.004. Thereafter, I iterated the logistic equation again 1000 times in a DO loop to obtain the new steady–state. The simulation terminated on $a \geq 4.0$. I avoided iterating the logistic equation when $a = 4.0$, since at this point, the logistic equation becomes unstable.

Finally, I exported the resulting trajectories of a and x into CTRL–C and plotted $x(a)$ using a point type plot.

[H10.8]* Gilpin's Bifurcation Map

Reproduce Fig.10.16. To create Fig.10.16, I wrote an ACSL program. In the initial segment, I set $k = 0.95$. I used an integer counter, which I initialized to zero. I limited the step size to 1.0 (using **maxterval**) and disabled the communication by setting arbitrarily $cint = 10^{12}$.

The DYNAMIC segment consisted of a DERIVATIVE block and a DISCRETE block. In the DERIVATIVE block, I solved the differential equations and scheduled a state–event to occur whenever the derivative of prey

x_1 crosses through zero. The state–event triggers execution of the DIS-CRETE block. The DISCRETE block is similar to the one of Hw.[H10.7], except that I incremented the competition factor k by 0.001 and did not iterate to determine the new steady–state. Also, I used the DISCRETE block to manually record the values of all variables using the call LOGD statement. It was necessary to place the termt statement inside the dis-crete block. Since we disabled communication, it won't be tested at the usual place.

Finally, I exported the resulting trajectories of k and x_1 into CTRL–C and plotted $x_1(k)$ using a point type plot.

Projects

[P10.1] Chaos in Gilpin's Model

We have seen that the chaotic range as a function of competition is very limited in Gilpin's model. Analyze the chaotic range as a function of other parameters. Design a method that will allow you to determine the bifur-cation points in Gilpin's model accurately and see whether you can find a relation between the bifurcation patterns of Gilpin's model as a function of the various model parameters.

[P10.2] Logistic Model

It turns out that energy–bound macroeconomies that exhibit logistic be-havior show a large recession at the point of inflection. We wish to analyze whether there exists a clearcut point of inflection in our global economy. For this purpose we start with the continuous logistic equation. Normalize the parameters a and b such that the final value of the logistic model is 1.0 and the gradient at the point of inflection is also 1.0. Determine the initial value of time such that $x = 0.01$ at $t = t_0$. Verify the correctness of your model by simulating the logistic model until a time t_f where x has reached 99% of its final value.

Choose 20 points equidistantly spaced over the simulated range. Disturb each value randomly by $\pm 1\%$ and reidentify a new logistic model using the "measured" data by means of linear regression. Simulate the newly iden-tified logistic model. Repeat 10 times with different random disturbances. Determine the variation in the final value and the time of inflection. Now throw the last "measured" value away before you reidentify a new logistic model, and repeat the analysis. Keep throwing more and more values away from the right. Obviously, the average variations of the final value and the time of inflection must grow as less and less data points are used for the identification. Plot the average variations over the time of the last data point used in the identification.

Use U.S. census data for the identification of a continuous logistic model. Start by using all data points up to 1990 and determine the final U.S. population and the year of inflection. Repeat by throwing data points away from the right. What is the earliest year in which a decent prediction of the point of inflection could have been made? Repeat with other energy-bound variables, such as world population, U.S. GNP divided by buying power per dollar, etc. Is there a strong correlation between the identified years of inflection among these various variables? When does the global recession occur? Is this maybe what is happening in eastern Europe (and in the U.S.!) right now?

Research

[R10.1] Evolutionary Games

Study the literature on evolutionary games and try to identify common patterns among the strategies. What are the innovators in these models, and how do the organizers work? Try to design a differential equation model that periodically generates new differential equations of a modified type and eliminates older differential equations that are not "fit" according to some chosen criterion. Analyze what conditions are necessary for self-organization to occur.

[R10.2] Chaotic Domains

Chaos is not *per se* a bad phenomenon. Without chaos, none of us would be alive. Our global ecology operates under chaotic steady–state conditions. Chaos ensures that all necessary minerals are constantly being recycled. A continuous steady–state of a variable (such as a mineral) would indicate that this variable has come to rest somewhere and no longer participates in the recycling process.

In world ecology, these chaotic conditions are seemingly quite robust, i.e., chaos occurs for a wide range of parameter values. Contrary to this, we found that in the Gilpin equations, chaos existed only for a small set of parameter values. It is of considerable interest to understand how chaotic regions in the parameter space are related to the system structure. We know that, in a continuous–time model, no chaotic behavior can result for systems of orders one or two. The smallest system order that can lead to chaos is three. However, at order three, the chaotic regions are small and vulnerable. I suspect that with increasing system order, the chaotic regions in the parameter space will become more and more dominant, i.e., more and more robust.

Analyze arbitrary Lotka–Volterra–type models, i.e., models of the type:

$$\dot{\mathbf{x}} = \big(\mathrm{diag}(\mathbf{a}) + \mathrm{diag}(\mathbf{x}) \cdot \mathbf{B}\big) \cdot \mathbf{x} \qquad (R10.2a)$$

Try to determine a general expression for the size of the chaotic regions in the parameter space as a function of the system order.

11

System Dynamics

Preview

In this chapter, we shall introduce a general strategy for inductive modeling, a strategy that will allow us to model systems with totally unknown meta–laws. Of course, models so constructed will not offer the same degree of validity as deductively constructed physical models and it will be important to discuss how the validity of these models can be assessed. While the methodology can be used to construct models in a completely inductive manner, it will allow us to incorporate in our model any physical insight that we may possess about the functioning of the process under investigation. The methodology has been coined *System Dynamics*. This is unfortunately somewhat of a misnomer. Haven't we been discussing the "dynamics of systems" in this book all along? Didn't I reference in Chapter 4 a book entitled *System Dynamics* that talks about simple electrical and mechanical systems? In this new context, "System Dynamics" denotes a specific semiphysical and semiinductive modeling methodology. To minimize the confusion, from now on, I shall always capitalize the term "System Dynamics" when I refer to this particular modeling methodology and not capitalize it when I refer to the dynamics of systems in general.

11.1 Introduction

If you were asked at this point to model the entire world to be able to speculate about the destiny of this planet, you would probably be at a loss. You know how to model electrical circuits, but the world? When you will have reached the end of this chapter, you will know how to "model the world."

The first exercise in every modeling endeavor is to choose the facets of the investigated system that we wish to capture in our model. Obviously, it would be unrealistic to assume that we can capture each and every facet of the system in a model unless we decide to duplicate the system itself. We make this selection by identifying a set of variables that will be the key players in our model.

In physical modeling, we usually did not pay much attention to this step of the modeling cycle since the choice of variables came about naturally. For example, in an electrical circuit, we started by deciding whether we were to capture the electrical phenomena only, or whether we were to include thermic phenomena as well. Thereafter, we applied the well–established meta–knowledge of electrical circuitry to generate the model. Only in the end, we chose our "key players," which at that time we called our *state variables*, simply as the outputs of every single integrator in the model. This means that we created our model first (on the basis of available meta-knowledge), and that we chose our state variables only later.

How are we going to choose our state variables when we model the world? State variables capture significant quantities of a system, which have the property that they can accumulate over time (which is, of course, just another way of saying that they are outputs of integrators). Typical candidates for state variables might be populations, money, frustration, love, tumor cells, inventory on stock, and knowledge. In System Dynamics, state variables are called either *levels* [11.6] or *stocks* [11.15], depending on which reference we use.

A first type of equation used in all System Dynamics models captures the fact that the change of each level (or stock) over time can be expressed as the difference between inflows and outflows of this level variable, for instance:

$$\dot{P} = BR - DR \qquad (11.1)$$

The derivative of a population P with respect to time in a closed system can be expressed as the difference between the births per unit time (the so–called birth rate, BR) and the deaths per unit time (the so–called death rate, DR). Table 11.1 lists inflows and outflows for the levels proposed earlier.

Table 11.1. Inflows and Outflows of Typical Level Variables.

Level	Inflow	Outflow
population	birth rate	death rate
money	income	expenses
frustration	stress	affection
love	affection	frustration
tumor cells	infection	treatment
inventory on stock	shipments	sales
knowledge	learning	forgetting

In System Dynamics, the inflows and outflows are called either *rates* [11.7] or *flows* [11.15], depending on which reference we use.

From Table 11.1, we can learn a number of things:

(1) The concepts of levels and rates are extremely general. Applications can be found everywhere.

(2) Rate variables could be levels at the same time (frustration). We shall see how we deal with this problem.

(3) System Dynamics modelers are notoriously "sloppy" with their nomenclature. Of course, treatment is not the outflow of tumor cells, but treatment leads to the death of tumor cells, and if we say that one unit of treatment can be equated with one unit of killed tumor cells at all times, we may not need to distinguish between these two variables in the model.

After we define all the rates, we need to formulate equations that relate the rates back to the levels. In a more familiar terminology, we need to derive a set of state equations. These equations can possibly be simplified by introducing auxiliary variables. In System Dynamics, these auxiliary variables are sometimes called *converters* [11.15].

So far, nothing extraordinary has been detected about the System Dynamics methodology. What is particular about System Dynamics is the way in which the state equations are formulated. This will be the next point on the agenda.

11.2 The Laundry List

In a first attempt to derive a set of state equations, we could try to enumerate all the factors that have an influence on the rate variables, for example:

$$\left.\begin{array}{r}\text{population}\\\text{material standard of living}\\\text{food ratio}\\\text{crowding ratio}\\\text{pollution}\end{array}\right\} \rightarrow \text{birth rate} \qquad (11.2)$$

In our world model, the birth rate is influenced by the population, the material standard of living, the food ratio, the crowding ratio, and the pollution. Such an enumeration is called a *laundry list* [11.15]. The influencing factors may be levels, rates, or converters.

Of course, we must be careful to avoid such dubious relations as:

$$\text{death rate} \rightarrow \text{birth rate} \qquad (11.3a)$$
$$\text{birth rate} \rightarrow \text{death rate} \qquad (11.3b)$$

which is just another way of saying that we should avoid the creation of algebraic loops among the rate variables.

Laundry lists are the first step on the way to deriving state equations.

11.3 The Influence Diagram

Once we have designed laundry lists for all rate variables, we can connect all these laundry lists in one flowchart. Figure 11.1 shows the flowchart of Gilpin's model [11.8].

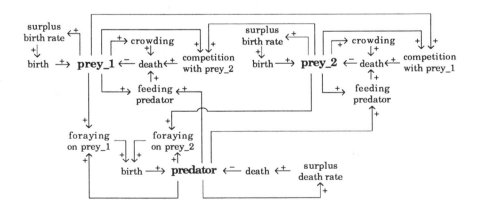

Figure 11.1. Influence diagram for Gilpin's model.

This type of flowchart is referred to as either an *influence diagram* [11.3] or a *causal loop diagram* [11.15], depending on which reference we use.

Notice the signs attached to all paths in an influence diagram. These signs describe whether the influence is positive or negative. For example, the birth rate influences the population in a positive sense, whereas the death rate influences the population in a negative sense.

Influence diagrams are somewhat similar to block diagrams and signal flow graphs. Yet, they are also quite different. We can draw an influence diagram at a much earlier instance of the modeling cycle than we can draw a block diagram or signal flow graph. While the influence diagram tells us which variable depends on which other variables, it does not reveal the nature of this dependency. An influence diagram is therefore much less formal than either a block diagram or a signal flow graph. This is a major strength, and at the same time a major weakness, of influence diagrams.

If we are told which state variables are to describe our "world model," we are ready to draw an influence diagram, but we would not know yet how to draw either a block diagram or a signal flow graph. This is clearly a strength. However, due to their informality, influence diagrams are somewhat unorganized. We tend to create too many dependencies — after all, ultimately, every event in this world is somehow related to every other event. The question is whether a proposed dependency contributes significantly to our modeling effort. A methodology that does not encourage us to distinguish between significant and insignificant relations, is certainly problematic. The lack of rigor is, therefore, a major weakness of influence diagrams, and it is, by the way, a major weakness of System Dynamics methodology as a whole. Using System Dynamics, we are all too easily seduced into creating models that are much too complicated, far more complicated than the quantity and quality of the available data are able to support and validate. We shall return to this point later in this text.

The signs that are attached to each influence path allow us to analyze the stability behavior of our model in qualitative terms. If we follow influence paths around a closed loop, we can count the negative signs that we meet along the way. If the total sum of negative signs is even, we have identified a *positive feedback loop*. Positive feedback loops are always unstable, i.e., they are responsible for unbounded growth in a model. For example, if we extract the feedback loop:

$$\text{Prey\_1} \rightarrow (+)\text{surplus birth rate} \rightarrow (+)\text{birth} \rightarrow (+)\text{Prey\_1} \qquad (11.4)$$

from the rest of the model, we observe that the level Prey_1 will grow exponentially beyond all bounds. If the total sum of negative signs is odd, we have identified a *negative feedback loop*. Negative feedback loops are more difficult to assess. If the number of level variables that we pass on our way around the loop is one or two, the negative feedback loop is certainly stable. For example, if we analyze the feedback loop:

$$\text{Predator} \rightarrow (+)\text{surplus death rate} \rightarrow (+)\text{death} \rightarrow (-)\text{Predator} \qquad (11.5)$$

we come across one negative sign, and therefore, this identifies a negative feedback loop. Along the way, we meet one level variable only (the predator), and thus the negative feedback loop is stable. If we extract this loop from the overall model, we observe that the predator population exhibits an exponential decay. Unfortunately, the influence diagram does not tell us which are level variables and which are not, which is another obvious shortcoming of influence diagrams.

If we meet more than two level variables along a negative feedback loop, we cannot tell whether the loop is stable. Stability will then depend on the total open–loop gain. This observation is related to the *Nyquist stability criterion* for feedback control systems. Most references in System Dynamics state incorrectly that "negative feedback loops are always stable." This is because the System Dynamics methodology is rarely applied to high–order models with very complex feedback mechanisms, and if it is, the stability properties of these models will not be analyzed in qualitative terms. Moreover, most researchers of System Dynamics are not versed in control theory.

We notice further that Fig.11.1 is less concise and less easy to read than the explicit state–space model itself:

$$\dot{x}_1 = x_1 - 0.001\ x_1^2 - 0.001\ x_1 x_2 - 0.01\ x_1 x_3 \qquad (11.6a)$$

$$\dot{x}_2 = x_2 - 0.0015\ x_1 x_2 - 0.001\ x_2^2 - 0.001\ x_2 x_3 \qquad (11.6b)$$

$$\dot{x}_3 = -x_3 + 0.005\ x_1 x_3 + 0.0005\ x_2 x_3 \qquad (11.6c)$$

which contains even more information than the influence diagram, since it explicitly states the nature of all dependencies. This is not a shortcoming of the influence diagram, but simply signifies that we have badly abused the System Dynamics methodology. System Dynamics is a modeling tool, not a model documentation tool. Once we have a working state–space model, it makes little sense to go back and construct a System Dynamics model after the fact. System Dynamics models are useful on the way to eventually producing state–space models for systems for which we lack applicable meta–knowledge, not the other way around. We never create a System Dynamics model for an electrical circuit. This would be pure nonsense.

Furthermore, System Dynamics is a poor methodology to describe strongly intermeshed systems, i.e., systems in which every state variable is tightly coupled with every other state variable, as this is the case in Gilpin's model. The reason for this statement will become clear in due course.

11.4 The Structure Diagram

To avoid some of the problems associated with influence diagrams, Forrester [11.5] suggested another representation, the structure diagram. *Structure diagrams* are similar to influence diagrams, but they distinguish clearly between levels, rates, and converters. Figure 11.2 shows the structure diagram of Gilpin's model.

Level variables are represented by square boxes. Most levels are bracketed by two little clouds that represent the sources and sinks of the level. *Sources* provide an infinite supply of the material which is stocked up (accumulated) in the level variable. *Sinks* provide an inexhaustible dumping place for the same material. The double lines from the source cloud to the level and further to the sink cloud symbolize the *flow of material*. Single lines symbolize the *flow of information*, i.e., they denote control signals. This is similar to the power bonds versus the signal paths in the bond graph modeling approach. However, in bond graph modeling, we concentrated on the flow of power, i.e., the power bonds were the dominant elements in the model. Here, we concentrate more on the flow of information, and therefore, the single lines are dominant.

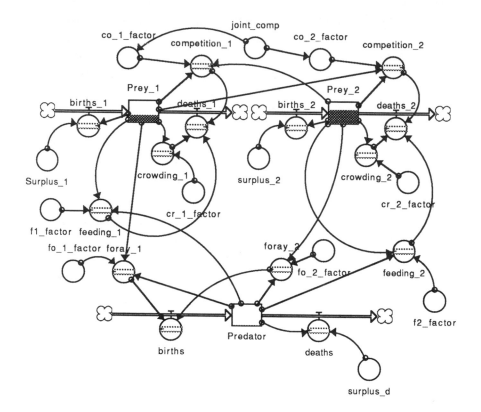

Figure 11.2. Structure diagram for Gilpin's model.

Rate variables are denoted by circles with an attached valve. The rate variables control the flow into and out of the storages (levels, stocks) symbolically by opening/closing the valve that they are responsible for. However, the rate variables do not themselves decide upon the amount of opening/closing of the valves. They are only the guardians of the valves, i.e., they are the paid laborers who operate the valves in accordance with what they are told to do.

Each rate variable has one or several masters that tell it how far to open/close its valve. This is symbolized by one or more control signals (single lines) ending in the rate variable. They can emanate from anywhere in the model, even from another rate variable. However, if rate variables are themselves used as masters, we have to be extremely careful that we avoid creating algebraic loops.

Converter variables are denoted by circles without an attached valve. They are messengers. They collect command information from one or several sources, preprocess it, and deliver it to a rate variable or another messenger. They are introduced strictly for convenience.

Sometimes, it is useful to maintain several state variables of the same type in the model. For example, if we compartmentalize an ecosystem, populations of the same species appear in each of the compartment models. In this case, migration of an animal from one compartment to the next is symbolized by material flow that emanates from one level variable and proceeds directly to the next level variable. A rate variable between the two levels controls the migration rate between the compartments.

If we were to express the populations of Gilpin's model in terms of their energy content, we could model the process of feeding as an energy (material) flow from the two prey levels to the predator level with a branch–off at the valves denoting the feeding efficiency factors. However, the System Dynamics methodology is poorly equipped to model energy flows in a decent manner. If that is what we have in mind, we are much better advised to construct a bond graph.

The structure diagram of Fig.11.2 was constructed using the modern and beautifully engineered System Dynamics modeling system STELLA [11.15]. During the simulation, STELLA animates the graph by showing the amount of stock currently accumulated in each of the levels and by depicting with "needle gauges" those rate and converter variables the values of which vary over time. Figure 11.2 was captured immediately after a simulation run was completed. It contains the animation information at final time.

STELLA has been nicely integrated into the Macintosh workbench environment. It makes optimal use of the object orientation of the Macintosh operating environment and it makes creating System Dynamics models a joy. Contrary to the old days when the graphical representation tools of System Dynamics were auxiliary tools designed to help the System Dynamics modeler develop his or her models using paper and pencil and then left it up to the user to convert the resulting models into a state–space description, in STELLA, all these design tools have been fully integrated into the software. The user designs her or his System Dynamics model on the screen using an object–oriented schematic capture program. Models are designed in a strictly top–down manner, i.e., the user starts with whatever is known about the system under investigation. Unknowns are automatically denoted with a question mark. The user refines his or her model in a stepwise fashion until all question marks have disappeared. At such time, STELLA can convert the model to a state–space model and simulate the resulting state–space model using traditional simulation techniques.

STELLA demonstrates drastically the distinction between modeling and simulation. *Modeling* is the art of capturing physical and other phenomena in a mathematical language. The *modeling software* supports the act of modeling, i.e., it helps the user to formulate and formalize his or her conceptions about the functioning of the real system in terms of a mathematical language. *Simulation* is the art of applying mathematical descriptions of stimuli to the mathematical description of the system and of performing manipulations on these mathematical descriptions in such a way that model behavior is being extracted that resembles the system behavior that we would experience if we were to apply the real stimuli to the real system. The *simulation software* supports the act of simulation, i.e., it helps us to transform the mathematical description of the model together with the mathematical description of the input stimuli into trajectory behavior.

STELLA provides us with an excellent modeling environment. STELLA also provides us with a (much less great) simulation environment. Why STELLA is less convincing in its simulation capabilities than in its modeling capabilities will be shown in due course. However, this problem can be fixed. The important issue is that we realize that the two methodologies, modeling and simulation, are fundamentally different and that it is possible to code them in the software separately in two distinct modules. Most modeling and simulation tools mix the modeling aspects with the simulation aspects. STELLA (and DYMOLA) are commendable exceptions to this commonly found confusion. As a consequence, it will be possible to fix what is wrong with STELLA's simulation engine in a manner that is entirely transparent to the user. This can furthermore be achieved without modifying a single line of code in the modeling software.

One of the strongest arguments in favor of STELLA is its carefully written manual, which I strongly recommend for further reading [11.15].

11.5 Structure Characterization

The next question is how do we go about determining the *structure* of our state equations, i.e., how do we determine the precise nature of the relations that influence the rate variables. This process is called *structure characterization*.

One of the real strengths of the System Dynamics methodology is that it allows us to blend deductive with inductive modeling techniques. Whenever we possess physical insight into how a rate variable is influenced by the system, we should by all means use it. This is the best thing that can happen to us. However, let us discuss the case where we lack such intuition. What shall we do about this problem?

Let us return to the world model. Our laundry list suggested that the birth rate *BR* depends on a number of factors, namely, the population *P*, the material standard of living *MSL*, the food ratio *FR*, the crowding ratio *CR*, and the pollution *POL*:

$$BR = \mathrm{f}(P, MSL, FR, CR, POL) \tag{11.7}$$

An important concept in System Dynamics is the exploitation of *small signal analysis*. If we don't know enough about a system as a whole, maybe we can at least capture the behavior of that system in the vicinity of its current system state. Accordingly, in each state equation, we capture the "normal" (i.e., known) behavior and we model alterations from the normal state as small signals. Applied to Eq.(11.7), we would write:

$$BR = BRN \cdot \mathrm{f}(P, MSL, FR, CR, POL) \tag{11.8}$$

where *BRN* denotes the normal birth rate, i.e., the birth rate observed at the time when the model was created, say 1970.

Physical intuition dictates that the birth rate must be proportional to the population, thus:

$$BR = BRN \cdot P \cdot \mathrm{f}(MSL, FR, CR, POL) \tag{11.9}$$

Now, we are at the end of our physical insight. We want to assume that Eq.(11.9) expresses a static relationship among all variables. If such an assumption is unjustified, we have chosen our levels incorrectly.

The next principle observed in most System Dynamics models is that the modeling task can be simplified if we postulate that all variables influence the state equation independently of each other, i.e.,

$$BR = BRN \cdot P \cdot \mathrm{f}_1(MSL) \cdot \mathrm{f}_2(FR) \cdot \mathrm{f}_3(CR) \cdot \mathrm{f}_4(POL) \tag{11.10}$$

Of course, this is a gross simplification that is in no way supported by physical evidence. Yet it makes our lives easier, and since we don't have anything better to go by, this may be preferable to having no model at all. At least it reduces drastically the amount of measurement data that we shall need in order to formulate models for the functions f_i. If we need n measurement points to decently identify any one of the functions f_i, we need $4n$ measurement points to identify all four functions as compared to n^4 points to identify the combined multivariable function.

Due to the assumption of small signal analysis, each of these functions must pass through the point $< 1.0, 1.0 >$, i.e., we normalize each of the influencing variables such that, in the year 1970, it assumes a value of 1.0. For example, the global material standard of living in the year 1970 is 1.0. The effect of the material standard of living on the birth rate in the year 1970 must thus also be 1.0, since *BRN* denotes the normal birth rate, i.e., the global birth rate in the year 1970.

If we define the absolute material standard of living as the yearly income of an individual, *MSL(t)* can be computed as the absolute material standard of living of the average inhabitant of this planet at time t divided by the absolute material standard of living of the average inhabitant of this planet in the year 1970.

How do we determine f_1 *(MSL)*? We compartmentalize our world in the year 1970. We find that the material standard of living in the year 1970 differs drastically from one country to another and so does the birth rate. Thus, we correlate the birth rate in different countries with the observed material standard of living in those countries and, voilà, the desired function f_1 has been identified. It turns out that the less money people have to raise children, the more they seem to enjoy having them.

This example shows the dangers of this type of modeling. The love story cliff is treacherously close. It is our responsibility to ensure a proper cause/effect relation between the variables that we correlate in this manner. Numerous reasons can be mentioned why people in the Third World produce more children:

(1) They have, on the average, a poorer education and don't understand birth control so well.

(2) They are too poor to buy a proper retirement policy and children are often the only way to ensure their material well–being when they are old.

(3) Children are often their only means of entertainment. In First

World countries, children are frequently experienced as hampering a successful career. Thus, potential parents must be truly convinced that children is what they really want. Otherwise, they may be better off without them.

All these pieces of physical insight get diluted in the global generalization of a relation between the material standard of living and the birth rate.

Moreover, a correlation between variables does not prove the existence of a direct causal relation at all. Maybe both variables are caused by yet another factor that influences both variables simultaneously. Just for fun, I once correlated the statistics of birth rates in Switzerland over the past 50 years with the statistics on stork populations — and indeed, a strong positive correlation could be observed among these two variables. I leave the conclusion to the reader. We shall talk more about causality in the next section of this chapter.

While we can leave static functional relationships among variables in a tabular form and simply interpolate from the table during the simulation, a technique introduced in Chapter 5, it is sometimes useful to determine an explicit formula that captures the functional relationship. Figure 11.3 shows a set of "measured" data that relates a variable y to a variable x.

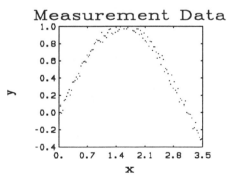

Figure 11.3. A set of "measured" data of y as a function of x.

These "measured" data were produced in CTRL–C (MATLAB) by adding 5% noise to both the input and output of a sine wave generator:

```
[>   x = 0 : 0.03 : 3.5;
[>   k1 = 0.1*RAND(x) − 0.05*ONES(x);
[>   k2 = 0.1*RAND(x) − 0.05*ONES(x);
[>   y = sin(x + k1) + k2;
```

Let us immediately forget where these data came from. The question of interest is whether or not we can identify the structure of the source from the measured data. We shall test the following three hypotheses:

$$y_a = \sin(x) \tag{11.11a}$$

$$y_b = a + b \cdot x + c \cdot x^2 \tag{11.11b}$$

$$y_c = a + b \cdot x + c \cdot x^2 + d \cdot x^3 \tag{11.11c}$$

We can identify an optimal set of parameters using regression analysis. Let me illustrate the concept by means of hypothesis #3.

Assuming that we have n measurements, we can plug all these n measurement data into the hypothetical equation:

$$y_{c_1} = a + b \cdot x_1 + c \cdot x_1^2 + d \cdot x_1^3 \tag{11.12a}$$

$$y_{c_2} = a + b \cdot x_2 + c \cdot x_2^2 + d \cdot x_2^3 \tag{11.12b}$$

$$\vdots \qquad\qquad\qquad\qquad \vdots$$

$$y_{c_n} = a + b \cdot x_n + c \cdot x_n^2 + d \cdot x_n^3 \tag{11.12n}$$

which can be rewritten using a matrix notation as follows:

$$
\begin{pmatrix}
1 & x_1 & x_1^2 & x_1^3 \\
1 & x_2 & x_2^2 & x_2^3 \\
\vdots & \vdots & \vdots & \vdots \\
1 & x_n & x_n^2 & x_n^3
\end{pmatrix}
\cdot
\begin{pmatrix}
a \\ b \\ c \\ d
\end{pmatrix}
=
\begin{pmatrix}
y_{c_1} \\ y_{c_2} \\ \vdots \\ y_{c_n}
\end{pmatrix}
\tag{11.13}
$$

or:

$$\mathbf{V}(\mathbf{x}) \cdot \mathbf{coef} = \mathbf{y} \tag{11.14}$$

where $\mathbf{V}(\mathbf{x})$ is a so–called Vandermonde matrix spanned over the vector $\mathbf{x}$. Assuming that $n > 4$, we are confronted with an overdetermined linear regression problem that we can solve in a least square's sense:

$$\mathbf{V}(\mathbf{x}) \cdot \mathbf{coef} - \mathbf{y} = \mathbf{res} \tag{11.15}$$

where **res** is the vector of residua. We try to determine the coefficient vector **coef** such that the $\mathcal{L}_2$ norm of **res**:

$$\|\mathbf{res}\|_2 = \sqrt{\sum_{\forall i} res_i^2} \qquad (11.16)$$

is minimized. This problem can be solved by computing the pseudoinverse of the rectangular matrix $\mathbf{V}(\mathbf{x})$:

$$\left(\mathbf{V}(\mathbf{x})' \cdot \mathbf{V}(\mathbf{x})\right) \cdot \mathbf{coef} = \mathbf{V}(\mathbf{x})' \cdot \mathbf{y} \qquad (11.17)$$

where $\mathbf{V}(\mathbf{x})' \cdot \mathbf{V}(\mathbf{x})$ is a square Hermitian matrix of size 4×4 that can be inverted:

$$\mathbf{coef} = \left(\mathbf{V}(\mathbf{x})' \cdot \mathbf{V}(\mathbf{x})\right)^{-1} \cdot \mathbf{V}(\mathbf{x})' \cdot \mathbf{y} \qquad (11.18)$$

where $\left(\mathbf{V}(\mathbf{x})' \cdot \mathbf{V}(\mathbf{x})\right)^{-1} \cdot \mathbf{V}(\mathbf{x})'$ is called the *pseudoinverse* of $\mathbf{V}(\mathbf{x})$. In MATLAB or CTRL–C, this problem can be conveniently coded using the "\" operator:

$$[> \quad coef = V \backslash y \qquad (11.19)$$

For the given data set, the following optimal functions were found:

$$y_a = \sin(x) \qquad (11.20a)$$
$$y_b = -0.0288 + 1.2511 \cdot x - 0.3938 \cdot x^2 \qquad (11.20b)$$
$$y_c = -0.0833 + 1.4434 \cdot x - 0.5325 \cdot x^2 + 0.0266 \cdot x^3 \qquad (11.20c)$$

Figure 11.4 shows the three fitted curves plotted as solid lines over the measured data set. Obviously, all three hypotheses can be easily defended on the basis of Fig.11.4. I also computed the $\mathcal{L}_2$ norms of the three residua vectors and found:

$$\|\mathbf{res}_1\|_2 = 0.3967 \qquad (11.21a)$$
$$\|\mathbf{res}_2\|_2 = 0.5395 \qquad (11.21b)$$
$$\|\mathbf{res}_3\|_2 = 0.4857 \qquad (11.21c)$$

The residuum of the sine wave fit is slightly better, but not sufficiently so to guarantee that this is the correct hypothesis. I also computed the cross–correlations between x and y for the three hypotheses and found that the largest cross–correlations are as follows:

$$\|\mathbf{corr}_1\|_\infty = 52.7961 \qquad (11.22a)$$
$$\|\mathbf{corr}_2\|_\infty = 52.6210 \qquad (11.22b)$$
$$\|\mathbf{corr}_3\|_\infty = 52.6761 \qquad (11.22c)$$

Again, the sine wave fit turned out to be slightly better, but this test shows even a less significant difference than the last one.

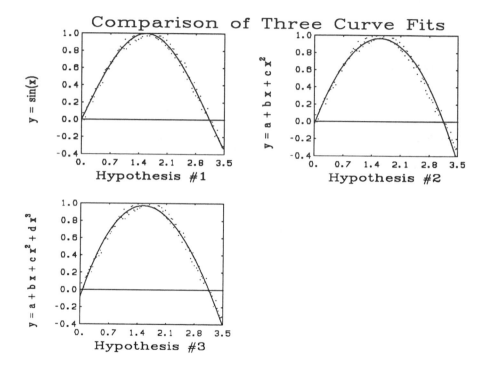

Figure 11.4. Curve fitting of "measured" data set.

Ivakhnenko and Lapa devised a technique that often allows us to distinguish between different hypotheses regarding functional relationships [11.11]. They suggested that we apply a transformation to both the data and the hypothesized curves such that, after the transformation, the hypothesized curves look sufficiently different, so that the measured data can be easily associated with one, but not with the other proposed hypotheses.

A good transformation for our case might be the following: We remember that $\sin^2(x) + \cos^2(x) = 1.0$. We therefore apply the following transformation to the data:

$$\mathbf{y_{new}} = \mathbf{y}^2 + \cos^2(\mathbf{x}) \tag{11.23}$$

The three hypotheses turn into:

$$y_{new_a} = 1.0 \tag{11.24a}$$

$$y_{new_b} = (a + b \cdot x + c \cdot x^2)^2 + \cos^2(x) \tag{11.24b}$$

$$y_{new_c} = (a + b \cdot x + c \cdot x^2 + d \cdot x^3)^2 + \cos^2(x) \tag{11.24c}$$

Figure 11.5 shows the transformed data.

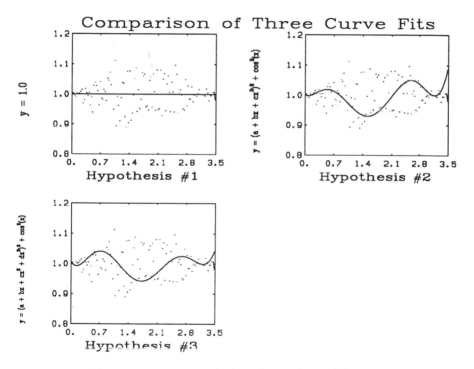

Figure 11.5. Curve fitting of transformed data set.

Clearly, the three curves can be easily distinguished from each other. Unfortunately, the "measured" data are spread so widely now that we still cannot associate them with certainty with any one of the three curves. I recomputed the residua and correlations and found:

$$\|\mathbf{res_1}\|_2 = 0.4922 \quad , \quad \|\mathbf{corr_1}\|_\infty = 117.0747 \tag{11.25a}$$

$$\|\mathbf{res_2}\|_2 = 0.6199 \quad , \quad \|\mathbf{corr_2}\|_\infty = 116.8677 \tag{11.25b}$$

$$\|\mathbf{res_3}\|_2 = 0.5829 \quad , \quad \|\mathbf{corr_3}\|_\infty = 116.9146 \tag{11.25c}$$

The situation is still unchanged. We have indications that the sine wave approximation is slightly better, but we cannot be sure that this hypothesis is physically correct. Had we added only 1% noise to our data instead of 5% noise, the applied transformation would indeed have disqualified the second and third hypotheses.

What this example teaches us is how little it takes before structural information gets lost in measurement data. This makes the inductive structure characterization a tough and often unsolvable problem.

11.6 Causality

Causality is a very natural notion in our everyday lives. An action causes a reaction. However, in the casual context, some time usually elapses between the action and the reaction, i.e., the concept of causality is more commonly applied to discrete–event systems than to continuous–time systems.

In engineering, we usually call a response of a system "causal" if it occurs simultaneously with the input stimulus or if it lags behind the input stimulus. We call it "noncausal" if it occurs prior to the input stimulus. Of course, noncausal responses cannot be generated by physical systems.

It is now interesting to discuss whether, given two different continuous signals, it can be decided that one of the two signals has been caused by the other. This is a favorite topic in the artificial intelligence literature, which is full of sometimes rather obscure notions about causality [11.12].

Let us analyze an electrical resistor characterized by its voltage and current. Does it make sense to say that the voltage across the resistor causes a current to flow, or does it make more sense to say that the current through the resistor causes a drop of potential? In the active form, *both* statements are correct. Yet given a measured voltage across and a measured current through a resistor, we cannot conclude after the fact which caused what. This is clearly a chicken–and–egg problem.

Let us apply the following experiment. Figure 11.6 shows three different circuit elements, a resistor, a capacitor, and an inductor connected to a noise voltage source.

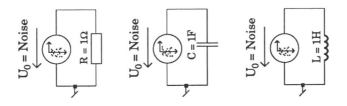

Figure 11.6. Three different circuit configurations.

In all three cases, we shall measure the current that flows through the element. We shall then compute the cross–correlation between the input signal u and output signal i in all three cases. In MATLAB or CTRL–C, this is easiest accomplished with several fast Fourier transforms. The following CTRL–C (or MATLAB) code produces the desired cross–correlation function between the input signal u and output signal i.

```
[>   u = [u,ZROW(u)];
[>   i = [i,ZROW(i)];
[>   sui =FFT(u).*CONJ(FFT(i));
[>   rui =IFFT(sui);
[>   corr = [rui(129 : 256), rui(1 : 128)];
```

We assume that both vectors u and i contain 128 measured data points. Zero–padding is applied to provide the required memory cells for the fast Fourier transform. The vector rui contains the correlation function, but not in the right sequence. The first half of the vector contains increasing positive values of delay between the input and output; the second half of the vector contains decreasing negative values of delay between the input and output. The last line of code puts the correlation function into the correct sequence.

Figure 11.7 shows the results obtained for the three circuit configurations. As expected, the cross–correlation assumes its peak value at $\delta t = 0.0$ for the resistor. It assumes its peak value for $\delta t > 0.0$ for the capacitor and it assumes its peak value for $\delta t < 0.0$ for the inductor. Consequently, we can say that the current "lags behind" the voltage in the inductor and "leads" the voltage in the capacitor.

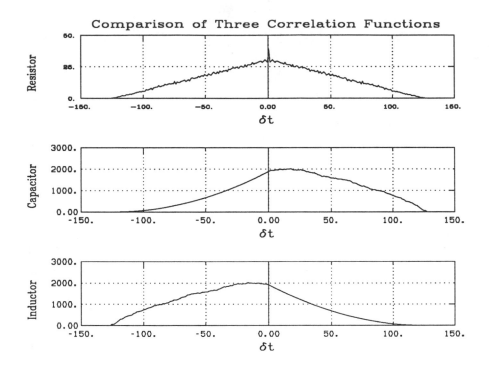

Figure 11.7. Cross–correlations for the three circuit configurations.

We might therefore be inclined to say that in an inductor the voltage causes current to flow, whereas in a capacitor, the current causes a drop of potential. The resistor can assume either of the two causalities. This definition is consistent with our earlier use of the term "causality" as we came across it in the context of bond graph modeling and from a computational point of view, this terminology makes a lot of sense.

Yet notice that in the preceding experiment, the voltage truly "caused" the current to flow in all three situations. In every single case, the voltage source was the physical origin of the observed phenomenon. Does this mean that the capacitor configuration is a "nonphysical system"? Of course not! Obviously, we can connect a voltage source to a capacitor. However, the system is indeed structurally singular. We are made to believe that the capacitor can store energy, but in the given configuration, it cannot. It simply attempts to differentiate the input.

Did we just discover that a physical system can indeed anticipate the input, i.e., that physical systems can have precognitions, or at least premonitions, of coming events? The answer is, in some way, yes. If we assume that before time $t = 0.0$, every single signal in the system was zero (a common notion in many areas of engineering design, such as control theory), our system has no way of knowing what is to come. If we try to apply a step voltage input to our capacitor, we have difficulties to do so since this would imply an infinitely large current at time zero, which would require an infinite amount of power to be put in the system. For sure, the system will not "accommodate" our desire by starting the current a little earlier. In this sense, the configuration is indeed nonphysical. Yet in a statistical sense, i.e., under statistical steady–state conditions, we can measure a cross–correlation function in which the current leads the voltage. In this context, the system can indeed anticipate future inputs. In other words, knowledge of the past behavior of a system allows us to anticipate future behavior, and in a statistical sense, this even includes future inputs. To obtain the capacitor curve of Fig.11.7, I cheated a little bit. I applied the same experiment as for the inductor and then simply swapped the input and output. However, I also tried the experiment with numerical differentiation. Due to the (even theoretical) impossibility of computing the derivative of a noise signal, the correlation is much smaller, but the current still leads.

We just learned that the discrimination of true causality from measurements alone is a hopeless undertaking. However, for practical purposes, we shall define the terms *causal* and *causality* as follows:

(1) Two signals are coupled through a *causal relation* if the (either positive or negative) cross–correlation between them is strong.

(2) In addition, two signals are coupled through a *causality relation* if one signal lags behind the other.

(3) If two signals show little or no cross–correlation, we call them *causally unrelated.*

(4) If two signals show a strong cross–correlation, but neither of the two signals lags behind the other, the signals are causally related, but their causality cannot be decided.

(5) If two signals show a significant amount of cross–correlation, and if signal #2 lags behind signal #1, we claim that signal #1 has caused signal #2. The stronger the cross–correlation of the two signals, the more pronounced is the causal relation among the two signals. The larger the lag time, the stronger is their causality relation.

11.7 Differential Versus Difference Equations

Traditionally, System Dynamics researchers have always formulated their state–space models as a set of difference equations. There are two reasons for this:

(1) Originally, System Dynamics had been closely linked with the DYNAMO simulation language [11.14], which solves only difference equations and no differential equations.

(2) It has been often claimed that the major application areas for System Dynamics are found in management and soft sciences. The quality of the data available in these fields does not justify the accuracy that numerical integration provides. Difference equations are cheaper to solve and they will serve the same purpose.

In DYNAMO, a difference equation for the change in population would be formulated in the following way:

$$P.K = P.J + (DT)(BR.JK - DR.JK) \qquad (11.26)$$

The index K denotes the next time step, whereas the index J indicates the current time step. JK denotes the interval between the current and the next time step. Obviously, we can reinterpret Eq.(11.26) as follows:

$$P_{k+1} = P_k + \Delta t \cdot (BR - DR) \qquad (11.27)$$

which is equivalent to our differential equation:

$$\dot{P} = BR - DR \qquad (11.28)$$

if the fixed–step forward Euler algorithm:

$$P_{k+1} = P_k + \Delta t \cdot \dot{P}_k \qquad (11.29)$$

is used for the numerical integration. DYNAMO operates on state equations, but forces the user to explicitly formulate her or his integration algorithm. Usually, this will be a fixed–step forward Euler algorithm. We can trick DYNAMO into the use of higher–order integration algorithms if we code them explicitly as difference equations. This is most unfortunate. Why should we have to code the integration algorithm manually if good simulation languages, such as ACSL, exist that will take care of the numerical integration for us? DYNAMO is simply a very old–fashioned language, and there is no excuse for using it any longer (as a matter of fact, there has not been an excuse for using it for quite some time already).

What about the second argument? It is indeed correct that most System Dynamics applications do not justify a sophisticated integration algorithm to be used from the point of view of a consistent data representation. However, we shall see in the companion book that the numerical accuracy of an integration scheme is dictated by three separate components: consistency, convergence, and numerical stability. A numerical scheme is called *consistent* if the analytical (i.e., infinitely accurate) solution of that scheme decently agrees with the analytical solution of the original problem. A scheme is called *convergent* if the local error of the scheme goes to zero if the discretization interval Δt approaches zero. The scheme is called *numerically stable* if the local errors do not accumulate over many steps, i.e., if the local error is a good indicator of the global error as well. From the point of view of consistency, it is perfectly legitimate to represent typical System Dynamics problems through a set of coarsely discretized difference equations. However, the requirement of numerical stability may still suggest using a sophisticated higher–order numerical integration scheme, even though the accuracy requirements of our model are low.

As I explained earlier, System Dynamics is a *modeling methodology*, which is totally unrelated to the underlying *simulation methodology*. STELLA does not use DYNAMO as its simulation engine. The developers of STELLA recognized rightly that System Dynamics models can (and should) be mapped into state–space models. STELLA currently offers three different numerical integration rules. The default method is forward Euler (like in DYNAMO), but in addition, STELLA offers two fixed–step Runge–Kutta algorithms, one of second order, and one of fourth order. However, this is insufficient. Variable–step algorithms are mandatory for many nonlinear problems in order to obtain accurate simulation results at a decent execution speed. Also, the manner in which the integration algorithms are implemented can have drastic effects on their execution speed. In fact, little reason exists why a developer of a modeling tool should have to reinvent the wheel and create his or her own simulation engine as well. It would have been equally easy to map the state–space model that is being created by STELLA's excellent modeling engine into a decent simulation language such as ACSL. ACSL has been developed over many years, and while ACSL still has its problems, its simulation engine is far better than anything that the average software designer could design on her or his own.

To prove my point, I ran the Gilpin model of Fig.11.2 in STELLA. Figure 11.8 shows the results for a joint competition factor of $k = 1.0$.

In order to obtain any meaningful results, I had to use the fourth–order Runge–Kutta algorithm with a step size of $\Delta t = 1.0$. The results shown in Fig.11.8 are still not very accurate. We would have to use a considerably smaller step size to, e.g., accurately determine the competition factor $k = 1.089$ at which the first prey x_1 dies out. Yet I could not reduce the step size any further since one simulation run required already an execution time of 30 min on my MacPlus.

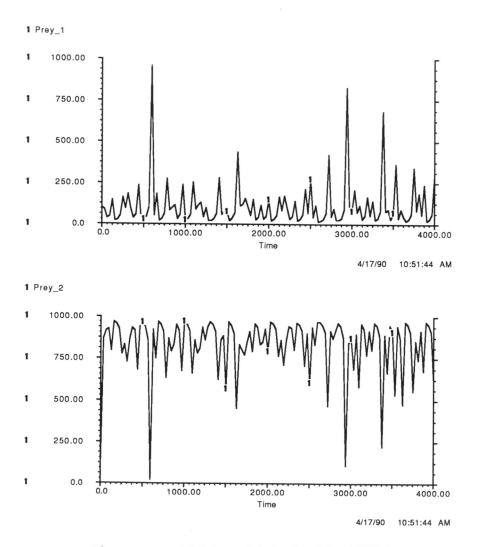

Figure 11.8. Gilpin's model simulated in STELLA.

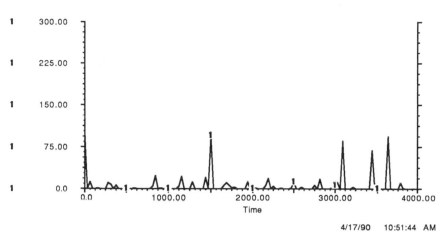

4/17/90   10:51:44 AM

Figure 11.8. Gilpin's model simulated in STELLA (continued).

The ACSL program employed in Chapter 10 required an execution time of 8 sec for a fourth–order Runge–Kutta algorithm with an average step size of $\Delta t_{avg} = 0.125$ on a VAX–8700. However, during each of the peaks, ACSL reduced the step size to a value of roughly $\Delta t_{min} = 10^{-6}$. Of course, the VAX–8700 is a substantially faster (and more expensive) machine than my Macintosh. However, the observed speed difference is primarily caused by the fact that STELLA *interprets* the state–space model at run–time rather than *compiles* it beforehand at least into threaded code (as DESIRE does). Consequently, STELLA can be used for fairly simple models only.

However, it would be an easy task to cure this deficiency. All that needs to be done is to turn STELLA into a program generator (similar to DYMOLA) that automatically generates an ACSL or DESIRE program from the state–space model and uses that for simulation. This would speed up the execution time of STELLA simulation runs by approximately a factor of 100.

Also, discrete interventions are common elements of System Dynamics models. We shall meet one such discrete intervention in a later section of this chapter in which we shall model the spread of an influenza epidemic. In STELLA, the user is forced to employ rather clumsy ways to implement such interventions. This is due to the fact that STELLA does not know the concept of discrete events. A "discrete intervention" is simply a time–event in terms of our previously used nomenclature. Since ACSL provides for explicit mechanisms to describe time–events, the resulting models could be made simpler, more elegant, and more robust if ACSL were used as the target language. The interpretive way in which STELLA's simulation engine currently operates is fine for model–debugging purposes and for simple models, i.e., it would make sense to retain this capability as an alternative and add the program generation capability to the code as a new feature.

11.8 The Larch Bud Moth Model

In Chapter 10, we discussed a population dynamics model of the larch bud moth, *Zeiraphera diniana* (Guenée), as observed in the upper Engadine valley of Switzerland. At that time, we used several simple two– (three–) species predator–prey models to describe the limit cycle behavior of the insect population.

Let us repeat this analysis now using the System Dynamics modeling methodology, which enables us to incorporate into the model as much physical knowledge about the dynamics of the system as we are able to gather.

It is known that (in the upper Engadine valley) the female insects deposit their eggs in the larch trees during the month of August every year, which then stay in an extended embryonic diapause that lasts until the spring of the following year. During the fall, the eggs are preyed upon by several species of *Acarina* and *Dermaptera*. During the winter, the eggs are naturally parasitized by a species of *Trichogramma*. The surviving eggs are ready for hatching in June. Extensive studies have shown the winter mortality to be a constant fraction of the deposited eggs:

$$Small\_Larvae = (1.0 - winter\_mortality) \cdot Eggs \qquad (11.30)$$

The average winter mortality is 57.28%. Many of the small larvae die from incoincidence, i.e., from a bad location, for example, if the branch where the egg was deposited dies during the winter. Measurements have shown that the incoincidence factor depends linearly on the raw fiber content of the needles of the larch tree:

$$incoincidence = 0.05112 \cdot Raw\,fiber - 0.17932 \qquad (11.31)$$

where the coefficients were found by linear regression. The raw fiber content is expressed in percentages. It varies between $Raw\,fiber_{min} = 12\%$ for healthy trees and $Raw\,fiber_{max} = 18\%$ for heavily damaged trees. From this, we can compute the number of larvae that survive to become large larvae:

$$Large\_Larvae = (1.0 - incoincidence) \cdot Small\_Larvae \qquad (11.32)$$

The large larvae move around and cause damage to the larch forest by eating the needles. Some of the large larvae die from starvation, i.e., if their food demand is not met. Others die due to physiological weakening. If the raw fiber content is too high, they may still survive the larvae state and cause more damage by eating, but they may die during their chrysalis state. We can thus compute the number of adult insects as follows:

$$Insects = (1.0 - weakening) \cdot (1.0 - starvation) \cdot Large\_Larvae \quad (11.33)$$

It was observed that the logarithm of the starvation factor depends linearly on the quotient of foliage and food demand:

$$starvation = \exp(-\frac{foliage}{food\_demand}) \qquad (11.34)$$

and that the weakening factor is again linearly dependent on the raw fiber content of the needles:

$$weakening = 0.124017 \cdot Raw\,fiber - 1.435284 \qquad (11.35)$$

The foliage is linear in the raw fiber content and the number of trees:

$$foliage = (-2.25933 \cdot Raw\,fiber + 67.38939) \cdot nbr\_trees \qquad (11.36)$$

where the number of trees was counted to be $nbr\_trees = 511147$. The food demand is linear in the number of large larvae:

$$food\_demand = 0.005472 \cdot Large\_Larvae \qquad (11.37)$$

Since only female insects contribute to the next generation, we compute:

$$Females = sex\_ratio \cdot Insects \qquad (11.38)$$

where the average percentage of female insects was found to be 44%, thus *sex_ratio* = 0.44. The number of eggs deposited for the next generation of insects can be computed as:

$$New\_Eggs = fecundity \cdot Females \qquad (11.39)$$

where the fecundity again depends on the raw fiber content of the needles. If the raw fiber content was high, the weight of the chrysalis will be low. Such insects may still survive, but they exhibit a lower fecundity than insects that were nourished well during their larvae state:

$$fecundity = -18.475457 \cdot Rawfiber + 356.72636 \qquad (11.40)$$

This concludes the life cycle of the larch bud moth.

Let us now analyze the life cycle of the trees. The damage caused by the insects can be expressed as follows:

$$defoliation = (1.0 - starvation) \cdot \left(\frac{food\_demand}{foliage}\right) \qquad (11.41)$$

The raw fiber recruitment, i.e., the change in needle quality is a complex function of the current needle quality and the current defoliation:

$$New\_Rawfiber = \text{Grecr}(defoliation, Rawfiber) \cdot Rawfiber \qquad (11.42)$$

The "Grecr" function is an experimental function that models the change in raw fiber content as a function of defoliation. If little defoliation takes place, the trees will slowly recover to their optimal raw fiber content of 12%. If heavy defoliation takes place, the raw fiber content slowly increases until it reaches its maximum value of 18%. Fischlin and Baltensweiler modeled this phenomenon in the following manner [11.4]:

if *defoliation* < 0.4
   then if *Rawfiber* < 11.99
     then *Grecr* $= 1.0$
     else $zRaw = 0.425 + \text{abs}\left(\frac{18.0 - Rawfiber}{Rawfiber - 11.99}\right)$
       if $zRaw > (Rawfiber - 11.99)$
         then $Grecr = \frac{11.99}{Rawfiber}$
         else $Grecr = 1.0 - \frac{zRaw}{Rawfiber}$
         end if
     end if
   else if *defoliation* < 0.8
     then $Grecr = 1.0 + \frac{(defoliation\ 0.4)\cdot(18.0 - Rawfiber)}{0.4 \cdot Rawfiber}$
     else $Grecr = \frac{18.0}{Rawfiber}$
     end if
   end if

This completes the description of the model. While regression analysis was used on several occasions to determine optimal parameter values, this is basically a physical model of the population dynamics system. Parameter fitting was applied in a strictly local manner to fit no more than two parameters at a time to a set of input/output measurements. I am therefore much more inclined to believe that I understand now what is going on in this system than I was after identifying the (much simpler) Lotka–Volterra model of Chapter 10.

Let us go ahead and try to capture this model in STELLA. Figure 11.9 shows the structure diagram of this model.

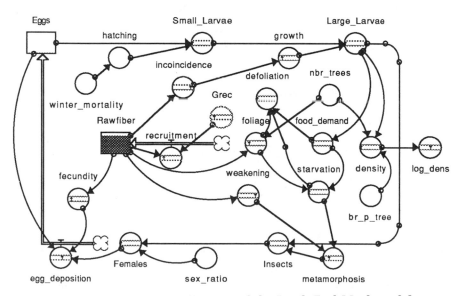

Figure 11.9. Structure diagram of the Larch Bud Moth model.

With the state variables, I had to cheat a little bit. STELLA is designed to solve differential equations rather than difference equations, thus, by modeling *Eggs* as a level variable and *egg_deposition* as a rate variable, STELLA forces us to accept the equation:

$$\frac{d}{dt}Eggs = egg\_deposition \tag{11.43}$$

or, by using the (standard) fixed–step Euler algorithm:

$$New\_Eggs = Eggs + \Delta t \cdot egg\_deposition \tag{11.44}$$

Comparing Eq.(11.44) with Eq.(11.39), we find that:

$$egg\_deposition = fecundity \cdot Females - \frac{Eggs}{\Delta t} \tag{11.45}$$

In the same manner, we find that:

$$recruitment = Grecr \cdot Rawfiber - \frac{Rawfiber}{\Delta t} \tag{11.46}$$

In our case, Δt was set to 1.0.

I had more problems with the "Grecr" function. STELLA does not provide us with procedural sections in the way ACSL does. This is a common disease of most graphically oriented modeling languages. I had to code the function in the following way:

$Grec = $ **if** $(defoliation < 0.4)$ **then** $Grec1$ **else** $Grec2$
$Grec1 = $ **if** $(Rawfiber < 11.99)$ **then** $Grec3$ **else** 1.0
$Grec2 = $ **if** $(defoliation < 0.8)$ **then** $Grec4$ **else** $Grec5$
$Grec3 = $ **if** $(zRaw < (Rawfiber - 11.99))$ **then** $Grec6$ **else** $Grec7$
$Grec4 = 1.0 + (defoliation - 0.4) * (18.0 - Rawfiber)/(0.4 * Rawfiber)$
$Grec5 = 18.0/Rawfiber$
$Grec6 = 11.99/Rawfiber$
$Grec7 = 1.0 - zRaw/Rawfiber$
$zRaw = 0.425 + $ abs$((18.0 - Rawfiber)/(Rawfiber - 11.99))$

Since I didn't want to clutter my nice structure diagram with all this detail, I used STELLA's *ghost* feature. I created ghosts of *Rawfiber* and *defoliation*, and computed "Grecr" in a separate structure diagram (within the same model), as shown in Fig.11.10.

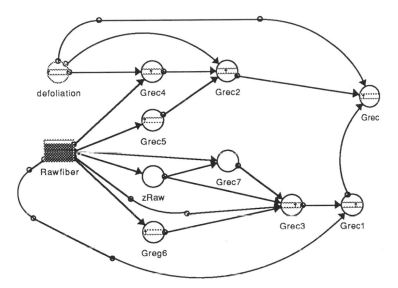

Figure 11.10. Structure diagram of the Grecr function.

Finally, I made a ghost of "Grecr" and copied it back into my original structure diagram. This is a way to create STELLA submodels. Of course, if so much detail is provided to and desired in a model such as expressed in the "Grecr" function, System Dynamics may not be the most appropriate modeling methodology any longer. It would have been equally easy to take the equations as given and code an ACSL program directly.

Finally, I added yet another separate structure diagram to capture the measurement data such that they can be plotted together with the simulated data on one graph. This is shown in Fig.11.11.

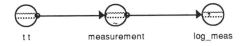

Figure 11.11. Structure diagram of the measurement data.

The *tt* block contains simply the simulation clock:

$$tt = TIME \tag{11.47}$$

and *measurement* is a tabular function that contains the measurement data. The measurement data are given in Table 11.2.

Table 11.2. Measurement Data for the Larch Bud Moth Model.

Time [year]	Larvae Density [#/kg]
1949	0.018
1950	0.082
1951	0.444
1952	4.174
1953	68.797
1954	331.760
1955	126.541
1956	21.280
1957	2.246
1958	0.085
1959	0.080
1960	0.371
1961	1.638
1962	22.878
1963	248.817
1964	184.272
1965	3.116
1966	0.019
1967	0.002
1968	0.059
1969	0.197
1970	1.068
1971	10.569
1972	173.932
1973	249.612
1974	176.023
1975	4.749
1976	0.014
1977	0.008
1978	0.056

Notice that STELLA allows us to maintain several unconnected structure diagrams in the same model.

Finally, we need to compute the output function of our model. The measurement data were expressed in number of larvae per kilogram dry needle biomass, i.e., we must compute the larvae density:

$$density = \frac{Large\_Larvae}{br\_p\_tree \cdot nbr\_trees} \tag{11.48}$$

where $br\_p\_tree = 91.3$ denotes the amount of dry needle biomass per tree expressed in kilograms.

Since STELLA does not provide for a logarithmic format of its graphs, we need to compute, in the model, the logarithms of the larvae density and our measurement data. This concludes the overall model description. Figure 11.12 shows the resulting graph, which looks similar to the one given in the paper by Fischlin and Baltensweiler [11.4].

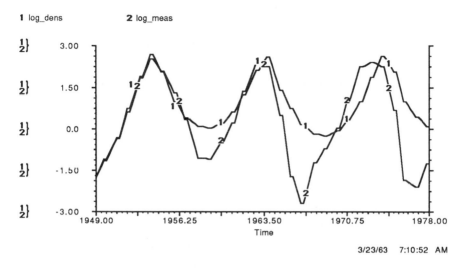

Figure 11.12. Simulation results of the Larch Bud Moth model.

The graph looks a little crooked. STELLA assumes all variables to be continuous over time. Since our state variables stay constant for a period of one year (our step size), the graph shows a staircase function.

11.9 The Influenza Model

Let us look at another system now. We wish to study the spreading of an influenza epidemic. This model has four state variables. We start out with a population of 10,000 noninfected individuals. At a given time, a stem of influenza bacteria is introduced into the system. Infection occurs spontaneously and some of the previously noninfected individuals become infected. The bacteria spread in the infected individuals and after some time, the individuals fall sick. The total number of contagious people (both those who are already sick and those who haven't broken down yet) spread the disease further and recruit new infected people among the noninfected population. The sick people eventually get cured due to the natural defenses of their immune systems. This creates a population of immune people. However, the body "forgets" the previous exposure after some time and the bacteria have a tendency to mutate. Thus, the immune people turn into noninfected people again who are susceptible to reinfection. The purpose of this example is to show how easy it is to create intuitive models using the System Dynamics methodology.

The structure diagram of this system is shown in Fig.11.13.

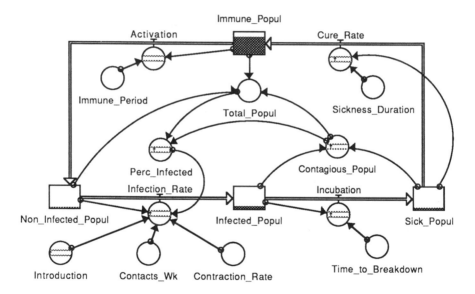

Figure 11.13. Structure diagram of the influenza model.

This diagram shows a new element. Outflows of one level can be used as inflows to another. The rate between two levels controls the flow from one level to the other, as a faucet controls the flow of water from the storage tank to the sink.

The time to breakdown (incubation period) is 4 weeks, the actual sickness lasts for 2 weeks, and the immune period lasts for 26 weeks. This allows us to write down a number of equations already:

$$Incubation = \text{INT}(\frac{Infected\_Popul}{Time\_to\_Breakdown}) \qquad (11.49a)$$

$$Cure\_Rate = \text{INT}(\frac{Sick\_Popul}{Sickness\_Duration}) \qquad (11.49b)$$

$$Activation = \text{INT}(\frac{Immune\_Popul}{Immune\_Period}) \qquad (11.49c)$$

A population of 10,000 is not sufficient to ignore the "quantization error." The INT function ensures that every one of the four populations will always contain an integer number of people.

The contagious population is the sum of the infected and the sick:

$$Contagious\_Popul = Infected\_Population + Sick\_Population \quad (11.50)$$

The total population is the sum of all four populations:

$$Total\_Popul = Non\_Infected\_Popul + Contagious\_Popul$$
$$+ Immune\_Popul \qquad (11.51)$$

which, of course, is constant (10,000) in our model. The percentage of currently infected people is:

$$Perc\_Infected = \frac{Contagious\_Popul}{Total\_Popul} \qquad (11.52)$$

The infection rate can be computed as the product of the noninfected population P_{NI}, the number of contacts that a noninfected person has per week C_{Wk}, the percentage of contagious people $Perc_C$, and the chance of contracting the disease in such a contact $Rate_C$:

$$Infection\_Rate = \text{INT}(P_{NI} \cdot C_{Wk} \cdot Perc_C \cdot Rate_C) \qquad (11.53)$$

where the average number of weekly contacts of one person with another is constant, $C_{Wk} = 15$, and also the contraction rate per contact is assumed to be constant, $Rate_C = 0.25$.

The introduction of the disease occurs as a discrete event during the eighth simulated week. STELLA does not provide for a mechanism to describe discrete events. Thus, we must model the introduction with a PULSE function:

$$Introduction = \text{PULSE}(1.0, 8.0, 1E3) \qquad (11.54)$$

introduces a pulse of height 1.0 at time 8.0 with a repetition frequency of 1000.0. This is simply added to the infection rate:

$$Infection\_Rate = \text{INT}(P_{NI} \cdot C_{Wk} \cdot Perc_C \cdot Rate_C + Introduction) \quad (11.55)$$

Finally, we must ensure that not more people are ever being infected than the entire pool of noninfected persons, thus:

$$
\begin{aligned}
Infection\_Rate = \text{MIN}(&\text{INT}(P_{NI} \cdot C_{Wk} \cdot Perc_C \cdot Rate_C \\
&+ Introduction), P_{NI})
\end{aligned}
\qquad (11.56)
$$

This concludes the description of the influenza model. Figure 11.14 shows the results of this model being simulated over a period of one year.

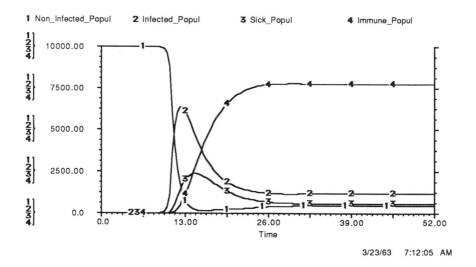

Figure 11.14. Simulation results for the influenza model.

After the introduction, the influenza epidemic spreads rapidly. Within 6 weeks, the percentage of sick people reaches its maximum of roughly 25%. A steady–state is reached about 20 weeks after the introduction. Obviously, the disease does not die out naturally. A certain percentage of the population loses immunity sufficiently fast to get reinfected before the bacteria stem has disappeared.

It might have been more realistic to model the four variables: *Contacts_Wk*, *Time_to_Breakdown*, *Sickness_Duration*, and *Immune_Period* through random functions rather than through constants, but let us save this modification for a homework problem.

11.10 Forrester's World Model

Let us now try to apply our methodology to a more ambitious problem. One of the most famous System Dynamics models ever developed and published was Forrester's world model [11.7]. How did this come about? In the late 1960s, a number of concerned scientists tried to decide whether there was a way to determine the destiny of the human race. They called themselves the "Club of Rome," and they started to investigate means to determine our future on this planet. In the sequence, Jay Forrester applied his System Dynamics methodology to this problem and in 1971 published his most famous book, *World Dynamics*.

Forrester decided that the "world" can be captured by five state variables (levels): population, pollution, nonrecoverable natural resources, capital investment, and percentage of the capital invested in the agricultural sector. His simulation starts in the year 1900 and the initial conditions for the five levels are:

$$Population = 1.65 \cdot 10^9 \qquad (11.57a)$$

$$Pollution = 2.0 \cdot 10^8 \qquad (11.57b)$$

$$Natural\_Resources = 9.0 \cdot 10^{11} \qquad (11.57c)$$

$$Capital\_Investment = 4.0 \cdot 10^8 \qquad (11.57d)$$

$$CIAF = 0.2 \qquad (11.57e)$$

He developed the structure diagram shown in Fig.11.15, which I redrew in STELLA.

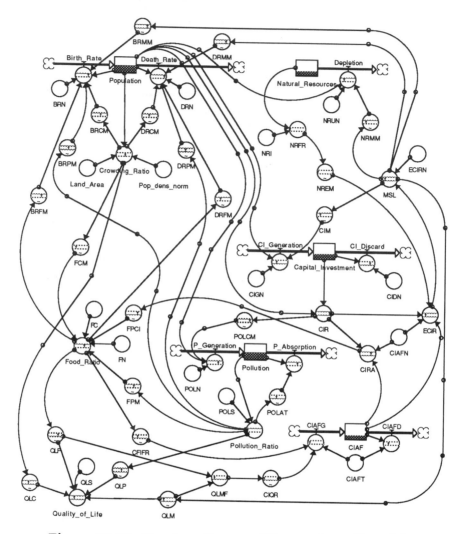

Figure 11.15. Structure diagram of Forrester's world model.

As usual, I captured the model at the end of the simulation. Consequently, those converters that do not contain a needle gauge are constants. Forrester used the following values for his constants:

$$BRN = 0.04 \quad (normal \ birth \ rate) \quad\quad (11.58a)$$

$$CIAFN = 0.3 \quad (CIAF \ normalization) \quad\quad (11.58b)$$

$$CIAFT = 15.0 \quad (CIAF \ time \ constant) \quad\quad (11.58c)$$

$$CIDN = 0.025 \quad (normal \ capital \ discard) \quad\quad (11.58d)$$

$$CIGN = 0.05 \quad (normal\ capital\ generation) \qquad (11.58e)$$

$$DRN = 0.028 \quad (normal\ death\ rate) \qquad (11.58f)$$

$$ECIRN = 1.0 \quad (capital\ normalization) \qquad (11.58g)$$

$$FC = 1.0 \quad (food\ coefficient) \qquad (11.58h)$$

$$FN = 1.0 \quad (food\ normalization) \qquad (11.58i)$$

$$Land\_Area = 1.35 \cdot 10^8 \quad (area\ of\ arable\ land) \qquad (11.58j)$$

$$NRI = 9.0 \cdot 10^{11} \quad (initial\ natural\ resources)\ (11.58k)$$

$$NRUN = 1.0 \quad (normal\ resource\ utilization) \qquad (11.58l)$$

$$POLN = 1.0 \quad (normal\ pollution) \qquad (11.58m)$$

$$POLS = 3.5999 \cdot 10^9 \quad (standard\ pollution) \qquad (11.58n)$$

$$Pop\_dens\_norm = 26.5 \quad (normal\ population\ density) \qquad (11.58o)$$

$$QLS = 1.0 \quad (standard\ quality\ of\ life) \qquad (11.58p)$$

Converters with a needle gauge but without a tilde denote algebraic relations. These were given as follows:

$$CIR = \frac{Capital\_Investment}{Population} \qquad (11.59a)$$

$$CIRA = CIR \cdot \frac{CIAF}{CIAFN} \qquad (11.59b)$$

$$Crowding\_Ratio = \frac{Population}{Land\_Area \cdot Pop\_dens\_norm} \qquad (11.59c)$$

$$ECIR = NREM \cdot CIR \cdot \frac{1.0 - CIAF}{1.0 - CIAFN} \qquad (11.59d)$$

$$Food\_Ratio = FPCI \cdot FCM \cdot FPM \cdot \frac{FC}{FN} \qquad (11.59e)$$

$$MSL = \frac{ECIR}{ECIRN} \qquad (11.59f)$$

$$NRFR = \frac{Natural\_Resources}{NRI} \qquad (11.59g)$$

$$Pollution\_Ratio = \frac{Pollution}{POLS} \qquad (11.59h)$$

$$QLMF = \frac{QLM}{QLF} \qquad (11.59i)$$

$$Quality\_of\_Life = QLS \cdot QLC \cdot QLF \cdot QLM \cdot QLP \qquad (11.59j)$$

where CIR denotes the capital investment ratio, $ECIR$ stands for the effective capital investment ratio, MSL is the material standard of living, and $NRFR$ is the fraction of the nonrecoverable natural resources remaining. Most of the equations are self–explanatory. A detailed rationale for these equations can be found in Forrester's book.

The converters with a needle gauge and a tilde are tabular functions of their respective inputs. The following tables list the tabular functions:

Table 11.3a. Tabular Function that Depends on *CIR*.

CIR	POLCM *P_Generation*
0.0	0.05
1.0	1.00
2.0	3.00
3.0	5.40
4.0	7.40
5.0	8.00

Table 11.3b. Tabular Function that Depends on *CIRA*.

CIRA	FPCI *Food_Ratio*
0.0	0.50
1.0	1.00
2.0	1.40
3.0	1.70
4.0	1.90
5.0	2.05
6.0	2.20

Table 11.3c. Tabular Functions that Depend on *Crowding_Ratio*.

Crowd_Rat	BRCM *Birth_Rate*	DRCM *Death_Rate*	FCM *Food_Ratio*	QLC *Qual_Life*
0.0	1.05	0.9	2.4	2.00
0.5				1.30
1.0	1.00	1.0	1.0	1.00
1.5				0.75
2.0	0.90	1.2	0.6	0.55
2.5				0.45
3.0	0.70	1.5	0.4	0.38
3.5				0.30
4.0	0.60	1.9	0.3	0.25
4.5				0.22
5.0	0.55	3.0	0.2	0.20

Table 11.3d. Tabular Functions that Depend on *Food_Ratio*.

Food_Ratio	BRFM Birth_Rate	DRFM Death_Rate	CFIFR CIAFG	QLF Qual_Life
0.00	0.0	30.0	1.00	0.0
0.25		3.0		
0.50		2.0	0.60	
0.75		1.4		
1.00	1.0	1.0	0.30	1.0
1.25		0.7		
1.50		0.6	0.15	
1.75		0.5		
2.00	1.6	0.5	0.10	1.8
3.00	1.9			2.4
4.00	2.0			2.7

Table 11.3e. Tabular Functions that Depend on *MSL*.

MSL	BRMM Birth_Rat	CIM CI_Gen	DRMM Death_Rat	NRMM Depletion	QLM Qual_Lif
0.0	1.20	0.1	3.00	0.00	0.2
0.5			1.80		
1.0	1.00	1.0	1.00	1.00	1.0
1.5			0.80		
2.0	0.85	1.8	0.70	1.80	1.7
2.5			0.60		
3.0	0.75	2.4	0.53	2.40	2.3
3.5			0.50		
4.0	0.70	2.8	0.50	2.90	2.7
4.5			0.50		
5.0	0.70	3.0	0.50	3.30	2.9
6.0				3.60	
7.0				3.80	
8.0				3.90	
9.0				3.95	
10.0				4.00	

Table 11.3f. Tabular Function that Depends on *NRFR*.

NRFR	NREM ECIR
0.00	0.00
0.25	0.15
0.50	0.50
0.75	0.85
1.00	1.00

Table 11.3g. Tabular Functions that Depend on *Pollution_Ratio*.

	BRPM	DRPM	FPM	Polat	QLP
Poll_Rat	*Birth_Rat*	*Death_Rat*	*Food_Rat*	*P_Absorp*	*Qual_Lif*
0.0	1.02	0.92	1.02	0.6	1.04
10.0	0.90	1.30	0.90	2.5	0.85
20.0	0.70	2.00	0.65	5.0	0.60
30.0	0.40	3.20	0.35	8.0	0.30
40.0	0.25	4.80	0.20	11.5	0.15
50.0	0.15	6.80	0.10	15.5	0.05
60.0	0.10	9.20	0.05	20.0	0.02

Table 11.3h. Tabular Function that Depends on *QLMF*.

	CIQR
QLMF	*CIAFG*
0.0	0.7
0.5	0.8
1.0	1.0
1.5	1.5
2.0	2.0

Some of the functions that depend on the same input variable have values specified at different sampling points, and some (more scary!) have values specified over inconsistent domains. However, the program runs without error, and thus, let us not be discouraged by such minor details.

Finally, I need to write down the equations for the rates:

$$Birth\_Rate = Population \cdot BRN \cdot BRCM \cdot BRFM \cdot$$
$$BRMM \cdot BRPM \tag{11.60a}$$

$$CIAFD = \frac{CIAF}{CIAFT} \tag{11.60b}$$

$$CIAFG = \frac{CFIFR \cdot CIQR}{CIAFT} \tag{11.60c}$$

$$CI\_Discard = CIDN \cdot Capital\_Investment \tag{11.60d}$$

$$CI\_Generation = CIGN \cdot CIM \cdot Population \tag{11.60e}$$

$$Death\_Rate = Population \cdot DRN \cdot DRCM \cdot DRFM \cdot$$
$$DRMM \cdot DRPM \tag{11.60f}$$

$$Depletion = Population \cdot NRUN \cdot NRMM \tag{11.60g}$$

$$P\_Absorption = \frac{Pollution}{POLAT} \tag{11.60h}$$

$$P\_Generation = Population \cdot POLN \cdot POLCM \tag{11.60i}$$

This completes the description of the model. Simulation results are shown in Fig.11.16.

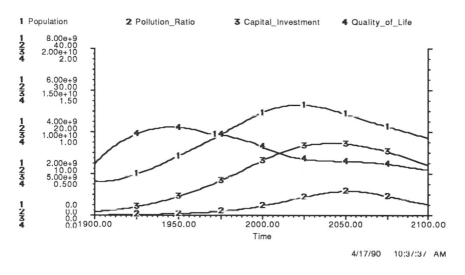

Figure 11.16. Simulation results for Forrester's world model.

These results agree with those that were presented in Forrester's book. Notice that I silently converted the former difference equations of Forrester's original text into a set of differential equations. In my simulation, I used a step size of $\Delta t = 0.125$ instead of $\Delta t = 1.0$ as suggested by Forrester in order to make the graphs look smoother.

Forrester drew quite a bit of heat for his publication from politicians since they didn't want to hear what he had to say and from colleagues since they considered his methods not sufficiently "solid" in scientific terms (and maybe since they were a little jealous of his unquestionable success). World modeling remained a fashionable topic for a number of years. Several other authors published their newest "findings," which I read with unbroken fascination and growing frustration. In order to avoid the criticisms and mockeries of their colleagues, all the later authors kept the details of their models for themselves and published only their "results." Forrester was maybe a little naïve, but he was at least honest. He played with open cards, and for this, he has my full respect.

What were these dramatic revelations that Forrester disclosed in his book? He discovered that a physical system with finite energy cannot exhibit a behavior in which any variable grows to infinity.

The population will have to stagnate, and so will the GNP. As a control engineer, I don't need a simulation model to determine that this is in fact a truism. There cannot be a question about the fact that we shall not see an annual increase of the GNP of 3% or whatever for an infinite period of time. However, what was really new (and frightening) about Forrester's model was the fact that the stagnation is not just an abstract concept, something that might happen to our descendants in the year 5000 A.D., but that it will happen soon, probably within the next 50 to 200 years. Our children may still see the day when recycling will no longer be an alibi exercise but will become a bitter necessity and when fines for causing environmental damage will no longer merely serve the purpose of ensuring the reelection of some politicians but will be so stiff that it will be more economical for potential offenders to avoid causing the damage in the first place. I am not talking here about accidents as they can, and always will, happen, but about the reckless and purposeful contamination of our scarce and irreplaceable living space.

11.11 Model Validation

Since we have moved away quite a bit from the rigid meta–laws of the hard physical sciences with which we started, it is now time to talk a little more about the process of model validation. How do we ensure that our models reflect reality? How do we verify, even if a given model reflects the already–observed behavior of the real system, that model predictions of future behavior do reflect the true future of the real system under investigation?

It is still too early to address this problem to its full extent. I shall do so in the companion book of this text when I also discuss the verification of simulation results. But now, I wish to at least point out some of the pitfalls of System Dynamics models in particular.

One of the real problems with System Dynamics models is the fact that the step of model validation is separated from the process of trajectory behavior generation. It is all too easy to forget the implicit (and explicit) assumptions behind a model once it has been properly "debugged" (i.e., it no longer produces any error messages!), and believe in simulation results just because they look elegant and maybe plausible. Certainly, Forrester succumbed to this temptation in the conclusions of his text, *World Dynamics*, and we all do at times.

One remedy that comes immediately to mind is the following: never extrapolate any variables of a semiphysical model (such as a System Dynamics model) far beyond the range of observed values. It is absurd to believe that we can predict the reaction of our planet to a pollution value that is 100 times higher than any value that we have ever physically observed.

In practice, I suggest that the STELLA designers add a feature to their software that allows the users to attach "demons" to their models that can watch over the integrity of inherent model assumptions during the execution of simulation runs. The simplest form of such a demon is a threshold indicator, which can be attached to any variable in the model. For example, when drawing a level variable denoting a population, we should be able to attach a demon to this variable that ensures that the population never decreases to a negative value. If this should ever happen, the simulation should immediately stop with an error message.

As can be noticed, a demon is nothing but an implementation of our well–known termination conditions (called **termt** in ACSL). I suggest implementing demons as a new icon of the structure diagram. Demons have one or several inputs, but no outputs. Connectors can terminate in demons, but they can never emanate from demons. When we double–click on a demon, a window pops up that enables us to formulate a termination condition in which all the variables connected to the demon must be used. If we don't wish to clutter up our structure diagram with the demons, we can simply create ghosts of your "demonized" variables and connect the demons to the ghosts in separate structure diagrams.

The second remedy that comes to mind is sensitivity analysis. The high "precision" (in terms of the number of displayed digits) that simulation results usually provide may be deceiving. It is quite obvious that parameter values are naturally associated with tolerance bands about a nominal value. Even in electrical circuitry, we know the value of a resistor only with an accuracy of 2%, 5%, or 10%, depending on the price we are willing to pay for the component. Consequently, it makes a lot of sense to investigate the sensitivity of the simulation results to parameter variations. If the sensitivity is large, we know that we must be extremely cautious in drawing conclusions about the system behavior. On the other hand, if the sensitivity is small, we can trust that our model exhibits a behavior that resembles the true behavior of the real system. If you are interested in learning more about how this can be accomplished, solve homework problems Hw.[H11.3] and Hw.[H11.4]. I shall return to this technique in more detail in the companion book of this text.

11.12 Summary

In this chapter, we introduced a new modeling methodology, called System Dynamics, which enabled us to create rather quickly semi-formal models of "soft" systems, i.e., systems as they are found in the biological and social sciences, as well as in economics and business administration. A new tool, STELLA, was also introduced, a tool that was specifically designed to facilitate the creation of System Dynamics models.

References

[11.1] François E. Cellier (1986), "Enhanced Run–Time Experiments for Continuous System Simulation Languages," *Proceedings SCS MultiConference on Languages for Continuous System Simulation* (F.E. Cellier, ed.), SCS Publishing, San Diego, Calif., pp. 78–83.

[11.2] François E. Cellier and C. Magnus Rimvall (1989), "Matrix Environments for Continuous System Modeling and Simulation," *Simulation*, **52**(4), pp. 141–149.

[11.3] R. G. Coyle (1977), *Management System Dynamics*, John Wiley & Sons, London, U.K.

[11.4] Andreas Fischlin and Werner Baltensweiler (1979), "Systems Analysis of the Larch Bud Moth System. Part 1: The Larch — Larch Bud Moth Relationship," Mitteilungen der Schweizerischen Entomologischen Gesellschaft, **52**, pp. 273–289.

[11.5] Jay W. Forrester (1964), *Industrial Dynamics*, MIT Press, Cambridge, Mass.

[11.6] Jay W. Forrester (1968), *Principles of Systems*, Wright–Allen Press, Cambridge, Mass.

[11.7] Jay W. Forrester (1971), *World Dynamics*, Wright–Allen Press, Cambridge, Mass.

[11.8] Michael E. Gilpin (1979), "Spiral Chaos in a Predator–Prey Model," *The American Naturalist*, **113**, pp. 306–308.

[11.9] William L. Heyward and James W. Curran (1988), "The Epidemiology of AIDS in the U.S.," *Scientific American*, October issue, pp. 72–81.

[11.10] James M. Hyman (1990), "Modeling The AIDS Epidemiology in the U.S.," private communication, Los Alamos National Laboratory, Los Alamos, NM.

[11.11] A. G. Ivakhnenko and Valentin G. Lapa (1967), *Cybernetics and Forecasting Techniques*, Series in Modern Analytic and Computational Methods in Science and Mathematics, **8**, American Elsevier Publishing, New York.

[11.12] Yumi Iwasaki and Herbert A. Simon (1986), "Causality in Device Behavior," *Artificial Intelligence*, **29**, pp. 3–32.

[11.13] Jonathan M. Mann, James Chin, Peter Piot, and Thomas Quinn (1988), "The International Epidemiology of AIDS," *Scientific American*, October issue, pp. 82–89.

[11.14] George P. Richardson and Alexander L. Pugh (1981), *Introduction to System Dynamics Modeling with DYNAMO*, Wright–Allen Series in System Dynamics, MIT Press, Cambridge, Mass.

[11.15] Barry Richmond, Steve Peterson, and Peter Vescuso (1987), *An Academic User's Guide to STELLA*, High Performance Systems, Inc., Lyme, N.H.

Bibliography

[B11.1] Jean D. Lebel (1982), "System Dynamics," in: *Progress in Modelling and Simulation* (F.E. Cellier, ed.), Academic Press, London, U.K., pp. 119–158.

Homework Problems

[H11.1] Influenza Model
Modify the influenza model given in this chapter in the following way:

$$Contacts\_Wk = U(5.0, 25.0) \qquad (H11.1a)$$
$$Time\_to\_Breakdown = N(4.0, 1.0) \qquad (H11.1b)$$
$$Sickness\_Duration = N(2.0, 1.0) \qquad (H11.1c)$$
$$Immune\_Period = N(26.0, 8.0) \qquad (H11.1d)$$

where $U(min, max)$ denotes a uniform distribution with a minimum value of min and a maximum value of max and $N(\mu, \sigma)$ denotes a normal distribution with a mean value of μ and a standard deviation of σ. Repeat the simulation and compare the results.

In a second simulation experiment, enhance the mean value of the immune period. Find the smallest mean value of the immune period necessary to ensure the natural extinction of the bacteria stem.

[H11.2] World Model

Jay Forrester determined that the stagnation in his world model was caused by the exhaustion of the nonrecoverable natural resources. As a remedy, he recommended reducing the normal use of natural resources (NRUN) to 25% of its former value in the year 1970.

Implement this change in STELLA using the STEP function. Simulate the modified system and check the results. Forrester discovered that the removal of this energy constraint just unchained another one. Now the pollution explodes and the population suffers a major breakdown due to an environmental disaster. He suggested reducing the normal pollution also to 25% of its former value in the year 1970.

Implement also this change in STELLA using the STEP function. Forrester liked the resulting curve much better. Extend the simulation period from the year 2100 to the year 2500, and compare the new curve with the original curve. Interpret the results.

I discovered that the energy constraint on the natural resources is very beneficial to a smooth transition from the growth phase to the stagnation phase. As a control engineer, I know that if I wish to reduce oscillations in a feedback control system, I should reduce the open–loop gain of the system, not enhance it. Thus, since Forrester's model suggests that we have already exceeded the steady–state value of the system, the conservation of nonrecoverable natural resources may be more detrimental than beneficial since ultimately, we shall have to learn to live without them anyway. I once "optimized" the behavior of the world model in terms of a smooth transition to the stagnation phase. The optimization suggested that I should pump several thousand barrels of oil *into* the ground every day, i.e., that I should not save the nonrecoverable natural resources, but spend them as quickly as I could. An early shortage of these resources will ensure that our system exhibits as little overshoot behavior as possible.

[H11.3] Sensitivity in the Large

Reimplement Forrester's world model in ACSL. Execute the simulation under control of either MATLAB or CTRL–C. Assume that the four parameters $CIAFT$, FC, $NRUN$, and $POLN$ are inaccurately known. Assume that the given values have a tolerance band of $\pm 25\%$ associated with them.

Rerun the simulation 16 times for all worst–case combinations of these parameters. Import into CTRL–C or MATLAB the population variable from the ACSL model, and store all resulting populations together in a matrix. Use MATLAB's (CTRL–C's) MAX and MIN functions to compute upper and lower bounds and plot the upper and lower envelopes of the population together on one graph as functions of time. What do you conclude?

[H11.4] Replication and Batch

This time, assume that the four parameters *CIAFT*, *FC*, *NRUN*, and *POLN* are uniformly distributed stochastic variables in the range nominal value ±25%. Perform 100 simulation runs with different seed values for the random–number generators. Import from ACSL the population trajectories and compute again upper and lower envelopes. What do you conclude?

[H11.5] Balancing Your Checkbook

A traveling salesperson makes an average net income of $2000 a month, more precisely, his or her income is $N(2000, 200)$, i.e., the income is normally distributed with a mean value of $2000 and a standard deviation of $200. He or she likes to spend 90% of his or her average income. The average income is computed as a moving average of the real income over six months.

Since STELLA does not provide us with a moving–average function, we need to construct this function the hard way. Make a ghost of the salesperson's income and delay it six times by one month in a separate structure diagram. Then add the six incomes and divide by six. This is the desired moving average. Make a ghost of the moving average and copy it back into the main diagram.

Of course, a cash constraint exists. The salesperson will not spend the desired amount of money unless she or he has sufficient cash in the bank. $4000 in the bank are considered sufficiently safe to spend what she or he likes to spend. Her or his real consumption is the product of the desired consumption and the cash constraint. The cash constraint is 1.0 if she or he has $4000 or more in the bank, otherwise, it is linear in her or his current savings.

At the beginning of the simulation, the salesperson has exactly $4000 in the bank. Four months into the first year, a competitor comes out with a new product that reduces the mean value of the poor salesperson's income by $500. The standard deviation remains the same. Simulate the system over 3 years. Plot on one graph the monthly net income, cash balance, and monthly consumption of our Schlemihl.

Research

[R11.1] The Epidemiology of AIDS

Study the October 1988 issue of *Scientific American*, a special issue devoted to the HIV infection, and in particular, the two articles on the epidemiology of AIDS by Heyward and Curran [11.9] describing the epidemiology of the disease in the U.S., and by Mann *et al.* [11.13] describing the international epidemiology of the disease. Scan the literature for further information to obtain as solid statistical material as possible on which to base your modeling efforts. Good sources of data are the Proceedings of the Annual International Conference on AIDS. Develop a System Dynamics model (a modification of our simple influenza model) that describes the spreading of the disease and reflects the observed data well. As usual with these types of studies, the crux is with the data. The available data are often speculative and largely inconsistent. It would be fairly easy to determine an inductive model describing the spreading of the disease fairly accurately, but this doesn't help us much since we haven't observed a saturation period yet, and this is exactly what our model is supposed to predict. Thus, we need a semiphysical model.

One of the most interesting facts about the epidemiology of AIDS is that the reported AIDS population did not grow exponentially during the early stages of the disease as would be expected, but instead, it grew polynomially with the third power of the time t. A very good inductive model of the reported AIDS cases in the U.S. between 1980 and 1988 can be described through the formula [11.10]:

$$A(t) = 175 \cdot (t - 1981.2)^3 + 340 \qquad (R11.1a)$$

If we try to retrofit a physical model to the observed behavior, we need a differential equation for $A(t)$. From Eq.(R11.1a), we can find immediately by differentiation:

$$\dot{A}(t) = 525 \cdot (t - 1981.2)^2 \qquad (R11.1b)$$

which can be rewritten as:

$$\dot{A}(t) = \frac{3 \cdot (A(t) - 340)}{(t - 1981.2)} \qquad (R11.1c)$$

Consequently, in order to obtain a polynomial growth, our model must contain some sort of $1/t$ factor. Such a $1/t$ factor makes sense for a disease that loses its virulence through natural mutation or diffusion or leads to a growing body of immunized and therefore nonsusceptible hosts. Unfortunately, neither of these two assumptions is plausible in the case of the AIDS disease. However, any decent semiphysical model of the epidemiology of

AIDS will have to come up with a plausible explanation of where this $1/t$ factor comes from, a hypothesis that should preferably be verifiable.

Develop a consistent System Dynamics model describing the epidemiology of AIDS in terms of the susceptible population $S(t)$, the infected population $I(t)$, and the sick population $A(t)$, possibly compartmentalized into various risk groups. It may also be necessary to divide the infected population into three separate subgroups denoting the group of freshly infected individuals (who are highly contagious, since no antibody has been developed yet), the group of symptom–free infected individuals (not very contagious, since the antibody constantly destroys the free retrovirus in the blood stream), and the group with early symptoms (highly contagious since the antibody is about to lose the battle).

Associate tolerance bands with the parameters of your model and perform a "sensitivity analysis in the large" (as described by the author in two previous articles [11.1,11.2]). From the sensitivity analysis, determine which will be the maximum and minimum values of reported AIDS infections in the years 1992, 1995, and 2000. When will the epidemic exhibit its peak value? What are the best and worst percentages of the infected population at that time? Will humanity survive the viral assault? How good is the confidence of your predictions?

[R11.2] Global Change

Create a System Dynamics model that describes the global change in Earth climate. Start by determining the state variables of the model. Clearly this must be an energy model. We must know how much energy arrives at our globe from the sun and what is the percentage of absorbed energy versus reflected energy. We must also know how much heat is produced by dissipative processes at the surface of our planet (such as burning fuel). What is the percentage of human–caused heat versus solar heat?

Obviously, the solar heat absorbed by the atmosphere depends on several factors such as the composition of our atmosphere, the percentage of cloud coverage, and the percentage of (dark) forests versus (clear) deserts. Other factors of importance are the salt content of our seas. Would a large–scale exploitation of solar energy change the situation? Could we influence the percentage of absorbed versus deflected energy by installing large mirrors in desert areas?

The composition of our atmosphere depends on many factors including human–caused air pollution, the percentage of forest coverage, and the evaporation of our seas. All these factors must somehow be expressed in our global change model.

In the global flow equilibrium, there must exist small–scale factors with large–scale effects; otherwise, there couldn't have occurred glacial epochs in the recent past of our global history. What are these influencing factors? One of the currently fashioned theories links the ice–ages with the flow of the gulfstream which in turn is said to be influenced by the salt content in the North–Atlantic. Can we verify this hypothesis by means of modeling? Can we exploit these factors to our advantage? What would happen if we were to place large sheets of swimming cloth (similar to pool covers) over a few critical areas in the North–Atlantic? This would certainly reduce the evaporation at these places. Thereby, the salt content of the water should slowly be reduced. Could this effect slow down the gulfstream sufficiently much to counter the effects of the warming trend?

12

Naïve Physics

Preview

Until this point, we have focused on a single question throughout this entire text: How can we get computer programs to mimic the behavior of physical systems. In this chapter, as well as in the following chapters, we shall deal with quite a different issue: We shall try to understand how humans model the behavior of physical systems in the absence of a computer, i.e., how they reason about the functioning of a device or process. In other words, we shall try to model the process of understanding itself. Naïve physics is one methodology that can address this question. Other methodologies will be discussed in due course.

12.1 Introduction

In the first 11 chapters of this text, we have tried to find the answer to a single question: How can we capture the dynamical behavior of a physical, or other, system in a mathematical model. It turned out that differential equations presented us with a powerful description mechanism to achieve this goal. How we go about solving these differential equations was of little concern to us. We used prefabricated simulation languages, such as ACSL, which hopefully would know how to handle the set of differential equations once it had been derived.

The rationale behind this approach was the following: the process of *simulation* can be fairly easily automated. Once we have a powerful simulation engine available to us we should not have to worry about how precisely this engine works, since we can trust that it can handle most simulation problems adequately and accurately on its own. However, the process of *modeling* is a very complex issue that requires all our human ingenuity and cannot be fully automated.

Unfortunately, this answer is not good enough — particularly in the context of Space applications. In Space, human labor will be a scarce resource for a long time to come. Thus, we cannot conceive the conquest of our solar system without the deployment of high-autonomy systems. In order to colonize our moon or the planet Mars, we must devise a technology in which autonomously operating robots can create a livable environment for humans. These robots will not be able to function properly unless they have the capability to make decisions on their own in a partially unknown environment.

How does the decision–making process work? Humans make decisions by envisioning a set of possible scenarios (experiments) and analyzing the effects of these scenarios using *mental models*. They "simulate" their mental models using *mental simulation*. The trajectories resulting from a mental simulation are referred to as *episodes*. Humans then go ahead with implementing the one scenario for which the mental simulation results in the most advantageous episode. Notice that modeling and simulation were not invented in the 1950s, they have been around since the first semisentient being tried to predict what the future would bring. If robots will ever be able to operate meaningfully on another planet, they must be able to mimic the human decision–making process. They must be able to create new models of their environment on the fly and use them immediately in mental simulation experiments. Thus, we need to design methodologies that allow us to model the human decision–making process, i.e., we need to model the mechanisms of human understanding itself.

This does not necessarily mean that robots should be prevented from solving differential equations simply because we aren't able to do so. The purpose of the exercise is not primarily to understand how our brain works, but to formulate an engineering design that either equals or surpasses our reasoning abilities. No (engineering) reason can be given why we should duplicate human inadequacies in our robots. Other (ethical) reasons could be given, which may suggest such a limitation, but these are not the topic of this book.

The straightforward solution to this problem is to provide the robot with an optimization algorithm aside from the simulation algorithm, i.e., the *optimization* algorithm is used to identify a state–space model, while the *simulation* algorithm is used to generate trajectory behavior of the identified state–space model. Unfortunately, several problems are intrinsic to this approach.

(1) Optimization of dynamical models is number crunching at its best. We would need very powerful onboard computers, since otherwise, each decision would take forever.

(2) The "smarter" (i.e., faster converging) we make the optimization algorithm, the more likely it will end up in a ditch. Thus, we need to choose between an optimization algorithm that converges very slowly, and one that converges faster — but more often than not converges to the wrong solution.

(3) In the section on structure characterization (Chapter 10), I explained that a good match of behavior does not guarantee that we have identified the right structure. Thus, together with the optimization algorithm, we need to automate the model validation process and this may be a yet more difficult task.

Contrary to the simulation algorithms, optimization algorithms do not lend themselves to full automation. Thus, while the differential equation description mechanism is a wonderful tool for capturing the behavior of a dynamical system under human supervision, it may not be the most appropriate tool when used as part of an automated decision–making process — unless we let go of the idea of optimization and try to mimic the human supervision process directly. Remember that while we were quite successful at modeling physical systems, we used optimization only in rare instances and very cautiously.

We shall return to this avenue later in this text. However, in the mean time, we shall check whether mechanisms other than differential equations exist that might be better suited to capture system behavior for the purpose of an automated decision maker.

If I hold a glass full of water in my hand and open my fingers, I know that the glass will fall down, and that, upon impact, it will break and spill the water all over my carpet. However, in my mental simulation, I don't solve any differential equations at all. I don't know when exactly the glass will hit the floor, and I don't know into how many pieces the glass will shatter, but often, these details are not important for proper decision making. Thus, the question is whether we can describe mental processes, that is, imprecise qualitative models, in precise mathematical terms.

The first approach we could try is to map out all possible behaviors of our system and formulate a recipe of appropriate actions for every imaginable situation. This approach is called *shallow reasoning*. A typical example for a shallow reasoner might be the "Grecr" function, which I introduced in Chapter 11. In general, a shallow reasoner consists of a usually very complex series of if–then–else clauses. As in the case of the optimizer, a number of problems are intrinsic to this approach. The first objection is the following: Even for relatively simple systems, the number of branches of our if–then–else tree can become very large. The tree usually grows exponentially with the behavioral complexity of the system. Several remedies have been proposed to reduce (but not overcome) this problem.

One remedy might be to artificially reduce the behavioral complexity of our system by lumping together state values that are different, but not to such an extent that it would bother us if we could no longer distinguish between them, i.e., we discretize our state–space. This is one way to cope with the imprecision of mental models. This state aggregation is done in all qualitative modeling systems, and in Chapter 1 of this text, I even used this fact to define the term "qualitative model." In both this and the next chapter, we shall look into the process of state discretization among other things.

Another remedy is to organize our tree a little better. It turns out that one cause of the phenomenal growth rate of our tree is the fact that the same branches are used over and over again. Instead of coding a hierarchical decision tree explicitly, we can formulate a number of parallel *rules* (another name for an if–then–else clause) and devise an algorithm that tells us in which sequence we should evaluate these rules. This implementation of the decision tree is called an *expert system*, and the search algorithm is called an *inference engine*.

However, these two techniques will only alleviate the complexity problem, not overcome it. The problem with this approach is that we operate on behavioral knowledge rather than on structural knowledge. In Chapter 10, I explained that the behavioral complexity of a system is usually very much larger than its structural complexity. For illustration, think of the simple logistic equation. If we wish to describe this system in terms of behavioral knowledge, we need to formulate a decision tree that branches in the same way as the bifurcation map does, i.e., we must distinguish between different behaviors depending on the value of the model parameter. Yet the structural knowledge (the difference equation) is extremely simple. No technique can help us overcome this problem if we insist on capturing the system behavior directly in terms of behavioral knowledge.

This problem was recognized by several researchers, such as de Kleer and Brown [12.8], Forbus [12.10], and Kuipers [12.12] who replaced the shallow reasoner with a deep reasoner. Unlike the shallow reasoner, the *deep reasoner* captures the knowledge about a system in terms of its structure rather than its behavior. A deep reasoner operates on something similar to a state–space model, but it avoids the problem of numerically integrating sets of differential equations. It operates on state variables with imprecisely determined values and it operates on imprecise structural operators such as:

$$y = \mathrm{M}^+(x) \tag{12.1}$$

meaning that the variable y is monotonically increasing with the variable x, and that $y = 0$ when $x = 0$. We could call this knowledge representation scheme a *qualitative state-space model* [12.5]. This approach, which is often referred to as the *naïve physics* approach to qualitative modeling, has received considerable attention lately, and it deserves to be treated in a separate chapter of this book. That is what we are going to do now.

12.2 Definitions

Unfortunately, the literature on qualitative modeling in general and naïve physics in particular is full of imprecisely defined, partially overlapping, and often even entirely redundant terminologies. Depending on the author, we meet terms such as *qualitative models*, *qualitative reasoning*, *qualitative physics*, *naïve physics*, and *common–sense reasoning*. All of these terms are used in very similar contexts, and they are hardly ever properly defined. I shall use the following terminology:

(1) *Qualitative variables* are variables that assume a finite ordered set of qualitative values, such as "minuscule," "small," "average," "large," and "gigantic." The literature on quantitative soft sciences is a little more precise on this definition than the literature on artificial intelligence. For instance, Babbie [12.1] distinguishes between:

   (a) *Nominal measures*, i.e., variables whose values have the only characteristics of exhaustiveness and mutual exclusiveness. Nominal measures are unordered sets. Typical nominal variables might be the religious affiliation or the hair color of a person. Such variables are not useful as state variables in a simulation. They can play a role as parameters.

   (b) *Ordinal measures*, i.e., variables that are nominal and rank–ordered. These variables are what I called qualitative variables earlier. However, sometimes we shall let go of the condition of mutual exclusiveness, for example, when we operate on *fuzzy sets*.

(c) *Interval measures*, i.e., variables that are ordinal and have the property that a distance measure can be defined between any two values, that is, interval variables can be added to and/or subtracted from each other. A typical candidate for a "soft" interval variable might be the intelligence quotient.

(d) *Ratio measures*, i.e., variables that are interval measures and have a true zero point.

However, Babbie does not distinguish clearly between discrete measures (integer numbers) and continuous measures (real numbers). Amazingly, this distinction seems unimportant to him.

(2) *Qualitative behavior* denotes a time–ordered set of values of a qualitative variable, i.e., an episode.

(3) *Qualitative models* are models that operate on qualitative states.

(4) A *qualitative simulation* is an episode generator that infers qualitative behavior from a qualitative model.

These are precisely defined terms that will prove useful in due course.

Why are we interested in qualitative modeling and simulation? A number of applications for this methodology can be named:

(1) *Incomplete system knowledge*: Some details about the system under investigation are missing. Without such detail, quantitative simulation (at least in the sense of a single trajectory generation) cannot work. For example, after an anomaly is detected in a flight, the pilot will usually switch off the autopilot since he or she cannot trust any longer that the model that is an inherent part of the autopilot still reflects the behavior of the modified system adequately. Since the qualitative model operates on more highly aggregated variables, it may be somewhat more robust, i.e., it may be less sensitive to system modifications.

(2) *Response to a class of experiments*: Until now, we always examined the response of a system to a single experiment. Sometimes it is more useful to examine the set of output trajectories resulting from applying an entire set of input trajectories to the system. Qualitative models are sometimes more adequate for this type of application. Sensitivity analysis in the large is an alternative [12.3,12.4].

(3) *Generalization for decision making*: It is sometimes hard to see the forest for the trees. Quantitative models generate quantita-

tive (that is, detailed) responses. It may be difficult to aggregate this detailed information in an automated system for purposes of knowledge generalization. It might be better not to generate this detailed knowledge in the first place. Qualitative modeling allows us to aggregate knowledge earlier in the game. This can sometimes be beneficial.

Evidently, a number of good reasons can be stated why we may wish to employ qualitative modeling and qualitative simulation. However, a number of bad reasons are also quite often quoted:

(4) "Qualitative simulation is cheaper than quantitative simulation. If quantitative simulation in a real–time situation cannot produce the results fast enough, qualitative simulation may be the answer to the problem." Wrong! Algorithms used for qualitative simulation are by no means faster than those used in quantitative simulation. In qualitative simulation, many alternative branches must generally be explored, whereas quantitative simulation usually produces one individual trajectory. Thus, quantitative simulation is normally faster than qualitative simulation if applicable. Thus, if your quantitative real–time simulation executes too slowly, don't go to qualitative simulation, go to a nearby computer store and buy yourself a faster computer.

(5) "Qualitative simulation requires a less profound understanding of the mechanisms that we wish to simulate. Therefore, if we don't fully understand the mechanisms that we wish to simulate, quantitative simulation is out of the question, whereas qualitative simulation may still work." Wrong again! Qualitative simulation has constraints as stringent as quantitative simulation, they are just a little different. A convenient user interface relieves the user from some of the intricacies of detailed understanding of the simulation mechanisms, not the modeling methodology *per se*. Today's languages for quantitative simulation (such as ACSL) are very user–friendly, more so than today's languages for qualitative simulation. This is due to the fact that quantitative simulation languages have been around much longer. Thus, if you don't understand what you are doing, don't go to qualitative simulation, go to an expert who does.

Qualitative modeling is not an *alternative* to quantitative modeling. When we have the knowledge available to produce a decent state–space model, it will in all likelihood work much better. Don't believe that because we can't solve differential equations in our heads our robots shouldn't do it either. Qualitative modeling presents us with an *enhancement* of our toolbox, and sometimes, this tool may be just the right one, but don't view (as unfortunately many researchers do) quantitative and qualitative modeling as in competition with each other. They are complementary rather than competitive techniques.

But let me get back to my definitions. I shall use the following terms:

(1) *Shallow models* of a system are models that capture the behavioral knowledge of the system directly, bypassing the structural knowledge entirely.

(2) *Deep models* of a system capture the structural knowledge of the system.

(3) *Shallow simulators* produce trajectory behavior from shallow models using a direct mapping of input/output behavioral patterns. A finite state machine is a typical example of a shallow simulator.

(4) *Deep simulators* produce trajectory behavior from deep models using a trajectory generation algorithm. Our state–space models are good candidates for deep models and the numerical integration algorithms are good candidates for deep simulators.

Notice that I didn't write anywhere that shallow models must be static models. If my finite state model predicts that today's output is a function of today's first input and yesterday's second input:

$$y(t) = \mathbf{f}(x_1(t), x_2(t - \Delta t)) \tag{12.2}$$

this might still be a shallow model, although it references time explicitly. The distinction is related to the properties of the function f. If f reflects the true structure of our system, it is a deep model, whereas if f reflects an inductively found approximate relationship between variables, it is a shallow model.

These are basically all the terms that we need. *Deep knowledge*, a term quite frequently used in the literature is thus equivalent to structural knowledge and *common–sense reasoning* is equivalent to qualitative simulation.

12.3 State Discretization and Landmarks

Naïve physics models are characterized by a very limited set of qualitative values. Variables usually assume only one of three values: +,

0, and −. Table 12.1 shows the mapping of a quantitative trajectory into a qualitative episode.

Table 12.1. Discretization of a Trajectory into an Episode.

Trajectory	Episode
18.73	+
9.22	+
4.01	+
−3.82	−
−5.06	−
−0.37	−
1.19	+

In order to maintain as much realism as possible within our highly aggregated state-space, we shall make the reasonable assumption that a continuous variable cannot jump from positive values to negative values without going through zero and vice versa. Thus, we shall expand the previous episode by adding values of 0 whenever the episode switches from + to − or from − to +. + and − are two *regions*, whereas 0 is a *landmark*. In all naïve physics systems, adjacent regions are always separated by landmarks. The {− 0 +} set is the simplest of all meaningful sets of regions and landmarks.

Different authors use different approaches to alleviate the problem of reducing the true behavior of a physical system to a trivial behavior in the process of discretization. Kuipers [12.12] has a mechanism to invent new landmarks on the fly and he includes the additional landmarks of −∞ and +∞ right from the beginning. Morgan [12.14] operates on the minimal set of regions and landmarks {− 0 +}, expanded only with two more elements: ? meaning *don't know* and ⊔ denoting an illegal or inconsistent value. (⊔ is a highly abstracted version of my Mac's trash can.) However, Morgan represents each variable through a *qualitative vector*:

$$x = (+ + -) \tag{12.3}$$

meaning that x and $\dot{x}$ are currently positive, while $\ddot{x}$ is currently negative. Morgan carries along with the current value of a state variable information about its first and second time derivatives.

In this chapter, I shall basically follow the approach of Morgan since it comes closest to our (by now well-introduced) concept of a state-space. Morgan's methodology can be easily modified to either include the first time derivative only, or to include higher time derivatives as well. We shall discuss the consequences of such a modification. However, for most applications, the suggested triple seems to be optimal.

12.4 Operations on Qualitative Variables

Two qualitative variables x and y can easily be added. The truth table for adding qualitative variables is as follows:

$$ADD = \begin{array}{c} x\backslash y \\ - \\ 0 \\ + \\ ? \end{array} \begin{array}{cccc} - & 0 & + & ? \\ \begin{pmatrix} - & - & ? & ? \\ - & 0 & + & ? \\ ? & + & + & ? \\ ? & ? & ? & ? \end{pmatrix} \end{array} \qquad (12.4)$$

I have written software that implements this technique called *QualSim*. QualSim is available either as a library of CTRL–C functions or as a MATLAB toolbox. The CTRL–C version of QualSim is made available by loading the QualSim library into CTRL–C. This is achieved with the command:

[>   **do** *qualsim : qualsim*

MATLAB recognizes all its toolboxes automatically.

In QualSim, $-$ is represented as 1, 0 as 2, $+$ as 3, ? as 4, and ⊔ as 5. Addition is represented through a constant matrix:

$$TADD = \begin{pmatrix} 1 & 1 & 4 & 4 \\ 1 & 2 & 3 & 4 \\ 4 & 3 & 3 & 4 \\ 4 & 4 & 4 & 4 \end{pmatrix} \qquad (12.5)$$

Two variables x and y can be added using the statement:

$$z = TADD(x, y) \qquad (12.6)$$

in a simple table look–up, exploiting the fact that x and y have integer values in the range from 1 to 4 and can therefore be interpreted as indices. If either x or y has a value of 5 (trash can), Eq.(12.6) will result in an error message. This is meaningful since we shouldn't use an illegal or inconsistent variable in any subsequent operation.

Of course, x and y are usually qualitative vectors. However, we know that:

$$\text{if } z = x + y$$
$$\rightarrow \dot{z} = \dot{x} + \dot{y} \tag{12.7}$$
$$\rightarrow \ddot{z} = \ddot{x} + \ddot{y}$$

Equation (12.7) has been implemented in a simple MATLAB (CTRL–C) function:

```
// [z] = QADD(x,y)
   for i = 1 : 3, ...
     z(i) = TADD(x(i),y(i)); ...
   end
return
```

Thus, two qualitative vectors can be added as follows:

$$z = QADD(x, y) \tag{12.8}$$

Similarly, we can define qualitative subtraction and multiplication operators with the following truth tables:

$$SUB = \begin{array}{c} x \backslash y \\ - \\ 0 \\ + \\ ? \end{array} \begin{array}{cccc} - & 0 & + & ? \\ \begin{pmatrix} ? & - & - & ? \\ + & 0 & - & ? \\ + & + & ? & ? \\ ? & ? & ? & ? \end{pmatrix} \end{array} \tag{12.9}$$

and:

$$MULT = \begin{array}{c} x \backslash y \\ - \\ 0 \\ + \\ ? \end{array} \begin{array}{cccc} - & 0 & + & ? \\ \begin{pmatrix} + & 0 & - & ? \\ 0 & 0 & 0 & 0 \\ - & 0 & + & ? \\ ? & 0 & ? & ? \end{pmatrix} \end{array} \tag{12.10}$$

In the multiplication, the corresponding vector function looks a little different since:

$$\text{if } z = x \cdot y$$
$$\rightarrow \dot{z} = x \cdot \dot{y} + \dot{x} \cdot y \tag{12.11}$$
$$\rightarrow \ddot{z} = x \cdot \ddot{y} + 2 \cdot \dot{x} \cdot \dot{y} + \ddot{x} \cdot y$$

which has been implemented as:

```
// [z] = CQMULT(x, y)
  z(1) = TMULT(x(1), y(1));
  z(2) = TADD(TMULT(x(1), y(2)), TMULT(x(2), y(1)));
  aux1 = TADD(TMULT(x(1), y(3)), TMULT(x(2), y(2)));
  z(3) = TADD(aux1, TMULT(x(3), y(1)));
return
```

The constant factor of 2 in the second derivative equation doesn't appear in the qualitative equations since $2 \cdot x$ is positive exactly if x is positive, etc. Thus, multiplication of a qualitative variable with a positive constant is a *do–nothing* operation.

The truth table for the qualitative division operator is not completely obvious. It has been defined as follows:

$$
\text{DIV} = \begin{array}{c} {}_{x}\backslash^{y} \\ - \\ 0 \\ + \\ ? \end{array} \begin{array}{cccc} - & 0 & + & ? \\ \left(\begin{array}{cccc} + & \sqcup & - & ? \\ 0 & ? & 0 & 0 \\ - & \sqcup & + & ? \\ ? & \sqcup & ? & ? \end{array} \right) \end{array} \tag{12.12}
$$

If you are interested to know what the corresponding $TDIV$ matrix and CQDIV function look like, solve Hw.[H12.1].

I already showed that multiplication of a qualitative variable with a positive constant is a do–nothing operation. Multiplication with a negative constant is the same as the negation of the qualitative variable. This can be written as:

$$ y = \text{QSUB}(CZERO, x) \tag{12.13a} $$

where $CZERO$ is a *constant* 0, i.e., the CTRL–C (MATLAB) vector:

$$ CZERO = [2, 2, 2] \tag{12.13b} $$

However, this has also been implemented as:

$$ y = \text{QMINUS}(x) \tag{12.14a} $$

which makes use of the qualitative negation vector $TMINUS$:

$$ TMINUS = [3, 2, 1, 4] \tag{12.14b} $$

Similarly to $CZERO$, also *constant* $+$ and *constant* $-$ vectors have been defined:

$$CPLUS = [3, 2, 2] \qquad (12.15a)$$
$$CMINUS = [1, 2, 2] \qquad (12.15b)$$

12.5 Functions of Qualitative Variables

In general, it is not so easy to define qualitative versions of functions. For example:

$$y = \sin(x) \qquad (12.16)$$

doesn't have a decent qualitative equivalent since the sign of y will depend on the actual quantitative value of x. However, some functions can be properly converted, for example:

$$y = CQEXP(x) \qquad (12.17)$$

can be easily defined since:

$$
\begin{aligned}
&\text{if } y = \exp(x) \\
&\rightarrow \dot{y} = \exp(x) \cdot \dot{x} \\
&\rightarrow \ddot{y} = \exp(x) \cdot (\ddot{x} + \dot{x}^2)
\end{aligned}
\qquad (12.18)
$$

and since $\exp(x)$ is positive for all values of x, we can define the CQEXP function as follows:

```
// [y] = CQEXP(x)
  y(1) = 3;
  y(2) = x(2);
  y(3) = TADD(x(3), TMULT(x(2), x(2)));
return
```

Other functions that are very useful are the M^+ and M^- functions. $M^+(x)$ is any monotonically increasing zero–crossing function of x, i.e.:

```
// [y] = MPLUS(x)
  y(1) = x(1);
  y(2) = x(2);
  y(3) = 4;
return
```

and $M^-(x)$ is any monotonically decreasing zero–crossing function of x, i.e.:

$$// \ [y] = MMINUS(x)$$
$$y(1) = TMINUS(x(1));$$
$$y(2) = TMINUS(x(2));$$
$$y(3) = 4;$$
return

As in the case of the exponential, we must apply the chain rule, since $y(2)$ denotes the time derivative of $y(1)$, not the derivative of $y(1)$ with respect to x.

12.6 Qualitative Simulation

Now we are ready to perform a qualitative simulation. Let me explain by means of an example how this works. Figure 12.1 shows a simple RC circuit:

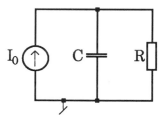

Figure 12.1. Simple RC circuit.

We wish to study the behavior of this circuit for a step function applied to the current source, assuming that the initial voltage across the capacitor is negative. This problem had originally been proposed by Williams [12.16].

A quantitative state–space description for this problem can be given as follows:

$$\dot{x} = \frac{-1}{R \cdot C} x + \frac{1}{C} u \qquad (12.19a)$$

$$y = x \qquad (12.19b)$$

where the input u is the current through the source and the output y is the voltage across the capacitor. The qualitative version of this state–space description is as follows:

$$// \ [xdot] = CQSTATE(x)$$
$$u = CPLUS;$$
$$xdot = QSUB(u, x);$$
$$\textbf{return}$$

and:

$$// \ [y] = QOUT(x)$$
$$y = x;$$
$$\textbf{return}$$

where CQSTATE models the *continuous qualitative state equations*, and QOUT models the *qualitative output equations*.

The initial conditions can be specified as:

$$// \ [x] = CQSTATEIC(dummy)$$
$$x = IMINUS;$$
$$\textbf{return}$$

The vector *IMINUS* denotes an *initial negative qualitative value*:

$$IMINUS = [1, 4, 4] \tag{12.20}$$

We happen to know that the initial capacitor voltage is negative, but we don't claim any knowledge about the initial value of the derivative of this voltage (remember: 4 denotes ?). Corresponding vectors have been defined for *IZERO* and *IPLUS*.

We can assign the initial value to x by calling CQSTATEIC. Thereafter, we can call CQSTATE to compute the initial value of the derivative. We find:

$$
\begin{array}{ccc}
t & x & \dot{x} \\
0 & ((-\ ?\ ?) & (+\ ?\ ?))
\end{array}
$$

However, since $x(2)$ must be equal to $\dot{x}(1)$, we can replace one of our question marks and iterate once more:

$$
\begin{array}{ccc}
t & x & \dot{x} \\
0 & ((-\ ?\ ?) & (+\ ?\ ?) \\
0 & (-\ +\ ?) & (+\ -\ ?))
\end{array}
$$

We repeat the same process a second time:

$$
\begin{array}{ccc}
t & x & \dot{x} \\
0 \\
0 \\
0
\end{array}
\left(
\begin{array}{cc}
(- \; ? \; ?) & (+ \; ? \; ?) \\
(- + \; ?) & (+ - \; ?) \\
(- + -) & (+ - +)
\end{array}
\right)
$$

and by now all question marks have disappeared. The relation between x and $\dot{x}$ is called a *consistency constraint*, and the mechanism to ensure consistency among all such relations is called the process of *constraint propagation* [12.7].

In QualSim, this has been implemented as follows:

```
ind = 1;
while ind = 1, ...
   xdot =CQSTATE(x); ...
   [x, ind] =ASSERT(x, xdot); ...
end
```

QualSim's ASSERT function propagates all constraints that apply to the x vector. If the x vector is modified by ASSERT, *ind* is returned as 1; if all constraints were satisfied without need to change x any further, *ind* is returned as 0; and if a constraint led to a conflict, *ind* is returned as -1.

Now, we are ready to *qualitatively integrate* the state vector to its next value, incrementing the *qualitative clock* to 1 [12.5]. Here, my methodology diverges from Morgan's [12.14]. I use a *qualitative forward Euler* algorithm to integrate the state equations. The quantitative version of this algorithm can be written as:

$$x_{k+1} = x_k + \Delta t \cdot \dot{x}_k \tag{12.21}$$

and its qualitative counterpart can thus be written as:

$$x_{k+1} = \text{QADD}(x_k, \dot{x}_k) \tag{12.22}$$

which has been made available as:

$$x_{k+1} = \text{QINT}(x_k, \dot{x}_k) \tag{12.23}$$

QINT is a little more general than QADD since it can integrate an entire qualitative state vector, whereas QADD operates on only one single qualitative variable (which is in itself a vector of length three). In our example, these two functions are identical since we analyze a first–order system.

The result of this integration is:

$$
\begin{array}{ccc}
t & x & \dot{x} \\
0 & (-\ ?\ ?) & (+\ ?\ ?) \\
0 & (-\ +\ ?) & (+\ -\ ?) \\
0 & (-\ +\ -) & (+\ -\ +) \\
1 & (?\ ?\ ?) & (?\ ?\ ?)
\end{array}
$$

Unfortunately, our previous method of constraint propagation will not help us any further at this point.

We need to *expand* our search, i.e., replace the ?s by all possible combinations of $-$, 0, and $+$. However, here we can apply a new set of constraints that we call *continuity constraints*. The first continuity constraint was mentioned previously. If x_k was $+$, x_{k+1} can only be $+$ or 0, but never $-$, since no variable can jump from $+$ to $-$ without passing through 0 on the way. Of course, the same holds true for $\dot{x}_k$ and $\ddot{x}_k$. But more continuity constraints exist. For example, if x_k was $+$ and $\dot{x}_k$ was either $+$ or 0, then x_{k+1} must be $+$ and cannot be 0. The constant matrix $TCONT$ tests for all continuity constraints:

$TCONT =$

$_l\backslash^n$	$--$	-0	$-+$	$-?$	$0-$	00	$0+$	$0?$	$+-$	$+0$	$++$	$+?$	$?-$	$?0$	$?+$	$??$
$--$	0	0	1	0	1	1	1	1	1	1	1	1	0	0	1	0
-0	0	0	0	0	1	1	1	1	1	1	1	1	0	0	0	0
$-+$	1	0	0	0	1	*	0	0	1	1	1	1	1	0	0	0
$-?$	0	0	0	0	1	1	0	0	1	1	1	1	0	0	0	0
$0-$	0	1	1	0	1	1	1	1	1	1	1	1	0	1	1	0
00	1	1	1	1	0	0	0	0	1	1	1	1	0	0	0	0
$0+$	1	1	1	1	1	1	1	1	1	1	0	0	1	1	0	0
$0?$	0	1	1	0	0	0	0	0	1	1	0	0	0	0	0	0
$+-$	1	1	1	1	0	*	1	0	0	0	1	0	0	0	1	0
$+0$	1	1	1	1	1	1	1	1	0	0	0	0	0	0	0	0
$++$	1	1	1	1	1	1	1	1	1	0	0	0	1	0	0	0
$+?$	1	1	1	1	0	1	1	0	0	0	0	0	0	0	0	0
$?-$	0	0	1	0	0	0	1	0	0	0	1	0	0	0	1	0
$?0$	0	0	0	0	0	0	0	0	0	0	0	0	0	0	0	0
$?+$	1	0	0	0	1	0	0	0	1	0	0	0	1	0	0	0
$??$	0	0	0	0	0	0	0	0	0	0	0	0	0	0	0	0

$$(12.24)$$

The left column shows combinations of either x_k and $\dot{x}_k$ or $\dot{x}_k$ and $\ddot{x}_k$. The top column shows the same combinations one time step into the future. A 0 element in the $TCONT$ matrix denotes a feasible transition, whereas a 1 element denotes a transition that is forbidden due to a conflict with one of the continuity constraints. The two elements marked by * are usually 1 except at time ∞ when they become 0 (to allow a smooth transition into steady–state).

QualSim's ASSERT function can operate on a single state vector x and its derivative vector $xdot$, but if x is a matrix with two rows, then $x(2,:)$ denotes the current state vector, while $x(1,:)$ is the last state vector. In this case, ASSERT will test for the continuity constraints in the following way:

$$c1 = 4 * xL(1) + xL(2) - 4;$$
$$c2 = 4 * x(1) + x(2) - 4;$$
$$ind = -TCONT(c1, c2);$$
if $ind = -1$, **return**, **end**
$$c1 = 4 * xL(2) + xL(3) - 4;$$
$$c2 = 4 * x(2) + x(3) - 4;$$
$$ind = -TCONT(c1, c2);$$
if $ind = -1$, **return**, **end**

where x denotes the current state variable, while xL denotes the state variable one time step back. $4 \cdot x + \dot{x} - 4$ maps the qualitative tuples into numbers between 1 and 16, which allow us to use the same trick as before, employing a simple (and fast) table look–up.

Of all 27 combinations of $-$, 0, and $+$ that would result from the three ?s in our state vector, only two combinations survive the propagation of the continuity and consistency constraints. These are:

$$
\begin{array}{c|cc}
t & x & \dot{x} \\
0 & (-\ ?\ ?) & (+\ ?\ ?) \\
0 & (-\ +\ ?) & (+\ -\ ?) \\
0 & (-\ +\ -) & (+\ -\ +) \\
1 & (?\ ?\ ?) & (?\ ?\ ?) \\
1a & (-\ +\ -) & (+\ -\ +) \\
1b & (0\ +\ -) & (+\ -\ +)
\end{array}
$$

Let us look at solution $1a$. We see that the state $1a$ is identical with the state 0. Here comes yet another constraint to play that we call the *steady–state constraint*. Obviously, since state $1a$ is identical to state 0, this state could be repeated forever. Is this physical? The answer is no. If $x = -$, and $\dot{x} = +$, then x must eventually become 0. This may take many steps, possibly infinitely many steps, but it must eventually happen. Thus, we can shrink this sequence into one single step and replace the x of state $1a$ by $x = (0\ +\ -)$, which we recognize as state $1b$. Thus, we can eliminate state $1a$ without loss of generality:

$$
\begin{array}{c c c}
t & x & \dot{x} \\
0 & (-\,?\,?) & (+\,?\,?) \\
0 & (-\,+\,?) & (+\,-\,?) \\
0 & (-\,+\,-) & (+\,-\,+) \\
1 & (?\,?\,?) & (?\,?\,?) \\
1a & (-\,+\,-) & (+\,-\,+) \\
1b & (0\,+\,-) & (+\,-\,+) \\
1 & (0\,+\,-) & (+\,-\,+)
\end{array}
$$

Now, we integrate again and find:

$$
\begin{array}{c c c}
t & x & \dot{x} \\
0 & (-\,+\,-) & (+\,-\,+) \\
1 & (0\,+\,-) & (+\,-\,+) \\
2 & (+\,?\,?) & (?\,?\,?)
\end{array}
$$

Only one solution survives the constraint propagation:

$$
\begin{array}{c c c}
t & x & \dot{x} \\
0 & (-\,+\,-) & (+\,-\,+) \\
1 & (0\,+\,-) & (+\,-\,+) \\
2 & (+\,+\,-) & (+\,-\,+)
\end{array}
$$

Integration leads to:

$$
\begin{array}{c c c}
t & x & \dot{x} \\
0 & (-\,+\,-) & (+\,-\,+) \\
1 & (0\,+\,-) & (+\,-\,+) \\
2 & (+\,+\,-) & (+\,-\,+) \\
3 & (+\,?\,?) & (?\,?\,?)
\end{array}
$$

and constraint propagation leads to:

$$
\begin{array}{c c c}
t & x & \dot{x} \\
0 & (-\,+\,-) & (+\,-\,+) \\
1 & (0\,+\,-) & (+\,-\,+) \\
2 & (+\,+\,-) & (+\,-\,+) \\
3 & (+\,+\,-) & (+\,-\,+)
\end{array}
$$

Thus, we need to apply the steady–state constraint again. Since $\dot{x} = +$ and $\ddot{x} = -$, $\dot{x}$ must become 0 after some time. Thus, we replace state 3 by:

$$
\begin{array}{c c c}
t & x & \dot{x} \\
0 & (-\,+\,-) & (+\,-\,+) \\
1 & (0\,+\,-) & (+\,-\,+) \\
2 & (+\,+\,-) & (+\,-\,+) \\
3 & (+\,0\,-) & (0\,-\,+)
\end{array}
$$

However, this state is in conflict with a consistency constraint. The second derivative must be 0, and therefore, the third derivative must also be 0:

$$
\begin{array}{c}
\begin{array}{ccc} t & x & \dot{x} \end{array} \\
\begin{array}{c} 0 \\ 1 \\ 2 \\ 3 \end{array}
\left(
\begin{array}{cc}
(-+-) & (+-+) \\
(0\ +-) & (+-+) \\
(++-) & (+-+) \\
(+\ 0\ \ 0) & (0\ \ 0\ \ 0)
\end{array}
\right)
\end{array}
$$

Integration leads to:

$$
\begin{array}{c}
\begin{array}{ccc} t & x & \dot{x} \end{array} \\
\begin{array}{c} 0 \\ 1 \\ 2 \\ 3 \\ 4 \end{array}
\left(
\begin{array}{cc}
(-+-) & (+-+) \\
(0\ +-) & (+-+) \\
(++-) & (+-+) \\
(+\ 0\ \ 0) & (0\ \ 0\ \ 0) \\
(+\ 0\ \ 0) & (0\ \ 0\ \ 0)
\end{array}
\right)
\end{array}
$$

We apply the steady–state constraint once more, but this time, everything is okay since 4 is indeed a valid steady–state solution.

QualSim provides a function:

$$[y, x] = \text{CQSIM}(nstp) \tag{12.25}$$

that computes the *continuous qualitative simulation* of the system described by the user–coded procedure files *CQSTATE.CTR*, *CQSTATEIC.CTR*, and *QOUT.CTR* (for CTRL–C; in MATLAB, the extensions are .*M*) over *nstp* clock steps, or until all branches have reached their steady–state values, whichever comes first. In our example, *y* will be returned as:

$$
y =
\left(
\begin{array}{ccc}
1 & 3 & 1 \\
2 & 3 & 1 \\
3 & 3 & 1 \\
3 & 2 & 2 \\
-1 & 0 & 0
\end{array}
\right)
\tag{12.26}
$$

The last row indicates that we have reached a *continuous steady–state*. The function:

$$\text{QPLOT}(y) \tag{12.27}$$

will produce the graph shown in Fig.12.2,

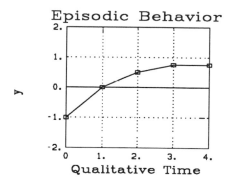

Figure 12.2. Episode of the RC circuit.

which is indeed the only possible physical solution to this problem. From the quantitative solution, we know that the transition from state 2 to state 3 requires infinitely many steps.

Notice that Morgan's [12.14] and my solutions don't exhibit the well–known "stuttering" behavior that has been observed by Forbus [12.10] and many others.

Encouraged by these nice results, let us try a somewhat more difficult problem:

$$\dot{x}_1 = +a \cdot x_2 \tag{12.28a}$$
$$\dot{x}_2 = -b \cdot x_1 \tag{12.28b}$$

which can be rewritten as:

$$\ddot{x}_1 = -(a \cdot b) \cdot x_1 = -\omega^2 \cdot x_1 \tag{12.29}$$

with the analytical solution:

$$x_1(t) = A \cdot \sin(\omega t) \tag{12.30a}$$
$$x_2(t) = (A\omega) \cdot \cos(\omega t) \tag{12.30b}$$

Let us check whether our qualitative simulation algorithm is able to reproduce this solution for us.

This problem is described in QualSim using the qualitative state–space model:

```
// [xdot] = CQSTATE(x)
   x1 = x(1 : 3);
   x2 = x(4 : 6);
   x1dot = x2;
   x2dot =QMINUS(x1);
   xdot = [x1dot, x2dot];
return
```

It becomes evident that x is indeed the entire state vector that must first be unpacked. The qualitative output equations are coded as:

```
// [y] = QOUT(x)
   y = x;
return
```

Since we wish to look at both state variables, we have no reason to unpack x in the first place. The initial conditions are assigned as follows:

```
// [x] = CQSTATEIC(dummy)
   x1 = IZERO;
   x2 = IPLUS;
   x = [x1, x2];
return
```

Thus, we start out with the following initial state:

$$
\begin{array}{ccccc}
t & x_1 & x_2 & \dot{x}_1 & \dot{x}_2 \\
0 & ((0\ ?\ ?) & (+\ ?\ ?) & (+\ ?\ ?) & (0\ ?\ ?))
\end{array}
$$

which is consolidated by propagating the consistency constraints to:

$$
\begin{array}{ccccc}
t & x_1 & x_2 & \dot{x}_1 & \dot{x}_2 \\
0 & ((0\ +\ 0) & (+\ 0\ -) & (+\ 0\ -) & (0\ -\ 0))
\end{array}
$$

This can be integrated twice:

$$
\begin{array}{ccccc}
t & x_1 & x_2 & \dot{x}_1 & \dot{x}_2 \\
0 & (0\ +\ 0) & (+\ 0\ -) & (+\ 0\ -) & (0\ -\ 0) \\
1 & (+\ +\ -) & (+\ -\ -) & (+\ -\ -) & (-\ -\ +) \\
2 & (+\ ?\ -) & (?\ -\ ?) & (?\ -\ ?) & (-\ ?\ +)
\end{array}
$$

The state vector contains three ?s. Thus, we need to consider 27 different combinations of $-$, 0, and $+$. Only eight of these 27 combinations survive the continuity constraints; only two of the eight survive the consistency constraints; and only one of the two survives the steady–state constraints:

t	x_1	x_2	$\dot{x}_1$	$\dot{x}_2$
0	(0 + 0)	(+ 0 −)	(+ 0 −)	(0 − 0)
1	(+ + −)	(+ − −)	(+ − −)	(− − +)
2	(+ 0 −)	(0 − 0)	(0 − 0)	(− 0 +)

If you are interested in studying the details of this process, solve Hw.[H12.2]. Now, we can integrate again over two steps, and we find:

t	x_1	x_2	$\dot{x}_1$	$\dot{x}_2$
0	(0 + 0)	(+ 0 −)	(+ 0 −)	(0 − 0)
1	(+ + −)	(+ − −)	(+ − −)	(− − +)
2	(+ 0 −)	(0 − 0)	(0 − 0)	(− 0 +)
3	(+ − −)	(− − +)	(− − +)	(− + +)
4	(? − ?)	(− ? +)	(− ? +)	(? + ?)

Again, we have three ?s, but as before, only one survives the process of constraint propagation. We can continue in the same manner over another four steps and find:

t	x_1	x_2	$\dot{x}_1$	$\dot{x}_2$
0	(0 + 0)	(+ 0 −)	(+ 0 −)	(0 − 0)
1	(+ + −)	(+ − −)	(+ − −)	(− − +)
2	(+ 0 −)	(0 − 0)	(0 − 0)	(− 0 +)
3	(+ − −)	(− − +)	(− − +)	(− + +)
4	(0 − 0)	(− 0 +)	(− 0 +)	(0 + 0)
5	(− − +)	(− + +)	(− + +)	(+ + −)
6	(− 0 +)	(0 + 0)	(0 + 0)	(+ 0 −)
7	(− + +)	(+ + −)	(+ + −)	(+ − −)
8	(0 + 0)	(+ 0 −)	(+ 0 −)	(0 − 0)

At this point, another steady–state constraint comes to play. Comparing state 8 with state 0, we find that those two states are exactly the same. Thus, we have identified a *periodic steady–state*, i.e., a limit cycle.

QualSim's qualitative plot function QPLOT generates the curves shown in Fig.12.3.

At this point in time, it makes sense to discuss how QPLOT works. The function:

$$y_p = \text{QPLOT}(y) \qquad (12.31)$$

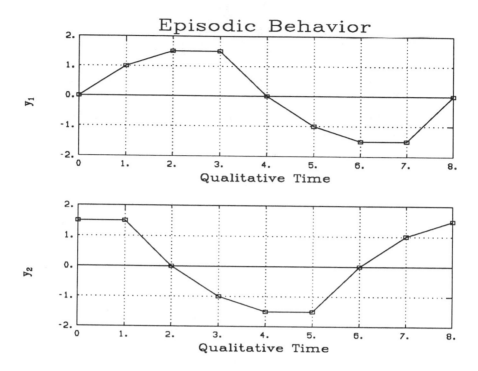

Figure 12.3. Episode of the sine wave model.

computes a qualitative output episode y_p on the basis of the qualitative output vector y using a set of simple rules such as:

$$\textbf{if } (y_k = +) \cap (\dot{y}_k = +) \cap (\ddot{y}_k = -)$$
$$\textbf{then } y_{p_{k+1}} = y_{p_k} + 0.5(y_{p_k} - y_{p_{k-1}})$$
$$\textbf{end} \tag{12.32}$$

with slight modifications for the startup period. At the end, QPLOT may have to iterate once more on the startup section of the episode to consolidate periodic steady–state conditions. This was required for the qualitative plot of Fig.12.3. In addition to computing the episode, QPLOT produces a graph. If you wish to learn more details about QPLOT, solve Hw.[H12.3].

It is interesting to note that this is the example for which Kuipers obtained additional spurious solutions that are nonphysical [12.12]. He claimed that the qualitative episode cannot distinguish between the true periodic sinusoidal solution and solutions of the types shown in Fig.12.4:

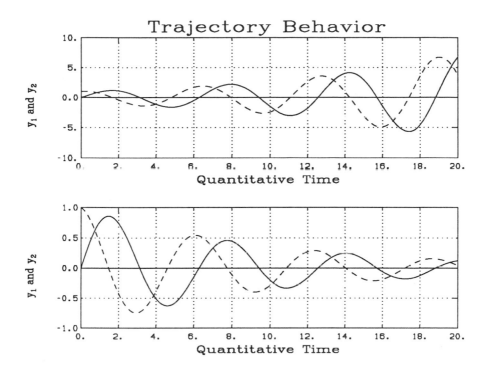

Figure 12.4. Trajectory behavior of unstable and stable oscillations.

Indeed, if we look at x_1 and x_2 or y and $\dot{y}$ alone, we can't tell the difference in terms of the qualitative variables. Yet if we include higher derivatives (as QualSim does), we can tell the difference. Our qualitative output vector $\boldsymbol{y}$ suggests that $\ddot{y}$ crosses through zero at exactly the same time when y crosses through zero, and $\dddot{y}$ crosses through zero exactly when $\dot{y}$ crosses through zero. This is only true for the periodic solution. In the two cases of Fig.12.4, one zero–crossing always lags behind or leads the other. Thus, the problems:

$$\ddot{y} - 2\sigma\omega\dot{y} + \omega^2 y = 0 \qquad (12.33a)$$
$$\ddot{y} + 2\sigma\omega\dot{y} + \omega^2 y = 0 \qquad (12.33b)$$

which will produce the trajectory behaviors shown in Fig.12.4, will lead to slightly different qualitative output vectors. If you don't believe me, solve Hw.[H12.4].

Unfortunately, QualSim is not yet smart enough to recognize the implications of the leading (lagging) derivatives. It will still suggest that the solution is periodic, only with a somewhat longer clock period (which has nothing to do with the quantitative period in the time domain). Well, we can't win them all.

What this example illustrates is the following. The inclusion of higher derivatives increases the computation time needed for the qualitative simulation. It may also lead to additional branches, branches that differ only in their higher derivatives. Yet it helps to filter out bogus solutions. For most practical purposes, the inclusion of the second derivative (i.e., the representation of the qualitative vectors through triples) is optimal.

Morgan suggested adding to the qualitative simulator a qualitative stability analyzer using the influence diagram technique described in Chapter 11 [12.14]. In this particular example, this technique would indeed work.

Let us now look at yet another example, which will exhibit true branching. The step response of the problem:

$$\dot{x}_1 = x_2 \tag{12.34a}$$
$$\dot{x}_2 = x_3 \tag{12.34b}$$
$$\dot{x}_3 = -c \cdot x_1 - b \cdot x_2 - a \cdot x_3 + u \tag{12.34c}$$
$$y = x_1 \tag{12.34d}$$

which can be expressed in matrix notation as follows:

$$\dot{\mathbf{x}} = \mathbf{A} \cdot \mathbf{x} + \mathbf{b} \cdot u$$
$$= \begin{pmatrix} 0 & 1 & 0 \\ 0 & 0 & 1 \\ -c & -b & -a \end{pmatrix} \cdot \mathbf{x} + \begin{pmatrix} 0 \\ 0 \\ 1 \end{pmatrix} \cdot u \tag{12.34e}$$
$$y = \mathbf{c}' \cdot \mathbf{x} + d \cdot u$$
$$= (1 \quad 0 \quad 0) \cdot \mathbf{x} + 0 \cdot u \tag{12.34f}$$

can be found quantitatively for different values of the positive constants a, b, and c. Figure 12.5 shows the trajectory behaviors resulting from arbitrary selections of valid combinations of the three parameters.

It is fairly easy to show that the four different types of behaviors shown in Fig.12.5 are indeed the only possible ones. We can look at the system analytically. The characteristic polynomial of this system is:

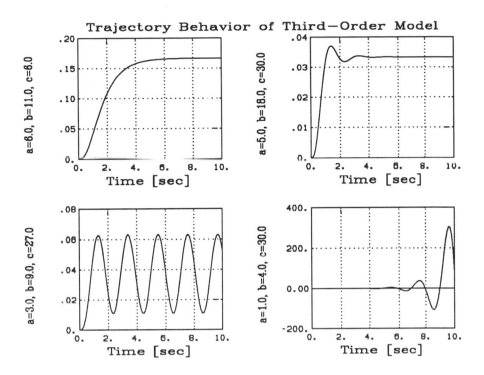

Figure 12.5. Trajectory behavior for various parameter combinations.

$$det(\lambda \cdot \mathbf{I}^{(n)} - \mathbf{A}) = \lambda^3 + a \cdot \lambda^2 + b \cdot \lambda + c = 0.0 \qquad (12.35)$$

The roots of the characteristic polynomial are the eigenmodi of the system. We can analyze the stability of the system by setting up a Routh–Hurwitz scheme, as shown in Fig.12.6.

S^3	1	b
S^2	a	c
S^1	$\dfrac{c - ab}{a}$	
S^0	c	

Figure 12.6. Routh–Hurwitz scheme for third–order model.

How the Routh–Hurwitz scheme is constructed is of no importance to our endeavor. Any introductory text on control theory (such as [12.9]) describes the procedure in detail. If all the elements in the first column of the Routh–Hurwitz scheme are positive, the system is stable, i.e., all three roots are in the left half λ plane. Otherwise, the number of sign changes in the first column of the Routh–Hurwitz scheme determines the number of unstable poles, i.e., the poles in the right half λ plane. It can be seen that, if $c > a \cdot b$, all three poles are in the left half plane. They can all be located on the negative real axis (exponential decay) or one can be located on the negative real axis while the other two form a conjugate complex pole pair (damped oscillation). If $c = a \cdot b$, one pole is still on the negative real axis, while the two dominant poles are now on the imaginary axis itself (undamped oscillation). If $c < a \cdot b$, one pole is still on the negative real axis, while the other two poles form a conjugate complex pole pair in the right half plane (excited oscillation). These are indeed all possible cases (for positive coefficients a, b, and c), as the following analysis shows.

Consider the case of the excited oscillation with:

$$\lambda_1 = -x \qquad\qquad (12.36a)$$
$$\lambda_2 = y + j \cdot z \qquad\qquad (12.36b)$$
$$\lambda_3 = y - j \cdot z \qquad\qquad (12.36c)$$

In this case, we can write the characteristic polynomial as:

$$(\lambda + x) \cdot (\lambda - y - j \cdot z) \cdot (\lambda - y + j \cdot z) = 0.0 \qquad (12.37)$$

which can be rewritten as:

$$\lambda^3 + (x - 2y)\lambda^2 + (y^2 + z^2 - 2xy)\lambda + x(y^2 + z^2) = 0.0 \qquad (12.38)$$

which has positive coefficients iff:

$$a = x - 2y > 0.0 \qquad\qquad (12.39a)$$
$$b = y^2 + z^2 - 2xy > 0.0 \qquad\qquad (12.39b)$$
$$c = x(y^2 + z^2) > 0.0 \qquad\qquad (12.39c)$$

whereby x, y, and z are all positive. The third condition is obviously always satisfied. The first condition can be rewritten as:

$$x > 2y \tag{12.40}$$

or:

$$2xy > 4y^2 \tag{12.41}$$

which we can plug into the second condition:

$$y^2 + z^2 > 2xy > 4y^2 \tag{12.42}$$

i.e., the borderline case can be determined as:

$$z^2 = 3y^2 \tag{12.43}$$

and therefore:

$$z = \sqrt{3} \cdot y \tag{12.44}$$

This is shown in Fig.12.7, in which the possible pole locations for positive coefficients a, b, and c are shaded.

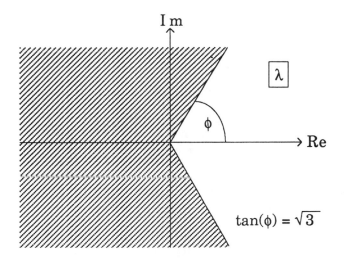

Figure 12.7. Pole locations for positive parameter combinations.

Therefore, it has been verified that this system does not exhibit any other modes of operation than the four modes shown in Fig.12.5.

Qualitative stability analysis will fail on this problem. Figure 12.8 shows the influence diagram of this set of equations.

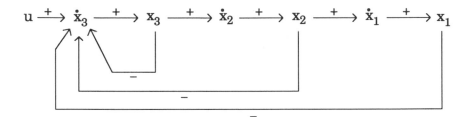

Figure 12.8. Influence diagram of third–order system.

The influence diagram shows three loops. All loops are negative feed-back loops. Thus, according to the literature on influence diagrams (such as [12.6]), this system would have to be stable. However, as the previous analysis and one of the sample runs show, this is not necessarily the case. The problem arises because one of the three feedback loops involves more than two integrators.

What does QualSim make of this problem? The problem can be specified as follows:

$$// \ [xdot] = CQSTATE(x)$$
$$x1 = x(1:3);$$
$$x2 = x(4:6);$$
$$x3 = x(7:9);$$
$$u = CPLUS;$$
$$x1dot = x2;$$
$$x2dot = x3;$$
$$aux = QADD(QADD(x1, x2), x3);$$
$$x3dot = QSUB(u, aux);$$
$$xdot = [x1dot, x2dot, x3dot];$$
$$\textbf{return}$$

Notice that CTRL–C (MATLAB) functions can be called recursively. The qualitative output equations are coded as:

$$// \ [y] = QOUT(x)$$
$$y = x(1:3);$$
$$\textbf{return}$$

The initial conditions are assigned as follows:

$$// \ [x] = CQSTATEIC(dummy)$$
$$x1 = IZERO;$$
$$x2 = IZERO;$$
$$x3 = IZERO;$$
$$x = [x1, x2, x3];$$
$$\textbf{return}$$

Already at step 2, two different solutions survive all constraints:

$$
\begin{array}{cccc}
t & x_1 & x_2 & x_3 & y \\
0 & (0\ 0\ 0) & (0\ 0\ +) & (0\ +\ -) & (0\ 0\ 0) \\
1 & (0\ 0\ +) & (0\ +\ +) & (+\ +\ -) & (0\ 0\ +) \\
2a & (0\ +\ +) & (+\ +\ +) & (+\ +\ -) & (0\ +\ +) \\
2b & (0\ +\ +) & (+\ +\ 0) & (+\ 0\ -) & (0\ +\ +)
\end{array}
$$

Notice that the two solutions are not different with respect to the output y. However, since the naïve physics modeling technique operates on the true system structure, we cannot reduce the behavior at this time. In fact, the branching factor is much larger than four in terms of all state variables. This makes the qualitative simulation quite a slow process. Only after the qualitative state vectors have been found are the qualitative output vectors computed (by calling QOUT). In this process, it may happen that several branches (logical columns) of the output are identical and can thus be merged into one. CQSIM will do this for us. It can also happen that two consecutive time steps (rows) of the output are identical. In that case, they will also be merged into one step, but only if the two rows are identical over all episodes.

This example shows that the naïve physics approach to qualitative simulation can lead to a formidable number of episodes. Already after seven steps, QualSim exhibits 29 different output episodes (and many more state episodes), and during the 10th step, QualSim crashes since CTRL–C runs out of stack space. MATLAB doesn't fare any better. Obviously, this approach is successful only in the simplest cases.

Tony Morgan had an excellent idea how to overcome this problem [12.15]. Rather than trying to compute all episodes explicitly, he suggested that all legal qualitative states of the system be determined and then be qualitatively integrated (or differentiated) over one step in order to find all legal successors of each of these legal states (which must, of course, also be among the legal states of the system).

Let me explain this concept by means of a second–order system with damping. In phase variables, this problem can be described as:

```
// [xdot] = CQSTATE(x)
  x1 = x(1 : 3);
  x2 = x(4 : 6);
  x1dot = x2;
  x2dot = QSUB(QMINUS(x1), x2);
  xdot = [x1dot, x2dot];
return
```

As in the case of the third–order system, this problem cannot be solved using the direct exploration of all episodes due to abundant branching.

We can evaluate all legal states by declaring the initial states as unknown. In this way, QualSim will determine the set of all feasible initial states, which is, of course, identical to the set of legal states. In QualSim, this can be accomplished in the following way:

```
// [x] = CQSTATEIC(dummy)
x1 = IWHAT;
x2 = IWHAT;
x = [x1, x2];
return
```

The global variable $IWHAT$ sets the qualitative vector to (? ? ?) (initial unknown). The global variable $CWHAT$ sets the qualitative vector to (? 0 0) (constant unknown).

The QualSim statement:

$$[> \quad [y, x] = \text{CQSIM}(1)$$

computes the set of legal states of this system, since it will accept all initial states that are consistent in themselves, i.e., that satisfy the consistency constraints.

Similarly, the statement:

$$[> \quad [y, x] = \text{CQSIM}(2)$$

computes the set of legal states and their immediate successors.

QualSim allows us to select the length of the qualitative vector. If we choose a length of two, i.e., if we operate on the state variables and their first derivatives only, we can still represent the output through a qualitative vector of length three since the system is specified in phase variables, thus:

$$\ddot{y} = \ddot{x}_1 = \dot{x}_2 \tag{12.45}$$

The results of the analysis can be summarized, as shown in Table 12.2.

Table 12.2. State Transition Table of Second–Order System.

State #	State	Successor State #s
(1)	(− − +)	(2)
(2)	(− 0 +)	(5)
(3)	(− + −)	(4), (8)
(4)	(− + 0)	(3), (8)
(5)	(− + +)	(4)
(6)	(0 − +)	(1)
(7)	(0 0 0)	(7)
(8)	(0 + −)	(13)
(9)	(+ − −)	(10)
(10)	(+ − 0)	(6), (11)
(11)	(+ − +)	(6), (10)
(12)	(+ 0 −)	(9)
(13)	(+ + −)	(12)

This is a so–called *finite state transition table*. While the information provided in the state transition table can be extracted from the x matrix generated by the CQSIM function, it is more convenient to use the statement:

$$[> \quad fst = \text{CQPERT}(0)$$

which calls CQSIM, and then postprocesses the data to generate the finite state transition table directly.

Notice that CQPERT requires the system to be specified in *phase variables*. This is necessary since CQPERT makes use of higher derivatives that can be automatically evaluated for systems specified in phase variables, but cannot easily be determined for systems specified in other types of state variables. CQPERT takes a dummy argument.

The state transition table does not include soft transitions, i.e., state transitions that can only occur as $t \to \infty$. These transitions are included in the state transition table when the following statements are issued:

$$[> \quad tcont = tcontsoft;$$
$$[> \quad fst = \text{CQPERT}(0)$$

By default, $tcont = tconthard$. The modified finite state transition table is shown in Table 12.3.

Table 12.3. Soft State Transition Table of Second–Order System.

State #	State	Successor State #s
(1)	$(--+)$	(2)
(2)	$(-0+)$	(5)
(3)	$(-+-)$	(4), (7), (8)
(4)	$(-+0)$	(3), (8)
(5)	$(-++)$	(4)
(6)	$(0-+)$	(1)
(7)	$(0\ 0\ 0)$	(7)
(8)	$(0+-)$	(13)
(9)	$(+--)$	(10)
(10)	$(+-0)$	(6), (11)
(11)	$(+-+)$	(6), (7), (10)
(12)	$(+0-)$	(9)
(13)	$(++-)$	(12)

This state transition table contains two additional transitions: $3 \to 7$ and $11 \to 7$. These transitions can only take place as $t \to \infty$.

It is possible to represent the finite state transition table graphically in the form of a so–called *PERT network*. This is sometimes also called a *finite state machine*. The PERT network of the system is shown in Fig.12.9.

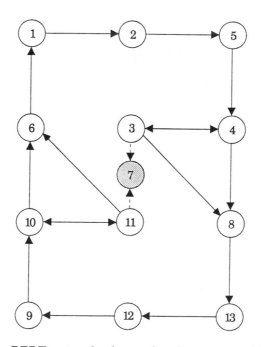

Figure 12.9. PERT network of second–order system with damping.

In a PERT network, each legal state is represented by a labeled node (a numbered circle). Transitions between states are indicated by directed paths. The two dashed paths in Fig.12.9 denote the soft transitions that lead to steady–state.

This method is indeed equivalent to exploring all episodes. Any given initial condition is represented by a legal state. Episodes starting from that initial condition can be found by simply following the directed paths through the PERT network. Whenever we pass through a node from which two or more transitions emanate, the qualitative behavior branches into several alternative episodes.

Two questions of considerable interest are the following: (i) Does this technique generate necessarily all legal episodes; and (ii) Does this technique generate legal episodes exclusively, or is it possible that the methodology sometimes produces bogus episodes that represent nonphysical qualitative behavior.

The first question can be answered affirmatively. By converting our state–space model from a quantitative to a qualitative form, we let go of a few constraints, i.e., we loosen our solution requirements. Obviously, we cannot eliminate correct solutions in this way. The set of qualitative constraints is an *outer envelope* (a majorante) of the set of all physical solutions. Since we explore all solutions that are compatible with the set of qualitative constraints, these must include all the physical solutions of the system.

Unfortunately, the second question cannot be answered affirmatively. Take any conservative mechanical system (i.e., a mechanical system without damping). Due to energy conservation requirements, the sum of all potential and kinetic energies stored in this system must remain constant at all times. In a quantitative simulation, it is not necessary to explicitly formulate this constraint. When we apply the d'Alembert principle to such a system and solve the resulting set of differential equations, we obtain a unique solution that implicitly satisfies the energy constraint. However, when we solve this same problem using qualitative simulation, the energy constraint is not necessarily always satisfied. It therefore makes sense to formulate the energy constraint explicitly in addition to the qualitative state equations since this can help us eliminate spurious solutions. This important result was first reported by Kuipers and Farquhar [12.13]. In QualSim, such additional constraints can be formulated in the user–supplied QCONSTR function:

$$// [ind] = QCONSTR(x, xdot)$$

$$\ldots$$

return

ind should be returned as 0 if all external constraints are satisfied, and it should be returned as 1 if at least one of the external constraints is violated. Function QCONSTR will be called by QualSim whenever the software attempts to assert a proposed state.

Does the PERT network of Fig.12.9 contain any spurious states or state transitions? To answer this question, let us repeat the analysis using qualitative vectors of length three. This allows us to determine the third derivative of the output. The resulting state transition table is shown in Table 12.4.

Table 12.4. Enhanced State Transition Table of Second–Order System.

State #	State	Successor State #s
(1)	$(- - + -)$	(2), (4)
(2)	$(- - + 0)$	(1), (4)
(3)	$(- - + +)$	(2)
(4)	$(- 0 + -)$	(9)
(5)	$(- + - -)$	(6), (14), (15)
(6)	$(- + - 0)$	(7), (16)
(7)	$(- + - +)$	(6), (15), (16)
(8)	$(- + 0 -)$	(5), (14)
(9)	$(- + + -)$	(8)
(10)	$(0 - + -)$	(1), (2)
(11)	$(0 - + 0)$	(1)
(12)	$(0 - + +)$	(2), (3)
(13)	$(0 \ 0 \ 0 \ 0)$	(13)
(14)	$(0 + - -)$	(23), (24)
(15)	$(0 + - 0)$	(25)
(16)	$(0 + - +)$	(24), (25)
(17)	$(+ - - +)$	(18)
(18)	$(+ - 0 +)$	(12), (21)
(19)	$(+ - + -)$	(10), (11), (20)
(20)	$(+ - + 0)$	(10), (19)
(21)	$(+ - + +)$	(11), (12), (20)
(22)	$(+ 0 - +)$	(17)
(23)	$(+ + - -)$	(24)
(24)	$(+ + - 0)$	(22), (25)
(25)	$(+ + - +)$	(22), (24)

We can now delete the third derivative and lump together all states that have become identical. This leaves us with the state transition table presented in Table 12.5.

Table 12.5. Reduced State Transition Table of Second–Order System.

State #	State	Successor State #s
(1)	(− − +)	(2)
(2)	(− 0 +)	(5)
(3)	(− + −)	(8)
(4)	(− + 0)	(3), (8)
(5)	(− + +)	(4)
(6)	(0 − +)	(1)
(7)	(0 0 0)	(7)
(8)	(0 + −)	(13)
(9)	(+ − −)	(10)
(10)	(+ − 0)	(6), (11)
(11)	(+ − +)	(6)
(12)	(+ 0 −)	(9)
(13)	(+ + −)	(12)

The resulting state transition table is identical to that of Table 12.2 except for the two omitted transitions: $3 \to 4$ and $11 \to 10$. These state transitions were obviously nonphysical. They were introduced in the process of qualitative integration using the qualitative forward Euler algorithm. These nonphysical transients are similar in nature to numerical integration errors occurring in a quantitative simulation. Expanding the length of our qualitative vectors from two to three has helped us eliminate some spurious solutions. This was obviously a good idea. While the second–order system exhibited some spurious state transitions, it did not exhibit any nonphysical "legal" states.

Among the remaining transitions, the two paths $4 \to 8$ and $10 \to 6$ are not exactly illegal, but they are boundary cases. In both these transitions, two elements of the qualitative vector change simultaneously. The transition $4 \to 8$ stands for the transition $(- + 0) \to (0 + -)$. However, since $(0 + 0)$ is not a legal state, this double transition can only take place via the intermediate state $(- + -)$ (which is the alternate path) or concurrently, which is obviously a borderline situation. A similar analysis can be made for the transition $10 \to 6$.

If the system is undercritically damped, its qualitative path will circle around the PERT network until, as $t \to \infty$, it finally moves into the steady–state node 7. If the system is overcritically damped, it will move to either state 3 or 11, whichever it reaches first, and stay in this node until, as $t \to \infty$, the system again moves into its steady–state node 7. These are obviously the only two behaviors that the second–order system with positive coefficients can exhibit. Thus, the qualitative simulation has indeed been able to capture the physical behavior of this system correctly.

Let us now return to our third–order model. We can easily determine the finite state transition table for this example. Using qualitative state vectors of length two, we can now compute a qualitative output vector of length four. The resulting state transition table contains 68 legal states, and up to eight paths emanate from a single node. No wonder that we were unable to explore the episodical behavior of this system explicitly. Even the resulting PERT network is a mess. The graph is far from planar and paths go in all directions. Thus, the best we can achieve here is to work with the finite state transition table directly. If you are interested in pursuing this question any further, solve Hw.[H12.6]).

12.7 Qualitative Discrete–Time Simulation

Analogy with other techniques that deal with discrete–time systems immediately leads to a meaningful definition of a qualitative discrete–time simulation [12.5]. In our qualitative vector, we simply replace the derivatives by values of the variable that are shifted into the future, i.e.,

$$x = (+ + -) \tag{12.46}$$

now means that x_k and x_{k+1} are positive, while x_{k+2} is negative. Since discrete–time systems can jump from positive values at one clock instant to negative values at the next, we must let go of all continuity constraints in this case.

Some of the operators are a little different now. For example:

$$
\begin{aligned}
\text{if } & z_k = x_k \cdot y_k \\
\rightarrow & z_{k+1} = x_{k+1} \cdot y_{k+1} \\
\rightarrow & z_{k+2} = x_{k+2} \cdot y_{k+2}
\end{aligned}
\tag{12.47}
$$

This operator has been implemented as:

```
// [z] = DQMULT(x, y)
  for i = 1 : 3, ...
    z(i) = TMULT(x(i), y(i)); ...
  end
return
```

Also, some constants are different. For example, a constant positive value is now:

$$DPLUS = [3, 3, 3] \qquad (12.48)$$

The qualitative integration is replaced by a shift operation:

$$x_{k_{new}} = x_{k+1_{old}} \qquad (12.49)$$

We can now try to solve the simple first–order problem once more, this time assuming this problem to represent the difference equation:

$$x_{k+1} = -a \cdot x_k + b \cdot u \qquad (12.50)$$

where u is a step input and x_0 is negative. This problem can be formulated in QualSim as follows:

```
// [xdot] = DQSTATE(x)
  u = DPLUS;
  xdot =QSUB(u, x);
return
```

and:

```
// [y] = QOUT(x)
  y = x;
return
```

and finally:

```
// [x] = DQSTATEIC(dummy)
  x = IMINUS;
return
```

Except for the replacement of $CPLUS$ by $DPLUS$, the problem formulation is exactly the same as in the continuous case. The QualSim routine:

$$[y, x] = \text{DQSIM}(nstp) \qquad (12.51)$$

performs the qualitative discrete–time simulation for us.

As we can verify immediately, this difference equation will lead to oscillatory behavior for $a > 0$. It will lead to a stable oscillation if $a < 1$, a marginally stable oscillation for $a = 1$, and an unstable oscillation if $a > 1$.

In the qualitative simulation, the oscillatory behavior is made possible by throwing the continuity constraints overboard. This problem leads immediately to heavy branching since QualSim tries desperately to distinguish between oscillations that become negative and others that don't. Of course, this is of no direct use to us. It might therefore be a little more natural to use a variable transformation:

$$\xi = x - \frac{b}{a+1} \tag{12.52}$$

which brings the steady–state value to zero. The quantitative difference equation can be rewritten as:

$$\xi_{k+1} = -a \cdot \xi_k \tag{12.53}$$

with the initial condition:

$$\xi_0 = x_0 - \frac{b}{a+1} < 0 \tag{12.54}$$

This problem can be solved conveniently in QualSim. It doesn't branch out and after only two steps we find a periodic steady–state:

$$
\begin{array}{ccc}
t & x_k & x_{k+1} \\
0 & (- \ ? \ ?) & (+ \ ? \ ?) \\
0 & (- + \ ?) & (+ - \ ?) \\
0 & (- + -) & (+ - +) \\
1 & (+ - +) & (- + -) \\
2 & (- + -) & (+ - +)
\end{array}
$$

Again, QualSim does not distinguish between a true periodic steady–state and stable or unstable oscillations.

What this example teaches us is that by a very small transformation of the posed problem, we can sometimes prevent the devastating desire of QualSim (and all other naïve physics tools) to branch out into impossibly many different episodes. However, this modification again requires quite a bit of human ingenuity and previous insight, exactly what we tried to gain from the qualitative simulation itself, and it doesn't lend itself easily to a full automation.

12.8 Pros and Cons

Unfortunately, the current state–of–the–art naïve physics approach to modeling shows considerably more adversities than advantages.

In this chapter, I demonstrated basically all that has been achieved. The results are not yet fully convincing.

Let me list the problems with which we are faced:

(1) The naïve physics approach did not live up to its promise to help us with the modeling process. We still needed to derive a state–space model manually. However, the approach saved us from the necessity of identifying model parameters through optimization [12.5].

(2) While we were able to reduce the naïve physics model to the familiar state–space description (or rather its qualitative counterpart), the qualitative simulation has a tendency to explode, i.e., to branch out immediately into an unmanageable multitude of different episodes. Thus, qualitative simulation is not robust at all. We cannot blindly take any state–space model, convert it to its qualitative form, run QualSim, and hope to get anything meaningful out of it. However, robustness was exactly what we were after in the first place. Morgan's new approach using the finite state transition table helps alleviate this problem to some extent [12.15].

(3) In order to get QualSim to work, we often require previous insight into the problem at hand, i.e., we must possess what we just try to gain — a most uncomfortable Catch–22 situation.

(4) Many state–space models involve trigonometric functions such as sin or cos. No qualitative counterpart to those functions can be defined unless we can limit the range of their inputs (small signal behavior).

(5) Naïve physics allows us to specify parameters as being positive or negative only. This is not of much practical use. In most engineering problems, we know the parameter values with a tolerance of, say, $\pm 10\%$. Unfortunately, the naïve physics approach does not allow us to specify a limited range for a parameter. Consequently, the episodes produced by the qualitative simulation are usually too rich since they also include the behavior of the system for values of the parameters outside their physically meaningful range. Kuipers' and Farquhar's QSIM program provides a partial answer to this problem [12.13]. QSIM operates on qualitative vectors of length two, but while the derivatives are always ternary variables of type $\{- 0 +\}$, the state variables themselves may contain additional landmarks and new landmarks can actually be discovered on the fly. This advantage was made possible (and is paid for) by forcing the user to formulate all constraints

and the results of all qualitative operations explicitly, rather than relying on a set of implicit constraints and computational rules as QualSim does. This makes the problem formulation a little more difficult to derive for QSIM than for QualSim, but it makes QSIM a little more flexible than QualSim.

Naïve physics is still a wide–open research area. While I am not yet convinced that this is really the right way to go, this is a fruitful area for research at the Master's or Ph.D. levels since so many open questions remain to be answered. Only the future will tell whether this technique as a whole will survive, or whether it will end up in the ⊔ of science.

12.9 Summary

In this chapter, we have introduced a highly advocated technique for qualitative modeling and simulation. We have shown that the technique is still in its infancy and that much research is needed to make it truly useful. The most comprehensive account of the naïve physics approach to qualitative modeling available today is the book edited by Daniel Bobrow [12.2]. However, also this book is somewhat limited in its perspective. For example, it would make a lot of sense to investigate the relations of the naïve physics approach to deep reasoning with other approaches to deep reasoning such as those used in cognitive psychology [12.11]. To my knowledge, such a comparison has never been made.

References

[12.1] Earl Babbie (1989), *The Practice of Social Research*, fifth edition, Wadsworth Publishing Company, Belmont, Calif.

[12.2] Daniel G. Bobrow, Ed. (1985), *Qualitative Reasoning about Physical Systems*, MIT Press, Cambridge, Mass.

[12.3] François E. Cellier (1986), "Enhanced Run–Time Experiments for Continuous System Simulation Languages," *Proceedings SCS MultiConference on Languages for Continuous System Simulation* (F.E. Cellier, ed.), SCS Publishing, San Diego, Calif., pp. 78–83.

[12.4] François E. Cellier and C. Magnus Rimvall (1989), "Matrix Environments for Continuous System Modeling and Simulation," *Simulation*, **52**(4), pp. 141–149.

[12.5] François E. Cellier and Nicolas Roddier (1991), "Qualitative State Spaces: A Formalization of the Naïve Physics Approach to Knowledge–Based Reasoning," *Proceedings AI, Simulation and Planning in High Autonomy Systems* (P.A. Fishwick, B.P. Zeigler, and J.W. Rozenblit, eds.), IEEE Computer Society Press, Los Alamitos, Calif.

[12.6] R. G. Coyle (1977), *Management System Dynamics*, John Wiley & Sons, London, U.K.

[12.7] Ernest Davis (1987), "Constraint Propagation with Interval Labels," *Artificial Intelligence*, **32**, pp. 281–331.

[12.8] Johan de Kleer and John S. Brown (1984), "A Qualitative Physics Based on Confluences," *Artificial Intelligence*, **24**, pp. 7–83.

[12.9] Richard C. Dorf (1989), *Modern Control Systems*, fifth edition, Addison–Wesley, Reading, Mass.

[12.10] Kenneth D. Forbus (1984), "Qualitative Process Theory," *Artificial Intelligence*, **24**, pp. 85–168.

[12.11] Dedre Gentner and Albert Stevens (1983), *Mental Models*, Lawrence Erlbaum Associates, Hillsdale, N.J.

[12.12] Benjamin Kuipers (1986), "Qualitative Simulation," *Artificial Intelligence*, **29**, pp. 289–338.

[12.13] Benjamin Kuipers and Adam Farquhar (1987), *QSIM: A Tool for Qualitative Simulation*, Internal Report: Artificial Intelligence Laboratory, The University of Texas, Austin.

[12.14] Antony J. Morgan (1988), *The Qualitative Behaviour of Dynamic Physical Systems*, Ph.D. dissertation, Wolfson College, The University of Cambridge, Cambridge, U.K.

[12.15] Antony J. Morgan (1990), "Accuracy in Qualitative Descriptions of Behaviour," *Proceedings Winter Simulation Conference*, New Orleans, La., pp. 520–526.

[12.16] Brian C. Williams (1984), "Qualitative Analysis of MOS Circuits," *Artificial Intelligence*, **24**, pp. 281–346.

Bibliography

[B12.1] Paul A. Fishwick (1988), "The Role of Process Abstraction in Simulation," *IEEE Trans. Systems, Man and Cybernetics*, **18**(1), pp. 18–39.

[B12.2]   Paul A. Fishwick and Paul A. Luker, Eds. (1991), *Qualitative Simulation, Modeling, and Analysis*, Springer–Verlag, New York.

[B12.3]   Paul A. Fishwick and Bernard P. Zeigler (1990), "Qualitative Physics: Towards the Automation of Systems Problem Solving," *Proceedings AI, Simulation and Planning in High Autonomy Systems* (B.P. Zeigler and J.W. Rozenblit, eds.), IEEE Computer Society Press, Los Alamitos, Calif., pp. 118–134.

[B12.4]   Patrick J. Hayes (1979), "The Naïve Physics Manifesto," in: *Expert Systems in the Micro–Electronic Age* (D. Michie, ed.), Edinburgh University Press, Edinburgh, Scotland, pp. 242–270.

[B12.5]   Raman Rajagopalan (1986), "Qualitative Modelling and Simulation: A Survey," *Proceedings AI Applied to Simulation* (E. Kerckhoffs, G.C. Vansteenkiste, and B.P. Zeigler, eds.), Simulation Series, 18(1), Springer–Verlag, Berlin, pp. 9–26.

[B12.6]   Lawrence E. Widman, Kenneth A. Loparo, and Norman R. Nielsen, Eds. (1989), *Artificial Intelligence, Simulation and Modeling*, John Wiley & Sons, New York.

[B12.7]   Bernard P. Zeigler and Jerzy W. Rozenblit, Eds. (1990), *AI, Simulation and Planning in High Autonomy Systems*, IEEE Computer Society Press, Los Alamitos, Calif.

Homework Problems

[H12.1] Qualitative Division

Find the qualitative division matrix *TDIV* and the qualitative division function CQDIV. Implement them as *myTDIV* and myCQDIV and compare the results of your functions with those available in QualSim.

[H12.2] Manual Qualitative Simulation

Simulate the sine wave problem manually over an entire limit cycle. Whenever you obtain ?s in your state vector, expand them in full and mark those that are in conflict with continuity constraints. Then reduce the set to those that satisfy all continuity constraints. Compute the time derivatives for the remaining set by evaluating the qualitative state–equations. Now mark those solutions that are in conflict with any consistency constraint. Reduce the set further and apply the steady–state constraints to the remaining set. Iterate until an entire limit cycle has been processed.

[H12.3] Qualitative Plots

Define a set of rules that describe the generation of episodes from qualitative vectors. Implement these rules in a function called myQPLOT and compare the results of your function with QualSim's implementation.

[H12.4] Qualitative Simulation Using QualSim

Simulate qualitatively the equations Eq.(12.33*a*) and Eq.(12.33*b*) using QualSim's CQSIM function. Apply QPLOT to graphically depict your results. Compare the resulting qualitative output vectors and episodes with those found for the sine wave function.

[H12.5] Spurious Solutions

Explain why the qualitative state vectors of length two were unable to detect the two spurious transitions of Table 12.2, whereas the qualitative state vectors of length three were able to eliminate these transitions.

[H12.6]* Finite State Transition Table

Use CQPERT to determine both the "hard" and "soft" finite state transition tables for the third–order model using qualitative vectors of length two. Repeat the analysis with vectors of length three, throw away the fourth derivative of the output, and compare the resulting state transition tables with the previously found tables. Do the original tables contain any bogus states or spurious transitions? Reduce the new state transition tables further by now throwing the third derivative of the output away. What is the number of remaining legal states? Draw a PERT network for the reduced state transition table. Make the graph as planar as possible. Explain how the circular paths in the PERT network relate to the feasible physical solutions.

Projects

[P12.1] DYMOLA Interface

Create a new interface to DYMOLA that will allow us to directly generate a qualitative simulation model from a DYMOLA program. In this way, the same model could be used alternatively to either quantitatively or qualitatively analyze a given problem.

[P12.2] Qualitative Integration

Currently, QualSim employs a qualitative forward Euler integration algorithm to advance the simulation clock from one qualitative time instant to the next. We may be able to get rid of some spurious state transitions if we employ a smarter integration scheme.

In quantitative numerical integration of differential equations, we sometimes employ so–called *predictor–corrector schemes*. You will learn more about those in the companion book on continuous system simulation. For the moment, let me propose the following (simple) predictor–corrector scheme. We compute a predictor of the next state using forward Euler:

$$x^P_{k+1} = x_k + \Delta t \cdot \dot{x}_k \qquad (P12.2a)$$

We then evaluate the state equations once to find a predictor for the state derivatives:

$$\dot{x}^P_{k+1} = f(x^P_{k+1}) \qquad (P12.2b)$$

We then improve our estimate of the next state by computing a corrector using backward Euler:

$$x^C_{k+1} = x_k + \Delta t \cdot \dot{x}^P_{k+1} \qquad (P12.2c)$$

We then use the corrected next state to continue with our simulation:

$$x_{k+1} = x^C_{k+1} \qquad (P12.2d)$$

Let us try to port this idea over to the qualitative simulation.

We start by using the qualitative forward Euler algorithm to find one or several next states that are compatible with the integration constraint. We treat this set of next states as predictors. All physical next states must be within this set, however, this set may contain a few nonphysical next states. We saw this in the second–order example with damping in which two bogus transitions survived the integration constraint.

We then evaluate a predictor for the qualitative state derivatives and we use a qualitative backward Euler algorithm as an additional integration constraint, i.e., we compute:

$$x^C_{k+1} = \text{QINT}(x_k, x^P_{k+1}) \qquad (P12.2e)$$

A physical solution must be within the set of predictors and the set of correctors, i.e., we continue our qualitative simulation with the common elements of both sets:

$$x_{k+1} = x_{k+1}^P \cap x_{k+1}^C \qquad\qquad (P12.2f)$$

Implement this modified integration algorithm in QualSim and check by means of the second–order example with damping whether the two non-physical state transitions that previously survived the forward Euler constraint are still around or whether they have now been eliminated.

Research

[R12.1] Bogus Solutions

Analyze the problem of bogus solutions. Formulate a general methodology that explains under what conditions the qualitative simulation produces bogus solutions and why.

[R12.2] Qualitative Stability

Analyze the problem of qualitative stability. Develop a general methodology that will allow us to distinguish between stable oscillations, marginally stable oscillations (periodic steady–state solutions), and unstable oscillations in a qualitative simulation.

[R12.3] Qualitative Control

Use the naïve physics approach to design a qualitative controller that can supervise a conventional controller of a process. For example, use the aircraft model of Chapter 4, apply arbitrary disturbances such as varying wind conditions to the aircraft, and let the qualitative controller train the conventional controller to operate the aircraft in the presence of these disturbances in a similar way as a human jet pilot cooperates with the low–level control circuitry of his or her aircraft.

13

Inductive Reasoning

Preview

In the last chapter, we discussed a crude approach to analyzing the behavior of a system by means of a coarse, qualitative, structural description of the real physical system. The claim was that this models the way humans reason about processes, and since humans are very adept at making correct decisions on the basis of incomplete knowledge, this approach may enable algorithms to duplicate such aptitude. It turned out that the results were not as promising as some researchers would like us to believe. Strong indicators can also be found that humans mostly assess the behavior of a system not on the basis of qualitative physical considerations, but on the basis of analogies with similar processes, the operation of which they have previously observed, i.e., that they use pattern recognition to analyze system behavior. In this chapter, we shall discuss one of several pattern recognition techniques that may be able to mimic how humans apply pattern recognition to reason about system behavior.

13.1 Introduction

In Chapter 12, I mentioned the following example: If I hold a glass with water in my hand and open my fingers, I know that the glass will fall to the ground, break into many pieces, and spill the water over my carpet. I don't need to solve a set of differential equations to come up with this assessment. I argued that it is sufficient to know that a positive force exists that pulls the glass down. The amount of that force is not important to the conclusion, except if I wish to know when exactly the glass will hit the floor. But is this really how I came to the correct conclusion? Or is it maybe because I remembered that my father, when I was 10 years old, let a two–liter bottle of Chianti wine slip through his fingers and our dog came and licked the entire contents of the bottle from the carpet and got himself completely drunk? But if this were true, how come my brain correlated the water glass event with a seemingly unrelated event that happened more than three decades ago, taking into consideration that I observe and store an unbelievable manifold of different episodes every day of my life? How come a physician whom I had visited only once four years earlier introduced himself to me (since his office had misplaced my previous patient card), but frowned after the third word and exclaimed: "But you *have* been here before, haven't you?" although the guy must be seeing 50 patients a day? In this chapter and the next, I shall try to bring us a little closer to unraveling these mysteries.

Let me start by explaining why I believe pattern recognition is a much more frequently used and much more powerful tool in assessing qualitatively the behavior of a system than knowledge–based reasoning. I still own a dog who loves to play ball. I kick the ball with the side of my foot (I usually wear sandals and a straight kick hurts my toes) and my dog runs after the ball as fast as he can. I was able to observe the following phenomenon: If I place my foot to the left of the ball, my dog will turn to the right to be able to run after the ball as soon as I hit it. He somehow knows that the ball will be kicked to the right. If I now change my strategy and place my foot to the right of the ball, my dog immediately swings around to be ready to run to the left. He obviously has some primitive understanding of the mechanics involved in ball kicking. However, I assure you that I never let my dog near my physics texts, and thus, he had no opportunity to study Newton's laws — not even in their naïve form.

A number of strong arguments can be mentioned in favor of the pattern recognition approach, but a number of strong arguments can also be brought forward that advise against it. What makes pattern recognition attractive?

(1) In Chapter 10, I demonstrated that structure characterization is a very tough problem. In the pattern recognition approach, we bypass this problem entirely by going directly for the behavior itself. It is a much simpler task to identify a pattern than to characterize a structure.

(2) Pattern recognition is qualitative by nature. We don't need to artificially modify our problem to make it qualitative.

(3) Pattern recognition algorithms lend themselves naturally to implementation on a parallel–processor architecture. Structure characterization algorithms don't share this property. Thus, pattern recognizers can be very fast.

What makes pattern recognizers unattractive for the purpose of reasoning?

(1) As I mentioned earlier, the behavioral complexity of a system is much larger than its structural complexity. Thus, we need to store much more information about a system if we represent it directly in a behavioral form. Consequently, pattern recognizers will usually not be able to characterize a system in general, but only for a limited set of input stimuli.

(2) In a reasoning system, we like to know why a system behaves in one way and not in another. Pattern recognition does not usually provide much insight into the *why*, only into the *how*. Consequently, it is very difficult to use knowledge extracted from a pattern recognizer in the process of knowledge generalization. However, this is also a weakness of humans. Only a few people are truly capable of effectively participating in the process of knowledge generalization.

Problem #1 can be overcome if we are able to make the pattern recognizer so fast that we can learn the behavior of our system on–line. In that case, we can identify the system behavior in the vicinity of the current operating point, make a decision, implement it, observe the effects of that implementation, determine a new operating point, and reidentify the process for this new operating point. Problem #2 is tough luck, but remember that the naïve physics approach hasn't helped us much with this problem either. This is simply a difficult problem to tackle.

The question remains: How did my physician recognize me after all these years? Somehow, we must be able to store *generic properties* of a system away, properties that are so generic that we can easily and quickly access them and compare them with the generic properties of a new system to find similar or equal patterns. In this chapter, I shall introduce one methodology that allows us to mimic this ability to a certain extent.

13.2 The Process of Recoding

In the last chapter, I discussed how continuous trajectories are discretized into episodes for the purpose of naïve physics modeling. Inductive reasoning, like all other qualitative modeling techniques, also requires a discretization of continuous phenomena.

This time, we shall allow more values than just $-$, 0, and $+$, and we shall concentrate on the regions themselves rather than on the landmarks that separate these regions from each other. The values that represent such regions can be symbolic (for example, 'tiny,' 'small,' 'average,' and 'big,' denoting four distinct regions) or integer numbers (for example, '1,' '2,' '3,' and '4,' denoting the same four regions). In an inductive reasoning system that is coded in LISP, symbolic names are probably preferred, whereas in an inductive reasoner coded in a predominantly numeric software such as MATLAB or CTRL–C, integers will be the representation of choice. From a practical point of view, it really doesn't matter which of the two representations is being used since one can easily be mapped into the other. The symbolic representation will make the code more readable, though.

In inductive reasoning, the regions are called *levels*. Notice the difference with System Dynamics. In System Dynamics, a "level" denotes a continuous state variable, whereas in inductive reasoning, it denotes one value of a discrete state variable. The process of discretizing continuous trajectories into discrete episodes is called *recoding*. Finally, a combination of legal levels of all state variables of a model is called a *state*. Thus, a model with n state variables, each of which is recoded into k levels has k^n legal states. An *episodical behavior* is a time history of legal states. The episodical behavior is the qualitative counterpart of the quantitative trajectory behavior.

Inductive reasoning is a technique invented in the 1970s by George Klir [13.5]. A first software system implementing Klir's ideas was SAPS [13.11]. Unfortunately, the original SAPS system was not sufficiently flexible to be of much practical use. We have therefore developed a new implementation, called SAPS–II, which is available as either a CTRL–C library or a MATLAB toolbox [13.4]. The CTRL–C version of SAPS–II can be accessed using the command:

> [>   **do** *saps* : *saps*

MATLAB recognizes all its toolboxes automatically.

In SAPS–II, levels are represented through positive integers. While QualSim is a very simple program package, SAPS–II is quite intricate, and I won't be able to discuss all the details of how SAPS has been implemented. However, this chapter can serve as a somewhat simplified user's guide.

The first question that we must ask ourselves is: How do we recode? How many levels should we select for each of our state variables? Where do we draw the borderline (i.e., where do we select the landmark) that separates two neighboring regions from each other?

Inductive reasoning is a completely inductive approach to modeling. It operates on a set of measured data points and identifies a model from previously made observations.

From statistical considerations, we know that in any class analysis, we would like to record each possible discrete state at least five times [13.7]. Thus, a relation exists between the possible number of legal states and the number of data points that we require to base our modeling effort upon:

$$n_{rec} \geq 5 \cdot n_{leg} = 5 \cdot \prod_{\forall i} k_i \qquad (13.1)$$

where n_{rec} denotes the total number of recordings, i.e., the total number of observed states, n_{leg} denotes the total number of different legal states, i is an index that loops over all variables, and k is an index that loops over all levels. If we postulate that each variable assumes the same number of levels, Eq.(13.1) can be simplified to:

$$n_{rec} \geq 5 \cdot (n_{lev})^{n_{var}} \qquad (13.2)$$

where n_{var} denotes the number of variables and n_{lev} denotes the chosen number of levels for each variable. The number of variables is usually given and the number of recordings is frequently predetermined. In such a case, we can find the optimum number of levels from Eq.(13.3):

$$n_{lev} = \text{ROUND}\left(\sqrt[n_{var}]{\frac{n_{rec}}{5}} \right) \qquad (13.3)$$

For reasons of symmetry, we often prefer an odd number of levels over an even number of levels. For example, the five levels 'much too low,' 'too low,' 'normal,' 'too high,' and 'much too high' might denote states of the heart beat of a patient undergoing surgery. By choosing an odd number of levels, we can group anomalous levels symmetrically around the normal state.

If the number of recordings is not predetermined, we might consider consulting with a human expert (in the preceding example, the surgeon) to determine a meaningful number of levels for a given variable.

The number of levels of our variables determines the expressiveness and predictiveness of our qualitative model. The *expressiveness* of a qualitative model is a measure of the information content that the model provides. Later in this chapter, I shall present formulae describing the information content of a qualitative model. The *predictiveness* of a qualitative model is a measure of its forecasting power, i.e., it determines the length of time over which the model can be used to forecast the future behavior of the underlying system [13.8].

If all variables are recoded into exactly one level, the qualitative model exhibits only one legal state. It is called a "null model." It will be able to predict the future behavior of the underlying system perfectly over an infinite time span (within the framework of its model resolution). Yet the prediction does not tell us anything useful. Thus, the null model is characterized by an infinitely high predictiveness and a zero expressiveness.

On the other hand, if we recode every variable into 1000 levels, the system exhibits myriad legal states. The expressiveness (i.e., resolution) of such a model will be excellent. Each state contains a large amount of valuable information about the real system. Yet the predictiveness of this model will be lousy unless we possess an extremely large base of observed data. In all likelihood, this model cannot be used to predict the behavior of the real system for even a single time step into the future.

Consequently, we must compromise. For most practical applications, we found that either three or five levels were about optimal [13.3,13.12].

Once we decide upon the number of levels for each variable, we must choose the landmarks that separate neighboring regions. Often, this is best done by consulting with a human expert. For example, we may ask a surgeon what he or she considers a 'normal' heart beat during surgery and when he or she would believe that the heart beat is definitely 'too low' or 'too high,' and when he or she would consider it to be 'critically too low' or 'critically too high.' If we are then able to predict the future behavior of the patient in terms of these qualitative variables, we may be able to construct a heart monitor that will warn the surgeon ahead of time about a predictable problem. This clearly sounds like a worthwhile research topic.

However, if the amount of observed data is limited, it may be preferable to maximize the expressiveness of the qualitative model. This demand leads to a clearly defined optimal landmark selection algorithm [13.8]. The expressiveness of the model will be maximized if each level is observed equally often. Thus, one way to find an optimal set of landmarks is to sort the observed trajectory values into ascending order, cut the sorted vector into n_{lev} segments of equal length, and choose the landmarks anywhere between the extreme values of neighboring segments. Let me demonstrate this process by means of an example. Figure 13.1 shows an observed trajectory of a continuous variable.

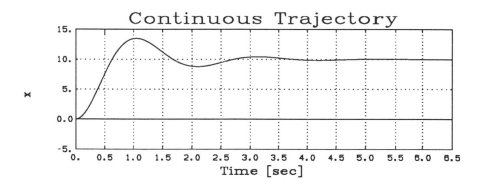

Figure 13.1. Trajectory behavior of a continuous variable.

We first discretize the time axis (how this is done in an optimal manner will be explained in due course). Let us say that this process leads to a trajectory vector of length 131. The observed values range from 0.0 to 13.5. Let us assume that we wish to recode this trajectory into the three distinct levels '1,' '2,' and '3.' If we would simply cut the domain into equal intervals of length 4.5, i.e.:

$$\text{`1'} \longleftrightarrow [0.0, 4.5]$$
$$\text{`2'} \longleftrightarrow (4.5, 9.0]$$
$$\text{`3'} \longleftrightarrow (9.0, 13.5]$$

the levels '1' and '2' would only occur very briefly and only during the initial phase of the episode. Thereafter, we would constantly observe a level of '3.' Figure 13.2 shows the recoded episode.

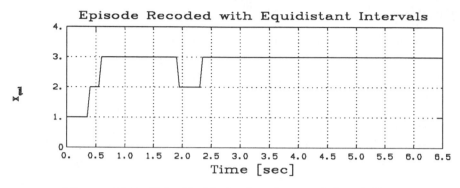

Figure 13.2. Episodical behavior of a recoded variable.

In the process of recoding, we throw away a lot of information. In our example, we lost most of the information regarding the oscillation frequency. However, if we use the previously described optimal algorithm, we sort the trajectory values in ascending order such that the first value is 0.0 and the last value is 13.5. We then cut the resulting vector of length 131 into three vectors of approximately equal length. The first vector contains the elements 1 to 43, the second vector contains the elements 44 to 86, and the third vector contains the elements 87 to 131. For the given example, the following values were found:

$$x_{sorted}(43) = \quad 9.8898$$
$$x_{sorted}(44) = \quad 9.8969$$
$$x_{sorted}(86) = 10.0500$$
$$x_{sorted}(87) = 10.0501$$

and by using the arithmetic mean values of neighboring observed data points in different segments as our landmarks LM_i, we find:

$$LM_1 = \frac{x_{sorted}(43) + x_{sorted}(44)}{2.0} = \quad 9.8934$$
$$LM_2 = \frac{x_{sorted}(86) + x_{sorted}(87)}{2.0} = 10.0500$$

Using these landmarks, we obtain the following three regions:

$$\text{‘1’} \longleftrightarrow [0.0, 9.8934]$$
$$\text{‘2’} \longleftrightarrow (9.8934, 10.05]$$
$$\text{‘3’} \longleftrightarrow (10.05, 13.5]$$

Figure 13.3 shows the same continuous trajectory as Fig.13.1 with the two new landmarks superimposed.

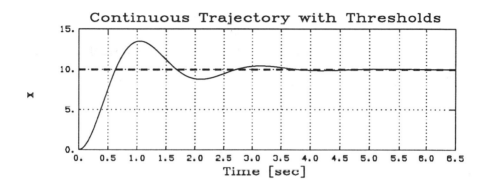

Figure 13.3. Continuous trajectory behavior with landmarks superimposed.

The band width of level ‘2’ is very narrow indeed. Figure 13.4 shows the recoded episodical behavior.

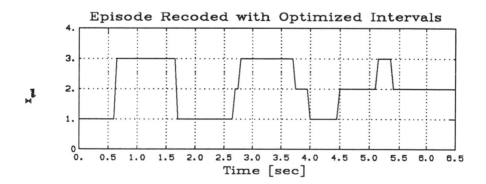

Figure 13.4. Episodical behavior of a recoded variable.

Clearly, the recoding of Fig.13.4 has preserved more information about the real system than the recoding of Fig.13.2.

Which technique will work best depends heavily on the application area. For the case of the heart surgeon, the "optimized" recoding would be meaningless. His or her goal is to receive an early warning when the heart beat is expected to become critical, not to observe each level equally often. Obviously, he or she wishes to observe level '3' (out of five levels) only, i.e., he or she wishes to keep his or her patient constantly within the normal range.

The following example, which was taken from Hugo Uyttenhove's Ph.D. dissertation [13.11], may serve as an illustration for the process of recoding. A heart monitor observes six different variables about a patient undergoing surgery. Each of these variables is being recoded into five different levels using the qualitative states:

$$'1' \longleftrightarrow \text{much too low}$$
$$'2' \longleftrightarrow \text{too low}$$
$$'3' \longleftrightarrow \text{normal}$$
$$'4' \longleftrightarrow \text{too high}$$
$$'5' \longleftrightarrow \text{much too high}$$

Table 13.1 lists the six variables together with their five ranges.

Table 13.1. Recoding of Heart Variables.

variable	'1'	'2'	'3'	'4'	'5'
Systolic Blood Pr.	< 75	[75, 100)	[100, 150)	[150, 180)	> 180
Mean Blood Pr.	< 50	[50, 65)	[65, 100)	[100, 110)	> 110
Central Venous Pr.		< 4	[4, 20)	> 20	
Cardiac Output	< 2	[2, 3)	[3, 7)	> 7	
Heart Rate	< 50	[50, 60)	[60, 100)	[100, 110)	> 110
Left Atrial Pr.	< 1	[1, 4)	[4, 20)	> 20	

Let us assume that the trajectory behavior of this six variable system has been recorded in the form of a trajectory behavior matrix *meas* with six columns denoting the six different variables and 1001 rows denoting different measurement instants (different time values). The following SAPS–II program segment can be used to recode these values:

```
[>  do saps:saps

[>  from = [   0.0      75.0     100.0    150.0    180.0
               75.0    100.0     150.0    180.0    999.9];
[>  to = 1 : 5;
[>  raw =RECODE(meas(:, 1),'domain', from, to);

[>  from = [   0.0      50.0      65.0    100.0    110.0
               50.0     65.0     100.0    110.0    999.9];
[>  r −RECODE(meas(:, 2),'domain', from, to);
[>  raw = [raw, r];

[>  from = [   0.0       4.0      20.0
                4.0      20.0     999.9];
[>  to = 2 : 4;
[>  r =RECODE(meas(:, 3),'domain', from, to);
[>  raw = [raw, r];

[>  from = [   0.0       2.0       3.0      7.0
                2.0       3.0       7.0    999.9];
[>  to = 1 : 4;
[>  r =RECODE(meas(:, 4),'domain', from, to);
[>  raw = [raw, r];

[>  from = [   0.0      50.0      60.0    100.0    110.0
               50.0     60.0     100.0    110.0    999.9];
[>  to = 1 : 5;
[>  r =RECODE(meas(:, 5),'domain', from, to);
[>  raw = [raw, r];

[>  from = [   0.0       1.0       4.0     20.0
                1.0       4.0      20.0    999.9];
[>  to = 1 : 4;
[>  r =RECODE(meas(:, 6),'domain', from, to);
[>  raw = [raw, r];
```

RECODE is one of the SAPS–II functions. As it is used in this example, it maps the regions (domains) specified in columns of the *from* matrix to the levels that are specified in the *to* vector. Thus, the *from* matrix and the *to* vector must have the same number of columns. In this example, each variable (column) is being recoded separately. The resulting episode r is then concatenated from the right to the previously found episodes, which are stored in *raw*. The code should be fairly self–explanatory otherwise.

One problem remains to be discussed: How big is big? Obviously, qualitative terms are somewhat subjective. In comparison to an adult, a 10–year–old usually has quite a different opinion about what an "old person" is. The concept of landmarks is a treacherous one. Is it really true that a systolic blood pressure of 100.1 is normal, whereas a systolic blood pressure of 99.9 is too low? Different physicians may have a different opinion altogether. My wife usually has a systolic blood pressure of about 90. Yet, her physician always smiles when he takes her blood pressure and predicts that she will have a long life. 90 is too low for what? What the previously mentioned heart surgeon probably meant when he or she declared 100 the borderline between 'normal' and 'low' was the following: If a 'normal' patient, i.e., a patient with an average 'normal' systolic blood pressure of 125 experiences a sudden drop of the blood pressure from 125 to 90 during surgery, then something is probably wrong. Does this mean that we should look at the time derivative of the systolic blood pressure in addition to the blood pressure itself? We probably should, but the surgeon couldn't tell us, because this is not the way he or she thinks. Most medical doctors aren't trained to think in terms of gradients and dynamic systems.

We have discovered two different problems.

(1) Asking an "expert" about which variables to look at can be a dubious undertaking. Experts in heart surgery aren't necessarily experts in expert system design. The surgeon won't understand the purpose of our question. Thus, we must use our own scrutiny and intelligence to interpret the answers of the so-called expert in an adequate way. The automation of the process of variable selection may be the toughest problem of all. We shall address this problem later in this chapter to some extent.

(2) Even if the question to the expert has been both properly formulated and properly understood, the determination of landmarks contains usually an element of subjectiveness. The crispness of a landmark may be deceiving. While a systolic blood pressure of 125 is clearly and undoubtedly a good and normal value, the matter becomes more confusing as we approach one of the neighboring landmarks.

Lotfi Zadeh tackled the latter problem [13.13,13.14,13.15]. He introduced fuzzy measures as a technique to deal with the uncertainty of landmarks. Instead of saying that the systolic blood pressure is normal for values above 100 and low for values below 100, a *fuzzy measure* allows us to specify that, as we pass the value 100 in the negative direction, the answer 'normal' becomes less and less likely, while the answer 'low' becomes more and more likely.

We can depict the fuzzy measure as shown in Fig.13.5.

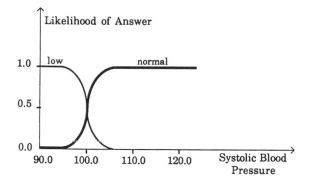

Figure 13.5. Fuzzy qualitative variable.

The sigmoidal curves are called *membership functions*. How precisely
these membership functions are shaped is up to the user. In SAPS–
II, we have implemented only one type of membership function: a
normal distribution, which is 1.0 at the arithmetic mean value μ_i of
any two neighboring landmarks and 0.5 at the landmarks themselves
[13.8]. This membership function can be easily calculated using the
equation:

$$Memb_i = \exp(-k_i \cdot (x - \mu_i)^2) \tag{13.4}$$

where x is the continuous variable that needs to be recoded, say the
systolic blood pressure, and k_i is determined such that the membership function $Memb_i$ degrades to a value of 0.5 at the neighboring
landmarks. Figure 13.6 shows the membership functions for the systolic blood pressure.

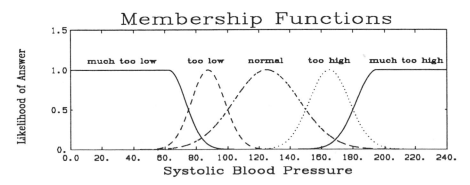

Figure 13.6. Membership functions of the systolic blood pressure.

The first and last membership functions are treated a little differently. Their shape (k_i value) is the same as for their immediate neighbors and they are semiopen.

If we wish to compute the membership functions for the heart monitor example, we need to modify the previously shown program segment in the following way:

```
[>    do saps:saps

[>    from = [  0.0     75.0    100.0    150.0    180.0
                75.0    100.0   150.0    180.0    999.9];
[>    to = 1 : 5;
[>    [raw, Memb, side] =RECODE(meas(:, 1),'fuzzy', from, to);

[>    from = [  0.0     50.0     65.0    100.0    110.0
                50.0    65.0    100.0    110.0    999.9];
[>    [r, m, s] =RECODE(meas(:, 2),'fuzzy', from, to);
[>    raw = [raw, r];   Memb = [Memb, m];   side = [side, s];

[>    from = [  0.0     4.0      20.0
                4.0     20.0    999.9];
[>    to = 2 : 4;
[>    [r, m, s] =RECODE(meas(:, 3),'fuzzy', from, to);
[>    raw = [raw, r];   Memb = [Memb, m];   side = [side, s];

[>    from = [  0.0     2.0      3.0      7.0
                2.0     3.0      7.0     999.9];
[>    to = 1 : 4;
[>    [r, m, s] =RECODE(meas(:, 4),'fuzzy', from, to);
[>    raw = [raw, r];   Memb = [Memb, m];   side = [side, s];

[>    from = [  0.0     50.0     60.0    100.0    110.0
                50.0    60.0    100.0    110.0    999.9];
[>    to = 1 : 5;
[>    [r, m, s] =RECODE(meas(:, 5),'fuzzy', from, to);
[>    raw = [raw, r];   Memb = [Memb, m];   side = [side, s];

[>    from = [  0.0     1.0      4.0      20.0
                1.0     4.0      20.0    999.9];
[>    to = 1 : 4;
[>    [r, m, s] =RECODE(meas(:, 6),'fuzzy', from, to);
[>    raw = [raw, r];   Memb = [Memb, m];   side = [side, s];
```

The *raw* matrix will be exactly the same as before (since RECODE will always pick the most likely answer), but we also obtain the fuzzy memberships of all our qualitative variables that are stored in the *Memb* matrix. The third matrix, *side*, contains a value of 0 whenever the measured data point coincides with the mean value of the neighboring landmarks, it assumes a value of -1 if the measured variable is smaller than the mean between the landmarks and it is $+1$ if the measured variable is larger than the mean between the landmarks.

In the process of recoding, a large amount of valuable information about our real system is discarded. The fuzzy membership retains some of this information, which will prove useful in due course. In fact, up to this point, no information has been lost at all. The original continuous signal can be regenerated accurately using the SAPS function:

[>   *meas* = REGENERATE(*raw*, *Memb*, *side*, *from*, *to*)

where the meaning of the *from* and *to* parameters is opposite from that before [13.8].

13.3 Input/Output Behavior and Masking

By now, we have recoded our trajectory behavior into a discrete episodical behavior. In SAPS–II, the episodical behavior is stored in a raw data matrix. Each column of the *raw data matrix* represents one of the observed variables and each row of the raw data matrix represents one time point, i.e., one recording of all variables or one recorded state. The values of the raw data matrix are in the set of legal levels that the variables can assume, i.e., they are all positive integers, usually in the range from '1' to '5.'

How does the episodical behavior help us identify a model of our system for the purpose of forecasting the future behavior of the system for any given input stream? Any model describes relationships between variables. That is its purpose. For example, in a state–space model, we describe the relationships between the state variables x_i and their time derivatives $\dot{x}_i$:

$$\dot{x}_i = f_i(x_1, x_2, \ldots, x_n) \tag{13.5}$$

If the state variables x_i have been recoded into the qualitative variables v_i and their time derivatives have been recoded into the qualitative variables w_i, we can write:

$$w_i = \tilde{f}_i(v_1, v_2, \ldots, v_n) \tag{13.6}$$

While $\tilde{f}_i$ will be a different function than f_i, the fact that (deterministic) relationships exist between the x_i and the $\dot{x}_i$ variables can be partially preserved in the process of recoding. The $\tilde{f}_i$ functions are (possibly deterministic) relationships between the v_i and the w_i variables.

The beauty of this transformation becomes evident when we try to identify these functional relationships. While the identification (characterization) of the f_i functions is a difficult task, the identification of the $\tilde{f}_i$ functions is straightforward. Since each of the v_i variables can assume only a finite set of values, we can characterize the $\tilde{f}_i$ functions through enumeration. Let me provide an example:

$$y = \sin(x) \tag{13.7}$$

is a quantitative relationship between two quantitative variables x and y. Let us recode the variable x into a qualitative variable v with four states, such that:

$$\begin{pmatrix} x & v \\ 1^{st} \text{ quadrant} & \text{`1'} \\ 2^{nd} \text{ quadrant} & \text{`2'} \\ 3^{rd} \text{ quadrant} & \text{`3'} \\ 4^{th} \text{ quadrant} & \text{`4'} \end{pmatrix} \tag{13.8}$$

If the angle x is anywhere between $0°$ and $90°$ plus or minus a multiple of $360°$, the qualitative variable v assumes a value of '1,' etc.; v simply denotes the quadrant of x. Let us recode the variable y into a qualitative variable w with two states, such that:

$$\begin{pmatrix} y & w \\ \text{negative} & \text{`1'} \\ \text{positive} & \text{`2'} \end{pmatrix} \tag{13.9}$$

Thus, w simply denotes the sign of y. This allows us to characterize the functional relationship between the two qualitative variables v and w as follows:

$$\begin{pmatrix} v & w \\ \text{`1'} & \text{`1'} \\ \text{`2'} & \text{`1'} \\ \text{`3'} & \text{`2'} \\ \text{`4'} & \text{`2'} \end{pmatrix} \tag{13.10}$$

Equation (13.10) is the qualitative counterpart of Eq.(13.7). Qualitative functions are finite automata that relate the qualitative variables to each other.

However, we have to be careful that we recode all variables in a consistent fashion. For example, if x had been recoded differently:

$$
\begin{matrix}
x & v \\
\begin{pmatrix}
-45°.. +45° & \text{'1'} \\
+45°.. +135° & \text{'2'} \\
+135°.. +225° & \text{'3'} \\
+225°.. +315° & \text{'4'}
\end{pmatrix}
\end{matrix}
\tag{13.11}
$$

with the same recoding for y, we would have obtained a nondeterministic relationship between v and w:

$$
\begin{matrix}
v & w & \text{prob} \\
\begin{pmatrix}
\text{'1'} & \text{'1'} & 50\% \\
\text{'1'} & \text{'2'} & 50\% \\
\text{'2'} & \text{'1'} & 100\% \\
\text{'3'} & \text{'1'} & 50\% \\
\text{'3'} & \text{'2'} & 50\% \\
\text{'4'} & \text{'2'} & 100\%
\end{pmatrix}
\end{matrix}
\tag{13.12}
$$

The third column denotes the relative frequency of observation, which can be interpreted as the conditional probability of the output w to assume a certain value, given that the input v has already assumed an observed value.

Equation (13.12) can be rewritten in a slightly different form:

$$
\begin{matrix}
v \backslash^w & \text{'1'} & \text{'2'} \\
\begin{matrix} \text{'1'} \\ \text{'2'} \\ \text{'3'} \\ \text{'4'} \end{matrix}
\begin{pmatrix}
0.5 & 0.5 \\
1.0 & 0.0 \\
0.5 & 0.5 \\
0.0 & 1.0
\end{pmatrix}
\end{matrix}
\tag{13.13}
$$

which is called a *state transition matrix* relating v to w. The values stored in the state transition matrix are the transition probabilities between a given level of v and and a certain level of w, i.e., they are the conditional probabilities of w given v:

$$
ST_{ij} = \mathrm{p}\{w = \text{'}j\text{'}|v = \text{'}i\text{'}\}
\tag{13.14}
$$

The element $< i, j >$ of the state transition matrix is the conditional probability of the variable w to become 'j,' assuming that v has a value of 'i.'

We don't want to fall into the same trap as in Chapter 12 where we assumed that we already knew the state–space model of our system.

The technique advocated in this chapter is totally inductive. Consequently, we are not going to assume that we know anything about our system, with the exception of the observed data streams, i.e., the measured trajectory behavior. Obviously, this says that we cannot know *a priori* what it means to recode our variables consistently. All we know is the following: For any system that can be described by a deterministic (yet unknown) arbitrarily nonlinear state–space model of arbitrary order, if we are lucky enough to pick all state variables and all state derivatives as our output variables, i.e., if all these variables are included in our trajectory behavior, and if we are fortunate enough to recode all these variables in a consistent fashion, then deterministic and static relationships will exist between the qualitative state variables and the qualitative state derivatives.

In the process of modeling, we wish to find finite automata relations between our recoded variables that are as deterministic as possible. If we find such a relationship for every output variable, we can forecast the behavior of our system by iterating through the state transition matrices. The more deterministic the state transition matrices are, the better the certainty that we will predict the future behavior correctly.

Let us now look at the development of our system over time. For the moment, I shall assume that our observed trajectory behavior was produced by quantitatively simulating a state–space model over time. Let us assume that the state–space model was coded in ACSL, that we integrate it using a fixed–step forward Euler algorithm, and that we log every time step in our trajectory behavior. In this case, we can write:

$$x_i(k+1) = x_i(k) + \Delta t \cdot \dot{x}_i(k) \tag{13.15}$$

The qualitative version of Eq.(13.15) is:

$$v_i(k+1) = \tilde{g}(v_i(k), w_i(k)) \tag{13.16}$$

Equation (13.15) is obviously a deterministic relationship between $x_i(k)$, $\dot{x}_i(k)$, and $x_i(k+1)$. Is $\tilde{g}$ a deterministic function? Let me assume that we choose our step size Δt very small in order to integrate accurately. In this case, the state variables will change very little from one step to the next. This means that after the recoding, $v_i(k+1)$ is almost always equal to $v_i(k)$. Only when x_i passes through a landmark will $v_i(k+1)$ be different from $v_i(k)$. Consequently, our qualitative model will have a tough time predicting when the landmark crossing will take place. Thus, it is not a good idea to include every tiny time step in our trajectory behavior. The time distance between two logged entries of our trajectory behavior δt should be chosen such that the two terms in Eq.(13.15) are of the same order of magnitude, i.e.:

$$||x_i|| \approx \delta t \cdot ||\dot{x}_i|| \tag{13.17}$$

Notice that δt is the communication interval, whereas Δt denotes the integration step size. These two variables have no direct relationship to each other. The data stream could as well be the output of a digital oscilloscope observing a real physical system. In such a case, Δt has no meaning, but δt still exists, and must be chosen carefully.

Let us now forget about state–space models. All we know is that we have a recorded continuous trajectory behavior available for modeling. We want to assume furthermore that we know which are the inputs into the real system and which are the outputs that we measure. Our trajectory behavior can thus be separated into a set of input trajectories u_i concatenated from the right with a set of output trajectories y_i, for example:

$$\begin{array}{c} time \\ 0.0 \\ \delta t \\ 2 \cdot \delta t \\ 3 \cdot \delta t \\ \vdots \\ (n_{rec} - 1) \cdot \delta t \end{array} \quad \begin{array}{ccccc} u_1 & u_2 & y_1 & y_2 & y_3 \\ \left(\begin{array}{ccccc} \dots & \dots & \dots & \dots & \dots \\ \dots & \dots & \dots & \dots & \dots \\ \dots & \dots & \dots & \dots & \dots \\ \dots & \dots & \dots & \dots & \dots \\ \vdots & \vdots & \vdots & \vdots & \vdots \\ \dots & \dots & \dots & \dots & \dots \end{array} \right) \end{array} \tag{13.18}$$

The trajectory behavior is recoded into an episodical behavior using the techniques described in the last section. Our modeling effort now consists in finding finite automata relations between the recoded variables that make the resulting state transition matrices as deterministic as possible. Such a relation could look like:

$$y_1(t) = \tilde{f}(y_3(t - 2\delta t), u_2(t - \delta t), y_1(t - \delta t), u_1(t)) \tag{13.19}$$

Equation (13.19) can be represented as follows:

$$\begin{array}{c} t \backslash^x \\ t - 2\delta t \\ t - \delta t \\ t \end{array} \quad \begin{array}{ccccc} u_1 & u_2 & y_1 & y_2 & y_3 \\ \left(\begin{array}{ccccc} 0 & 0 & 0 & 0 & -1 \\ 0 & -2 & -3 & 0 & 0 \\ -4 & 0 & +1 & 0 & 0 \end{array} \right) \end{array} \tag{13.20}$$

The negative elements in this matrix denote inputs of our qualitative functional relationship. This example has four inputs. The sequence in which they are enumerated is immaterial. I usually enumerate them from left to right and top to bottom. The positive value is the output. Thus, Eq.(13.20) is a matrix representation of Eq.(13.19). In inductive reasoning, such a representation is called a mask. A *mask* denotes a dynamic relationship between qualitative variables. In SAPS–II, masks are written as either MATLAB or CTRL–C matrices. A mask has the same number of columns as the episodical behavior to which it should be applied and it has a certain number of rows. The number of rows of the mask matrix is called the *depth* of the mask. The mask can be used to flatten a dynamic relationship out into a static relationship. We can shift the mask over the episodical behavior, pick out the selected inputs and outputs, and write them together in one row. Figure 13.7 illustrates this process.

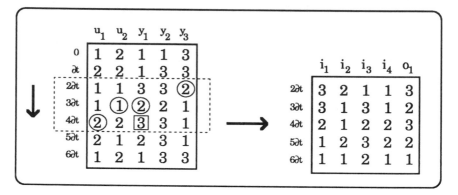

Figure 13.7. Flattening dynamic relationships through masking.

After the mask has been applied to the raw data, the formerly dynamic episodical behavior has become static, i.e., the relationships are now contained within single rows. in SAPS–II, this operation can be performed using the IOMODEL function:

$$[> \quad io = \text{IOMODEL}(raw, mask)$$

IOMODEL will translate the raw data matrix on the left side of Fig.13.7 into the flattened data matrix on the right side of Fig.13.7.

We still haven't discussed how δt is picked in practice. Experience has shown that the equality of Eq.(13.17) can be translated into the following general rule: The mask should cover the largest time constant that we wish to capture in our model. If the trajectory behavior stems from measurement data, we should measure a Bode diagram of the system that we wish to model. This allows us to determine the band width ω_{3dB} of the system. The largest time constant (i.e., the settling time) of the system can be computed from Eq.(13.21):

$$t_s \approx \frac{2\pi}{\omega_{3dB}} \tag{13.21}$$

If our chosen mask depth is 3, the mask spans a time interval of $2\delta t$, thus:

$$\delta t = \frac{t_s}{2} \tag{13.22}$$

The mask depth should be chosen as the ratio between the largest and smallest time constant that we wish to capture in our model, but this ratio should not be larger than 3 or 4. Otherwise, the inductive reasoner won't work very well, since the computing effort grows exponentially with the size of the mask.

13.4 Inductive Modeling and Optimal Masks

An inductive reasoning model is simply a set of masks that relate the input variables and previous values of the output variables to the current values of the outputs. We shall usually forbid relations between various outputs at the current time, since if:

$$y_1(t) = \tilde{f}_1(y_2(t)) \tag{13.23a}$$
$$y_2(t) = \tilde{f}_2(y_1(t)) \tag{13.23b}$$

we have an algebraic loop.

The question remains: How do we find the appropriate masks? The answer to this question was already given. We need to find the masks that, within the framework of the allowable masks, present us with the most deterministic state transition matrix since this matrix will optimize the predictiveness of our model. In SAPS–II, we have introduced the concept of a mask candidate matrix. A *mask candidate matrix* is an ensemble of all possible masks from which we choose the best one by a mechanism of exhaustive search. The mask candidate matrix contains -1 elements where the mask has a potential input, it contains a $+1$ element where the mask has its output, and it contains 0 elements to denote forbidden connections. Thus, the mask candidate matrix for our previous five–variable example will be:

$$
\begin{array}{c}
t \backslash x \\
t - 2\delta t \\
t - \delta t \\
t
\end{array}
\begin{array}{ccccc}
u_1 & u_2 & y_1 & y_2 & y_3 \\
\left(\begin{array}{ccccc}
-1 & -1 & -1 & -1 & -1 \\
-1 & -1 & -1 & -1 & -1 \\
-1 & -1 & +1 & 0 & 0
\end{array} \right)
\end{array}
\qquad (13.24)
$$

The SAPS–II program segment:

```
[>   mcan = -ONES(3,5);
[>   mcan(3, 3:5) = [1, 0, 0];
[>   maxcompl = 5;
[>   mask =OPTMASK(raw, mcan, maxcompl)
```

determines the optimal mask from the set of candidate masks. *raw* is the raw data matrix, and *mcan* is the mask candidate matrix. OPTMASK will go through all possible masks of complexity two, i.e., all masks with one input, and find the best. It will then proceed and try all masks of complexity three, i.e., all masks with two inputs, and find the best of those, etc. The third parameter *maxcompl* enables us to limit the maximum complexity, i.e., the largest number of nonzero elements that the mask may contain. This is a useful feature. In all practical examples, the quality of the masks will first grow with increasing complexity, then reach a maximum, and then decay rapidly. Thus, by setting *maxcompl*, we can reduce the time that the optimization takes. A good value for *maxcompl* is usually five. In order to disable this feature, *maxcompl* can be set to zero.

How do we determine the quality of a mask? Let me explain the process by means of a simple example. Let us assume that we have found the following raw data matrix (episodical trajectory) of a three–variable system:

$$
raw = \left[\begin{array}{ccc}
2 & 1 & 1 \\
1 & 2 & 3 \\
1 & 1 & 4 \\
1 & 1 & 3 \\
1 & 1 & 2 \\
2 & 2 & 2 \\
2 & 1 & 2 \\
2 & 2 & 3 \\
1 & 2 & 1 \\
1 & 2 & 1 \\
2 & 1 & 4
\end{array} \right]
$$

Each mask will lead to a different state transition matrix. For example, the mask:

$$mask = [\begin{matrix} -1 & -2 & 0 \\ 0 & 0 & +1 \end{matrix}]$$

will lead to the following input/output model:

$$io = [\begin{matrix} 2 & 1 & 3 \\ 1 & 2 & 4 \\ 1 & 1 & 3 \\ 1 & 1 & 2 \\ 1 & 1 & 2 \\ 2 & 2 & 2 \\ 2 & 1 & 3 \\ 2 & 2 & 1 \\ 1 & 2 & 1 \\ 1 & 2 & 4 \end{matrix}]$$

The *basic behavior* of this input/output matrix is a lexical listing of all observed states together with their observation frequencies:

$$b = [\begin{matrix} 1 & 1 & 2 \\ 1 & 1 & 3 \\ 1 & 2 & 1 \\ 1 & 2 & 4 \\ 2 & 1 & 3 \\ 2 & 2 & 1 \\ 2 & 2 & 2 \end{matrix}] \quad p = [\begin{matrix} 0.2 \\ 0.1 \\ 0.1 \\ 0.2 \\ 0.2 \\ 0.1 \\ 0.1 \end{matrix}]$$

In SAPS–II, the basic behavior can be computed using the statement:

$$[> \quad [b, p] = \text{BEHAVIOR}(io)$$

This gives rise to the following state transition matrix:

in\out	'1'	'2'	'3'	'4'
'11'	0.000	0.667	0.333	0.000
'12'	0.333	0.000	0.000	0.667
'21'	0.000	0.000	1.000	0.000
'22'	0.500	0.500	0.000	0.000

which shows on the left side the combined values of the two inputs, on the top row the values of the output, and in the table itself, the conditional probabilities. In out example, the two inputs are binary variables, whereas the single output has four levels. In addition, we need the absolute probabilities (observation frequencies) of the input states. For our example, the following values are found:

$$
\begin{array}{cc}
\text{input} & \text{prob} \\
\text{`11'} & 0.3 \\
\text{`12'} & 0.3 \\
\text{`21'} & 0.2 \\
\text{`22'} & 0.2
\end{array}
$$

In SAPS–II, the state transition matrix and input probability vector can be computed using the statement:

$$[> \quad [st, ip] = \text{STMATRIX}(io, 2)$$

where the second input argument denotes the number of input variables of the input/output matrix. In our example, the input/output matrix contains two inputs and one output.

Now we can compute the *Shannon entropy* [13.10] of the state transition matrix, which is a measure of the information content of the state transition matrix. The Shannon entropy is computed with the following formula:

$$
HM = -\sum_{\forall i} (p\{\text{inp} = \text{`}i\text{'}\}
$$
$$
\cdot \sum_{\forall j} (p\{\text{out} = \text{`}j\text{'}|\text{inp} = \text{`}i\text{'}\} \cdot \log_2(p\{\text{out} = \text{`}j\text{'}|\text{inp} = \text{`}i\text{'}\}))) \quad (13.25)
$$

In our example:

$$
\begin{aligned}
-HM =\ & 0.3 \cdot [0.667 \cdot \log_2(0.667) + 0.333 \cdot \log_2(0.333)] \\
& + 0.3 \cdot [0.667 \cdot \log_2(0.667) + 0.333 \cdot \log_2(0.333)] \\
& + 0.2 \cdot [1.0 \cdot \log_2(1.0)] \\
& + 0.2 \cdot [0.5 \cdot \log_2(0.5) + 0.5 \cdot \log_2(0.5)] \\
=\ & -0.275 - 0.275 + 0.0 - 0.2 \\
=\ & -0.75
\end{aligned}
$$

and thus:

$$HM = 0.75$$

The state transition matrix is completely deterministic if it contains one $+1$ element in every row, while all other elements are 0. In that case, the Shannon entropy is:

$$HM_{\min} = 0.0$$

The worst case occurs if all outcomes are equally probable, i.e., if the state transition matrix contains only elements of the same magnitude, in our case: 0.25 (since the output has four levels). For this case, we find the following Shannon entropy:

$$HM_{\max} = 2.0$$

The maximum entropy depends on the number of rows and columns of the state transition matrix. We can introduce an *uncertainty reduction measure*, which is defined as follows:

$$HR = 1.0 - \frac{HM}{HM_{\max}} \tag{13.26}$$

In SAPS–II, the Shannon entropy and uncertainty reduction measure of a state transition matrix can be determined using the statement:

$$[> \quad [HM, HR] = \text{ENTROPY}(st, ip)$$

HR can be used as a *quality measure*. In the worst case, HR is equal to 0.0, while in the best case, HR is equal to 1.0.

However, a problem remains with this approach. If we increase the complexity of the mask, we find that the state transition matrix becomes more and more deterministic. With growing mask complexity, more and more possible input states (combinations of levels of the various input variables) exist. Since the total number of observations n_{rec} remains constant, the observation frequencies of the observed states will become smaller and smaller. Very soon, we shall be confronted with the situation where every state that has ever been observed has been observed precisely once. This leads obviously to a completely deterministic state transition matrix. Yet the predictiveness of the model may still be very poor, since already the next predicted state has probably never before been observed, and that will be the end of our forecasting. Therefore, we must include this consideration in our quality measure.

I mentioned earlier that from a statistical point of view we would like to make sure that every state is observed at least five times [13.7]. Therefore, we introduce an *observation ratio* [13.8]:

$$OR = \frac{5 \cdot n_{5\times} + 4 \cdot n_{4\times} + 3 \cdot n_{3\times} + 2 \cdot n_{2\times} + n_{1\times}}{5 \cdot n_{leg}} \tag{13.27}$$

where:

n_{leg} = number of legal input states;
$n_{1\times}$ = number of input states observed only once;
$n_{2\times}$ = number of input states observed twice;
$n_{3\times}$ = number of input states observed thrice;
$n_{4\times}$ = number of input states observed four times; and
$n_{5\times}$ = number of input states observed five times or more.

If every legal input state has been observed at least five times, OR is equal to 1.0. If no input state has been observed at all (no data), OR is equal to 0.0. Thus, OR also qualifies for a quality measure.

We define the *quality of a mask* as the product of its uncertainty reduction measure and its observation ratio:

$$Q = HR \cdot OR \tag{13.28}$$

The *optimal mask* is the mask with the largest Q value.

The OPTMASK function can be used to compute all these quantities. The full syntax of this function is as follows:

$$[> \quad [mask, HM, HR, Q, mhis] = \text{OPTMASK}(raw, mcan, maxcompl)$$

mask is the optimal mask found in the optimization. *HM* is a row vector that contains the Shannon entropies of the best masks for every considered complexity. *HR* is a row vector that contains the corresponding uncertainty reduction measures. Q is a row vector that contains the corresponding quality measures, and *mhis* is the mask history matrix. The *mask history matrix* contains, concatenated to each other from the right, the best masks at each of the considered complexities. One of these masks is the optimal mask, which for reasons of convenience is also returned separately.

Until now, we haven't used our fuzzy membership functions. Remember that the fuzzy membership associated with the value of a qualitative variable is a *measure of confidence*. It specifies how confident we are that the assigned value is correct. If we compute the input/output matrix, we can assign a confidence to each row. The *confidence* of a row of the input/output matrix is the joint membership of all the variables associated with that row [13.8].

Let me demonstrate this concept by means of our simple three–variable example. Assume that the following fuzzy membership matrix accompanies our raw data matrix:

$$
\begin{array}{ll}
raw = \begin{bmatrix}
2 & 1 & 1 \\
1 & 2 & 3 \\
1 & 1 & 4 \\
1 & 1 & 3 \\
1 & 1 & 2 \\
2 & 2 & 2 \\
2 & 1 & 2 \\
2 & 2 & 3 \\
1 & 2 & 1 \\
1 & 2 & 1 \\
2 & 1 & 4
\end{bmatrix}
&
Memb = \begin{bmatrix}
0.61 & 1.00 & 0.83 \\
0.73 & 0.77 & 0.95 \\
0.73 & 0.88 & 1.00 \\
0.51 & 0.91 & 1.00 \\
0.55 & 0.92 & 0.92 \\
0.71 & 0.77 & 0.78 \\
0.63 & 0.91 & 0.69 \\
0.86 & 0.83 & 0.83 \\
0.77 & 0.97 & 0.70 \\
0.78 & 0.93 & 0.75 \\
0.89 & 0.81 & 1.00
\end{bmatrix}
\end{array}
$$

The joint membership of i membership functions is defined as the smallest individual membership:

$$
Memb_{joint} = \bigcap_{\forall i} Memb_i = \inf_{\forall i}(Memb_i) \stackrel{\mathrm{def}}{=} \min_{\forall i}(Memb_i) \tag{13.29}
$$

SAPS–II's FIOMODEL function computes the input/output matrix together with the confidence vector:

$$
[> \quad [io, conf] = \text{FIOMODEL}(raw, mask)
$$

Applied to our raw data matrix and using the same mask as before:

$$
mask = \begin{bmatrix}
-1 & -2 & 0 \\
0 & 0 & +1
\end{bmatrix}
$$

we find:

$$
\begin{array}{ll}
io = \begin{bmatrix}
2 & 1 & 3 \\
1 & 2 & 4 \\
1 & 1 & 3 \\
1 & 1 & 2 \\
1 & 1 & 2 \\
2 & 2 & 2 \\
2 & 1 & 3 \\
2 & 2 & 1 \\
1 & 2 & 1 \\
1 & 2 & 4
\end{bmatrix}
&
conf = \begin{bmatrix}
0.61 \\
0.73 \\
0.73 \\
0.51 \\
0.55 \\
0.69 \\
0.63 \\
0.70 \\
0.75 \\
0.78
\end{bmatrix}
\end{array}
$$

The *conf* vector indicates how much confidence we have in the individual rows of our input/output matrix. We can now compute the basic behavior of the input/output model. Rather than counting the observation frequencies, we shall accumulate the confidences. If a state has been observed more than once, we gain more and more confidence in it. Thus, we sum up the individual confidences. In SAPS–II, this can be achieved using the statement:

$$[> \quad [b, c] = \text{FBEHAVIOR}(io, conf)$$

Applied to our simple example, we find:

$$b = [\ \begin{matrix} 1 & 1 & 2 \\ 1 & 1 & 3 \\ 1 & 2 & 1 \\ 1 & 2 & 4 \\ 2 & 1 & 3 \\ 2 & 2 & 1 \\ 2 & 2 & 2 \end{matrix} \ ] \qquad c = [\ \begin{matrix} 1.06 \\ 0.73 \\ 0.75 \\ 1.51 \\ 1.24 \\ 0.70 \\ 0.69 \end{matrix} \ ]$$

Notice that the c vector is no longer a probability. The c_i elements no longer add up to 1.0.

This leads now to a modified state transition matrix. The SAPS–II statement:

$$[> \quad [st, ic] = \text{FSTMATRIX}(io, conf, 2)$$

produces the following results:

in \ out	'1'	'2'	'3'	'4'
'11'	0.00	1.06	0.73	0.00
'12'	0.75	0.00	0.00	1.51
'21'	0.00	0.00	1.24	0.00
'22'	0.70	0.69	0.00	0.00

which shows on the left side the combined values of the two inputs, in the top row the values of the output, and in the table itself, the confidence values. The total input confidence vector is:

input	conf
'11'	1.79
'12'	2.26
'21'	1.24
'22'	1.39

The total input confidences are computed by summing the individual confidences of all occurrences of the same input in the basic behavior. Notice that in all these computations, the actual qualitative variables are exactly the same as before, only their assessment has changed. The previously used probability measure has been replaced by a fuzzy measure.

The optimal mask analysis can use the fuzzy measure as well. The statement:

$$[> \ [mask, HM, HR, Q, mhis] = \text{FOPTMASK}(raw, Memb, mcan, maxcompl)$$

uses the fuzzy measure to evaluate the optimal masks. In order to be able to still use the Shannon entropy, we normalize the row sums of the state transition matrices to 1.0. It can happen that FOPT-MASK picks another mask as its optimal mask than the previously used OPTMASK routine. Since we use more information about the real system, we shall obtain a higher mask quality in most cases. Notice that the concept of applying the Shannon entropy to a confidence measure is somewhat dubious on theoretical grounds since the Shannon entropy was derived in the context of probabilistic measures only. For this reason, some scientists prefer to replace the Shannon entropy by other types of performance indices [13.6,13.9], which have been derived in the context of the particular measure chosen. However, from a practical point of view, numerous simulation experiments performed by my students and me have shown the Shannon entropy to also work satisfactorily in this context.

13.5 Forecasting Behavior

Once we have determined the optimal mask, we can compute the input/output model resulting from applying the optimal mask to the raw data and we can compute the corresponding state transition matrix. Now we are ready to forecast the future behavior of our system, i.e., we are ready to perform a qualitative simulation.

Forecasting is a straightforward procedure. We simply loop over input states in our input/output model and forecast new output states by reading out from the state transition matrix the most probable output given the current input. Let me explain this procedure by means of the previously used example. Given the raw data matrix:

$$raw = \begin{bmatrix} 2 & 1 & 1 \\ 1 & 2 & 3 \\ 1 & 1 & 4 \\ 1 & 1 & 3 \\ 1 & 1 & 2 \\ 2 & 2 & 2 \\ 2 & 1 & 2 \\ 2 & 2 & 3 \\ 1 & 2 & 1 \\ 1 & 2 & 1 \\ 2 & 1 & 4 \end{bmatrix}$$

which is assumed to consist of two input variables and one output variable. The future inputs over the next four steps are:

$$inp = \begin{bmatrix} 1 & 1 \\ 1 & 1 \\ 2 & 2 \\ 2 & 1 \end{bmatrix}$$

and we wish to forecast the output vector over the same four steps. Let me assume furthermore that the optimal mask is the one used earlier:

$$mask = \begin{bmatrix} -1 & -2 & 0 \\ 0 & 0 & +1 \end{bmatrix}$$

For this case, we have already computed the input/output model and the state transition matrix. After the input/output model has been computed, the mask covers the final two rows of the raw data matrix. In order to predict the next output, we simply shift the mask one row further down. The next input set is thus: '2 1.' From the state transition matrix, we find that this input leads in all cases to the output '3.' Thus, we copy '3' into the data matrix at the place of the next output. We then shift the mask one row further down. At this time, the mask reads the input set '1 1.' From the state transition matrix, we find that the most probable output is '2,' but its probability is only 66.7%. We continue in the same manner. The next input set is again '1 1.' Since this input set is assumed to be statistically independent of the previous one (an unreasonable but commonly made assumption), the joint probability is the product of the previous cumulative probability with the newly found probability, thus $p = (2/3) \cdot (2/3) = 4/9 = 44.4\%$. The next input set is '22.' For this case, we find that the outcomes '1' and '2' are equally likely (50%). Thus, we arbitrarily pick one of those. The cumulative output probability has meanwhile decreased to 22.2%.

This is exactly how, in SAPS–II, the FORECAST routine predicts future states of a recoded system.

$$[> \quad [f2, p] = \text{FORECAST}(f1, mask, nrec, minprob)$$

forecasts the future behavior of a given system $f1$ where $f1$ contains the raw data model concatenated from below with the future inputs filled from the right with arbitrary zero values, thus:

$$[> \quad f1 = [raw; inp, \text{ZROW}(nstp, nout)]$$

where $nstp$ denotes the number of steps to be forecast and $nout$ denotes the number of output variables in the raw data model. $mask$ is the optimal mask to be used in the forecasting; $nrec$ denotes the number of recorded past data values, i.e., $nrec$ tells the forecasting routine how many of the rows of $f1$ belong to the past and how many belong to the future; and $minprob$ instructs SAPS to terminate the forecasting process if the cumulative probability decreases below a given value. This feature can be disabled by setting $minprob$ to zero.

Upon return, $f2$ contains the same information as $f1$ but augmented by the forecast outputs, i.e., some or all of the ZROW values have been replaced by forecasts. p is a column vector containing the cumulative probabilities. Of course, up to row $nrec$, the probabilities are all 1.0 since these rows contain past, i.e., factual, information.

If during the forecasting process an input state is encountered that has never before been recorded, the forecasting process comes to a halt. It is then the user's responsibility to either collect more data, reduce the number of levels, or pick an arbitrary output and continue with the forecasting.

How can we make use of the fuzzy memberships in the forecasting process? The procedure is very similar to our previous approach. However, in this case, we don't pick the output with the highest confidence. Instead, we compare the membership and side functions of the new input with the membership and side functions of all previous recordings of the same input and pick as the output the one that belongs to the previously recorded input with the most similar membership [13.8]. For this purpose, we compute a cheap approximation of the regenerated continuous signal:

$$d = 1 + side * (1 - Memb) \tag{13.30}$$

for every input variable of the new input set and store the regenerated d_i values in a vector. We then repeat this reconstruction for all previous recordings of the same input set. We finally compute the $\mathcal{L}_2$ norms of the difference between the d vector of the new input and the d vectors of all previous recordings of the same input, and pick the one with the smallest $\mathcal{L}_2$ norm. Its *output* and *side* values are then used as forecasts for the *output* and *side* values of the current state. We proceed a little differently with the membership values. Here, we take the five previous recordings with the smallest $\mathcal{L}_2$ norms and compute a distance–weighted average as the forecast for the fuzzy membership values of the current state.

In SAPS–II, this is accomplished by use of the function:

$$[> \quad [f2, Memb2, side2] = \text{FFORECAST}(f1, Memb1, side1, mask, nrec)$$

The fuzzy forecasting function will usually give us a more accurate forecast than the probabilistic forecasting function. Also, if we use fuzzy forecasting, we can retrieve pseudocontinuous output signals with a relatively high quality using the REGENERATE function.

13.6 A Linear System – An Example

Let us once more analyze the same example that we discussed in Chapter 12:

$$\dot{\mathbf{x}} = \mathbf{A} \cdot \mathbf{x} + \mathbf{b} \cdot u$$
$$= \begin{pmatrix} 0 & 1 & 0 \\ 0 & 0 & 1 \\ -2 & -3 & -4 \end{pmatrix} \cdot \mathbf{x} + \begin{pmatrix} 0 \\ 0 \\ 1 \end{pmatrix} \cdot u \qquad (13.31a)$$

$$y = \mathbf{C} \cdot \mathbf{x} + \mathbf{d} \cdot u$$
$$= \begin{pmatrix} 1 & 0 & 0 \\ 0 & 1 & 0 \\ 0 & 0 & 1 \end{pmatrix} \cdot \mathbf{x} + \begin{pmatrix} 0 \\ 0 \\ 0 \end{pmatrix} \cdot u \qquad (13.31b)$$

This time we shall use fixed values for the parameters and the entire state vector as output. Figure 13.8 shows the three Bode diagrams of this multivariable system superposed onto a single graph.

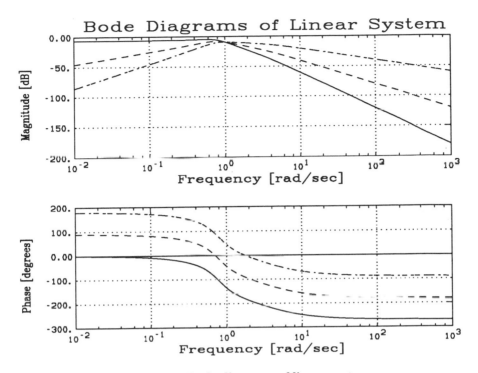

Figure 13.8. Bode diagrams of linear system.

In this example, I computed the Bode diagrams directly in CTRL–C. However, for any stable physical system, we can measure the Bode diagram; thus, this does not compromise the generality of the approach. We see that the band width of this system is $\omega_{3dB} \approx 1$ sec^{-1}. Therefore, the settling time is $t_s \approx 6$ sec. We wish to use a mask with a depth of three, and therefore, the communication interval should be $\delta t \approx 3$ sec.

In order to exert all frequencies of this system in an optimal manner, we shall simulate this system (directly in CTRL–C or MATLAB) applying a binary random sequence as the input signal [13.2]. We decided to recode each of the output states into three levels (the input is already binary), and therefore, the number of legal states can be computed as:

$$n_{leg} = \prod_{\forall i} k_i = 2 \cdot 3 \cdot 3 \cdot 3 = 54 \tag{13.32}$$

and the required number of recordings is:

$$n_{rec} = 5 \cdot n_{leg} = 270 \qquad\qquad (13.33)$$

Let us simulate the system over 300 communication intervals. This is accomplished as follows:

```
[>   t = 0 : 3 : 900;
[>   u =ROUND(RAND(t));
[>   x0 =ZROW(3,1);
[>   SIMU('ic', x0);
[>   y =SIMU(a, b, c, d, u, t);
```

Figure 13.9 shows the results of the continuous–time simulation.

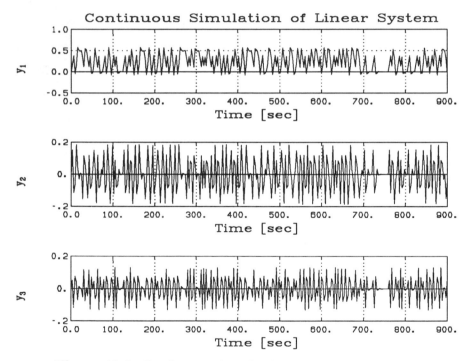

Figure 13.9. Continuous–time simulation of linear system.

Notice that we use the continuous–time simulation here as we could have used a digital oscilloscope. We shall not make any use of the fact that we know the structure of our system.

We shall use our optimal recoding algorithm to discretize our three output variables:

```
[>   do saps:saps
[>   meas = [u', y'];
[>   m = meas;
[>   for i = 2 : 4, ...
         [indx, mi] =SORT(meas(:, i)); ...
         m(:, i) = mi; ...
     end
[>   LM = m(1, :);
[>   LM = [LM; 0.5 * (m(100, :) + m(101, :))];
[>   LM = [LM; 0.5 * (m(200, :) + m(201, :))];
[>   LM = [LM; m(300, :);
[>   raw = meas;
[>   to = 1 : 3;
[>   for i = 2 : 4, ...
         from = [LM(1 : 3, i), LM(2 : 4, i)]'; ...
         r =RECODE(meas(:, i), 'domain', from, to); ...
         raw(:, i) = r; ...
     end
```

This code segment sorts each trajectory (column) vector separately, then subdivides the sorted vector into three segments of equal size to determine the optimal landmark values (*LM*). Thereafter, the measurement data are recoded separately for each trajectory. At the end of the code segment, *raw* contains the recoded raw data matrix.

We are now ready to search for the optimal masks. We operate on three separate mask candidate matrices, one for each of the three outputs. We shall compute the quality vector and mask history matrix since it turns out that we shall also put the suboptimal masks to good use. We shall keep the three best masks and sort them in order of decreasing quality. The following code segment shows the optimal mask analysis for the first output. The other two masks are computed accordingly. Notice that we shall use the first 270 rows of the raw data matrix only.

```
[>   rraw = raw(1 : 270, :);
[>   mcan = −ONES(3, 4);
[>   mcan(3, 2 : 4) = [1, 0, 0];
[>   [mask, hm, hr, q1, mhis1] =OPTMASK(rraw, mcan, 5);
[>   indx =SORT(q1);
[>   m1a = mhis1(:, 4 * (indx(1) − 1) + 1 : 4 * indx(1));
[>   m1b = mhis1(:, 4 * (indx(2) − 1) + 1 : 4 * indx(2));
[>   m1c = mhis1(:, 4 * (indx(3) − 1) + 1 : 4 * indx(3));
[>   m1 = [m1a, m1b, m1c];
```

At the end of this code segment, $m1$ contains the three best masks concatenated to each other from the right.

We are now ready to forecast. During the forecasting process, it will happen from time to time that an input state is encountered that has never before been recorded. In this case, the forecasting routine will come to a halt and leave it up to the user what to do next.

We decided to try the following strategy: since the input state depends on the masks, we simply repeat the forecasting step with the next best mask hoping that the problem goes away. If this doesn't help, we try the third mask. This is the reason why I saved the suboptimal masks.

I coded the forecasting in a separate routine, called FRC, which is called from the main procedure as follows:

```
[>   deff frc
[>   inpt = raw(271 : 300, 1);
[>   pred = FRC(rraw, inpt, m1, m2, m3);
```

At this point, it should have become clear why I simulated over 300 steps although I needed only 270 recordings. The final 30 steps of the continuous–time simulation will be used to validate the forecast.

The FRC routine operates in the following way: We loop over the 30 steps of the forecast. In each step, we call the SAPS–II routine FORECAST three times, once with each of the three optimal masks to forecast one value only. At the end of the step, we concatenate the new row (forecast) to the raw data from below and repeat.

If FORECAST isn't able to predict a value since the input state has never been seen before, it returns the raw data unchanged, i.e., the number of rows upon output is the same as upon input. In that case, we repeat the FORECAST with the next best mask. If none of the three best masks is able to predict the next step, we pick a value at random. The following code segment shows how FRC works:

```
//[frcst] = FRC(raw,inpt,m1,m2,m3);
   m1a = m1(:,1 : 4);  m1b = m1(:,5 : 8);  m1c = m1(:,9 : 12);
   m2a = m2(:,1 : 4);  m2b = m2(:,5 : 8);  m2c = m2(:,9 : 12);
   m3a = m3(:,1 : 4);  m3b = m3(:,5 : 8);  m3c = m3(:,9 : 12);
   r = raw;
   [row, col] = SIZE(raw);
   [n, m] = SIZE(inpt);
   for i = 1 : n, ...
      in = inpt(i); ...
      fc = [in,ZROW(1,3)]; ...
      fcc = [r; fc]; ...
      ff1 =FORECAST(fcc, m1a, row + i - 1, 0); ...
      [rf, cf] =SIZE(ff1); ...
      if rf <> row + i, ...
         ff1 =FORECAST(fcc, m1b, row + i - 1, 0); ...
         [rf, cf] =SIZE(ff1); ...
         if rf <> row + i, ...
            ff1 =FORECAST(fcc, m1c, row + i - 1, 0); ...
            [rf, cf] =SIZE(ff1); ...
            if rf <> row + i, ...
               ff1 = [ff1;ROUND(RAND(1,4))]; ...
            end, ...
         end, ...
      end, ...
      ... //
      ... // Same code for ff2 and ff3
      ... //
      ff = [in, ff1(row + i, 2), ff2(row + i, 3), ff3(row + i, 4)]; ...
      r = [r; ff]; ...
   end
   frcst = r;
return
```

I then compared the data from the simulation with the forecast data:

```
[>   simdat = raw(271 : 300, :);
[>   frcdat = pred(271 : 300, :);
[>   error = simdat - frcdat;
```

and these are the data that I found:

$$
simdat = \begin{bmatrix}
0 & 3 & 1 & 1 \\
0 & 1 & 1 & 3 \\
0 & 1 & 2 & 2 \\
0 & 1 & 2 & 2 \\
0 & 1 & 2 & 2 \\
1 & 2 & 3 & 3 \\
0 & 3 & 1 & 1 \\
1 & 2 & 3 & 3 \\
0 & 2 & 1 & 1 \\
0 & 1 & 1 & 3 \\
0 & 1 & 2 & 2 \\
1 & 2 & 3 & 3 \\
1 & 3 & 2 & 1 \\
0 & 2 & 1 & 2 \\
1 & 1 & 3 & 3 \\
0 & 3 & 1 & 1 \\
0 & 1 & 1 & 3 \\
1 & 2 & 3 & 3 \\
0 & 3 & 1 & 1 \\
1 & 2 & 3 & 3 \\
1 & 3 & 3 & 1 \\
0 & 2 & 1 & 1 \\
0 & 1 & 2 & 3 \\
1 & 2 & 3 & 2 \\
0 & 3 & 1 & 1 \\
1 & 2 & 3 & 3 \\
1 & 3 & 3 & 1 \\
1 & 3 & 2 & 2 \\
1 & 3 & 2 & 2 \\
0 & 2 & 1 & 1
\end{bmatrix}
\qquad
frcdat = \begin{bmatrix}
0 & 3 & 1 & 1 \\
0 & 1 & 1 & 3 \\
0 & 1 & 2 & 2 \\
0 & 1 & 2 & 2 \\
0 & 1 & 2 & 2 \\
1 & 2 & 3 & 3 \\
0 & 3 & 1 & 1 \\
1 & 2 & 3 & 3 \\
0 & 2 & 1 & 1 \\
0 & 1 & 1 & 3 \\
0 & 1 & 2 & 2 \\
1 & 2 & 3 & 3 \\
1 & 3 & 2 & 1 \\
0 & 2 & 1 & 2 \\
1 & 1 & 3 & 3 \\
0 & 3 & 1 & 1 \\
0 & 1 & 1 & 3 \\
1 & 2 & 3 & 3 \\
0 & 3 & 1 & 1 \\
1 & 2 & 3 & 3 \\
1 & 3 & 3 & 1 \\
0 & 2 & 1 & 1 \\
0 & 1 & 2 & 3 \\
1 & 2 & 3 & 2 \\
0 & 3 & 1 & 1 \\
1 & 2 & 3 & 3 \\
1 & 3 & 3 & 1 \\
1 & 3 & 2 & 2 \\
1 & 3 & 2 & 2 \\
0 & 2 & 1 & 1
\end{bmatrix}
$$

It can be seen that the 30 data rows do not contain even a single error. The forecasting procedure worked beautifully.

It is very easy to replace the RECODE, OPTMASK, and FORE-CAST functions with their fuzzy counterparts. Also in this case, the forecasting works without a single error. However, now we can use the REGENERATE routine to obtain a forecast of the continuous–time signals.

Figure 13.10 displays the true signals with the regenerated ones superimposed.

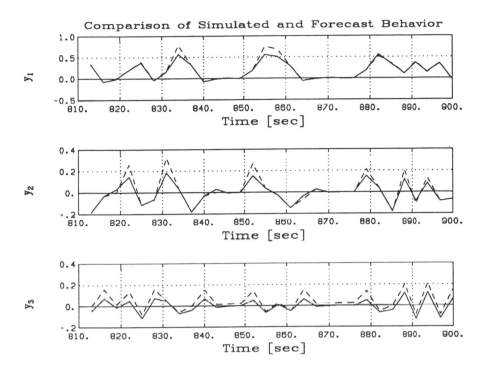

Figure 13.10. True and regenerated continuous–time signals.

The solid lines represent the results from the continuous simulation. They look discontinuous because of the large communication interval of 3 sec used in the simulation. CTRL–C's (MATLAB's) plot routine uses linear interpolation between communication points. The dashed lines are the pseudocontinuous signals that were regenerated from the forecast using the fuzzy membership functions. If you are interested in how Fig.13.10 was produced, solve Hw.[H13.1].

Since this example gave us sensational results, let us check whether we can capitalize on this idea.

13.7 Gambling the Stock Market

Figure 13.11 shows the end–of–day prices of a stock of a particular company recorded over a period of 775 days.

Figure 13.11. End–of–day prices of a company.

These are real data. The name of the company is unimportant, especially since the company obviously didn't fare too well over these two years.

We notice immediately a problem with these data. Even if the data would follow a straight decreasing line, SAPS would have problems predicting anything since each new value is "new" (has never occurred before). Before we can apply SAPS to these data, it is important that we manipulate the data in such a way that they become stationary, yet the original data can be reconstructed at any time. This process is called *detrending* of the data.

Many algorithms exist for the detrending of data. In some cases, we may decide to compute an *informal derivative*:

$$y_k = x_k - x_{k-1} \tag{13.34}$$

which, in SAPS–II, can be computed using the function:

$$[> \quad y = \text{DIFF}(x, 1)$$

In the world of finances, it is more common to compute the *daily return*, which is defined as:

$$ret_k = \frac{pr_k - pr_{k-1}}{pr_{k-1}} \tag{13.35}$$

The return on day k is defined as the end–of–day price on day k minus the end–of–day price on day $k - 1$ divided by the end–of–day price on day $k - 1$. SAPS–II doesn't provide for a function to

compute the daily return directly, but it is a trivial task to define
such a function:

```
// [y] = RETRN(x)
   [n, m] =SIZE(x);
   for = 1 : m, ...
      y(:, i) =DIFF(x, 1) ./ x(1 : n − 1, i); ...
   end
return
```

Figure 13.12 shows the daily return for the same company.

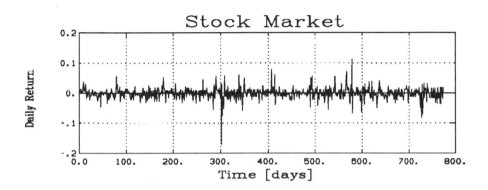

Figure 13.12. Daily return of a company.

The data are now clearly detrended. It looks pretty much like a noise
signal. Since we have nothing better to go by, let us try to predict
future values of return from earlier values of return, i.e., we use a
sort of recursive filter for prediction.

All we are interested in is whether the value of the stock will
increase or decrease. Thus, we use our optimal recoding algorithm
to recode the daily return data into the three levels 'up,' 'stationary,'
and 'down,' such that each of these levels is recorded equally often.
In SAPS–II, we shall represent 'down' as '1,' 'stationary' as '2,' and
'up' as '3.' A natural gambling strategy would then be to sell stock
when the indicator is 'down,' to buy stock when it is 'up,' and to do
nothing when it is 'stationary.'

I decided on a mask depth of 10, performed an optimal mask
analysis using the first 500 data points (just like in the last section),
and tried to forecast the next 200 values. Then I computed the
error by subtracting the forecast values from the true values and I
computed a bar chart of the errors, which is shown in Fig.13.13.

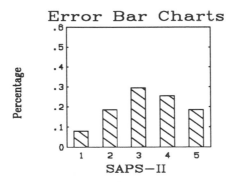

Figure 13.13. Bar chart of the SAPS–II forecasting error.

CTRL–C's bar chart routine doesn't place correct labels on the x–axis. The values shown can be interpreted as:

$$x\_value = true\_value - forecast\_value + 3$$

Thus, a value of '3' indicates that the forecast was correct, a value of '1' indicates that the true value had been 'down' while the predicted value had been 'up,' etc. Obviously, SAPS predicts the correct value in roughly 33% of all cases — which is not overly impressive taking into account that we operate with three levels.

To prove my point, I performed the following experiment. Instead of using an optimal mask analysis for forecasting, I simply picked, at random, a number between '1' and '3,' and called this my forecast. The error bar chart for this case is shown in Fig.13.14.

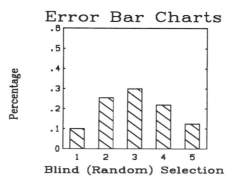

Figure 13.14. Bar chart of the blind forecasting error.

Clearly, the forecasting power of SAPS was nil in this case. SAPS performed about as well as the blind algorithm, i.e., we simply coded a very expensive random number generator. Since the sample size chosen was fairly small (200 samples), a large variability remains in the data, and the theoretical values of 33% correct answers, 22% answers off by one in either direction, and 11% answers off by two in either direction are not reflected accurately by our limited statistical experiment.

Can anything be done at all? Let us first look at the most obvious forecasting strategy: the return tomorrow will be the same as today. Figure 13.15 shows the error bar chart for this case.

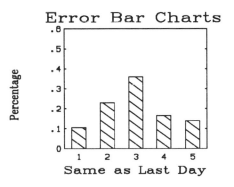

Figure 13.15. Bar chart of the persistent forecasting error.

Again, the results are similar, that is, little positive correlation exists within the data. Let us therefore try a more refined technique: linear regression. I postulate that the return at day k can be predicted from a weighted sum of the returns of the previous 10 days, i.e.:

$$x_k = \sum_{i=1}^{10} a_i \cdot x_{k-i} \qquad (13.36)$$

If we apply Eq.(13.36) to a number of days, e.g., one month and write all the resulting equations below each other, we obtain:

$$\begin{pmatrix} x_{k-1} \\ x_{k-2} \\ \vdots \\ x_{k-30} \end{pmatrix} = \begin{pmatrix} x_{k-2} & x_{k-3} & \cdots & x_{k-11} \\ x_{k-3} & x_{k-4} & \cdots & x_{k-12} \\ \vdots & \vdots & \ddots & \vdots \\ x_{k-31} & x_{k-32} & \cdots & x_{k-40} \end{pmatrix} \cdot \begin{pmatrix} a_1 \\ a_2 \\ \vdots \\ a_{10} \end{pmatrix} \qquad (13.37)$$

that is:

$$
\begin{pmatrix} a_1 \\ a_2 \\ \vdots \\ a_{10} \end{pmatrix} = \begin{pmatrix} x_{k-2} & x_{k-3} & \cdots & x_{k-11} \\ x_{k-3} & x_{k-4} & \cdots & x_{k-12} \\ \vdots & \vdots & \ddots & \vdots \\ x_{k-31} & x_{k-32} & \cdots & x_{k-40} \end{pmatrix}^{-1} \begin{pmatrix} x_{k-1} \\ x_{k-2} \\ \vdots \\ x_{k-30} \end{pmatrix} \tag{13.38}
$$

The matrix to be inverted is a so–called Hankel matrix (a matrix that is constant along its antidiagonals), and the inverse is, in fact, a pseudoinverse since the matrix is rectangular.

By using the pseudoinverse of the Hankel matrix in Eq.(13.38), we solve the overdetermined linear equation system in a least squares sense. This is called the linear regression problem. CTRL–C's (or MATLAB's) "\" operator can be used to formulate the linear regression problem.

The following CTRL–C code forecasts the return of days 501 to 700 on the basis of the available data. We recompute a new regression vector a for every new day on the basis of the previous month of measured data. The regression vector is then used to predict the actual (continuous) return for the next day only. The following code segment implements this algorithm:

```
[>    pdata = data
[>    for i = 501:700,...
          r = data(i − 2 : −1 : i − 40);...
          m = [r(1:30),r(2:31),r(3:32),r(4:33),r(5:34),...
              r(6:35),r(7:36),r(8:37),r(9:38),r(10:39)];...
          x = data(i − 1 : −1 : i − 30);...
          a = m\x;...
          pdata(i) = data(i − 1 : −1 : i − 10)' * a;...
      end
```

where m is the Hankel matrix as described in Eq.(13.38).

I then used my optimal recoding algorithm to recode the predicted data into three levels and compared the recoded prediction with the true data. Figure 13.16 shows the resulting error bar chart for this case.

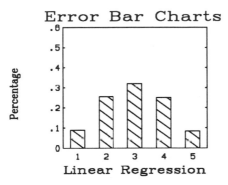

Figure 13.16. Bar chart of the regression forecasting error.

Obviously, the regression algorithm didn't fare much better. However, we have a number of parameters that can be varied: (i) the length of the regression vector, in our case 10; (ii) the number of columns of the Hankel matrix, in our case 30; and (iii) the frequency of recomputing the regression vector, in our case 1 day. The performance of the regression algorithm depends somewhat on the selection of these three parameters. I optimized these parameters and soon discovered that the best results are obtained when a square Hankel matrix is used and when the regression vector is recomputed every day. The results for a 2×2 Hankel matrix are shown in Fig.13.17.

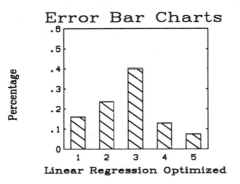

Figure 13.17. Bar chart of optimized regression forecasting error.

This time, we had some success. We now predict roughly 40% of the values correctly in comparison to 33% from before. Somewhat promising results were obtained for Hankel matrices of sizes 2×2 up to 7×7. For larger Hankel matrices, the predictive power is again reduced to noise level. The best results were found for the 2×2 case. This proves that the system exhibits at least some eigendynamics that can be captured and exploited, but the results are not very impressive.

Why did SAPS perform so poorly in this case? Two major reasons are to be mentioned. The first problem is the following: Every system's behavior is dictated by two components: (i) its eigendynamics and (ii) its inputs. This is true also for the stock market. Unfortunately, we have not captured the effects of the inputs at all. All our models were simple data filters that tried to predict future values of return from previous values of return. Obviously, the influence of external events on the stock market is formidable. However, we weren't able to capture those since we have no clear indicator as to what these inputs are. Since very little positive correlation exists between neighboring data points, it is insufficient to estimate these inputs indirectly by measuring their effects on the stock market. We are badly in need of direct measurements of these inputs which, at this point in time, we even don't know how to characterize. The second problem is the following: SAPS requires many data points to find a decent optimal mask. Since we didn't capture the full eigendynamics of the system but only a very limited excerpt of those, the system seems to change constantly. We cannot rely on 500 past data points (i.e., 1.5 years worth of data) to generate a model since the behavior of the system, as experienced through the daily return data, changes so quickly. This is what really broke our neck in the case of SAPS. The more successful regression model operated on a few days of data only, whereas SAPS needed 1.5 years worth of data in order to come up with a model. The regression model will not work either if we extend the size of the Hankel matrix or don't recompute the regression vector in regular intervals.

I am quite convinced that SAPS could work. The dominant time constant of the stock market is on the order of one day, thus the sampling rate is quite appropriate. We don't really need to rely on interday data in order to predict the stock market. Secondly, I am convinced that the behavior of the stock market is not as time–varying as it appears to be. This is only the case because we haven't yet determined which state variables to use. With the proper choice of inputs and state variables, SAPS could clearly outperform the regression model. The need to rely on so much past data shouldn't be a problem if we were able to reduce the apparent time dependence of the system. This is clearly a worthwhile research topic.

The true (and unsolvable) problem is the following: We shall never be able to predict the stock market far into the future since the influence of external inputs is at least half of the game. In order to retain the observed (mild) success with the regression model, we were able to predict only one day ahead. The probability of correctness of any longer predictions rapidly decay to the noise level of 33%. Within a short time span, however, the market price of an individual stock will not grow very much. Due to the commission that we have to pay in order to buy stock, we can't really exploit the system unless we are able to operate on a very large scale (since the commission to be paid depends on the trade volume). Thus, even with the best model in the world, small savers like you and me will have to rely on professional investors if we want our "gambling" to turn into an investment and not merely be a gamble.

13.8 Structure Characterization

We have seen that the mechanism of optimal mask analysis can provide us with an effective means to describe the behavior of a system. Yet since the methodology relies on an exhaustive search of possible masks from a set of mask candidates, this search will become prohibitively expensive if we make the set of mask candidates too large. Experience has shown that the number of variables among which the optimal mask is to be found should not be much larger than five and that the mask depth should usually be limited to three or four. This can pose a severe limitation of the technique since we are often confronted with systems with large numbers of candidate variables among which we should choose an appropriate subset for the optimal mask analysis. The stock market illustrates this problem very well. What are adequate inputs and which are appropriate state variables? The choices are manifold.

In this section, we shall study the problem of selecting subsets of variables for optimal mask analyses. We call this the *structure characterization* problem. Contrary to Chapter 11 where structure characterization had been defined as the process of identifying a particular function that relates one or several inputs to one or several outputs, we shall now define the structure characterization as the process of determining which variables are related to which other variables in a multivariable system.

This problem is more closely related to what we called *causality* in Chapter 11, and the classical statistical technique to decide this question is *correlation analysis*. If a strong (positive or negative) correlation exists between two variables, they belong to the same subsystem, i.e., they are causally related to each other. However, a problem remains with this concept. Let us assume that we measure two variables *a* and *b* related to each other through the equation:

$$a = b$$

Obviously, these variables have the largest positive correlation possible. Yet we should not select both *a* and *b* for our optimal mask analysis since once we select one of these variables the other does not add any additional information to our analysis. It is redundant.

In this section, we shall introduce a new technique called *reconstruction analysis* which can be used as an alternative to the previously introduced correlation analysis. However, while correlation analysis works well on continuous signals, reconstruction analysis has been devised to operate on recoded, i.e., discrete, signals.

SAPS–II distinguishes between three different types of structure representation. A *causal structure* lists the variables that form the subsystems as a row vector whereby subsequent subsystems are separated by a zero. Below this row vector, masks are coded that show the causal relation between the variables of the subsystems. For example, the structure shown in Fig.13.18

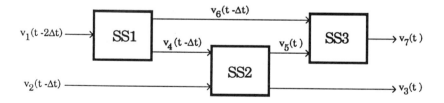

Figure 13.18. A causal structure.

can be coded as:

$$
istc = \begin{bmatrix}
1 & 4 & 6 & 0 & 2 & 3 & 4 & 5 & 0 & 5 & 6 & 7 \\
-1 & 0 & 0 & 0 & 0 & 0 & 0 & 0 & 0 & 0 & 0 & 0 \\
0 & 1 & 2 & 0 & -1 & 0 & -2 & 0 & 0 & 0 & -1 & 0 \\
0 & 0 & 0 & 0 & 0 & 1 & 0 & 2 & 0 & -2 & 0 & 1
\end{bmatrix}
$$

As indicated in the first row, one subsystem consists of the variables 1, 4, and 6. The mask written underneath this structure shows that variable 1 is an input (negative entry in the mask), while variables 4 and 6 are two outputs (positive entries in the mask).

By eliminating the masks from the causal structure, we obtain a *composite structure*. Obviously, the composite structure contains less information than the causal structure. The direction information is lost and the timing information is gone. In SAPS–II, the composite structure is represented by the first row of the causal structure:

$$istr = [\ 1\ \ 4\ \ 6\ \ 0\ \ 2\ \ 3\ \ 4\ \ 5\ \ 0\ \ 5\ \ 6\ \ 7\ ]$$

which can stand for the structure shown in Fig.13.19a,

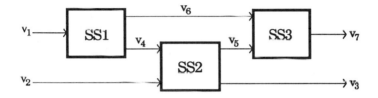

Figure 13.19a. Possible representation of a composite structure.

but which can also represent the quite different structure shown in Fig.13.19b.

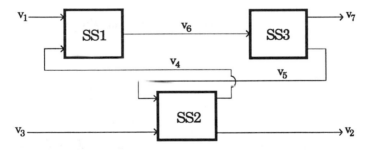

Figure 13.19b. Another representation of the same composite structure.

It can be easily verified that both systems contain the same three subsystems and are thus identified by the same composite structure. The composite structure no longer contains information about which variables are inputs and which are outputs.

The third structure representation available in SAPS–II is the binary structure. A *binary structure* is an ordered list of all binary relations between variables belonging to the same subsystem. The SAPS–II function:

$$[> \quad istb = \text{BINARY}(istr)$$

generates the binary structure out of the composite structure. For our example, the resulting binary structure is:

$$istb = \begin{bmatrix} 1 & 4 \\ 1 & 6 \\ 2 & 3 \\ 2 & 4 \\ 2 & 5 \\ 3 & 4 \\ 3 & 5 \\ 4 & 5 \\ 4 & 6 \\ 5 & 6 \\ 5 & 7 \\ 6 & 7 \end{bmatrix}$$

The binary structure can be easily obtained from the composite structure by drawing imaginary lines between any two variables of each subsystem, writing down the variable names (numbers) at the two ends of each such line as a pair, sorting all these pairs alphabetically, and eliminating all the redundant pairs.

Again, the binary structure contains less information than the composite structure it represents. For example, the system shown in Fig.13.20

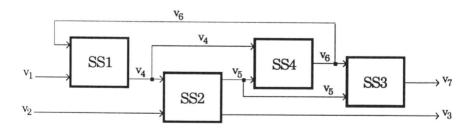

Figure 13.20. Another realization of the same binary structure.

with the composite structure:

$$istr = [\ 1\ 4\ 6\ 0\ 2\ 3\ 4\ 5\ 0\ 5\ 6\ 7\ 0\ 4\ 5\ 6\ ]$$

possesses the same binary structure as our previous system, although it contains an additional subsystem (SS4), and thus a different composite structure.

In SAPS–II, the command:

$$[> \quad istr = \text{COMPOSE}(istb)$$

produces the minimal composite structure among all possible composite structures representable by the same binary structure, i.e., it produces the composite structure with the smallest number of subsystems.

Similar to the optimal mask analysis, we wish to perform an *optimal structure analysis*, which we shall present with a set of candidate structures out of which the qualitatively best structure is selected. This means that we need to formulate a quality measure for structures. This measure is based on the composite structure.

Given the behavior model of a set of raw (recoded) data:

$$[> \quad [b, p] = \text{BEHAVIOR}(raw)$$

Notice that the behavior model does not imply knowledge of which variables are inputs and which are outputs. In the past, we always applied the BEHAVIOR function to input/output models, but this is not necessary. If we extract a particular subsystem from the behavior model, we simply throw away those columns that represent variables that are not included in the subsystem, we then merge rows that have become indistinguishable, add up their probabilities (or confidences), and sort the resulting behavior model in alphabetical order. In SAPS–II, this can be accomplished using the function:

$$[> \quad [b2, p2] = \text{EXTRACT}(b1, p1, istr)$$

where *istr* is a *primitive composite structure*, i.e., a composite structure that contains only one subsystem.

Given the behavior model:

$$b = \begin{bmatrix} 0 & 0 & 0 \\ 0 & 0 & 1 \\ 0 & 1 & 2 \\ 1 & 0 & 0 \\ 1 & 1 & 1 \\ 1 & 1 & 2 \end{bmatrix} \qquad p = \begin{bmatrix} 0.2 \\ 0.1 \\ 0.1 \\ 0.2 \\ 0.1 \\ 0.3 \end{bmatrix}$$

We can extract the first two variables using the command:

$$[> \quad [b12, p12] = \text{EXTRACT}(b, p, [1, 2])$$

which results in the behavior model:

$$b12 = [\; 0 \; 0 \qquad p12 = [\; 0.3$$
$$0 \; 1 \qquad\qquad 0.1$$
$$1 \; 0 \qquad\qquad 0.2$$
$$1 \; 1 \;] \qquad\qquad 0.4 \;]$$

We can alternatively extract the second and third variable using the command:

$$[> \quad [b23, p23] = \text{EXTRACT}(b, p, [2, 3])$$

which results in the behavior model:

$$b23 = [\; 0 \; 0 \qquad p23 = [\; 0.4$$
$$0 \; 1 \qquad\qquad 0.1$$
$$1 \; 0 \qquad\qquad 0.1$$
$$1 \; 1 \;] \qquad\qquad 0.4 \;]$$

We can then combine the two substructures using the command:

$$[> \quad [bb, pp] = \text{COMBINE}(b12, p12, b23, p23, [1, 2, 0, 2, 3])$$

where the fifth input parameter is again a composite structure that describes the nature of the recombination of substructures. The probability of a recombined state is computed as the probability of the particular state of the first subsystem multiplied with the conditional probability of the particular state of the second subsystem. For example, the probability of the state $[0, 0, 0]$ is 0.24 since the probability of the state $[0, 0]$ of the first subsystem is 0.3, and the conditional probability of the state $[0, 0]$ of the second subsystem is 0.8 (if the variable '2' has assumed a value of 0, variable '3' assumes a value of 0 in 80% of the cases and a value of 1 in 20% of the cases). The resulting behavior model is:

$$bb = [\; 0 \; 0 \; 0 \qquad pp = [\; 0.24$$
$$0 \; 0 \; 1 \qquad\qquad 0.06$$
$$0 \; 1 \; 1 \qquad\qquad 0.02$$
$$0 \; 1 \; 2 \qquad\qquad 0.08$$
$$1 \; 0 \; 0 \qquad\qquad 0.16$$
$$1 \; 0 \; 1 \qquad\qquad 0.04$$
$$1 \; 1 \; 1 \qquad\qquad 0.08$$
$$1 \; 1 \; 2 \;] \qquad\qquad 0.32 \;]$$

When we compare the recombined system with the original system, we find that two additional states have been introduced, but both of these have low probabilities of occurrence. We can compute the differences between the probabilities of the original behavior model and the recombined behavior model and compute the $\mathcal{L}_2$ norm of the difference vector. This is defined as the *reconstruction error* of the particular composite structure.

In SAPS–II, the reconstruction error of a particular composite structure can be computed directly using the function:

$$[> \quad crr = \text{STRUCTURE}(b, p, istr)$$

The optimal structure analysis is a *minimax problem*. Obviously, the totally connected structure has always a reconstruction error of 0.0 since nothing is extracted and nothing is recombined. Thus, we cannot simply minimize the reconstruction error. Instead, we try to maximize the number of subsystems (i.e., enhance the complexity of the internal structure of the system) while keeping the reconstruction error below a user–specified maximum allowed error. For any given complexity, we choose the structure that minimizes the reconstruction error.

How do we enumerate possible structures and how do we define the complexity of a structure? The structure search algorithm used in the optimal structure analysis bases on the binary structure representation of the system. The "complexity" of a structure is simply defined as the number of rows of the binary structure. For example, the binary structure shown earlier in this section has a complexity of 12. Notice that this definition of complexity is not congruent with the previously used term "complexity of the internal structure." The most "complex" system is the totally connected system that has no internal structure at all.

Three different structure optimization algorithms have currently been implemented. The *structure refinement algorithm* starts with a totally interlinked composite structure in which all possible binary relations are present. This structure, of course, shows no reconstruction error at all as there is nothing to be reconstructed. Binary relations are canceled one at a time and the structure is selected that exhibits the smallest reconstruction error. The iteration goes on by taking the selected binary structure and again canceling one binary relation at a time and picking the structure with the smallest reconstruction error. The search proceeds from the highest complexity to lower and lower complexities. At each level, the resulting reconstruction error will be either larger than before or the same. The search continues until the reconstruction error becomes too large.

The *structure aggregation algorithm* starts with a system in which each variable forms a substructure of its own that is not linked to any other variable. No binary relations are thus initially present and the reconstruction error of this structure is very large. Binary relations are added one at a time and at each level the structure is selected that shows the largest reduction of the reconstruction error. The iteration goes on until the reconstruction error has become sufficiently small.

The *single refinement algorithm* starts similar to the structure refinement algorithm. However, instead of canceling one binary relation only, all binary relations that exhibit a sufficiently small reconstruction error are canceled at once, and only one step of refinement is performed.

All three algorithms are suboptimal algorithms, since none of them investigates all possible structures. Therefore, in a sufficiently complex system, the three algorithms may well suggest three different structures, and it often pays to try them all. The single refinement algorithm is much cheaper than the other two algorithms, and yet it performs amazingly well. Thus, in a real–time (on–line) structure identification, this will probably be the algorithm of choice.

In SAPS–II, an optimal structure analysis is performed using the function:

$$[> \quad istr = \text{OPTSTRUC}(b, p, errmax, group, algor)$$

where $[b, p]$ is the behavior relation of the system to be analyzed, and *errmax* is the largest reconstruction error tolerated. The *group* parameter allows aiding the optimal structure algorithm by providing *a priori* knowledge about the structure to be analyzed. For example, the grouping information:

$$group = [\ 1 \quad 2 \quad 3 \quad 1 \quad 0 \quad 4\ ]$$

tells the optimization algorithm that, in a six–variable system, the first and the fourth variable appear in any subsystem either together or not at all, whereas the fifth variable is certainly disassociated, i.e., it appears only in one subsystem in which no other variable is represented. Thus, the six–variable system is effectively reduced to a four–variable system. If no *a priori* knowledge about the structure exists, the grouping information should be coded as:

$$group = 1 : 6$$

The *algor* parameter finally tells the analysis which of the three algorithms to use. Possible values are:

$$algor = \ 'REFINE'$$
$$algor = \ 'AGGREGATE'$$
$$algor = \ 'SINGLEREF'$$

istr is the resulting composite structure of the system.

One additional feature has been built into SAPS–II. After an optimization has taken place, the resulting structure can be postoptimized by applying the single refinement algorithm once more to each of its substructures. This algorithm is executed using the command:

$$[> \quad istp = \text{SINGLEREF}(istr, b, p, errmax)$$

As an example, let us once more consider the open heart surgery problem discussed earlier in this chapter. This problem consists of a six–variable model recoded as shown in Table 13.1. Expert knowledge was used to determine the landmarks between the five different levels for each variable. The recoded raw data model was stored away on a file using the CTRL–C command:

$$[> \quad \textbf{save } raw > saps{:}heart.dat$$

The following CTRL–C macro performs an optimal structure analysis on the previously saved raw data model:

```
[>   do saps:saps
[>   repo = 1;
[>   load < demo:heart.dat
[>   [b, p] =BEHAVIOR(raw);
[>   igr = 1 : 6
[>   ermax = 0.016
[>   is1 =OPTSTRUC(b, p, ermax, igr, 'refine')
[>   isr1 =SINGLEREF(is1, b, p, ermax)
[>   is2 =OPTSTRUC(b, p, ermax, igr, 'aggregate')
[>   isr2 =SINGLEREF(is2, b, p, ermax)
[>   is3 =OPTSTRUC(b, p, ermax, igr, 'singleref')
[>   isr3 =SINGLEREF(is3, b, p, ermax)
```

The analysis starts by applying the structure refinement algorithm to the behavior relation of the system. No *a priori* knowledge about the structure is yet available. The resulting composite structure is:

$$is1 = (1, 2, 4)(1, 3, 4)(2, 4, 5)(2, 4, 6)$$

A postoptimization of this structure leads to:

$$isr1 = (1, 2, 4)(1, 3, 4)(2, 4, 6)(5)$$

Obviously, the fifth variable (heart rate) is only weakly related to the other variables in the system.

Next, the structure aggregation algorithm is applied. This algorithm suggests the composite structure:

$$is2 = (2,4,6)(1,2,3,4)(5)$$

and postoptimization leads to:

$$isr2 = (3,4)(1,2,4)(2,4,6)(5)$$

Finally, I tried the single refinement algorithm. This time, the following composite structure is found:

$$is3 = (3,4)(4,6)(1,2,4)(5)$$

which is not modified by postoptimization.

Analyzing the different suggested structures, it becomes obvious that the heart rate does not need to be considered at all. It also shows that a very strong link exists between variables one and four. Thus, those two variables can easily be grouped together. The analysis is now repeated, applying the appropriate grouping information:

$$[> \quad igr = [\ 1 \ \ 2 \ \ 3 \ \ 1 \ \ 0 \ \ 4 \ ]$$

This time, all three algorithms suggest the same composite structure:

$$istr = (1,2,4)(1,3,4)(1,4,6)(5)$$

which seems to be a good working hypothesis for a continuation of the system analysis.

13.9 Causality

While the concept of a causal structure was introduced in the last section, we haven't made use of this concept yet. While the optimal structure analysis was able to reveal causal relations among variables, it has not helped us to determine causality relations between those variables.

We could employ the correlation techniques that were suggested in Chapter 11 for this purpose, i.e., after we have determined which variables belong together, we could compute cross–correlations pairwise between them and check whether one of the variables of any such pair lags significantly behind the other. Yet correlation analysis works better on continuous variables and it might be more appropriate to tackle the recoded data directly.

Two separate routes are basically open to solve this problem:

(1) We can apply an optimal mask analysis repetitively in a loop, i.e., we compute an optimal mask separately for every true output at time t (as we have done in the past) and then interpret some of the inputs to these optimal masks as outputs from another optimal mask shifted further back in time. If an input to an optimal mask is a true system input, no further mask needs to be evaluated to explain it. If it is a true output (evaluated at an earlier point in time), we already know the optimal mask — it is the same mask as for the current time simply shifted back a number of steps. However, if the input to the optimal mask is an auxiliary variable that is neither a true input nor a true output, we make it an output and compute an optimal mask for it.

(2) We realize that the optimal structure analysis is flat. The time dependence is not preserved. Remember that the structure analysis operates on the behavior model rather than on the raw data model. The behavior model does not retain the time flow information. In order to remedy this problem, we can duplicate the raw data model, shift it by one row, and write the shifted data to the side of the original data as additional variables. We can repeat this process for a number of times. In SAPS–II, this can be achieved by applying the IOMODEL function to the raw data with a mask in which all elements are inputs except for the last, which we shall call an output. If the mask has n_{depth} rows and n_{var} columns, we obtain an input/output model with $n_{var} \times n_{depth}$ columns. We can then apply an optimal structure analysis to the enhanced input/output model and reinterpret the resulting structure in terms of the original time–dependent variables.

Both techniques may be used to obtain a causal structure of our system. However, the resulting structures are not necessarily the same, and it is not obvious at this point what is the precise relationship between the resulting structures and when which of the two techniques might be more more appropriate to use than the other. Moreover, many "correct" structures may exist that faithfully represent a given data set. How can we discriminate among them? How can we gain additional evidence that will help us increase our faith in one of the possible structures and discard others? Many questions haven't been answered yet in this context. This is therefore an excellent area for research.

13.10 Summary

In this chapter, we introduced a number of pattern recognition techniques that can be used for qualitative simulation or forecasting of system behavior. Optimal mask analysis allows us to determine qualitative causality relations among a set of causally related variables. Optimal masks can be viewed as a sort of feature extractor. Similar patterns will lead to the same optimal mask, and therefore, the optimal mask can help us recognize similarities between patterns. These patterns can be either temporal patterns (such as time signals) or static patterns (such as images). Optimal mask analysis is still a far cry from automating the ability of my physician to remember faces after a long time, but it may be a first step toward mimicking this capability in a computer program. Optimal structure analysis allows us to reveal causal relations among a set of time–related variables. This technique can be useful as a means to identify important factors that are related to a particular event.

SAPS–II was introduced as a tool to qualitatively analyze systems using these techniques. SAPS–II is available as a CTRL–C function library and as a Pro–Matlab (PC–Matlab) toolbox. In this text, I reproduced the CTRL–C solutions to the presented algorithms. The MATLAB solutions look very similar.

The research described in this chapter can be attributed largely to one scientist and colleague, George Klir, a great thinker. I wish to acknowledge my gratitude for his constructive comments and frequent encouragements.

References

[13.1]   François E. Cellier (1987), "Prisoner's Dilemma Revisited — A New Strategy Based on the General System Problem Solving Framework," *International J. of General Systems*, **13**(4), pp. 323–332.

[13.2] François E. Cellier (1987), "Qualitative Simulation of Technical Systems Using the General System Problem Solving Framework," *International J. of General Systems*, **13**(4), pp. 333–344.

[13.3] François E. Cellier (1991), "General System Problem Solving Paradigm for Qualitative Modeling," in: *Qualitative Simulation, Modeling, and Analysis* (P.A. Fishwick and P.A. Luker, eds.), Springer–Verlag, New York, pp.51–71.

[13.4] François E. Cellier and David W. Yandell (1987), "SAPS–II: A New Implementation of the Systems Approach Problem Solver," *International J. of General Systems*, **13**(4), pp. 307 322.

[13.5] George J. Klir (1985), *Architecture of Systems Problem Solving*, Plenum Press, New York.

[13.6] George J. Klir (1989), "Inductive Systems Modelling: An Overview," *Modelling and Simulation Methodology: Knowledge Systems' Paradigms* (M.S. Elzas, T.I. Ören, and B.P. Zeigler, eds.), Elsevier Science Publishers B.V. (North–Holland), Amsterdam, The Netherlands.

[13.7] Averill M. Law and W. David Kelton (1990), *Simulation Modeling and Analysis*, second edition, McGraw–Hill, New York.

[13.8] DongHui Li and François E. Cellier (1990) "Fuzzy Measures in Inductive Reasoning," *Proceedings 1990 Winter Simulation Conference*, New Orleans, La., pp. 527–538.

[13.9] Glenn Shafer (1976), *A Mathematical Theory of Evidence*, Princeton University Press, Princeton, N.J.

[13.10] Claude E. Shannon and Warren Weaver (1964), *The Mathematical Theory of Communication*, University of Illinois Press, Urbana.

[13.11] Hugo J. Uyttenhove (1979), *SAPS — System Approach Problem Solver*, Ph.D. dissertation, SUNY Binghampton, N.Y.

[13.12] Pentti J. Vesanterä and François E. Cellier (1989), "Building Intelligence into an Autopilot — Using Qualitative Simulation to Support Global Decision Making," *Simulation*, **52**(3), pp. 111–121.

[13.13] Lotfi A. Zadeh (1985), "Syllogistic Reasoning in Fuzzy Logic and Its Application to Usuality and Reasoning with Dispositions," *IEEE Trans. Systems, Man, and Cybernetics*, **SMC**–15(6), pp. 754–763.

[13.14] Lotfi A. Zadeh (1986), "A Simple View of the Dempster–Shafer Theory of Evidence and Its Implication for the Rule of Combination," *The AI Magazine*, Summer Issue, pp. 85–90.

[13.15] Lotfi A. Zadeh (1987), "A Computational Theory of Dispositions," *International J. of Intelligent Systems*, **2**, pp. 39–63.

Bibliography

[B13.1] Russell L. Ackoff (1978), *The Art of Problem Solving*, Wiley–Interscience, New York.

[B13.2] Ludwig von Bertalanffy (1969), *General System Theory: Foundations, Development, Applications*, G. Braziller Publishing, New York.

[B13.3] G. Broekstra (1978), "On the Representation and Identification of Structure Systems," *International J. of Systems Science*, 9(11), pp. 1271–1293.

[B13.4] Brian R. Gaines (1979), "General Systems Research: Quo Vadis," *General Systems Yearbook*, **24**, pp. 1–9.

[B13.5] George J. Klir, Ed. (1978), *Applied General Systems Research*, Plenum Press, New York.

[B13.6] Allan Newell and Herbert A. Simon (1972), *Human Problem Solving*, Prentice–Hall, Englewood Cliffs, N.J.

[B13.7] John L. Pollock (1990), *Monic Probability and the Foundations of Induction*, Oxford University Press, New York.

[B13.8] Joseph E. Robertshaw, S. J. Mecca, and M. N. Rerick (1979), *Problem Solving: A Systems Approach*, McGraw–Hill, New York.

[B13.9] Herbert A. Simon (1972), "Complexity and the Representation of Patterned Sequences of Symbols," *Psychological Reviews*, **79**, pp. 369–382.

[B13.10] Herbert A. Simon (1977), *Models of Discovery and Other Topics in the Methods of Science*, Reidel Publishing, Boston, Mass.

[B13.11] G. Towner (1980), *The Architecture of Knowledge*, University Press of America, Washington, D.C.

Homework Problems

[H13.1] Fuzzy Forecasting

Implement the algorithms as presented in Section 13.6 and reproduce the probabilistic forecasting results shown in that section.

Implement a fuzzy forecasting routine:

$$[> \quad [frcst, Mf, sf] = \text{FFRC}([raw, Mr, sr], [inpt, Mi, si], m1, m2, m3)$$

It is necessary to concatenate the membership and side matrices from the right to the raw data matrix and the input matrix since CTRL–C limits the number of formal arguments of any function to no more than 10. FFRC is very similar to FRC, except that the calls to FORECAST are replaced by corresponding calls to FFORECAST.

Modify the main routine by replacing the 'domain' parameter with the 'fuzzy' parameter in RECODE and by replacing the OPTMASK calls by corresponding FOPTMASK calls.

Add at the end a call to REGENERATE to retrieve the continuous–time signals and plot the true signals (from the CTRL–C or MATLAB simulation) together with the retrieved ones as shown in Fig.13.10.

[H13.2] Political Modeling [13.5]

We wish to investigate the relationship between the political situation in this country and the long–range stock market trends. The political situation is captured in terms of three variables: P_P denotes the party affiliation of the President, C_H denotes the House control, and C_S denotes the Senate control. Each of these three variables is of the enumerated type with values 'R' for Republican and 'D' for Democrat. The stock market index is represented through the variable I_{SM}, which is also enumerated with the values 'u' for up and 'd' for down.

Table H13.2 lists the values of these four variables between 1897 and 1989.

Table H13.2. Political/Marketing Variables [13.5].

Period [years]	P_P	C_H	C_S	I_{SM}
1897 – 1901	R	R	R	u
1901 – 1905	R	R	R	u
1905 – 1909	R	R	R	u
1909 – 1913	R	R	R	u
1913 – 1917	D	D	D	d
1917 – 1921	D	D	D	d
1921 – 1925	R	R	R	d
1925 – 1929	R	R	R	u
1929 – 1933	D	R	R	u
1933 – 1937	D	D	D	u
1937 – 1941	D	D	D	d
1941 – 1945	D	D	D	d
1945 – 1949	D	D	D	u
1949 – 1953	D	D	D	d
1953 – 1957	R	R	R	u
1957 – 1961	R	R	D	u
1961 – 1965	D	D	D	u
1965 – 1969	D	D	D	u
1969 – 1973	R	D	D	d
1973 – 1977	R	D	D	d
1977 – 1981	D	D	D	d
1981 – 1985	R	D	R	u
1985 – 1989	R	D	D	u

It is assumed that the party affiliation of the President is an *exogenous variable*, i.e., it depends only on quantities that have not been included in the model (in particular, the personality of the candidate). However, the other three variables are *endogenous variables*, i.e., they are determined internally to the model. In terms of our normal nomenclature, the party affiliation of the President is an input variable, while the other three variables are output variables.

We wish to use optimal mask analysis to make predictions about the future of this political system. Start by computing three optimal masks for the three output variables. House and Senate control may depend upon past values of all four variables one and two steps back only. The stock market index may depend on these same variables plus the current value of the President's party affiliation. Forecast the future of the system over 12 years using these optimal masks. Repeat your forecast for all eight possible combinations of the input variable.

[H13.3] Model Validation

We wish to validate the model that was derived for the political system of Hw.[H13.2]. For this purpose, we recompute optimal masks using only the first 75% of the data. Forecast the behavior of the system to the current date using the previously found optimal masks, and compare the outcome with the real data. Repeat the forecast with some of the suboptimal masks in the mask history that have a similar quality and check which of these suboptimal masks produces the smallest forecasting error.

[H13.4] Boolean Logic

Given the memoryless logical circuit shown in Fig.H13.4.

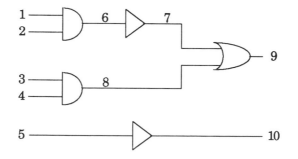

Figure H13.4. Circuit diagram of a logical circuit.

The circuit contains two logical AND gates, two logical inverters, and one logical OR gate. The fact that the circuit is not completely connected is of no concern.

Write a logical table of all possible states of this circuit. Enumerate all possible combinations of inputs and compute the set of resulting outputs. The enumeration results in a table containing 32 rows (possible states) and 10 columns (variables). Represent the logical false state as '0,' and the logical true state as '1.' This table can be interpreted as a raw data model.

We now forget where the raw data came from. We wish to relearn the structure of our circuit from the given raw data model using optimal structure analysis and optimal mask analysis.

Let us start with optimal structure analysis. We know that this circuit has no memory. Thus, every input/output relation is immediate. Therefore, we can apply the optimal structure analysis directly to the behavior model of the given raw data matrix. Try all three optimization algorithms. No grouping information is to be used. Set the maximum error *errmax* to the machine resolution *eps*. This is reasonable since we know that the data have been produced by a completely deterministic logical circuit. Thus, we cannot tolerate any reconstruction error at all. Set the SAPS system variable *repo* (internal output) to 1, so that you can see how the software reasons about your data. Interpret the results.

Let us now try to solve the same problem using optimal mask analysis. This time, we want to assume that we know that variables 1 to 5 are true system inputs, while variables 9 and 10 are true system outputs that furthermore do not depend directly on each other. Since we know that all input/output relations are immediate, we choose for each output a mask candidate matrix of depth one in which the five inputs and the three auxiliary variables are possible inputs. If we find that the optimal masks depend on any of the auxiliary variables, we repeat the analysis by making this auxiliary variable our new output. It can depend on the five inputs and all other auxiliary variables. We repeat the analysis until all optimal mask inputs have been reduced to true system inputs. Interpret the results.

The second technique provides us immediately with a causal structure. Manually derive the resulting composite structure. Compute the behavior model of the overall system and extract one after the other each of the subsystems. Compare the behavior models of the subsystems with the logical truth tables of the hardware components.

[H13.5] Boolean Logic

Given the memoryless logical circuit shown in Fig.H13.5.

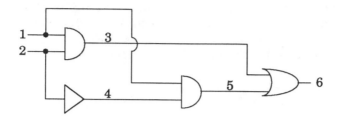

Figure H13.5. Circuit diagram of a logical circuit.

Repeat the analyses proposed in Hw.[H13.4] to this new model. This time, the raw data model has six columns (variables) but only four rows (legal states). What do you conclude?

[H13.6] Logic Circuit with Memory

Given the logical circuit with memory shown in Fig.H13.6.

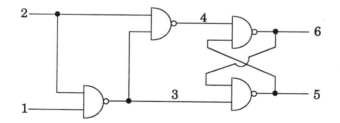

Figure H13.6. Circuit diagram of a logical circuit with memory.

This circuit consists of four NAND gates; however, contrary to the previous examples, this circuit is not memoryless.

This time, the approach used earlier won't work since not all input/output relations are single-valued. Instead, start with the signals '1,' '2,' and '5' in their false (i.e., '0') state. Compute the resulting values for the signals '3,' '4,' and '6.'

Then apply pseudorandom binary signals to the two inputs, i.e., at each "clock," each of the two inputs can either toggle or stay the same. For each clock, compute the resulting values of the two outputs '5' and '6,' and of the two auxiliary variables '3' and '4.' Repeat for 200 steps.

At this point, we have produced our raw data matrix. Notice that while the overall circuit is able to memorize, each of the four subcircuits is still memoryless. The memory is achieved by means of the feedback loops.

Apply the same techniques as in the last two homework problems to check whether you can relearn the structure of this system.

Now we wish to try yet another approach. This time, let me assume that we know for a fact that the circuit to be identified was built in NAND logic. Use the EXTRACT function of SAPS–II to associate each of the outputs with any two other signals, extract the three–variable subsystem, and check whether the resulting behavior model corresponds to the truth table of a NAND gate. If one or the other of the identified gate inputs is an auxiliary variable, repeat the analysis with this variable as the new output.

Since quite a bit of testing needs to be performed, we better write a general–purpose CTRL–C (MATLAB) function that can perform this analysis for an arbitrary behavior model.

[H13.7]* Logic Circuit with Faults

We want to repeat Hw.[H13.6]. This time, we wish to study the influence of faults in the raw data matrix.

We generate the raw data matrix in exactly the same manner as in Hw.[H13.6]. Then we disturb the raw data matrix by introducing a certain percentage of erroneous elements. For instance, 1%, 2%, 5%, or 10% of the elements in the raw data matrix are being assigned the wrong value.

Repeat the same analyses as in Hw.[H13.6] using the disturbed raw data matrix for the four degrees of fault severity suggested earlier. When using optimal structure analysis, it may now be necessary to raise the *errmax* parameter beyond the machine resolution *eps* in order to accommodate for the digital noise. When using the EXTRACT function, you need to allow for a somewhat inaccurate truth table. In each step, try all combinations of gate input variables and pick the one that gives you the smallest least squares error when compared to the truth table of the NAND gate.

Critically evaluate the capability of the three techniques to function properly under the influence of digital noise.

Projects

[P13.1] TicTacToe

Design a CTRL–C (MATLAB) routine that can play TicTacToe.

$$[> \quad new = \mathrm{TicTacToe}(old)$$

plays one step of TicTacToe. *old* refers to the board before the step. *old* is represented through a 3×3 matrix consisting of integer elements with the values 0, 1, and 2. A value of 0 indicates that the particular field is unoccupied, 1 indicates that the field is occupied by the program, and 2 indicates that the field is occupied by the human player. *new* refers to the board after the step has been executed. If the game starts with a zero matrix, the program will make the first move; if it starts with one field already occupied by a 2 element, the human player begins.

Begin by designing a set of logical rules that describe both the (syntactic) rules of the game (such as: 'only those fields can be occupied which are currently unoccupied,' or: 'only one field can be altered within one step'), as well as the (semantic) strategies (such as: 'if the opponent has currently occupied two fields in a line, and the third field in that line is currently unoccupied, then occupy that field immediately'). It is possible to describe the entire TicTacToe game by a set of seven logical rules.

Design a CTRL–C (MATLAB) routine that can interrogate the set of rules (the so–called rule base), and therefore, can play TicTacToe. This program is a so–called expert system.

Design a data collection routine that can observe the game and stores the data (the board) away in a raw data matrix. Design another MATLAB (CTRL–C) routine which, using SAPS–II functions, can analyze the raw data matrix by either observing how the program plays against you, or by observing how you play against the program. It can thereby relearn a set of logical rules (a behavior model) that describe both the syntactic rules and the semantic strategies of the game. In all likelihood, this automatically generated rule base will be considerably longer than the one that you created manually. When I tried it, the new rule base contained 124 rules.

Notice that you need to operate on an input/output model. If the board before the TicTacToe program plays is stored as input and the board after the TicTacToe program has played is stored as output, SAPS–II will learn the strategy of the TicTacToe program. On the other hand, if the board after the TicTacToe program has played is stored as input and the board before the TicTacToe program plays again is stored as output, SAPS–II will learn the strategy of the human player.

Let the expert system now play using the rule base that was automatically generated by SAPS–II rather than the manually generated rule base. If all goes well, the new program should be as good at playing TicTacToe as the original program was.

Notice the significance of this result. Instead of generating a rule base manually, the data collection routine could also have been used to observe two human players playing one against the other. Thus, we have found a methodology to automatically synthesize knowledge bases from observed data.

Replace TicTacToe by another game, such as the Prisoner's Dilemma problem [13.1]. The data collection routine must be modified to interface correctly with the new problem. However, the rule base synthesizer (the SAPS–II routine) and the expert system shell (the play routine) should be able to learn and play the new game without a single modification.

[P13.2] Intelligent Autopilot

Apply the techniques presented in Section 13.6 to the continuous simulation of a Boeing 747 jetliner in high–altitude horizontal flight, i.e., to the simulation program designed in Pr.[P4.1]. The purpose of this study is to check whether we can apply the proposed techniques as reliably to a highly nonlinear system as to a linear system.

Enhance the ACSL program designed in Pr.[P4.1] by a set of faulty operational modes such as heavy ice on the wings or loss of one of the four engines.

For each of the fault modes, identify a set of optimal masks that characterizes the fault.

Modify the ACSL program once more. This time, one of the faults should be chosen arbitrarily at a randomly selected point in time during the simulation.

Use the forecasting routine to identify when the accident has happened and discriminate the correct fault by comparing the recoded continuous data after the accident occurred and the transients resulting from the accident have died out with forecasts obtained using all of the stored fault masks. The forecast with the smallest deviation identifies (most likely) the type of fault that has occurred.

Research

[R13.1] Rule Base Minimization

Start with the two automatically generated rule bases of Pr.[P13.1]. The purpose of this research is to formulate an automated procedure that can reduce the synthesized rule base back to a minimal rule base. For this purpose, you might have to design a statistical extension to the Morgan rules of logic, or alternatively, a statistical extension to the Karnaugh diagram. If applied to the synthesized rule base of the TicTacToe problem, the result should be a rule base that looks similar to the rule base that had been manually designed to start with.

The solution of this problem is important for the understanding of automated knowledge generalization or abstraction mechanisms.

[R13.2] Heart Monitor

In collaboration with a heart surgeon, establish a set of variables to be monitored during open heart surgery. For a number of months, collect data during actual surgery and record what actions the surgeon took when and why on the basis of the data that he or she observed. Use the collected physical data as inputs and the actions of the surgeon as outputs. Use inductive reasoning to formulate a model that behaves similarly to the surgeon.

Design an apparatus that can thereafter observe data during actual surgery, reason about this data on–line in a similar fashion as the surgeon did before, and provide consulting advise for the same or another surgeon. If asked why the particular advice was given, the apparatus should be able to relate the decision back to the reasoning performed by the original surgeon. It is important that the model has some power of prediction, i.e., that it can forecast potential problems before they actually occur and provide, together with the actual advice, a verbal prediction of the physical state on which that advice is based.

[R13.3] Anesthesiology

A major problem in anesthesiology is the assessment of the depth of the anesthesia. If too little drug is applied, the anesthesia is not sufficiently deep, the patient feels pain, and the risk of a postoperative shock syndrome is enhanced. However, the drug itself is highly toxic and it is therefore important to apply the minimum amount required for safe surgery.

In collaboration with an anesthesiologist, decide on a number of secondary factors that she or he uses to indirectly assess the depth of anesthesia. Similarly to the heart monitor problem, develop a consultation system that, on the basis of physical measurements suggests an appropriate dosage of the drug and optimal timing for its administration.

[R13.4] Stock Market

Decide on a number of input parameters (such as the Dow–Jones index) and state variables (such as daily returns of companies) to be monitored. Choose the variables such that the time dependence of the resulting optimal masks is minimized (i.e., for any subset of data over time, the resulting optimal masks should be the same or at least very similar).

Design a SAPS forecasting model with optimal prediction power and compare the results with an optimized linear regression model.

[R13.5] Causal Structure Characterization

Design a general–purpose routine that is able to derive a causal structure using several different techniques, which can then compare the resulting causal structures with each other, and decide which of the resulting structures is the most likely candidate.

14

Artificial Neural Networks and Genetic Algorithms

Preview

In Chapters 12 and 13, we have looked at mechanisms that might lead to an emulation of human reasoning capabilities. We approached this problem from a macroscopic point of view. In this chapter, we shall approach the same problem from a microscopic point of view, i.e., we shall try to emulate learning mechanisms as they are believed to take place at the level of neurons of the human brain and evolutionary adaptation mechanisms as they are hypothesized to have shaped our genetic code.

14.1 Introduction

In this chapter, we shall discuss how artificial neural networks can indeed emulate the human capability of *association* (remembering of similar events) and of *learning* (the autonomous organization of knowledge based on stochastic input stimuli). We shall then proceed to discuss yet another "learning" mechanism: the encoding of hereditary knowledge in our genetic code.

Artificial neural networks were invented several decades ago. Much of the early research on neural networks was accomplished by McCulloch and Pitts [14.22] and by Hebb [14.11]. This research was consolidated in the late 1950s with the conceptualization of the so-called *perceptron* [14.31].

Unfortunately, a single devastating report by Minsky and Papert discredited artificial neural network research in the late 1960s [14.24].

In this report, the authors demonstrated that a perceptron cannot even learn the "behavior" of a simple exclusive–or gate. As a consequence of this publication, research in this area came to a grinding halt. Remarkably, an entire branch of research can be aborted because of a single influential individual, especially in the United States where research funds are not provided automatically, but must be requested through research proposals. For almost 20 years, funding for artificial neural network research was virtually nonexistent. Only the last couple of years have seen a renaissance of this methodology due to the relentless efforts and pioneering research of a few individuals such as Grossberg [14.8,14.9,14.10], Hopfield [14.15], and Kohonen [14.18]. Meanwhile, artificial neural network research has been fully rehabilitated, and virtually thousands of research results are published every year in dozens of different journals and books. Impressive results have been obtained in many application areas, such as image processing and robot grasping systems [14.30].

Artificial neural networks still suffer somewhat from their pattern recognition heritage. Classical pattern recognition techniques were always concerned primarily with the recognition of static images. For this reason, most research efforts in neural networks were also concentrated around the analysis of static information. For quite some time, it was not recognized that humans more often than not base their decisions on *temporal patterns* or *cartoons*, i.e., series of sketchy static images that comprise an entire episode. Notice that graphical trajectory behavior is called a *film*, whereas graphical episodical behavior is called a *cartoon*. If I drive my car through the city of Tucson, and I suddenly see a ball rolling onto the street from behind a parked vehicle, I immediately engage the brake because I expect a child to run after the ball. I would be much less alarmed if the ball were simply lying on the street. Thus, my decision is based on an entire cartoon rather than on a single static image. The cartoon of the rolling ball is a temporal pattern that is matched with stored temporal patterns of my past. The prediction of the future event is based on the association and replay of another temporal pattern that once followed the rolling ball pattern in time, namely, that of a running child.

Artificial neural networks are well suited to identify temporal patterns. All that needs to be done is to store the individual images of the cartoon below each other in a large array and treat the entire cartoon as one pattern. Due to the inherent parallelism in artificial neural network algorithms, the size of a network layer (the length of the pattern array) does not have to increase the time needed for its processing.

Recently, artificial neural network researchers have begun to investigate temporal patterns. Promising results have been reported in the context of speech recognition systems [14.37]. Other important applications are related to dynamical systems. They include the design of adaptive controllers for nonlinear systems [14.26] and the development of fault diagnosers for manufacturing systems [14.33]. However, much remains to be done in these areas.

Finally, it is important to distinguish between research in engineering design of artificial neural networks [14.29,14.35], and scientific analysis (modeling) of the mechanisms of signal processing in the human cerebellum [14.3,14.17] and information processing in the human cerebrum [14.7,14.34]. Engineering design concentrates on very simple mathematical models that provide for some learning capabilities. All artificial neurons are identical models of simple motor neurons, and the interconnections between artificial neurons are structured in the form of a well organized, totally connected, layered network. The biological analogy is of minor significance. The goal of engineering design research is to create algorithms that can be implemented in robots and that provide these robots with some, however modest, reasoning capabilities. Scientific analysis focuses on modeling the activities of the human brain. These models are usually considerably more sophisticated. They distinguish between different types of neurons and attempt to represent correctly the interconnections between these different neurons as they have been observed in various zones of the human brain. However, the applicability of the research results for engineering purposes is of much less importance than the scientific understanding itself.

In this chapter, we want to pursue both research directions to some extent. However, the primary accent is clearly on engineering design, not on neurobiological modeling.

14.2 Artificial Neurons

Figure 14.1 shows a typical motor neuron of the human cerebellum. The motor neuron is the most studied and best understood neuron of the human brain. Altogether, the human brain contains on the order of 10^{11} neurons.

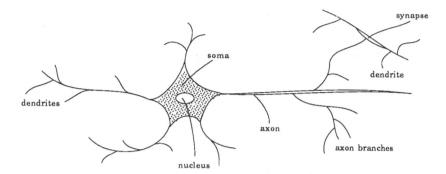

Figure 14.1. Motor neuron of the human cerebellum.

When a neuron "fires," it sends out an electrical pulse from its soma through its single efferent axon. The axon branches out into several subaxons and passes the electrical pulse along all of its branches. Each neuron has on the order of 1000 axon branches. These branches terminate in the vicinity of dendrites of neighboring neurons. The pulse that is sent through the axon is of a fixed amplitude known as the action potential, and it travels along the axon at a high speed of approximately 20 m sec^{-1} [14.3]. The pulse has a duration of approximately 1 msec. Once a neuron has fired, it needs to rest for at least 10 msec before it can fire again. This period is called the refractory period of the neuron [14.3].

When a pulse arrives at the end of an axon branch, it is transmitted to a dendrite of a neighboring neuron in a synaptic contact by molecules known as neurotransmitters. The synapses can be of the excitatory (positive) or inhibitory (negative) type. In the synapse, the signal is converted from a digital to an analog signal. The amplitude of the signal transmitted by the synapse depends on the strength of the synapse. In the afferent dendrites, the transmitted signal travels at a much lower speed toward the soma of its neuron. The typical dendrite attenuation time is on the order of 10 msec [14.3]. However, the attenuation time varies greatly from one neuron to another, because some synaptic connections are located far out on an afferent dendrite, while others are located directly at the base of the primary dendrite, or at the soma itself, or even at the axon hillock (the base of the efferent axon).

If the sum of the analog signals arriving at a soma from its various dendrites is sufficiently high, the neuron fires and sends out a digital pulse along its axon. Consequently, the soma converts the arriving analog signals back to digital signals using frequency modulation. In this way, signals are propagated through the brain by means of consecutive firings of neurons. During their voyage through the brain, the signals are constantly converted back and forth between a digital and an analog form.

Figure 14.2 shows a typical artificial neuron as used in today's artificial neural networks. Typical artificial neural networks contain several dozens to several thousands of individual artificial neurons.

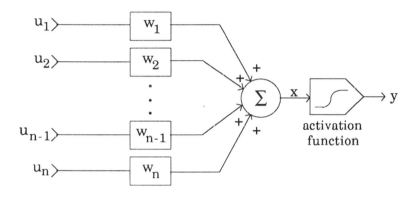

Figure 14.2. Neuron of an artificial neural network.

The input signal u_i symbolizes the digital pulses arriving from the ith neuron. The gains w_{ji} are the synaptic weights associated with the digital–to–analog (D/A) conversion of the neurotransmission from the ith neuron to the jth neuron across the synaptic cleft. The state of the jth neuron x_j is computed as the weighted sum of its inputs:

$$x_j = \sum_{\forall i} w_{ji} \cdot u_i = \mathbf{w_j}' \cdot \mathbf{u} \tag{14.1}$$

The output of the jth neuron y_j is computed as a nonlinear function of its state:

$$y_j = \mathbf{f}(x_j) \tag{14.2}$$

Figure 14.3 shows some typical activation functions.

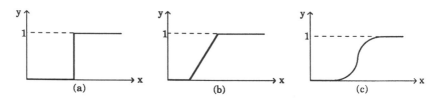

Figure 14.3. Typical activation functions of artificial neurons.

Most artificial neural networks ignore the frequency modulation of real neural systems, i.e., the output corresponding to a constant input is constant rather than pulsed. They also ignore the low–pass characteristics of real neurons, i.e., the time delay of the signal transmission across the true neuron is neglected. While the former omission may be harmless, the latter causes formidable problems when feedback loops are present in the artificial neural network.

Engineering–oriented artificial neural networks use the artificial neuron as the basic building block for constructing networks. All artificial neurons are functionally identical and all synapses are excitatory. Inhibitory synapses are simulated by allowing weighting factors to become negative. These artificial neural networks are symmetrically structured, totally connected, layered networks. Many engineering–oriented artificial neural networks are *feedforward networks* or *cascade networks*. Their response is immediate, since the time constants of the neurons are not modeled. Feedforward networks can be considered elaborate nonlinear function generators. Due to the lack of feedback loops, these networks have no "memory," i.e., they don't store any signals. Information is "stored" only by means of weight adjustments. *Recurrent networks* or *reentrant networks* contain feedback loops. Feedback loops are mandatory if the network is supposed to learn cartoons directly, i.e., if the individual images of temporal patterns are not compressed in time to one instant and stored underneath each other in a long pattern array (as proposed earlier), but are to be fed sequentially into the network as they arrive. A serious problem with most recurrent networks is their tendency to become unstable as the weights are adjusted. Global stability analysis of a recurrent network is a very difficult problem due to the inherent nonlinear activation functions at each node.

Neurobiologically oriented neural network models take into consideration the fact that the human brain contains various different types of neurons that are interconnected in a few standard patterns and form local neuronic "unit circuits." For example, the cerebellar unit circuit by Green and Triffet distinguishes between granule, Golgi, basket, stellate, and Purkinje neurons, which are interconnected as shown in Fig.14.4 [14.7,14.34].

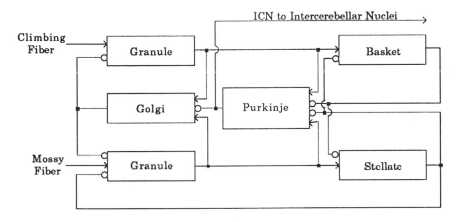

Figure 14.4. Unit circuit of the cerebellum.

Arrowheads indicate excitatory synaptic connections, whereas small circles indicate inhibitory synaptic connections. Notice the local feedback loops within the unit circuit. The unit circuits cannot work properly unless the dynamics of the individual neurons are modeled. Triffet and Green do indeed model the time and frequency response characteristics of signals traveling through the cerebellum. In fact, they even take into consideration the frequency modulation of neuronic signals, i.e., they model the action potential, the refractory period, and the resting period of the neuron.

14.3 Artificial Neural Engineering Networks

While the model of an individual neuron bears a certain similarity to the real world, today's artificial neural networks as they are used in engineering are totally artificial. This is due to the fact that we must enforce a very unnatural order among interconnections between artificial neurons to guarantee decent execution times of our artificial neural network programs. Most artificial neural network programs rely on matrix manipulations for communications between neurons. To this end, it is necessary to enforce a strict topology among the neuronic connections.

The most common artificial neural networks operate on layers of n artificial neurons. Within a given layer, none of the artificial neurons are connected to each other, but each artificial neuron is connected to every neuron of the next layer of m neurons. Figure 14.5 shows a typical network configuration.

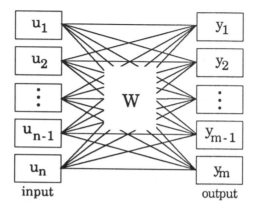

input   output

Figure 14.5. Simple artificial neural network.

The computational advantage of this structure becomes evident when we apply Eqs.(14.1–2) to this network. In matrix form, we can write:

$$\mathbf{x} = \mathbf{W} \cdot \mathbf{u} \tag{14.3a}$$

$$\mathbf{y} = \mathbf{f}(\mathbf{x}) \tag{14.3b}$$

where $\mathbf{x}$ and $\mathbf{y}$ are vectors of length m and $\mathbf{u}$ is a vector of length n. $\mathbf{W}$ is the interconnection matrix between the first and second layer. It has m rows and n columns. One single matrix multiplication and one vector function suffice to simulate the entire network with all its $n \times m$ connections.

What is this infamous problem discovered by Marvin Minsky that disgraced neural networks for more than a decade? A *perceptron* is an artificial neuron with a threshold output function as shown in Fig.14.3a. Let us look at a single perceptron with two inputs u_1 and u_2 and one output y. We wish to make this perceptron learn an exclusive–or function, i.e., the function whose truth table is presented in Table 14.1.

Table 14.1. Truth Table of Exclusive–Or Gate.

u_1	u_2	y
0	0	0
0	1	1
1	0	1
1	1	0

In our convention, '0' represents the logical false state, while '1' represents the logical true state. The output of the exclusive–or gate is true if the two inputs are in the opposite state and it is false, if both inputs are in the same state. "Learning" in an artificial neural network sense simply means to adjust the values of the synaptic weights. Figure 14.6 depicts our perceptron.

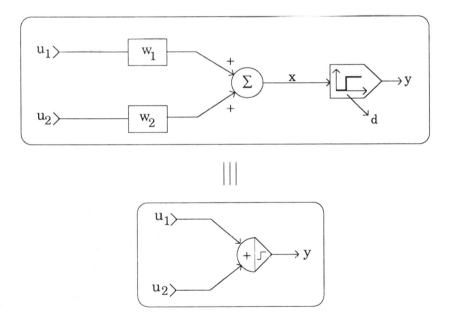

Figure 14.6. Single perceptron for exclusive–or gate.

The state of this perceptron can be written as:

$$x = w_1\, u_1 + w_2\, u_2 \tag{14.4}$$

The output is '1' if the state is larger than a given threshold, d, and it is '0' otherwise. Thus, we can represent the perceptron as shown in Fig.14.7.

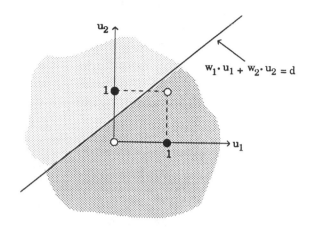

Figure 14.7. Perceptron solution of exclusive–or problem.

Figure 14.7 shows the output of the perceptron in the $< u_1, u_2 >$ plane. The empty circles represent the desired '0' outputs, whereas the filled circles represent the desired '1' outputs. The slanted line separates the $< u_1, u_2 >$ plane into a half–plane for which the perceptron computes a value of '0' (below the line) and a half–plane for which it computes a value of '1' (above the line). By adjusting the two weight factors w_1 and w_2 and the threshold d, we can arbitrarily place the slanted line in the $< u_1, u_2 >$ plane. Obviously, no values of w_1, w_2, and d can be found that will place the two filled circles on one side of the line and the two empty circles on the other. This simple truth was enough to bring most artificial neural network research to a grinding halt.

Figure 14.8 shows a slightly more complex artificial neural network that can solve the exclusive–or problem.

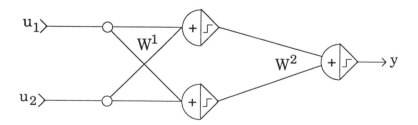

Figure 14.8. Two–layer solution of exclusive–or problem.

The enhanced network contains three perceptrons. Figure 14.9 shows how this network solves the exclusive–or problem.

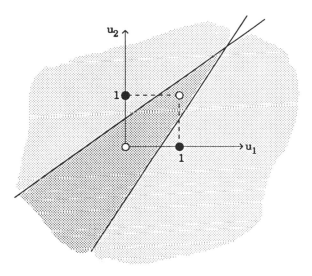

Figure 14.9. Two–layer solution of exclusive–or problem.

The darker shaded area of the $< u_1, u_2 >$ plane is the area for which the network computes a '1.' Thus, while the single–layer network failed to solve the exclusive–or problem, a simple two–layer network will do the trick. Multilayer networks can mask arbitrary subspaces in the input space. These subspaces can be concave as well as convex.

Complex architectures of arbitrarily connected networks are still fairly seldom used. This is because, until recently, we didn't understand how to adjust the weighting coefficients of an arbitrarily connected network to learn a particular pattern (such algorithms will be introduced toward the end of this chapter).

Backpropagation and counterpropagation are two training algorithms that have become popular since they provide systematic ways to train particular types of multilayered feedforward networks.

However, before we discuss artificial neural network learning in more detail, let me explain why artificial neural networks are attractive as tools for pattern recognition.

14.4 The Pattern Memorizing Power of Highly Rank–Deficient Matrices

We have seen that a one–layer perceptron network multiplies the input vector **u** from the left with a weighting matrix **W** to determine the state **x** of the network. Let us assume that the input of our network is the ASCII code for a character, say Z:

$$\mathbf{u} = \text{ASCII}(Z) = (0 \quad 1 \quad 0 \quad 1 \quad 1 \quad 0 \quad 1 \quad 0)' \qquad (14.5)$$

We choose our weighting matrix as the outer product of the input with itself except for the diagonal elements, which are set equal to '0' [14.29]:

$$\mathbf{W} = \mathbf{u} \cdot \mathbf{u}' - \text{DIAG}(\text{DIAG}(\mathbf{u} \cdot \mathbf{u}')) = \begin{pmatrix} 0 & 0 & 0 & 0 & 0 & 0 & 0 & 0 \\ 0 & 0 & 0 & 1 & 1 & 0 & 1 & 0 \\ 0 & 0 & 0 & 0 & 0 & 0 & 0 & 0 \\ 0 & 1 & 0 & 0 & 1 & 0 & 1 & 0 \\ 0 & 1 & 0 & 1 & 0 & 0 & 1 & 0 \\ 0 & 0 & 0 & 0 & 0 & 0 & 0 & 0 \\ 0 & 1 & 0 & 1 & 1 & 0 & 0 & 0 \\ 0 & 0 & 0 & 0 & 0 & 0 & 0 & 0 \end{pmatrix} \qquad (14.6)$$

W is a dilute (sparsely populated) symmetric matrix of rank 4. If we feed this network with the correct input, the output is obviously the same as the input, since:

$$\mathbf{x} = \mathbf{W} \cdot \mathbf{u} \approx (\mathbf{u} \cdot \mathbf{u}') \cdot \mathbf{u} = \mathbf{u} \cdot (\mathbf{u}' \cdot \mathbf{u}) = k \cdot \mathbf{u} \qquad (14.7)$$

The output threshold eliminates the k factor. The elimination of the diagonal elements is unimportant except for input vectors, which contain one single '1' element only.

Now let us perform a different experiment. We feed into the network a somewhat different input, say:

$$\mathbf{u} = (0 \quad 1 \quad 0 \quad 0 \quad 1 \quad 0 \quad 0 \quad 1)' \qquad (14.8)$$

A comparison to the previous input shows that three bits are different. Amazingly, the network will still produce the same output as before. This network is able to recognize "similar" inputs, or equivalently, is able to filter out digital errors in a digital input signal. Hopfield has shown that this is a general property of such highly rank–deficient sparsely populated matrices [14.15]. He showed that the technique works for input signals that have a smaller or equal number of '1' elements than '0' elements. Otherwise, we simply invert every bit of the input.

One and the same network can be used to recognize several different inputs u_i with and without distortions. For this purpose, we simply add the W_i matrices. Hopfield showed that a matrix of size $n \times n$ can store up to $0.15 \cdot n$ different symbols [14.15]. Thus, in order to store and recognize the entire upper–case alphabet, each letter should be represented by a vector of at least 174 bits in length. The ASCII code is completely inadequate for this purpose. It does not contain enough redundancy. Notice that the code sequence of Eq.(14.8) was recognized as the character Z, while in fact, it is the ASCII representation of another character, namely, the character I.

A good choice may be a pixel representation of a 14×14 pixel field, as shown in Fig.14.10.

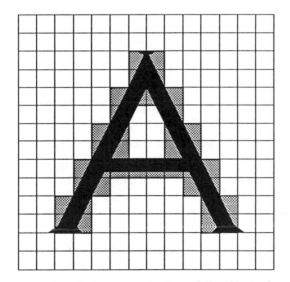

Figure 14.10. Pixel representation of a written character.

We simply number the pixels from left to right, top to bottom. White pixels are represented as '0,' while grey pixels are represented as '1.' We then write all the pixels into a vector of length 196. Minor variations in the writing of characters can be filtered out by the artificial neural network. This technique can be used to recognize handwritten characters. Some preprocessing will be necessary to center the character in the pixel field, normalize its size, and correct for slanting.

The same method also works if the desired output is symbolically different from the given input. For example, it may be desirable to map the pixel representation of a handwritten, upper–case character into its ASCII code. In this case, we use the following weighting matrix:

$$\mathbf{W} = \sum_{\forall i}(\mathbf{y} \cdot \mathbf{u}') \tag{14.9}$$

where $\mathbf{u}$ is the 196–bit–long pixel vector and $\mathbf{y}$ is the corresponding 8–bit–long ASCII code vector. Thus, $\mathbf{W}$ is now a rectangular matrix of the size 8 × 196. This is how we believe the brain maps visual pixel information into symbols (not necessarily in ASCII format, of course).

Auditory information can be treated similarly. Here, we could sample and digitize the audio signal and store an entire series of values sampled at successive sampling intervals as an input vector. Each time value is represented by a number of bits, perhaps eight. If we assume that the minimum required band width of the ear is 10 kHz and that one spoken word lasts 1 sec, we need to store 8 × 10,000 bits in the input vector. According to Hopfield [14.15], 80,000 bits allow us to recognize 12,000 different words, which should be sufficient for most purposes. We can then map the input vector of length 80,000 into an output vector of length 14, since 14 bits suffice to store 12,000 words. In this way, we should be able to comfortably recognize the spoken language. Of course, we need some additional preprocessing to "center" the spoken word in the time window, i.e., we need to determine when one word ends and the next begins. We should also normalize the spoken word in length and altitude, i.e., compensate for fast versus slow speakers as well as for female versus male voices. Also, it may be better to operate on frequency signals instead of time signals. Among different speakers, the FFT of the spoken word varies less than the corresponding time signal.

Much more has been written about this topic. Several authors have discussed these types of weight matrices from a statistical point of view. They showed that optimal separation between different patterns is achieved if the input patterns are orthogonal to each other. The optimal $\mathbf{W}$ matrix turns out to be the input correlation matrix. A good overview is given in Hecht–Nielsen's new book [14.12].

14.5 Supervised and Unsupervised Learning

We are now ready to discuss mechanisms of artificial neural network learning. In the last section, we knew how to set the synaptic weights. In a more general situation, this will not be the case. The question thus is: How do we modify the weights of our interconnection matrices such that the artificial neural network learns to recognize particular patterns?

Traditionally, two modes of neural network learning have been distinguished: supervised learning and unsupervised learning. In *supervised learning*, the artificial neural network "attends school." A "teacher" provides the network with an input and asks the network for the appropriate output. The teacher then provides the network with the correct output and uses either positive reinforcement or punishment to enhance the chance of a correct answer next time around.

Supervised (interactional) learning has been observed within humans and other living organisms at various levels. At the lowest level of learning, we notice *behavior modification schemes* (Skinnerism). These schemes do not assume that the person or animal has any insight regarding what she or he is taught. Behavior modification programs are prevalent in animal training and the rehabilitation of the severely mentally retarded. In normal human education, behavior modification plays a minor role, yet this is the only form of learning that is being imitated by today's artificial neural networks.

At the next higher level, we should mention the mechanisms of social learning [14.1]. *Social learning* operates on concepts such as social modeling and shaping. The child learns through mechanisms of vicarious learning and imitation. Learning occurs in a feedback loop. The child reacts to external stimuli, but the teacher also modifies his or her own behavior on the basis of the reactions he or she observes in the child, and the child, in return, will notice the effects that his or her behavior has on the teacher's behavior, and so on. The problems and patterns presented to the child are initially very simple and become more and more intricate as the child develops. This is a fruitful concept that could (and should) be adapted for use in artificial neural networks. Today's artificial neural network research focuses on network training and ignores the importance of the training pairs. Optimization of training pairs for accelerated network learning would be a worthwhile research topic. Maybe we can create a "teacher network" that accepts arbitrary training pairs and preprocesses them for learning by a "student network." Perhaps one type of student network could be trained to become a next generation teacher network. Notice that social learning (like the previously discussed Skinnerism) focuses on a purely phenomenological view of learning. Social learning analyzes the interaction between the teacher and the student but ignores what happens inside the student and/or the teacher.

At an even higher level, psychologists focus on the mechanisms of cognition. They discuss the role of *cognitive functions* (language, attention span, learning abilities, reasoning, and memorization) [14.27]. At this level, we no longer focus on *what* is learned, but rather on *how* learning occurs, that is, how we learn to learn. In analogy to artificial neural network terminology, this corresponds to using a second artificial neural network in conjunction with the original network. The output of the second network is the algorithm that the first network uses to adapt its weights. We then use a third network on top of the second one, and so forth, until we come to a point where the highest–level network has become so abstract and general that it doesn't require any training at all, but can be used to bootstrap itself. Presently, our state–of–the–art neural networks are far from such degrees of complexity. All that we have achieved with our artificial neural networks is the capability to emulate only the lowest level of behavioral modification schemes.

Supervised training of artificial neural networks is attractive because it is fairly easy to implement. It is unattractive because the network is kept in an unproductive learning phase for a long time before the learned knowledge can be applied to solve real problems. In a very simple analogy, our children have to grow up and attend school for many years before they can be integrated into the work force and "make money" on their own. However, the analogy is not truly appropriate since the child is extremely productive during his or her training period except for one particular aspect: the economic one.

In *unsupervised learning*, the network can be used immediately, but it will produce increasingly better results as time passes. Learning is accomplished by comparing the current input/target pair with previous similar input/target pairs. The network is trained to produce consistent results [14.35]. Many biologists insist that at least low–level learning occurs in a basically unsupervised mode [14.35]. I am not sure this is correct. Even motoric functions are learned through mechanisms of reinforcement and punishment (behavior modification) and mechanisms of imitation (social learning). Walking is learned by falling many times and by trying to imitate walking adults. Yet the child indeed does not learn to walk in a "dry run" mode, and walks only after mastering the problem in theory. The dichotomy between supervised and unsupervised learning is an artifact. Usually, artificial neural networks are studied in isolation. Humans do not learn in isolation. They live in an environment to which they react and which reacts to them in return. Real learning is similar to adaptive control. The network is constantly provided with real inputs from which it computes real outputs, and it is also provided with desired outputs with which it compares them. It then tries to modify its behavior until the real outputs resemble the desired outputs. However, while the network learns, it is constantly in use. Thus, it would make sense to replace the term unsupervised learning by adaptive learning.

The real problem with artificial neural network learning lies somewhere else. Artificial neural networks can observe only the inputs and outputs of the real system, but not its internal states. The question thus is: How does the network modify the weights of internal hidden layers? This dilemma is at the origin of the (artificial) distinction between supervised and unsupervised learning.

The answer to this question is quite simple: Artificial neural network training is an extremely slow and tedious process. Modifying the weights of the internal network layers is accomplished either by trial and error or gradient propagation. The real question thus is: How do we guarantee the convergence of the weight adaptation algorithm in use?

We shall address this question in greater generality toward the end of this chapter. For now, we shall restrict ourselves to the analysis of a few simple feedforward network topologies and develop learning algorithms for those networks only.

Single–Layer Networks

Let us compute the difference between the jth desired output $\hat{y}_j$ and the jth real output y_j:

$$\delta_j = \hat{y}_j - y_j \tag{14.10}$$

In the case of a perceptron network, δ_j will be either -1, 0, or $+1$.

In the case of other networks, δ_j can be a continuous variable. We then compute the change in the weight from the ith input to the jth output as follows:

$$\Delta_{ji} = g \cdot \delta_j \cdot u_i \qquad (14.11)$$

The rationale for this rule is simple. If the jth real output is too small, δ_j is positive. In this case, we need to reinforce those inputs that are strong by making them even stronger. This will raise the output level. However, if the jth real output is too large, δ_j is negative. In this case, we need to weaken those inputs that are strong. This will reduce the output level.

The constant multiplier g is the gain value of the algorithm. For larger values of g, the algorithm converges faster if it does indeed converge, but it is less likely to converge. For smaller values of g, the chance of convergence is greater, but convergence will take longer.

Finally, we update the weight w_{ji} in the following manner:

$$w_{ji_{new}} = w_{ji_{old}} + \Delta_{ji_{old}} \qquad (14.12)$$

In a matrix form, we can summarize Eqs.(14.10–12) as follows:

$$\mathbf{W}_{k+1} = \mathbf{W}_k + g \cdot (\hat{\mathbf{y}} - \mathbf{y}_k) \cdot \mathbf{u}_k{}' \qquad (14.13)$$

Equation (14.13) is commonly referred to as the *delta rule* [14.36].

Notice the following special case. If we choose the initial weighting matrix as zero, $\mathbf{W}_0 = 0$, then the initial output will also be zero, $\mathbf{y}_0 = 0$. If we furthermore choose a gain factor of $g = 1.0$, we can compute the weighting matrix after one single iteration as follows:

$$\mathbf{W}_1 = \hat{\mathbf{y}} \cdot \mathbf{u}_0{}' \qquad (14.14)$$

which is identical to the explicit weight assignment formula, Eq.(14.9), that we used in the last section.

For most applications, we shall choose a considerably smaller gain value such as: $g = 0.01$ and we shall start with a small random weighting matrix:

$$\mathbf{W}_0 = 0.01 * \text{RAND}(m, n) \qquad (14.15)$$

Backpropagation Networks

Backpropagation networks are multilayer networks in which the various layers are cascaded. Figure 14.11 shows a typical three–layer backpropagation network.

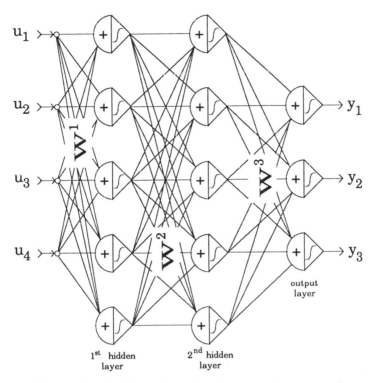

Figure 14.11. Three–layer backpropagation network.

In order for the backpropagation learning algorithm to work, we must assume that the activation function of each artificial neuron is differentiable over its entire input range. Thus, perceptrons cannot be used in backpropagation networks.

The most commonly used activation function in a backpropagation network is the sigmoid function:

$$y = \text{sigmoid}(x) = \frac{1.0}{1.0 + \exp(-x)} \qquad (14.16)$$

The shape of the sigmoid function is graphically shown in Fig.14.12.

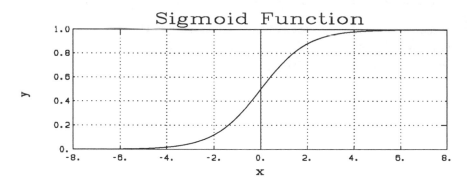

Figure 14.12. Sigmoid function.

The sigmoid function is particularly convenient because of its simple partial derivative:

$$\frac{\partial y}{\partial x} = y \cdot (1.0 - y) = \text{logistic}(y) \qquad (14.17)$$

The partial derivative of the output y with respect to state x does not depend on x explicitly. It can be written as a logistic function of the output y.

We shall train the output layer in basically the same manner as in the case of the single–layer network, but we shall modify the formula for δ_j. Instead of simply using the difference between the desired output $\hat{y}_j$ and the true output y_j, we multiply this difference by the activation gradient:

$$\delta_j = \frac{\partial y}{\partial x} \cdot (\hat{y}_j - y_j) = y_j \cdot (1.0 - y_j) \cdot (\hat{y}_j - y_j) \qquad (14.10^{alt})$$

Therefore, the matrix version of the learning algorithm for the output layer can now be written as:

$$\mathbf{W}_{k+1}^n = \mathbf{W}_k^n + g * \left(\mathbf{y}_k^n \cdot * (\text{ONES}(\mathbf{y}_k^n) - \mathbf{y}_k^n) \cdot * (\hat{\mathbf{y}} - \mathbf{y}_k^n)\right) * \mathbf{u}_k^{n\prime} \quad (14.13^{alt})$$

The subscript k denotes the kth iteration, whereas the superscript n denotes the nth stage (layer) of the multilayer network. I assume that the network has exactly n stages. Equation (14.13^{alt}) is written in a pseudo–CTRL–C (pseudo–MATLAB) style. The '*' operator denotes a regular matrix multiplication, whereas the '.*' operator denotes an elementwise multiplication. The vector $\mathbf{u}_k^n$ is obviously identical to $\mathbf{y}_k^{n-1}$. Let:

$$\vec{\delta}_k^n = \mathbf{y}_k^n \,.\ast\, (\mathrm{ONES}(\mathbf{y}_k^n) - \mathbf{y}_k^n) \,.\ast\, (\hat{\mathbf{y}} - \mathbf{y}_k^n) \qquad (14.18)$$

denote the kth iteration of the $\vec{\delta}$ vector for the nth (output) stage of the multilayer network. Using Eq.(14.18), we can rewrite Eq.(14.13^{alt}) as follows:

$$\mathbf{W}_{k+1}^n = \mathbf{W}_k^n + g \,\ast\, \vec{\delta}_k^n \,\ast\, \mathbf{u}_k^{n\prime} \qquad (14.19)$$

Unfortunately, this algorithm will work for the output stage of the multilayer network only. We cannot train the hidden layers in the same fashion since we don't have a desired output for these stages. Therefore, we replace the gradient by another (unsupervised) updating function. The $\vec{\delta}$ vector of the ℓth hidden layer is computed as follows:

$$\vec{\delta}_k^\ell = \mathbf{y}_k^\ell \,.\ast\, (\mathrm{ONES}(\mathbf{y}_k^\ell) - \mathbf{y}_k^\ell) \,.\ast\, (\mathbf{W}_k^{\ell+1\prime} \,\ast\, \vec{\delta}_k^{\ell+1}) \qquad (14.20)$$

Instead of weighing the $\vec{\delta}$ vector with the (unavailable) difference between the desired and the true output of that stage, we propagate the weighted $\vec{\delta}$ vector of the subsequent stage back through the network. We then compute the next iteration of the weighting matrix of this hidden layer using Eq.(14.19) applied to the ℓth stage, i.e.:

$$\mathbf{W}_{k+1}^\ell = \mathbf{W}_k^\ell + g \,\ast\, \vec{\delta}_k^\ell \,\ast\, \mathbf{u}_k^{\ell\prime} \qquad (14.21)$$

In this fashion, we proceed backward through the entire network.

The algorithm starts by setting all weighting matrices to small random matrices. We apply the true input to the network and propagate the true input forward to the true output, generating the first iteration on all signals in the network. We then propagate the gradients backward through the network to obtain the first iteration on all the weighting matrices. We then use these weighting matrices to propagate the same true input once more forward through the network to obtain the second iteration on the signals and then propagate the modified gradients backward through the entire network to obtain the second iteration on the weighting matrices. Consequently, the $\mathbf{u}^\ell$ and $\mathbf{y}^\ell$ vectors of the ℓth stage are updated on the forward path, while the $\vec{\delta}^\ell$ vector and the $\mathbf{W}^\ell$ matrix are updated on the backward path. Each iteration consists of one forward path followed by one backward path.

The backpropagation algorithm was made popular by Rumelhart *et al.* [14.32]. It presented the artificial neural network research community with the first systematic (although still heuristic) algorithm for training multilayer networks. The backpropagation algorithm has a fairly benign stability behavior. It will converge on many problems provided the gain g has been properly selected. Unfortunately, its convergence speed is usually very slow. Typically, a backpropagation training session may require several hundred thousand iterations for convergence.

Several enhancements of the algorithm have been proposed. Frequently, a bias vector is added, i.e., the state of an artificial neuron is no longer the weighted sum of its inputs alone, but is computed using the formula:

$$\mathbf{x} = \mathbf{W} \cdot \mathbf{u} + \mathbf{b} \tag{14.22}$$

Conceptually, this is not a true enhancement. It simply means that the neuron has an additional input, which is always '1.' Consequently, the bias term is updated as follows:

$$\mathbf{b_{k+1}} = \mathbf{b_k} + g \cdot \vec{\delta}_k \tag{14.23}$$

Also, a small momentum term is frequently added to the weights in order to improve the convergence speed [14.19]:

$$\mathbf{W_{k+1}} = (1.0 + m) \cdot \mathbf{W_k} + g \cdot \vec{\delta}_k \cdot \mathbf{u_k}' \tag{14.24a}$$

The momentum should obviously be added to the bias term as well:

$$\mathbf{b_{k+1}} = (1.0 + m) \cdot \mathbf{b_k} + g \cdot \vec{\delta}_k \tag{14.24b}$$

The momentum m is usually very small, $m \approx 0.01$.

Other references add a small percentage of the last change in the matrix to the weight update equation [14.12]:

$$\mathbf{\Delta W_k} = g \cdot \vec{\delta}_k \cdot \mathbf{u_k}' \tag{14.25a}$$
$$\mathbf{W_{k+1}} = \mathbf{W_k} + \mathbf{\Delta W_k} + m \cdot \mathbf{\Delta W_{k-1}} \tag{14.25b}$$

Finally, it is quite common to limit the amount by which the $\vec{\delta}$ vectors, the $\mathbf{b}$ vectors, and the $\mathbf{W}$ matrices can change in a single step. This often improves the stability behavior of the algorithm.

Counterpropagation Networks

Robert Hecht–Nielsen introduced a two–layer network that can be trained much more quickly than the backpropagation network [14.12]. In fact, counterpropagation networks can be trained instantaneously. We can provide an explicit algorithm for generating the weighting matrices of counterpropagation networks. The idea behind counterpropagation is fairly simple. The problem of teaching a two–layer network to map an arbitrary set of input vectors into another arbitrary set of output vectors can be decomposed into two simpler problems:

(1) We map the arbitrary set of inputs into an intermediate (hidden) digital layer in which the kth input/target pair is represented by the kth unit vector, $\mathbf{e_k}$:

$$\mathbf{e_k} = [\, 0\, ,\, 0\, ,\, \ldots\, ,\, 0\, ,\, 1\, ,\, 0\, ,\, \ldots\, ,\, 0\,]' \qquad (14.26)$$

The kth unit vector, $\mathbf{e_k}$ is a vector of length k that contains only '0' elements except in its kth position where it contains a '1' element.

(2) We map the intermediate (hidden) digital layer into the desired arbitrary output layer.

Obviously, the length of the hidden layer must be as large as the number of different input/target pairs with which we wish to train the network.

The map from the intermediate layer to the output layer is trivial. Since each output is driven by exactly one '1' source and since this source doesn't drive any other output, the output layer weighting matrix consists simply of a horizontal concatenation of the desired outputs:

$$\mathbf{W^2} = [\, \mathbf{y_1^2}\, ,\, \mathbf{y_2^2}\, ,\, \ldots\, ,\, \mathbf{y_n^2}\,] \qquad (14.27)$$

where $\mathbf{y_i^2}$ denotes the ith output vector of the second stage, which is driven by the ith unit input vector of the second stage, which is identical to the ith output vector of the first stage, $\mathbf{y_i^1}$.

Thus, the interesting question is: Can we train a single–layer perceptron network (the first stage of the counterpropagation network) to map each input into an output such that exactly one of the output elements is '1' while all other output elements are '0' and such that no two output vectors are identical if their input vectors are different?

Teuvo Kohonen [14.18] addressed this question. In a way, the two layers of the counterpropagation network are inverse to each other. The optimal weighting matrix can be written as follows:

$$\mathbf{W}^1 = [\, \mathbf{u}_1^1 \,,\, \mathbf{u}_2^1 \,,\, \ldots \,,\, \mathbf{u}_n^1 \,]' \qquad (14.28)$$

The weighting matrix of the first stage is the transpose of the matrix that consists of a horizontal concatenation of the input vectors.

The threshold of the perceptrons d is automatically adjusted such that only one of the states of the first stage is larger than the threshold.

Let us apply this technique to the problem of reading handwritten characters. The purpose is to map the pixel representation of handwritten upper–case characters to their ASCII code. Counterpropagation enables us to solve the character reading problem in one single step without need for any training. Let us assume that each input character is resolved in a 14×14 pixel matrix. Thus, each character is represented by a pixel vector of length 196. Since the alphabet contains 26 different upper–case characters, the hidden Kohonen layer must be of length 26. Each of the outputs is of length eight. Thus, the total network contains 34 artificial neurons: 26 perceptrons and 8 linear output neurons. The dimension of the weighting matrix of the first stage is 26×196. We store the pixel representation of the ideally written character A as the first–row vector of $\mathbf{W}^1$. We then concatenate from below the pixel representation of the ideally written character B, etc. This will map the pixel representation of the kth character of the alphabet into the kth unit vector $\mathbf{e}_k$. The dimension of the weighting matrix of the second stage is 8×26. We store the ASCII representation of the character A as the first–column vector of $\mathbf{W}^2$ and concatenate from the right the ASCII representation of the character B, etc. This will map the unit vector representation of the kth character of the alphabet into its ASCII representation.

Several unsupervised training algorithms have been devised that make the network adaptive to variations in the input vectors. A typically used updating rule is the following:

$$\mathbf{w}\mathbf{w}_{k+1}^{1}{}' = \mathbf{w}\mathbf{w}_k^{1}{}' + g \cdot (\mathbf{u}_k^{1}{}' - \mathbf{w}\mathbf{w}_k^{1}{}') \qquad (14.29)$$

where $\mathbf{w}\mathbf{w}^{1}{}'$ denotes the "winning" row vector of the $\mathbf{W}^1$ matrix. According to Eq.(14.29), we don't update the entire $\mathbf{W}^1$ matrix at once. Instead, we update only one row at a time, namely, the row that corresponds to the "winning" output of the Kohonen layer, i.e., the one output that is '1' while all others are '0.' Equation (14.29) is frequently referred to as the *Kohonen learning rule*. Notice that Kohonen learning does not always work. Improved learning techniques have been proposed. However, we shall refrain from discussing these updating algorithms here in more detail. Additional information is provided in Wasserman [14.35].

In general, counterpropagation networks work fairly well for digital systems, i.e., mappings of binary input vectors into binary output vectors. They don't work very well for continuous systems since the hidden layer must be digital and must enumerate all possible system states. Backpropagation networks work fairly well for continuous systems. They don't work well for systems with digital output, because their activation functions must be continuously differentiable.

Notice that the counterpropagation networks introduced in this section are somewhat different from those traditionally found in the artificial neural network literature. While the general idea behind the counterpropagation architecture is the same, the explicit algorithm for generating the weighting matrices is more appealing. Classical counterpropagation networks randomize the first (Kohonen) layer, then use Eq.(14.29) to train the Kohonen layer. They use a so-called Grossberg outstar as the second layer. However, my algorithm is more attractive since it doesn't require any iterative learning and works reliably for all types of digital systems, whereas Eq.(14.29) often leads to convergence problems.

14.6 Neural Network Software

Until now, we have written down all our formulae in this chapter in a pseudo–MATLAB (pseudo–CTRL–C) format. This was convenient since the reader should be familiar with this nomenclature by now.

Let us use this approach to solve an example problem. We wish to design a counterpropagation network to solve the infamous exclusive–or problem. Since this is a digital system, we expect the counterpropagation network to work well. Figure 14.13 depicts the counterpropagation network for this problem.

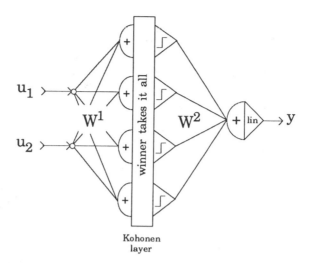

Figure 14.13. Counterpropagation network for XOR.

This system has two inputs and one output. Since the truth table contains four different states, the hidden layer must be of length four. Until now, we have always used the state '0' to denote false and the state '1' to denote true. This is not really practical for most artificial neural networks. Therefore, we shall use the real value -1.0 to denote the logical false state and the real value $+1.0$ to denote the logical true state in both the input vector and the output. The hidden layer will still use '0' and '1.'

With this exception, the counterpropagation network functions as described in the previous section. The code for this network is shown here:

```
// This procedure designs a counterpropagation network for XOR
//
deff winner -c
//
// Define the input and target vectors
//
inpt = [ −1 − 1    1    1
         −1    1 − 1    1 ];
target = [ −1    1    1 − 1 ];
//
// Set the weighting matrices
//
W1 = inpt';
W2 = target;
```

```
// Apply the network to evaluate the truth table
//
y = ZROW(target);
for nbr = 1:4, ...
    u1 = inpt(:, nbr); ...
    x1 = W1 * u1; ...
    y1 = WINNER(x1); ...
    u2 = y1; ...
    x2 = W2 * u2; ...
    y2 = x2; ...
    y(nbr) = y2; ...
end
//
// Display the results
//
y
//
return
```

This procedure is fairly self–explanatory. The WINNER function determines the largest of the perceptron states and assigns a value of +1.0 to that particular output in a "winner–takes–it–all" fashion. The WINNER function is shown here:

```
// [y] = WINNER(x)
//
// This procedure computes the winner function
//
[n, m] = SIZE(x);
ind = SORT(x);
y = ZROW(n, m);
y(ind(n)) = 1;
//
return
```

The WINNER function sorts the input vector x in increasing order. The vector *ind* is not the sorted array, but an index vector that shows the position of the various elements in the original vector. Thus, $x(ind(1))$ is the smallest element of the x vector, and $x(ind(n))$ is the largest element of the x vector. We then set the output vector y to 0.0, except for the winning element, which is set equal to +1.0.

The performance of the counterpropagation network is flawless. However, the network is not error–tolerant, since the input is digital and does not contain any redundancy. Consequently, a single–bit error converts the desired input vector into another undesired, yet equally legal, input vector. The Hamming distance between two neighboring legal input states is exactly one bit.

Let us now try to solve the same problem using a backpropagation network. We expect problems since the system is digital. Figure 14.14 shows the resulting backpropagation network.

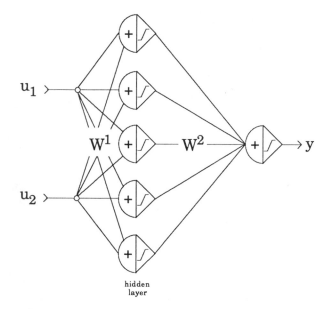

Figure 14.14. Backpropagation network for XOR.

The length of the hidden layer is arbitrary. In our program, we made this a parameter, *lhid*, which can be chosen at will. The program is shown here:

```
// This procedure designs a backpropagation network for XOR
// Select the length of the hidden layer (LHID) first
//
deff limit -c
deff tri -c
//
// Define the input and target vectors
//
inpt = [ −1 −1   1   1
         −1   1 −1   1 ];
target = [ −1   1   1 −1 ];
```

```
// Set the weighting matrices and biases
//
W1 = 0.1 * (2.0*RAND(lhid, 2)− ONES(lhid, 2));
W2 = 0.1 * (2.0*RAND(1, lhid)− ONES(1, lhid));
b1 = ZROW(lhid, 1);   b2 = ZROW(1);
WW1 = ZROW(lhid, 2);   WW2 = ZROW(1, lhid);
bb1 = ZROW(lhid, 1);   bb2 = ZROW(1);
//
// Set the gains and momenta
//
g1 = 0.6;   g2 = 0.3;
m1 = 0.06;   m2 = 0.03;
//
// Set the termination condition
//
crit = 0.025;   error = 1.0;   count = 0;
//
// Learn the weights and biases
//
while error > crit, ...
    count = count + 1; ...
    ... //
    ... // Loop over all input/target pairs
    ... //
    error = 0; ...
    for nbr = 1:4, ...
        u1 = inpt(:, nbr); ...
        y2h = target(nbr); ...
        ... //
        ... // Forward pass
        ... //
        x1 = WW1 * u1 + bb1; ...
        y1 = LIMIT(x1); ...
        u2 = y1; ...
        x2 = WW2 * u2 + bb2; ...
        y2 = LIMIT(x2); ...
        ... //
        ... // Backward pass
        ... //
        e = y2h − y2; ...
        delta2 = TRI(y2) .* e; ...
        W2 = W2 + g2 * delta2 * (u2') + m2 * WW2; ...
        b2 = b2 + g2 * delta2 + m2 * bb2; ...
        delta1 = TRI(y1) .* ((WW2') * delta2); ...
        W1 = W1 + g1 * delta1 * (u1') + m1 * WW1; ...
        b1 = b1 + g1 * delta1 + m1 * bb1; ...
        error = error+ NORM(e); ...
    end, ...
```

```
... // Update the momentum matrices and vectors
... //
WW1 = W1;   WW2 = W2; ...
bb1 = b1;   bb2 = b2; ...
end
//
// Apply the learned network to evaluate the truth table
//
y = ZROW(target);
for nbr = 1:4, ...
    u1 = inpt(:, nbr); ...
    x1 = WW1 * u1 + bb1; ...
    y1 = LIMIT(x1); ...
    u2 = y1; ...
    x2 = WW2 * u2 + bb2; ...
    y2 = LIMIT(x2); ...
    y(nbr) = y2; ...
end
//
// Display the results
//
y
//
return
```

It took some persuasion to get this program to work. The first difficulty was with the activation functions. The sigmoid function is no longer adequate since the output varies between -1.0 and $+1.0$, and not between 0.0 and 1.0. In this case, the sigmoid function is frequently replaced by:

$$y = \frac{2}{\pi} \cdot \tan^{-1}(x) \qquad (14.30)$$

which also has a very convenient partial derivative:

$$\frac{\partial y}{\partial x} = \frac{2}{\pi} \cdot \frac{1.0}{1.0 + x^2} \qquad (14.31)$$

However, this function won't converge for our application either. Since we wish to obtain outputs of exactly $+1.0$ and -1.0, we would need infinitely large states, and therefore infinitely large weights.

Without the $2/\pi$ term, the network does learn, but converges very slowly. Therefore, we decided to eliminate the requirement of a continuous derivative and used a limit function as the activation function:

```
// [y] = LIMIT(x)
//
// This procedure computes the limit function
//
[n, m] = SIZE(x);
for i = 1:n, ...
    y(i) = MIN([MAX([x(i), -1.0]), 1.0]); ...
end
//
return
```

In this case, we cannot backpropagate the gradient. Instead, we make use of the fact that we know that all outputs must converge to either $+1.0$ or -1.0. We therefore punish the distance of the true output from either of these two points using the tri function [14.19]:

```
// [y] = TRI(x)
//
// This procedure computes the tri function
//
[n, m] = SIZE(x);
y = ONES(n, m)- ABS(x);
//
return
```

We call this type of network a *pseudobackpropagation network*.

In addition to the weighting matrices, we need biases and momenta. The optimization starts with a zero–weight matrix, but adds small random momenta to the weights and biases. After each iteration, the momenta are updated to point more toward the optimum solution.

The program converges fairly quickly. It usually takes less than 20 iterations to converge to the correct solution. The program is also fairly insensitive to the length of the hidden layer. The convergence is equally fast with $lhid = 8$, $lhid = 16$, and $lhid = 32$.

This discussion teaches us another lesson. The design of neural networks is still more an art than a science. We usually start with one of the classical textbook algorithms ... and discover that it doesn't work. We then modify the algorithm until it converges in a satisfactory manner for our application. However, there is little generality in this procedure. A technique that works in one case may fail when applied to a slightly different problem. The backpropagation algorithm, as presented in this section, was taken from

Korn's new book [14.19]. This book contains a wealth of little tricks and ideas for improving the convergence speed of neural network algorithms.

Notice the gross difference between backpropagation and the functioning of our brain. Backpropagation learning is a gradient technique. Such optimization techniques have a tendency to either diverge or converge on a local minimum. In comparison, our brain learns slowly but reliably and doesn't exhibit any such convergence problems. Thus, gradient techniques, besides not being biologically plausible, may not be the best of all learning techniques. In an artificial neural network, the robustness of the optimization technique is much more important than the convergence speed. Unfortunately, these two performance parameters are always in competition with each other, as I shall demonstrate in the companion book of this text.

I explained earlier that most artificial neural network programs (such as backpropagation networks) will need many iterations for convergence. However, one thing that we certainly don't want to do is to rerun a MATLAB or CTRL-C procedure several hundred thousand times. The efficiency of such a program would be incredibly poor. Thus, MATLAB and CTRL-C are useful for documenting neural network algorithms, not for using them in an actual implementation. (Because the previously demonstrated exclusive–or problem is trivial from a computational point of view, CTRL-C was able to solve this problem acceptably fast.)

Special artificial neural network hardware is currently under development. Neural network algorithms lend themselves to massive parallel processing, thus a hardware solution is clearly indicated. However, today's neural network chips are still exorbitantly expensive and not sufficiently flexible. For the time being, we must therefore rely on software simulation tools.

Granino Korn recently developed a new DESIRE dialect, called DESIRE/NEUNET [14.19]. This code has been specially designed for the simulation of artificial neural networks. It has been optimized for fast compilation and fast execution. The DESIRE/NEUNET solution of the exclusive–or backpropagation network is given here:

−− ARTIFICIAL NEURAL NETWORK
−− Exclusive-Or Backpropagation

−− Constants
$lhid = 8$ | $g1 = 0.6$ | $g2 = 0.3$
$m1 = 0.06$ | $m2 = 0.03$ | $crit = 0.025$
−− Declarations
ARRAY $u1[2]$, $y1[lhid]$, $y2[1]$
ARRAY $W1[lhid, 2]$, $W2[1, lhid]$, $b1[lhid]$, $b2[1]$
ARRAY $WW1[lhid, 2]$, $WW2[1, lhid]$, $bb1[lhid]$, $bb2[1]$
ARRAY $delta1[lhid]$, $delta2[1]$, $e[1]$
ARRAY $inpt[4, 2]$, $target[4, 1]$
−− Read Constant Arrays
data $-1, -1; -1, 1; 1, -1; 1, 1$ | **read** $inpt$
data $-1; 1; 1; -1$ | **read** $target$
−− Initial conditions
for $i = 1$ **to** $lhid$
  $W1[i, 1] = 0.1*\text{ran}(0)$ | $W1[i, 2] = 0.1*\text{ran}(0)$
  $W2[1, i] = 0.1*\text{ran}(0)$
  $WW1[i, 1] = 0.0$ | $WW1[i, 2] = 0.0$
  $WW2[1, i] = 0.0$
  $b1[i] = 0.0$ | $bb1[i] = 0.0$
next
$b2 = 0.0$ | $bb2 = 0.0$ | $error = 0.0$
$min = -1.0$ | $max = 1.0$ | $sr = 4$

$TMAX = 20.0$ | $t = 1$ | $NN = 20$

−− scaling
$scale = 5$
drun $TEACH$
drun $RECALL$

DYNAMIC

---

label $TEACH$
$iRow = t$ | **VECTOR** $u1 = inpt\#$
VECTOR $y1 = WW1 * u1 + bb1$; min, max
VECTOR $y2 = WW2 * y1 + bb2$; min, max
VECTOR $e = target\# - y2$
VECTOR $delta2 = e * tri(y2)$
VECTOR $delta1 = WW2\% * delta2*tri(y1)$
LEARN $W1 = g1 * delta1 * u1 + m1 * WW1$
LEARN $W2 = g2 * delta2 * y1 + m2 * WW2$
UPDATE $b1 = g1 * delta1 + m1 * bb1$
UPDATE $b2 = g2 * delta2 + m2 * bb2$
DOT $e2 = e * e$ | $error = error + e2$

```
- - - - - - - - - - - - - - - - - - - - - - - - - - - - - -
SAMPLE sr
term crit − error
error = 0.0
MATRIX WW1 = W1  |  MATRIX WW2 = W2
VECTOR bb1 = b1  |  VECTOR bb2 = b2
dispt error
- - - - - - - - - - - - - - - - - - - - - - - - - - - - -
label RECALL
iRow = t  |  VECTOR u1 = inpt#
VECTOR y1 = WW1 ∗ u1 + bb1;  min, max
VECTOR y2 = WW2 ∗ y1 + bb2;  min, max
in1 = u1[1]  |  in2 = u1[2]  |  out1 = y2[1]
type in1, in2, out1
- - - - - - - - - - - - - - - - - - - - - - - - - - - - -
/ − −
/PIC 'xor.prc'
/ − −
```

This DESIRE program exhibits a number of new features that we never before met. DESIRE is able to handle vectors and matrices in a somewhat inflexible but extremely efficient manner. Contrary to CTRL–C (or MATLAB), DESIRE does not allow us to easily manipulate individual elements within matrices or vectors. DESIRE's matrix and vector operators deal with the whole data structure at once. Thus, some algorithms, such as the genetic algorithms described later in this chapter, can be elegantly programmed in CTRL–C, while they are almost impossible to implement in DESIRE. Yet DESIRE works very well for many classical neural network algorithms such as backpropagation and counterpropagation.

In DESIRE, all vectors and matrices must be declared using an ARRAY statement. ARRAY declarations can make use of previously defined constants.

Within the DYNAMIC block, vector assignments can be made using the VECTOR statement. The '∗' operator in a vector assignment denotes either the multiplication of a matrix with a vector, or the multiplication of a scalar with a vector, or the elementwise multiplication of two vectors. Notice that the VECTOR statement has been explicitly designed for the simulation of artificial neural networks. The statement:

$$\text{VECTOR } y1 = WW1 \ast u1 + bb1; \quad min, max \tag{14.32}$$

computes the state vector of a neural network and simultaneously its output using a hard limiter as the activation function. DE-SIRE/NEUNET also offers most other commonly used activation functions and their derivatives as system–defined functions, such as the tri function used in the preceding program. Remember that DESIRE is case–sensitive. The VECTOR assignment:

$$\textbf{VECTOR } y = y + x \tag{14.33}$$

can be abbreviated as:

$$\textbf{UPDATE } y = x \tag{14.34}$$

Matrix assignments can be made using the MATRIX statement. The '*' operator in a MATRIX statement denotes the outer (Hadamard) product of two vectors or the multiplication of a scalar with a matrix. The MATRIX assignment:

$$\textbf{MATRIX } A = A + B \tag{14.35}$$

can be abbreviated as:

$$\textbf{LEARN } A = B \tag{14.36}$$

The DOT statement computes the inner product of two vectors. The result is a scalar.

The '%' operator denotes a matrix transpose. *WW2%* is the transpose of matrix *WW2*. The '#' operator is a row vector extraction operator. In the preceding program, *inpt* is a matrix. *inpt#* extracts one particular row from that matrix, namely, the row indicated by the system variable *iRow*. '#' is a modulo operator. If *iRow* is larger than the number of rows of *inpt*, '#' starts counting the rows anew. During the first step, $t = 1$ and therefore $iRow = 1$. Thus, *inpt#* extracts the first–row vector from the *inpt* matrix. During the second step, $iRow = 2$ and therefore the second–row vector is used. In this way, the network gets to use all four input/target pairs. During the fifth step, $iRow = 5$ and since *inpt* has only four rows, the first row is extracted again. This language construct is much less elegant than CTRL–C's (and MATLAB's) wild–card feature. It is the price we pay for DESIRE's ultrafast compilation and execution. Remember that DESIRE was designed for optimal efficiency, not for optimal flexibility.

The SAMPLE block is similar to a DISCRETE block with an **interval** statement in ACSL. It is normally used to model difference equations. The argument of the SAMPLE statement denotes the frequency of execution of the SAMPLE block. In our example, the SAMPLE block will be executed once every four communication intervals. This construct is somewhat awkward, because it isn't necessarily meaningful to link the sampling rate with the communication interval. It may be desirable to sample a signal much more often than we wish to store results for output. The argument of the SAMPLE block should therefore refer to an arbitrary time interval rather than a multiple of the communication interval. Moreover, we may wish to simulate several discrete blocks sampled at different frequencies. Unfortunately, DESIRE allows only one SAMPLE block to be specified in every program.

Finally, we notice that several different DYNAMIC blocks can be coded in a single DESIRE program. Labels can precede sections of DYNAMIC code. These labels can be referenced in the **drun** statement.

Although I am quite critical of some of the details of the DESIRE language specification, DESIRE/NEUNET is clearly the best tool currently available for neural network simulations. On a 386–class machine, the exclusive–or backpropagation program compiles and executes in considerably less than 1 sec.

ACSL also offers matrix manipulation capabilities, but they are not useful for neural network simulation. Neither DESIRE nor ACSL provide easy access to individual matrix or vector elements. Yet the reasons are different. In ACSL, matrices were only an afterthought. They were implemented as *generic macros* (ACSL's "macro macro"). Consequently, matrix operations must be coded in inverse polish notation. For example, the statement:

$$\mathbf{x} = \mathbf{W} \cdot \mathbf{u} + \mathbf{b} \tag{14.37}$$

would have to be coded in ACSL as:

$$\mathrm{MADD}(x = \mathrm{MMUL}(W, u) \; , \; b) \tag{14.38}$$

The compilation of generic macros is fairly slow. Furthermore, ACSL does not offer any nonlinear vector functions, as they are needed to describe the nonlinear activation function of artificial neurons.

14.7 Neural Networks for Dynamical Systems

Artificial neural feedforward networks are basically nonlinear multivariable function generators. They statically map a set of inputs into a set of outputs.

For the state–space representation:

$$\dot{\mathbf{x}} = \mathbf{f}(\mathbf{x}, \mathbf{u}, t) \tag{14.39}$$

at any given time t, Eq.(14.39) maps the state vector $\mathbf{x}$ and the input vector $\mathbf{u}$ statically into the state derivative vector $\dot{\mathbf{x}}$. Thus, an artificial neural feedforward network should be able to "identify" the state–space model, i.e., to learn the system behavior. Unlike classical identification techniques, we need not provide the neural network with the explicit structure of the state–space model. In this respect, the neural network operates like the inductive reasoners discussed in Chapter 13.

Figure 14.15 shows a typical configuration of an adaptive (self–tuning) controller of a plant. The fast inner loop controls the inputs of a plant using a controller. The optimal controller parameters $\mathbf{p}_c$ depend on the current model parameters $\mathbf{p}_m$. In the slow outer loop, the model parameters $\mathbf{p}_m$ are identified from measurements of the plant input u and the plant output y. The optimal controller parameters $\mathbf{p}_c$ are then computed as a nonlinear function of the model parameters $\mathbf{p}_m$.

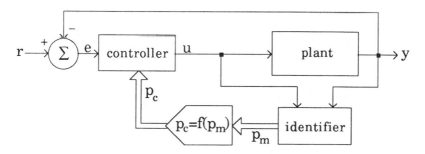

Figure 14.15. Plant with self–tuning regulator.

Classical adaptive controllers work only if the model can be parameterized, i.e., if the uncertainties of the model can be represented through a model parameter vector $\mathbf{p}_m$. Moreover, most of the known parameter identification techniques require the plant itself to be linear. This is true for both self–tuning regulators and model–reference adaptive controllers.

It is feasible to replace both the identification stage and the map from the model parameters to the controller parameters with neural networks. An excellent current review of research in this area was recently published by Narendra and Parthasarathy [14.26]. They show how the linear plant requirement can be eliminated. In this chapter, we shall not pursue the identification of adaptive controllers any further. Instead, we shall restrict our discussion to a very simple example of the identification of a dynamical system. Given the system:

$$\dot{\mathbf{x}} = \mathbf{A} \cdot \mathbf{x} + \mathbf{b} \cdot u$$
$$= \begin{pmatrix} 0 & 1 & 0 \\ 0 & 0 & 1 \\ -2 & -3 & -4 \end{pmatrix} \cdot \mathbf{x} + \begin{pmatrix} 0 \\ 0 \\ 1 \end{pmatrix} \cdot u \qquad (14.40a)$$

$$y = \mathbf{C} \cdot \mathbf{x} + \mathbf{d} \cdot u$$
$$= \begin{pmatrix} 1 & 0 & 0 \\ 0 & 1 & 0 \\ 0 & 0 & 1 \end{pmatrix} \cdot \mathbf{x} + \begin{pmatrix} 0 \\ 0 \\ 0 \end{pmatrix} \cdot u \qquad (14.40b)$$

We wish to train an artificial neural network to behave like the state–space model of this system. We shall use a backpropagation network similar to the one used before to replace the state equations, Eq.(14.40a). For this problem, we simulated the linear system in CTRL–C and stored the resulting *inpt* and *target* matrices for off–line (supervised) learning. The CTRL–C code for the simulation is given here:

```
// Define the system
//
a = [   0    1    0
        0    0    1
       -2  - 3  - 4 ];
b = [ 0 ; 0 ; 1 ];
c = EYE(a);
d = ZROW(b);
```

```
// Simulate the system in CTRL – C
//
t = 0:3:900;
u = ROUND(RAND(t));
x0 = ZROW(3,1);
SIMU('ic', x0)
[y, x] = SIMU(a, b, c, d, u, t);
//
// Postprocess the results
//
xdot = ZROW(x);
for i = 1:301, ...
    xdot(:, i) = a * x(:, i) + b * u(i); ...
end
//
inpt = [ u' , x' ];
target = xdot';
//
// Save the results
//
save inpt target   > linear.dat
//
return
```

The system to be learned contains four inputs (u and **x**) and three targets ($\dot{\mathbf{x}}$). There are 301 different input/target pairs available for training. The CTRL–C version of the backpropagation network is given here:

```
// This procedure designs a backpropagation network for
// learning the behavior of a linear system
// Select the length of the hidden layer (LHID) first
//
deff fatan -c
//
// Load the input and target vectors
//
load inpt target   < linear.dat
[npair, ninpt] = SIZE(inpt);
[npair, ntarg] = SIZE(target);
```

```
// Initialize the weighting matrices and biases
//
W1 = 0.01 * (2.0*RAND(lhid, ninpt)− ONES(lhid, ninpt));
W2 = 0.01 * (2.0*RAND(ntarg, lhid)− ONES(ntarg, lhid));
b1 = ZROW(lhid, 1);   b2 = ZROW(ntarg, 1);
WW1 = ZROW(lhid, ninpt);   WW2 = ZROW(ntarg, lhid);
bb1 = ZROW(lhid, 1);   bb2 = ZROW(ntarg, 1);
//
// Set the gains
//
g1 = 0.1;   g2 = 0.005;
//
// Set the termination condition
//
crit = 3.0;   error = 10.0;   count = 0;
//
// Learn the weights and biases
//
while error > crit, ...
   count = count + 1; ...
   ... //
   ... // Loop over all input/target pairs
   ... //
   error = 0; ...
   for nbr = 1:npair, ...
      u1 = inpt(nbr, :)'; ...
      y2h = target(nbr, :)'; ...
      ... //
      ... // Forward pass
      ... //
      x1 = WW1 * u1 + bb1; ...
      y1 = (2.0/pi)*ATAN(x1); ...
      u2 = y1; ...
      x2 = WW2 * u2 + bb2; ...
      y2 = (2.0/pi)*ATAN(x2); ...
      ... //
      ... // Backward pass
      ... //
      e = y2h − y2; ...
      delta2 = (2.0/pi)*FATAN(y2) .* e; ...
      W2 = W2 + g2 * delta2 * (u2'); ...
      b2 = b2 + g2 * delta2; ...
      delta1 = (2.0/pi)*FATAN(y1) .* ((WW2') * delta2); ...
      W1 = W1 + g1 * delta1 * (u1'); ...
      b1 = b1 + g1 * delta1; ...
      error = error+ NORM(e); ...
   end, ...
```

```
    ... // Update the momentum matrices and vectors
    ... //
    WW1 = W1;   WW2 = W2; ...
    bb1 = b1;   bb2 = b2; ...
    err(count, 1) = error; ...
end
//
// Save the learned network weights and biases for later
//
save WW1 WW2 bb1 bb2 err  > linear_2.dat
//
return
```

Since the system is continuous, we replaced the LIMIT/TRI function pair by an ATAN/FATAN function pair, where FATAN is the partial derivative of ATAN. To obtain convergence, the gains had to be considerably smaller than those used for the exclusive–or problem. The momentum terms had to be eliminated for this network. We looped over all available input/target pairs before we updated the weighting matrices and bias vectors. Thus, each iteration of the weight learning algorithm contained $npair = 301$ forward and backward passes through the network.

All desired target variables are approximately 0.1 units in amplitude. Thus, if our backpropagation network computes a real target of similar amplitude, but with arbitrary direction, the error is also approximately 0.1. Since we accumulate the errors of all input/target pairs, we expect an initial total error of about 30. This error should be reduced by one order of magnitude before we can claim that our network has learned the system. Thus, we set $crit = 3.0$.

The learning required 774 iterations and consumed more than 7 hours of CPU time on a VAX–11/8700. Obviously, this is not practical. If you wish to know how much time DESIRE/NEUNET requires to train the same network, solve Hw.[H14.2]. It turns out that DESIRE/NEUNET solves this problem within a few minutes on a 386–class machine. Moreover, notice that I gave you good gain values to start with. In most cases, we shall need to rerun the same optimization many times in order to determine decent gain values. CTRL–C and MATLAB are obviously not useful for practical neural network computations. However, these languages are excellent tools for documenting neural network algorithms.

Figure 14.16 displays the total error as a function of the iteration count.

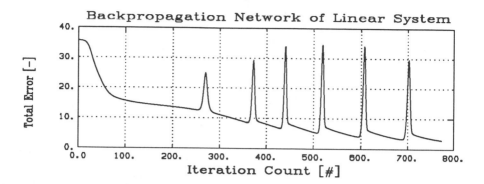

Figure 14.16. Learning a backpropagation network for a linear system.

The results are rather interesting. The weight learning problem can be interpreted as a discrete–time dynamical system, i.e., the updating of the weights is like solving a highly nonlinear high–dimensional set of difference equations. This dynamical system exhibits an oscillatory behavior. About once every 95 iterations, the network goes through a short phase in which the error temporarily grows. However, the network settles on a lower error level after each successive temporary instability. Notice that this dynamical system is chaotic. Each peak is slightly different from every other. The system behaves similarly to the Gilpin model for a competition factor of 1.0.

Let us analyze the behavior of our linear system at a particular point in time, $t = 813$ sec. The input and the state vector at that time have values of:

$$u = 0.0; \quad \mathbf{x} = \begin{pmatrix} -0.0464 \\ -0.0646 \\ 0.0756 \end{pmatrix}$$

and the true state derivative vector is:

$$\dot{\mathbf{x}}_{\mathbf{true}} = \begin{pmatrix} -0.0646 \\ 0.0756 \\ -0.0157 \end{pmatrix}$$

The approximated state derivative vector computed by the backpropagation network is:

$$\dot{\mathbf{x}}_{\mathbf{A.N.N.}} = \begin{pmatrix} -0.0606 \\ 0.0728 \\ -0.0170 \end{pmatrix}$$

Thus, the backpropagation network has indeed successfully learned the multivariable function. The approximated state derivative values are only slightly smaller than the true values.

Notice that the true system itself may be interpreted as an artificial neural network (A.N.N.):

$$\dot{\mathbf{x}} = \mathbf{A} \cdot \mathbf{x} + \mathbf{b} \cdot u = (\mathbf{A} \quad \mathbf{b}) \cdot \begin{pmatrix} \mathbf{x} \\ u \end{pmatrix} = \mathbf{W} \cdot \mathbf{u_1} \qquad (14.41)$$

Thus, we could have identified a single–layer network with four inputs and three outputs using a linear activation function. We could have trained this network much more easily, and the results would have been even better ... but this would have been no fun. The purpose of the exercise was to show that we can train an arbitrary artificial neural network to blindly learn the behavior of an arbitrary multivariable function.

We chose to learn the network off–line in a supervised learning mode. However, we could have easily learned the network on–line in an adaptive learning mode. We would simply have used input/target pairs as they arrive to train our $\mathbf{W^i}$ matrices and $\mathbf{b^i}$ vectors and occasionally update the $\mathbf{WW^i}$ matrices and $\mathbf{bb^i}$ vectors.

Encouraged by these nice results, let us now close the loop. Figure 14.17a shows the true system, which we shall now replace with the approximated system of Fig.14.17b.

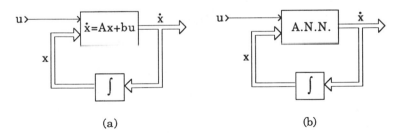

(a)   (b)

Figure 14.17a–b. Approximation of a linear system by an A.N.N.

We close the loop around the backpropagation network using vector integration. As in Chapter 13, we shall assume that forecasting begins at time $t = 813$ sec. We shall predict the future state trajectories of the system using the A.N.N. Figure 14.18 shows the results.

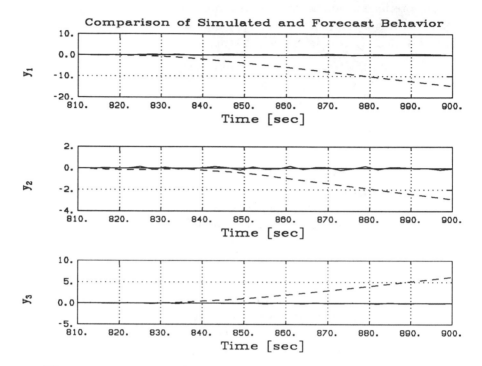

Figure 14.18. Comparison of true and approximated trajectories.

The solid lines represent the true trajectory behavior, while the dashed lines represent the approximated trajectory behavior. This was obviously not such a brilliant idea after all. The A.N.N. approximation and the true trajectory behavior vary greatly. The A.N.N. systematically underestimates the first and second state derivatives, whereas it systematically overestimates the third state derivative. The errors in each individual step are fairly small, but these errors accumulate, and the approximated solution quickly drifts away from the true solution. Unlike the results obtained in Chapter 13, errors accumulate if we close an integration loop around a neural feedforward network trained to approximate the behavior of an open–loop system.

Let us now try another approach. Remember that in Chapter 13 we chose a mask depth of three, i.e.:

$$\mathbf{x}_{k+1} = \mathbf{f}(u_{k-1}, \mathbf{x}_{k-1}, u_k, \mathbf{x}_k, u_{k+1}) \tag{14.42}$$

Thus, we could try to train an A.N.N. to approximate the closed–loop behavior of our dynamical system using Eq.(14.42). This time, we have nine inputs and three outputs. Since I have become impatient and don't want to spend several more hours training yet another backpropagation network, I shall use a counterpropagation network instead. Since counterpropagation networks work much better for digital systems, let us try the idea shown in Fig.14.19.

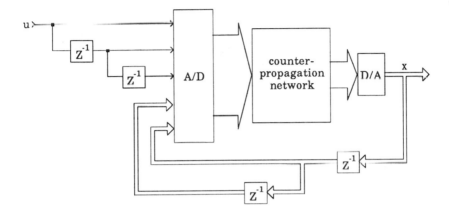

Figure 14.19. Counterpropagation network for linear system.

We generate the weighting matrices of the feedforward counterpropagation network by converting the nine analog inputs to 90 digital inputs and the three analog targets to 30 digital targets using 10–bit analog–to–digital (A/D) converters. We set up the counterpropagation network using the first 269 input/target pairs. Thus, the counterpropagation network has 90 inputs, a hidden layer of length 269, and an output layer of length 30. The off or false state of the digital inputs and targets is represented by -1.0, whereas it is represented by '0' in the hidden layer. The on or true state of all digital signals is represented by $+1.0$.

During recall, we convert the analog input vectors of length nine to digital input vectors of length 90 using the same 10–bit A/D converters. For each digital input vector, the counterpropagation network predicts a digital target vector of length 30. We then convert the resulting 30 digital targets back to three analog targets using 10–bit digital–to–analog (D/A) converters.

The algorithm can be slightly improved by using so–called Gray–

code converters in place of the usual binary–code converters [14.13]. Thereby, the Hamming distance between neighboring signals is always 1 bit which reduces the risk of incorrect predictions.

In our example, we use this configuration to predict the trajectory behavior of the linear dynamical system over the last 87.0 time units or 30 input/target pairs. Figure 14.20 shows the results of this effort.

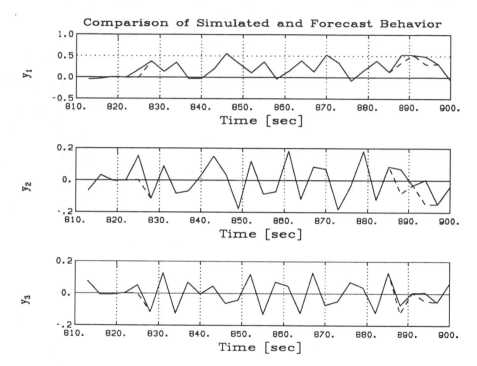

Figure 14.20. Comparison of true and approximated trajectories.

This time, we were right on target. Within the forecasting sequence, only a few predictions were incorrect. There are two possible sources of errors:

(1) Since the input/target map is no longer strictly deterministic, the same input may produce several different targets. In that case, we may subsequently recall the wrong target.

(2) A particular input may never have been observed during the setup period. In this case, the prediction will be arbitrary and probably incorrect.

As in the solution presented in Chapter 13, errors don't accumulate. An error that occurs once is not propagated through the network.

Compared to the SAPS method, the counterpropagation network allows us to operate with much better resolution. In SAPS, we had to discretize our continuous signals into no more than five different levels. In the counterpropagation network, we discretized our continuous signals into 1024 different levels. The price that we pay for this luxury is an increased memory requirement. The SAPS masks store their knowledge about the system in a much more compact fashion. The weighting matrices of the counterpropagation network have dimensions 269×90 and 30×269, and thus, they are fairly large. If you wish to learn how Fig.14.20 was produced, solve Hw.[H14.4].

Notice that this algorithm can also be implemented in an adaptive learning mode. To do this, we begin with a hidden layer of length one and store the first input/target pair as the two weighting matrices $\mathbf{W}^1$ and $\mathbf{W}^2$. We then process the second input. The second input will probably differ from the first. In this case, the counterpropagation network will forecast the wrong target. If this happens, we concatenate the new input as an additional row to the $\mathbf{W}^1$ matrix from below and we concatenate the true target as an additional column to the $\mathbf{W}^2$ matrix from the right, thereby incrementing the length of the hidden layer. If the target is predicted correctly, we leave the A.N.N. as is. After many iterations, most of the inputs will have been seen earlier, and eventually, the network will stabilize. If you wish to study this algorithm in more detail, solve Hw.[H14.5].

Yet another technique to emulate dynamical systems is direct identification of a *recurrent network*, i.e., identification of an A.N.N. with built–in feedback loops. This approach was first proposed by Hopfield [14.15]. The most popular and widely studied recurrent network is therefore referred to as a *Hopfield net*. Hopfield nets have only one layer and require a much smaller number of neurons than the previously proposed counterpropagation networks. For our (trivial) example, three artificial neurons would suffice. The Hopfield net would simply be the sampled–data version of our continuous system. Other commonly used variations of recurrent networks are the multi-layered *bidirectional associative memories* and Grossberg's *adaptive resonance circuits (ARTs)*. Recurrent networks can be problematic with respect to their stability behavior. We won't pursue this avenue any further. However, notice that the explicitly dimensioned counterpropagation network of Fig.14.19 is in fact also a recurrent network. It does not exhibit stability problems.

14.8 Global Feedback Through Inverse Networks

As I mentioned earlier, today's A.N.N.s bear little resemblance to actual neural networks found in mammals. Let us now return to the question of how our brain works, how it processes information, and how it learns. Obviously, my remarks must become a little more philosophical at this point and I won't be able to support my hypotheses with short CTRL–C or DESIRE/NEUNET programs.

We wish to discuss how our brain is believed to process various types of sensory input signals. Let us pursue the processing of information related to the animal dog. Its symbol in my brain will be denoted as $\mathcal{DOG}$, the sound "dog" will be written as $\approx dog \approx$, and a picture of a dog will be coded as $\wedge\!\wedge^{\overset{\bigcirc\!\prec}{}}$.

My brain has a *visual neural network* that maps $\wedge\!\wedge^{\overset{\bigcirc\!\prec}{}}$ into $\mathcal{DOG}$ and an *auditory neural network* that maps $\approx dog \approx$ into $\mathcal{DOG}$. Thus, the brain can map different sensory input signals into the same symbol.

If somebody says $\approx dog \approx$ to me, an image of my dog, $my\!-\!\wedge\!\wedge^{\overset{\bigcirc\!\prec}{}}$, appears before my inner eye. How is this possible and what is an inner eye? It seems that the brain is not only capable of mapping pictures into symbols, but it can also map symbols back into pictures. Somehow, these pictures can superpose real images captured by the eye. We call this internal seeing the *inner eye*. I shall denominate the neural network that maps symbols back into pictures the *inverse visual neural network*.

Similarly, my brain also has an *inverse auditory neural network*. In fact, my brain has an inverse neural network for each of my senses. If I watch old slides from one of my trips to the Middle East, I suddenly "smell" the characteristic spices in the bazaars (such as sumak), and I "hear" the muezzin call the believers to prayer. These sensations are produced by the inverse neural networks, i.e., by global feedback loops that connect the symbols right back to the senses from which they originated. Usually, these feedback loops are fairly rigorously suppressed. We have been taught to ignore these feedback signals most of the time. They are constantly present, however, and play an important role in our lives.

Figure 14.21 depicts these feedback loops for the visual and auditory networks.

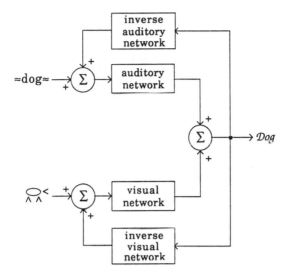

Figure 14.21. Global feedback through inverse neural networks.

I am convinced that this global feedback mechanism is essential to our brain's ability to learn. Without such feedback, the brain could mindlessly react to stimuli, but it could never achieve higher levels of cognition. We would not be able to truly think. Let me elaborate on this idea a little further in the next section.

14.9 Chaos and Dreams

It turns out that while the neural networks themselves are massive parallel processors at the symbolic or conscious level, we can think only one thought at a time. This means that at the output summer of Fig.14.21, where the symbols are formed and ultimately rise to the level of consciousness, we have a winner–takes–it–all situation. We are faced with competition.

Remember our lessons from Chapter 10:

high–dimensional system
+ feedback loops
+ competition
$\Longrightarrow$ chaos

If we take a high–dimensional system like our brain, introduce some feedback loops (hopefully without making the system unstable), and introduce competition, the result is almost invariably chaos. I therefore argue that, due to the existence of global feedback loops, our brain operates permanently under conditions of a chaotic steady–state.

Normally, the external inputs are much more powerful than the feedback loops. Thus, they will usually win the competitive battle at the output summer. However, during the night, when the external inputs are reduced to a minimum, the feedback loops take over. The behavior of our brain is then dictated by its own chaotic steady–state behavior. We call this mental state *dreaming*. In our dreams, colorful sceneries appear before our inner eyes. Since the visual input is usually dominant, this is what we most likely will remember in the morning. However, with our "inner ears," we also hear people speak to us and I am convinced that we also smell with our "inner nose," taste with our "inner taste buds," and touch with our "inner skin."

Recall another lesson from Chapter 10. For self–organization to occur in a system, we need an innovator and an organizer. In our brain, the *innovator* is the chaotic feedback. Without this feedback, we would never create new and original ideas. Instead, we would react to our environment like mindless machines [14.23]. The *organizer* is the reward mechanism. It is responsible for the survival of the "good ideas." Thus, without chaos, no learning or self–organization could occur. No wonder today's robots are "mindless." We haven't yet figured out how to introduce chaos into their "brains." This certainly is a worthwhile research topic.

14.10 Internalization Processes and Control Mechanisms

In our brains, the relative weights of the external versus internal inputs are controlled by various mechanisms. The control is exerted by two higher–level functions that have been coined our *ego* and our *superego*. The intensity of the feedback loops is controlled by mechanisms usually referred to as our will (attributed to the *ego*), and by societal taboos (attributed to the *superego*). The latter mechanism prevents us from pursuing particular thoughts beyond a danger level, a threshold coded into our *superego*. When a dangerous idea pops up, the *superego* automatically reduces the weights of the feedback loops and we divert our attention to another topic [14.23]. Moreover, we all want to be liked. Since strangely behaving people are not liked, our *ego* reduces the impact of the internal feedback loops to an extent where we react to external stimuli reliably and coherently.

In order to understand these mechanisms better, we need to expand the block diagram of Fig.14.21 to that in Fig.14.22.

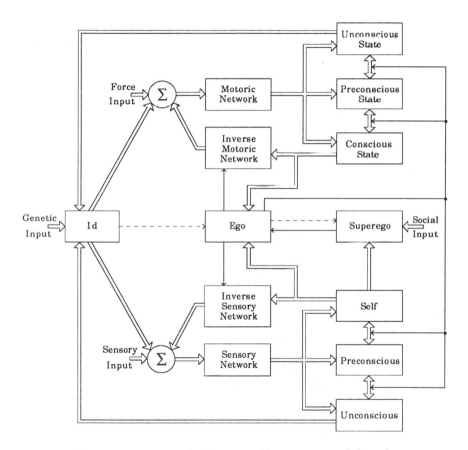

Figure 14.22. Block diagram of human mental functions.

From a psychodynamical point of view, we can distinguish between three major subjective functions (we could call them programs), the *id*, the *ego*, and the *superego*. They operate on three major objective functions (we could call them data bases), the *unconscious*, the *preconscious*, and the *conscious self*. The *id* is the earliest of our programs. It is purely hereditary, i.e., we are born with it. Initially, the *id* performs all of the control functions. During the first three years of our lives, the *ego* develops and subsequently assumes most of the control functions. However, since the *ego* is flexible and comparatively easy to reprogram (since it is predominantly conscious), we need yet another mechanism to ensure the overall stability of the system. This is the *superego*. It sets the limits for the *ego*. The *superego* develops last among the three programs, around age four or five. The *superego* is much more difficult to reprogram than the *ego* since it is mostly unconscious. The *superego* holds our unconscious beliefs and commands. It knows what is inherently right and wrong. It is the basis of our conscience, our superstitions, and all sorts of societal and familial taboos.

From the perspective of a control engineer, our brain contains two major model–reference adaptive control (MRAC) loops. On the sensory side, the innermost automatic feedback control loop contains the sensory network (forward path), which feeds the conscious self, the preconscious, and the unconscious. It also contains the inverse sensory network (feedback path). The innermost control loop is supervised by the *ego*, our major adaptive controller. It receives information mostly from the conscious self, but also from the preconscious and unconscious. The *ego* decides whether the system performs adequately. If the system reacts too slowly, the feedback gains of the innermost control loop are increased. If it starts to become unstable, the feedback gains are decreased. The *id* can perform similar functions. It receives information mostly from the unconscious, but also from the preconscious and conscious self. However, it is mostly content to let the *ego* do its job. It has more–direct means to intervene. It can provide strong input signals directly to our sensory system. If the *id* decides that I should eat, it provides my stomach sensors with the strong sensation of hunger. Notice that the *id* does not directly tamper with my data bases. It does not make me fall upon the idea that I might wish to eat. It stimulates my sensory input and lets the sensory network process the sensation of hunger, which will ultimately arrive at my self, will be forwarded as useful information to my *ego*, which instructs the second (motoric) control loop to do something about it. The *superego* is the external model of the MRAC. It provides the *ego* with the needed set values and ensures the overall stability of the system. The *ego* can be reprogrammed, but only to the extent that the *superego* permits. The motoric MRAC is a mirror image of the sensory MRAC. The innermost loop contains the motoric network (forward path), which feeds the three objective functions or data bases that maintain positional and velocity information. The inverse motoric network feeds back the perception of my current position, velocity, and force vectors. The *ego* is again the adaptive controller that puts purpose into my motions. The *id* can perform similar functions, but it usually doesn't. It has more–direct means to influence my motoric behavior, through reflexes. The *superego* knows "my place in the world." It prevents the *ego* from requesting motoric actions that are considered indecent.

From the perspective of a software engineer, we usually distinguish between data, and programs that act on data. We may classify the conscious self, the preconscious, and the unconscious, as well as their motoric counterparts as data, and the *id*, *ego*, and *superego* as programs that act upon these data. The four network boxes have not been named in the psychological literature. They are also programs, but hardwired ones. They are basically data filters. However, as software engineers, we know that the distinction between programs and data is not very crisp. An ACSL program is a data filter that maps input trajectories (input data) into output trajectories (output data). However, the ACSL preprocessor is yet another data filter that maps the ACSL program (input data) into a FORTRAN program (output data). Thus, what constitutes data and what constitutes programs depends on our perspective. In LISP, we notice this fusion even more clearly. There is really no difference at all between a data item and an instruction operating on a data item. In an expert system, we usually distinguish between the knowledge base that contains the facts and the rule base that contains instructions for how to process the facts. Yet both are data bases that are syntactically identical. This similarity applies to our brain as well. We possess only one brain that stores both the facts (objective functions) and the rules (subjective functions). When we look at our brain under a microscope, we cannot distinguish between them, because the microscope shows us only the syntax, not the semantics of our brain.

In Fig.14.22, data flow is shown as double lines. Control signals are shown as single lines. The basic programs of our brain are the four network boxes. The *ego* acts as an incremental compiler that constantly modifies the four basic programs, but it tampers more with the feedback networks than with the feedforward networks. The *ego* constantly modifies parameters in our feedback networks. The dashed lines in Fig.14.22 represent control signals of a deeper type. Initially, there is no *ego*. The *id* represents the desire for an *ego* to develop. Thus, the *id* controls the creation of the *ego* program. After the *ego* is fully developed, the *id* even delegates this function to the *ego*, i.e., the *ego* starts reprogramming itself.

Even this capability is not unheard of in software engineering. ELLPACK [14.28] is a simulation language for solving elliptic partial differential equations. Like in ACSL, a preprocessor translates ELLPACK programs into FORTRAN. However, the ELLPACK language is not static. It is meant to be user–modifiable. Therefore, the preprocessor is not hand–coded. Instead, the system comes with a compiler generator that can automatically generate a new version of the preprocessor from a data template file that contains an abstract description of the ELLPACK syntax (grammar) and semantics (code to be generated). Then the software designers decided that this data template file was too difficult to create manually. Consequently, they designed a template processor that generates the required data template file from a yet more abstract description. Naturally, they didn't want to manually code the template processor either. Instead, they described the syntax and semantics of the template processor and generated it using the same compiler generator that they had used before to generate the ELLPACK preprocessor. Finally, they described the syntax and semantics of the compiler generator itself in terms of the syntax and semantics of the compiler generator, and they now can feed this description into one version of the compiler generator and use it to generate the next version of itself. They actually started out with a very simple compiler generator and bootstrapped it by iteratively processing it through itself.

The *ego* and *superego* together perform the function of a highly effective MRAC. If the feedback gains are set too low, we lose our creativity and inspiration. If they are set too high, the system becomes unstable and we end up in a mental institution. Yet the set values of the adaptive controller are highly individual.

People who effectively suppress their feedback loops are perceived as reliable and predictable, but not very imaginative, possibly as compulsive, and in extreme cases as neurotic. These people are the bureaucrats of our society. They desire predictable routine lives. They feel uncomfortable changing jobs. They usually prefer the apparent safety of an existing situation, even if it is unbearable, to the uncertainty of change. Neurotic people have a small *ego* and an overpowering *superego* that leaves the *ego* little latitude for improvement. The feedback loops are heavily suppressed. Much of the conscious self is intentionally repressed into the preconscious and unconscious. Threatening thoughts are swept under the carpet rather than confronted.

People who allow their feedback gains to be at a higher level are considered inspirational and imaginative, but also somewhat impulsive and incalculable. These people are the artists and Bohemians of our society. They are considered somewhat egotistic (which simply means that they have a strong *ego*). They will change a bad situation rather than suffer. Since their *ego* is strong, they don't need an overpowering *superego* to keep the system stable. Artists allow themselves to be influenced by their unconscious. They allow information to flow from their unconscious into their conscious self. They consider their unconscious their best friend rather than a threatening enemy. In psychodynamic terms, this is called the ability to experience a "controlled regression."

Psychotic people, finally, can be described as people whose *ego* is disintegrating. The *ego* is not stable, the *ego* boundaries dissolve, and the feedback loops are out of control. Whatever exists of the *superego* contributes to the disintegration of the *ego*. The weakened *ego* perceives the flow from the unconscious as extremely threatening, but cannot stop it. The feedback loops take over. The psychotic person reacts incoherently. The individual cannot concentrate on any one topic, but changes the subject frequently during the course of a conversation. The person suddenly "hears voices" (through the inverse auditory network). She or he is in a constant state of panic because of being overwhelmed by the flow of incomprehensible undigested information from the unconscious into the conscious self and back to the sensory level through the inverse networks.

Some psychedelic drugs, such as LSD, have the tendency to increase both the external and internal input gains. LSD users have described that they "see the colors much more intensively." Since the feedback gains are increased simultaneously, the system becomes less stable. Consequently, some drug users have been described as exhibiting psychotic behavior. Since drug use disables the control mechanisms of the *ego*, this is a very dangerous proposition, even independently of the chemical side effects.

Until now, we have only discussed the "end product," the fully developed adult person. We have described the control mechanisms of the strong and healthy. We have pointed out how some mental disabilities, neurosis and psychosis, reflect upon functional deficiencies of some of the control mechanisms of our personality. We have not yet analyzed how and why these deficiencies have occurred in the first place.

From a phenomenological perspective, it may be observed that some of us grow up under adverse conditions. The adults around us don't provide us with sufficient affection (psychologists call this phenomenon *neglect*) or they suffocate us with too much of it (psychologists call this phenomenon *overprotection*). Input signals may be "scrambled" [14.20], i.e., words imply something, but mean something else (psychologists call this the *double bind*).

Children who grow up under such adverse conditions may eventually lose their interest in the external input, because it is either too painful (neglect or overprotection) or too unreliable (double bind). They turn into day dreamers ... because they consider their internal input less threatening than input from the environment [14.4]. Notice that day dreaming is not a negative phenomenon *per se*. It is the sole source of our creativity and inspiration. It is the wood from which our geniuses are carved. Yet under more severe circumstances, such children may become psychotic. They hear voices, they react incoherently to their environment, and external objects such as their parents or partners may no longer be cathected [14.5]. They are caught in a world in which they are the "only actors on a stage which encompasses the entire world." All other people can only be perceived as either threats or properties. Finally, if the trauma is experienced sufficiently early in life (within the first few months after birth), and if it is sufficiently strong, they simply switch off the external input altogether and become autistic.

However, a much more comprehensive picture has been painted by Otto Kernberg [14.16]. He identifies several pre–Oedipal development phases. The first phase, called the *primary undifferentiated autistic phase*, lasts for a few months after birth. During this phase, the child slowly develops a symbiotic relationship with his or her mother. The child cannot yet distinguish between itself and the mother. If this symbiotic relationship is traumatized, the child remains autistic. When a strong and stable symbiotic relationship has been established, the child enters the second phase, called the *phase of primary undifferentiated self–object images*. An all–good mother–self image enables the child to slowly learn to differentiate between the self–images and the object–images. This is the time when every child plays the "peekaboo" game. The child covers its head with a blanket and is "gone" ... but of course, the mother is gone as well. Within a short time span, the tension becomes unbearable and the blanket is removed. The child is "back" ... and so of course is the mother. If the mother is not able to maintain an embracing, allowing, and loving relationship with her child during this critical developmental phase, the child experiences an extreme aggression against the loved object, which can only be overcome by re–fusing the self–images with the object–images. The child cannot properly define its ego boundaries and this is the seed that will eventually lead to a psychosis, because a strong *ego* is needed for the development of a healthy *superego*. In the last pre–Oedipal phase, the child learns to integrate libidinally determined and aggressively determined self–images and object–images. It learns to integrate love and hate and to acknowledge the coexistence of both in the self and in others. The child learns that nobody is all good or all evil, and yet it is integrated enough to accept this fact without being threatened.

Notice that Kernberg's description does not contradict the phenomenological description given earlier. It only provides us with a more profound analysis of the internal mechanisms of the higher–level mental functions of the human personality.

Most neurophysiologists tend to attribute psychoses to chemical problems with neurotransmitters. Several hypotheses have been formulated relating various types of chemical substances to the occurrence of psychoses, none of which has yet been proven. Although this approach seems quite different from the psychodynamic explanations given earlier, these hypotheses do not contradict each other. Just as muscles in our body shrink when they are not in use, so are the synaptic strengths between neighboring neurons believed to weaken when the involved axon is not frequently fired. This is precisely the biological hypothesis behind our weight adjustment algorithms as expressed in today's artificial neural networks. Thus, the "disconnection of inputs" will ultimately be electrochemically implemented in our brain in the form of weakened synaptic strengths. This may be what neurophysiologists attempt to confirm with their measurements. The only remaining question is: What came first, the chicken or the egg?

14.11 Genetic Learning

Let us now return to the mechanisms of learning. I had mentioned earlier that gradient techniques are dangerous because of potential stability problems, besides the fact that they are not biologically plausible.

In this section, I shall introduce another optimization technique that does not exhibit the stability problems characteristic of gradient techniques, and while this approach is not biologically plausible in the context of neural learning, it has at least been inspired by biology. Genetic algorithms were first developed by John Holland in the late

1960s [14.14]. As with the neural networks, the basic idea behind genetic algorithms encompasses an entire methodology. Thus, many different algorithms can be devised that are all variations of the same basic scheme.

The idea behind genetic algorithms is fairly simple. Let me describe the methodology by means of a particular dialect of the genetic algorithms applied to the previously introduced linear system backpropagation network. In that problem, we started out by initializing the weighting matrices and bias vectors to small random numbers. The randomization was necessary in order to avoid stagnation effects during startup. Yet we have no reason to believe that the initial choice is close to optimal or even that the weights remain small during optimization. Thus, the initial weights (parameters) may differ greatly from the optimal weights, causing the optimization to require many iterations. Also, since backpropagation learning is basically a gradient technique, the solution may converge on a local rather than a global minimum, although this didn't happen in this particular example.

Genetic algorithms provide us with a means to determine optimal parameter values more reliably even in "rough terrain," i.e., when applied to systems with a cost function that has many "hills" and "valleys" in the parameter space.

Let us assume that we already know approximate ranges for the optimal weights. In our case, the optimal weights belonging to the $\mathbf{W}^1$ matrix assume values between -2.0 and $+2.0$, those belonging to the $\mathbf{W}^2$ matrix assume values in the range -0.5 to 0.5, those from the $\mathbf{b}^1$ vector are between -0.05 and 0.05, and those from $\mathbf{b}^2$ are bounded by -0.005 and 0.005. I am cheating a little. Since I solved the backpropagation problem already, I know the expected outcome. The more we can restrict the parameter ranges, the faster the genetic algorithm will converge.

We can *categorize* the parameter values by classifying them as 'very small,' 'small,' 'large,' and 'very large,' respectively. In terms of the terminology used in Chapter 13, we transform the formerly quantitative parameter vector into a qualitative parameter vector. A semiquantitative meaning can be associated with the qualitative parameters using fuzzy membership functions as shown in Fig.14.23 for the parameters stored in $\mathbf{W}^1$.

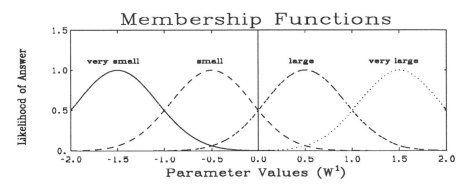

Figure 14.23. Fuzzy membership functions for parameter values.

The number of levels can, of course, be chosen freely. In our example, we decided to use four levels, $nlev = 4$. Let us now denote each class by a single upper–case character:

$$A \Leftrightarrow \text{very small}$$
$$B \Leftrightarrow \text{small}$$
$$C \Leftrightarrow \text{large}$$
$$D \Leftrightarrow \text{very large}$$

Thus, each qualitative parameter can be represented through a single character. We may now write all qualitative parameter values into a long character string such as:

$$ABACCBDADBCBBADCA$$

where the position in the string denotes the particular parameter and the character denotes its class. The length of the string is identical to the number of parameters in the problem. This is our qualitative parameter vector.

Somehow, this string bears a mild resemblance to our genetic code. The individual parameters mimic the amino acids as they alternate within the DNA helix. Of course, this is an extremely simplified version of a genetic code.

In our example, let us choose a hidden layer of length $lhid = 8$. Consequently, the size of $\mathbf{W}^1$ is 8×4, since our system has four inputs, and the size of $\mathbf{b}^1$ is 8. The size of $\mathbf{W}^2$ is 3×8, since the system has three targets, and the size of $\mathbf{b}^2$ is 3. Therefore, the total number of parameters of our problem *npar* is 67. Thus, the parameter string must be of length 67 as well.

The algorithm starts out with a genetic pool. We arbitrarily generate *nGenSt* = 100 different genetic strings and write them into a matrix of size 100 × 67. In CTRL–C (or MATLAB), it may be more convenient to represent the genes by integer numbers than by characters. The genetic pool can be created as follows:

$$GenPool = \text{ROUND}(nlev * \text{RAND}(nGenSt, npar)$$
$$+ 0.5 * \text{ONES}(nGenSt, npar))$$

Initially, we pick 10 arbitrary genetic strings (row vectors) from our genetic pool. We assign quantitative parameter values to them using their respective fuzzy membership functions by drawing random numbers using the fuzzy membership functions as our distribution functions. Next we generate weighting matrices and bias vectors from them by storing the quantitative parameters back into the weighting matrices in their appropriate positions. Finally, we evaluate our feedforward network 301 times using the available input/target pairs for each of these 10 parameter sets. The result will be 10 different *figures of merit*, which are the total errors, the sums of the individual errors for each training pair, found for the given weighting matrices. We sort the 10 performance indices and store them in an array. This gives us a vague first estimate of network performance.

We then arbitrarily pick two genetic strings (the parents) from our pool, draw an integer random number *k* from a uniform distribution between 1 and 67, and simulate a *crossover*. We pick the first *k* characters of one parent string (the head), and combine them with the remainder (the tail) of the other parent string. In this way, we obtain a new qualitative genetic string called the child. We then generate quantitative parameter values for the child using the fuzzy membership functions and simulate again. If the resulting performance of the child is worse than the fifth of the ten currently stored performance indices, we simply throw the child away. If it is better than the fifth string in the performance array, but worse than the fourth, we arbitrarily replace one genetic string in the pool with the child and place the newly found performance index in the performance array. The worst performance index is dropped from the performance array. If it is better than the fourth, but worse than the third stored performance, we duplicate the child once and replace two genetic strings in the pool with the two copies of the child. Now 2 of the 100 strings in the pool are (qualitatively) identical twins. If it is better than the third but worse than the second performance, we replace four genetic strings in the pool with the child. If it is better than the second but worse than the first, we replace six genetic strings in the pool with the child. Finally, if the child is the all–time champion, we replace 10 arbitrary genetic strings in the genetic pool with copies of our genius.

We repeat this algorithm many times, deleting poor genetic material while duplicating good material. As time passes, the quality of our genetic pool hopefully improves.

It could happen that the very best combination cannot be generated in this way. For example, the very best genetic string may require an A in position 15. If (by chance) none of the randomly generated 100 genetic strings had an A in that position or if those genes that had an A initially got purged before they could prove themselves, we will never produce a child with an A in position 15. For this reason, we add yet another rule to the genetic game. Once every $nmuta = 50$ iterations, we arbitrarily replace one of the characters in the combined string with a randomly chosen new value, simulating a *mutation*. Eventually, this mutation will generate an A in position 15.

Obviously, this algorithm can be applied in an adaptive learning mode. Our genetic pool will hopefully become better and better, and with it, our forecasting power will increase.

Of course, this algorithm can be improved. For instance, if we notice that a particular parameter stabilizes into one class, we can recategorize the parameter by taking the given class for granted and selecting new subclasses within the given class. In our example, we might notice that the parameter 27, which belongs to $\mathbf{W}^1$, always assumes a qualitative value of C, i.e., its quantitative value is in the range between 0.0 and 1.0. In this case, we can subdivide this range. We now call values between 0.0 and 0.25 'very small' and assign a character of A to them. Values between 0.25 and 0.5 are now called 'small' and obtain a character value of B, etc. I decided to check for recategorization once every 50 iterations, whenever I simulated a mutation. I decided that a recategorization was justified whenever 90% of the genes in one column of *GenPool* had assumed the same value, $nperc = 0.9$.

I ran my genetic algorithm over 800 iterations, which required roughly 2 hours of CPU time on our VAX–11/8700. The execution time was less than that of the backpropagation program since each iteration contains only the forward pass and no backward pass and the length of the hidden layer was reduced from 16 to 8. Thus, for a fair comparison between the two techniques, I should have allowed the genetic algorithm to iterate 2800 times. The results of this optimization are shown in Fig.14.24a.

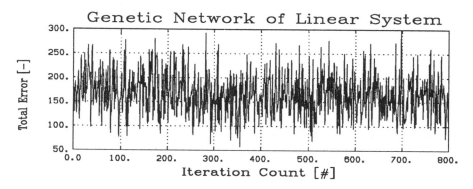

Figure 14.24a. Optimization of linear system with genetic algorithm.

Obviously, this optimization didn't work too well. Figure 14.24b shows a moving average computed over 100 iterations. The first value in Fig.14.24b is the average of the first 100 values of Fig.14.24a, the second value is the average of values 2 through 101 of Fig.14.24a, etc. I computed the moving average using the AVERAGE function of SAPS-II.

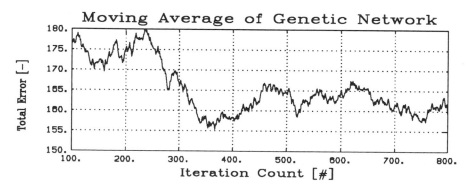

Figure 14.24b. Moving average of linear system.

The genetic algorithm does indeed learn. However, progress is painfully slow. My interpretation of these results is as follows: The terrain (in the parameter space) is very rough. Therefore, since we decided to use only four levels, each level contains both high mountains and deep gulches. Since we only retain the class values but not the quantitative values themselves, we throw away too much information. Consequently, I decided to rerun the optimization with $nlev = 16$. Since there are now more possible outcomes, I decided to consider 30% a solid majority vote, and thus, I reduced $nperc$ to 0.3. Another 1.2 CPU–hours later, I obtained the results for the modified algorithm. The simulation required less time because the optimization was terminated after 445 iterations. The results are shown in Fig.14.25.

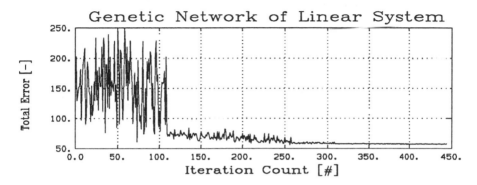

Figure 14.25. Optimization of linear system with genetic algorithm.

This time, the genetic algorithm learned the weights much faster. Unfortunately, good genetic material was weeded out too quickly, and the algorithm ended up in a ditch.

Montana and Davis designed another genetic algorithm specifically for the purpose of training neural feedforward networks [14.25]. They argue against eliminating useful information by coding fuzzy information into our genetic strings. Indeed, both the crossover operator and the mutation operator can be applied to both quantitative (real) and qualitative (fuzzy) parameters. In addition, they designed a set of interesting more–advanced genetic operators. They claim that networks function due to the synergism between weights associated with individual nodes. Thus, instead of applying the crossover algorithm blindly, they keep all the incoming weights of a node intact, and use either those of the father or those of the mother. Also, they consider multiple crossovers. Each node with all its incoming weights is arbitrarily taken from either the father or the mother. Thus, they simulate multiple crossovers of entire features. This makes a lot of sense. Montana and Davis also developed a very interesting concept of node assessment. They evaluate the quality (error) of a network in exactly the same manner that I use, i.e., they add the errors of the network over all training pairs. Then they remove an individual node from the network, i.e., they lobotomize all incoming and outgoing connections of that node by setting the corresponding weights equal to zero and recompute the quality of the modified network. They repeat the same procedure over and over, each time lobotomizing exactly one node. Using this information, they define the node whose presence has the least effect on the overall quality as the *weakest node*. Their mutation algorithm influences all incoming and outgoing weights of the weakest node in the hope of thereby improving the quality of the overall network. Again, this algorithm makes a lot of sense from an engineering point of view. They use a different distribution function for randomizing the initial weights of the network. Initially, they evaluate the quality of the entire genetic pool. However, in each generation, they pair up only one couple (as I do) and produce only one child, which replaces the worst genetic string in the genetic pool (unless it is even worse). The parents are chosen randomly, but with a distribution function such that the second–best genetic string is chosen 0.9 times as often as the best, and the third–best string is chosen 0.9 times as often as the second best, etc. Also this rule makes a lot of sense. Figure 14.26 shows the results of a simulation of the same problem that was discussed earlier, but now using the algorithm by Montana and Davis [14.25].

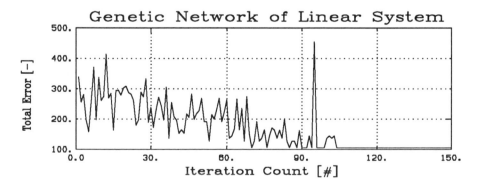

Figure 14.26. Montana–Davis optimization of linear system.

The algorithm is very efficient. The error is reduced quickly and the system learns fast. Unfortunately, it stagnates. Because the real parameter values are stored, no new information is entered into the system except through mutation. The system finds the smallest error among all the combinations of parameters present in the initial genetic pool reliably and quickly, but then it is stuck. The local superman wipes out his competition effectively and efficiently and becomes a tyrant ... unfortunately, he is but a midget in global terms.

This algorithm suffers from the same disease as mine. In both algorithms we were greedy and tried to retain as much good genetic material as possible. We never let a good genetic string die. This is the seed of stagnation.

Even the fittest among us must die for progress to survive.

Goldberg suggested using a genetic algorithm closer to a biological model [14.6]. He proposed the following genetic dialect: We start out with a randomly chosen qualitative genetic pool (as in my algorithm). We evaluate the quality of the entire genetic pool (as in the case of the algorithm by Montana and Davis). We rank the genetic strings according to their quality. We define the *fitness* of a genetic string as:

$$\text{fitness} = \frac{1.0}{\text{total error}} \qquad (14.43)$$

We then add up the fitnesses of all genetic strings in the genetic pool and define the *relative fitness* of a genetic string as:

$$\text{relative fitness} = \frac{\text{fitness}}{\text{sum over all fitnesses}} \qquad (14.44)$$

We then replace the entire genetic pool by a new pool in which each genetic string is represented never, once, or multiple times proportional to its relative fitness. Poor genetic strings are removed, while excellent genetic strings are duplicated many times. We then pair the genetic strings up arbitrarily. Each pair produces exactly two offspring, one consisting of the head of the first string concatenated with the tail of the second and the other consisting of the head of the second string concatenated with the tail of the first. We then let the old generation die and replace the entire genetic pool by the new generation. The algorithm is repeated until convergence.

This algorithm grants fit adults many children with varying sex partners, potentially including twin siblings, and deprives unfit adults of the right to reproduce. The algorithm enforces strict birth control.

An obvious disadvantage of this genetic dialect is the need to evaluate the fitness of the entire genetic pool once per generation. Thus, we can optimize this algorithm over 16 generations only if we wish to compare it to the previously advocated dialect. However, I decided to compute 100 iterations anyway. Figure 14.27a shows the results of this optimization. I plotted the mean value of the total errors of all genetic strings in the genetic pool.

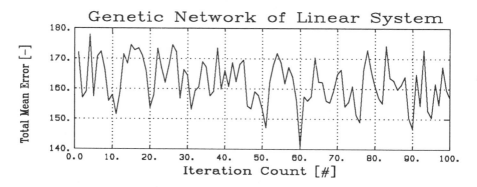

Figure 14.27a. Optimization of linear system with Goldberg's algorithm.

The results are disappointing. If the algorithm has learned anything, the improvement is lost in the noise. I then computed a moving average of the previously displayed mean values over 50 generations. The results are shown in Fig.14.27b.

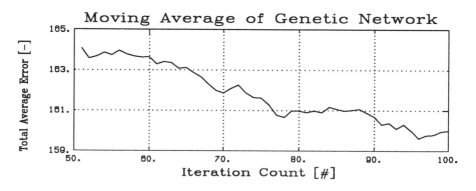

Figure 14.27b. Moving average of Goldberg's algorithm.

Notice that the algorithm does indeed learn. However, the progress is unbelievably slow. I ran this program in batch. It required just over 12 CPU–hours. Obviously, I cannot determine whether this algorithm will stagnate or find the true minimum, but I believe that it will eventually find the true minimum.

The problem with Goldberg's algorithm is the following. The entire idea of the genetic crossover operator bases on the naïve belief that the child of two fit parents is, at least in a statistical sense, a fit child. This belief is justified in nature since the genetic parameters reflect features and the child will inherit an entire feature either from the father or the mother. The overall fitness of a person is defined as the cumulative quality of all of his or her features. Thus, by inheriting features from both parents, fit parents will indeed have fit children. However, in our case, the individual parameters don't represent features. Each parameter influences all features, and each feature is influenced by all parameters. There is no compelling reason to believe that the crossover child of two fit parents is more fit than the average genetic string. Amazingly, the simulation results showed that such a child is indeed statistically more fit than the average genetic string ... but only by a narrow margin. This is why progress was so incredibly slow. It might have been worthwhile to combine Goldberg's algorithm with the previously proposed algorithm by Montana and Davis by combining the genetic operators of the latter (crossover of features and mutation of the weakest node) with the social behavior of the former (replacement of the entire population once per generation), but I was afraid that the director of our computer center would knock me over my head if I continued in this way.

We have just demonstrated the power of evolutionary development. We learned a lesson: For evolution to work, we must permit all individual genetic strings to die, irrespective of their quality. Retaining any individual string invariably leads to stagnation and the evolutionary process comes to a halt. It is the power of ever–changing nonrepetitive variations — we call this phenomenon a chaotic steady–state — which enables the evolutionary process to continue.

> In the beginning, there was Chaos.
> Chaos nurtures Progress.
> Progress enhances Order.
> Order tries to defy Chaos at all cost.
> ... But the day Order wins the final battle
> against Chaos, there will be mourning.
> 'Cause Progress is dead.

Genetic algorithms are a class of simple stochastic optimization techniques. Their behavior was demonstrated here by means of a neural network learning problem. However, no direct relationship exists between the two. Genetic algorithms can be interpreted as one particular implementation of a Monte Carlo optimization technique and can be applied to arbitrary optimization problems. We shall return to this discussion in the companion book of this text in the context of general–purpose parameter–estimation methods. It made sense to introduce the genetic algorithms here due to their inspirational biological foundation.

In the context of artificial neural networks, the genetic algorithm provides us with a systematic and stable technique to optimize arbitrarily constructed networks. This idea is fairly new and hasn't yet been exploited to its full potential. The idea is fruitful, because it removes configuration constraints on artificial neural networks. For instance, it allows us to optimize arbitrarily connected perceptron networks in a general, systematic, and robust (though fairly inefficient) way.

14.12 Neurobiological Learning

We have discussed various techniques for learning the weights of a neural network. The explicit counterpropagation network, in both its original feedforward and its derived recurrent forms, is very attractive since it does not require any training at all. Yet the algorithm requires that the coded (symbolic) targets be known *a priori* and this is certainly not how our brain works. A freshly born human child does not have a notion of a $\mathcal{DOG}$. It learns the symbol itself from observation. In this sense, human learning is indeed unsupervised. It is not even obvious that different human beings use the same symbolic representation for $\mathcal{DOG}$. If we were able to connect the *ego* of one person to the self of another (what an atrocious idea!), we might discover that the *ego* doesn't understand a word of what it reads in the self ... since the self uses a foreign alphabet.

The same difficulty holds true for the backpropagation algorithm, aside from the fact that the backpropagation algorithm learns far too slowly to be a realistic model of what happens in our brain.

Genetic algorithms are not plausible at all in the context of neurobiological learning, but at least they add a stochastic component to the learning process. I am convinced that this stochastic component exists in our brain, but it is introduced through the mechanism of chaotic feedback loops.

How does the brain learn? We don't really know yet. All we can say is that the brain learns reliably (no stability problems) and with amazing efficiency.

However, first attempts to shed light on this mystery have been made. Green and Triffet [14.7,14.34] have modeled "unit circuits" of the human brain (the allocortex, the cerebellum, and the cerebrum). Unit circuits are themselves organized in minizones (columns) which are connected to macrozones (rows), comprising a matrix structure. The unit circuits are interconnected in various ways. For example, in the cerebellum, the granule cells connect to the Purkinje cells of all four neighboring unit circuits (excitatory synapses), and the basket cells connect to the Purkinje cells of all four neighboring unit circuits (inhibitory synapses).

Triffet and Green represent their artificial neurons in a way that resembles the biological reality much more closely than any other artificial neural networks. Their artificial neurons simulate the frequency modulation of real neurons. The potential of each neuron can assume one of ten discrete levels. Levels '−6' to '−1' represent refractory states, level '0' represents the resting state, levels '+1' and '+2' represent excited states, and at level '+3,' the neuron fires, thereby returning to level '−6' for the next refractory period. The neuron may immediately return to a higher refractory level. This will happen if the sum of the incoming weights of the firing neuron is sufficiently large.

The proposed model is a discrete–time model. Each clock impulse represents 2 msec. During the refractory period, the potential is incremented by one level once every clock impulse. Consequently, it will take 12 msec for a neuron to return from the lowest refractory level to the resting level. During the refractory period, all other input is ignored. This procedure simulates the compulsory refractory period of biological neurons.

Once a neuron has reached its resting level or an excited level, it is susceptible to both synaptic and extracellular input. Whether the potential of a resting or excited neuron will change during one clock impulse depends on three factors:

(1) Neurons have a natural tendency to return to their resting state. Thus, if no input is applied to the neuron, it will gradually return to its resting state. This is called the *potential relaxation* mechanism.

(2) Neurons are influenced by firings of neighboring neurons located both within the same unit circuit and within neighboring unit circuits. The synaptic weights of the (partly excitatory and partly inhibitory) inputs will influence the potential of the neuron. If the neuron is exposed to a strong inhibitory input, the potential will be decremented. If it is exposed to a strong excitatory input, the potential will be incremented.

(3) Neurons are influenced by global electromagnetic wave transmissions (extracellular excitation or inhibition). Neurobiological measurements confirm that "intention potentials" sweep across specific regions of animal brains in advance of any motoric action.

In terms of our standard nomenclature, we can describe the potential (state) updating algorithm in the following way:

$$
\begin{aligned}
x_{t+\delta t}^{\ell} =& x_t^{\ell} + \text{relax}(x_t^{\ell}) + \left(1 - \text{relax}(x_t^{\ell})\right) \\
& \cdot \left(\sum_{\forall j} w_t^{\ell j} \cdot y_t^{j} + b_t^{\ell} - \text{relax}(-x_t^{\ell})\right)
\end{aligned}
\tag{14.45}
$$

where y_t^j is the output of the jth neuron at time t, and relax(.) is the relaxation function. It is '1' if its argument is negative and '0' otherwise. Consequently, the (negative) state of a refractory neuron is simply incremented by '1' once every clock impulse. The (nonnegative) state of a resting neuron is exposed to synaptic input from all incoming neurons. The synaptic strength of the connection from the jth to the ℓth neuron at time t is represented by the weight $w_t^{\ell j}$. The extracellular input to the ℓth neuron at time t is represented through the bias b_t^{ℓ}. The third term ensures that excited neurons relax to their resting state when left alone.

The output function is simply:

$$
y_{t+\delta t}^{\ell} = \text{firing}(x_{t+\delta t}^{\ell})
\tag{14.46}
$$

The firing function is '1' if its argument is '+3' or larger and it is '0' otherwise. That is, the output of the neuron simply registers the fact that the neuron just fired.

Finally, we need to return the fired neuron to a refractory state. The proposed rule is simply:

$$
\text{if } y_{t+\delta t}^{\ell} = 1 \text{ then } x_{t+\delta t}^{\ell} = \min(x_{t+\delta t}^{\ell} - 9, -4)
\tag{14.47}
$$

Thus, if the state is at the lowest firing level (level +3), it returns immediately to the lowest refractory level (level −6). If the state is elevated one level beyond the lowest firing level, it returns to the refractory level −5. If it is elevated two levels beyond the lowest firing level, it returns to the refractory level −4, but this is the highest level to which we allow a fired neuron to return. Such a neuron will require 8 msec to relax to its resting level.

Both the notation and the precise logic of the described algorithm deviate slightly from those used by Triffet and Green [14.7,14.34] in order to fit smoothly into the framework of this chapter.

The neural network learns both temporal and static patterns using three separate mechanisms:

(1) The synaptic weights, which are represented as integers, are incremented or decremented using a modified Hebb rule. However, unlike the traditional artificial neural networks, the optimal weights of a temporal pattern are themselves functions of time. They may constantly be modified (even after learning is completed) while a temporal pattern is processed by the brain. Green and Triffet [14.7,14.34] call these temporal sequences "programs."

(2) The circuit learns particular features of temporal patterns simply as a consequence of the memorizing power of the feedback loops in the circuit.

(3) The circuit is prepared for learning by the electromagnetic "intention wave" that precedes the actual neuronic signals.

The weight updating algorithm is straightforward. Whenever firing one neuron leads to firing another, the weight connecting these two neurons is incremented or decremented depending on the nature of the synaptic connection. If the connection is excitatory, the (positive) weight is incremented; if it is inhibitory, the (negative) weight is decremented. In both cases, the weight becomes stronger. If a weight has not been modified for a given period of time, it can be automatically weakened, i.e., positive weights are decremented while negative weights are incremented. This mechanism is called *weight relaxation*. Weight relaxation simulates the process of forgetting. Notice that this simple algorithm solves the problem of updating the hidden layers of traditional neural networks. Implementation of the frequency modulation of biological neurons provides us with an algorithm to train neurons arbitrarily located anywhere in the network efficiently and reliably.

Biases are updated by a discrete implementation of the (distributed) wave equation. The input to the wave equation is the (distributed) firing pattern of the Purkinje cells. Whenever a Purkinje cell fires, the bias of that unit circuit is incremented. The time constant of the wave equation is chosen so that the electromagnetic wave travels slightly faster through the network than the neuronic signals.

Green and Triffet [14.7,14.34] claim that their networks can learn patterns much more quickly than traditional networks. All weights are learned in parallel. No backpropagation is necessary. While the authors haven't yet convinced me that they have really modeled true neurobiological learning, their conceptual framework is revolutionary, and it certainly deserves further pursuit.

14.13 Summary

In this chapter, we discussed how biological research can inspire us in solving engineering problems and how engineering methodologies can help us understand biological processes better. It has helped us gain a better understanding of how our brain and body functions achieve optimal performance and it has brought us a step closer to answering the question of how engineering systems, such as robots, can be equipped with a modest amount of decision–making capability and responsibility. This entire research area is new and many problems still await a solution.

References

[14.1] Albert Bandura (1977), *Social Learning Theory*, Prentice–Hall, Englewood Cliffs, N.J.

[14.2] Rodney A. Brooks (1986), "A Robust Layered Control System for a Mobile Robot," *IEEE J. of Robotics and Automation*, Vol. RA2, pp.14–23.

[14.3] Rodney M. J. Cotterill, Ed. (1988), *Computer Simulation in Brain Science*, Cambridge University Press, Cambridge, Mass.

[14.4] William R. D. Fairbairn (1954), *An Object–Relations Theory of the Personality*, Basic Books, New York.

[14.5] Paul Federn (1952), *Ego Psychology and the Psychoses*, Basic Books, New York.

[14.6] David E. Goldberg (1989), *Genetic Algorithms in Search, Optimization, and Machine Learning*, Addison–Wesley, Reading, Mass.

[14.7] Herbert S. Green and Terry Triffet (1989), "A Zonal Model of Cortical Functions," *J. Theoretical Biology*, **136**, pp. 87–116.

[14.8] Stephen Grossberg (1982), *Studies of Mind and Brain: Neural Principles of Learning, Perception, Development, Cognition, and Motor Control*, D. Reidel Publishing, Hingham, Mass.

[14.9] Stephen Grossberg (1987), *The Adaptive Brain*, North–Holland Publishing, Amsterdam, The Netherlands.

[14.10] Stephen Grossberg (1988), *Neural Networks and Natural Intelligence*, MIT Press, Cambridge, Mass.

[14.11] Donald O. Hebb (1949), *The Organization of Behavior: A Neuropsychological Theory*, John Wiley & Sons, New York.

[14.12] Robert Hecht–Nielsen (1990), *Neurocomputers*, Addison–Wesley, Reading, Mass.

[14.13] Eugene Hnatek (1976), *A User's Handbook of D/A and A/D Converters*, Wiley–Interscience, New York.

[14.14] John Holland (1975), *Adaptation in Natural and Artificial Systems*, University of Michigan Press, Ann Arbor, Mich.

[14.15] John J. Hopfield (1982), "Neural Networks and Physical Systems with Emergent Collective Computational Abilities," *Proceedings of the National Academy of Sciences, USA*, **79**, pp. 2554–2558, National Academy of Sciences, Washington, D.C.

[14.16] Otto Kernberg (1975), *Borderline Conditions and Pathological Narcissism*, Aronson, New York.

[14.17] Christof Koch and Idan Segev, Eds. (1989), *Methods in Neuronal Modeling: From Synapses to Networks*, A Bradford Book, MIT Press, Cambridge, Mass.

[14.18] Teuvo Kohonen (1989), *Self–Organization and Associative Memory*, third edition, Series in Information Sciences, **8**, Springer Verlag, Berlin.

[14.19] Granino A. Korn (1991), *Neural–Network Experiments on Personal Computers*, MIT Press, Cambridge, Mass.

[14.20] Ronald D. Laing (1969), *The Divided Self*, Pantheon Books, New York.

[14.21] Richard P. Lippmann (1987), "An Introduction to Computing with Neural Nets," *IEEE ASSP Magazine*, April, pp. 4–22.

[14.22] Warren S. McCulloch and Walter Pitts (1943), "A Logical Calculus of the Ideas Immanent in Nervous Activity," *Bulletin of Mathematical Biophysics*, **5**, pp. 115–133.

[14.23] Marvin Minsky (1985), *The Society of Mind*, Simon and Schuster, New York.

[14.24] Marvin L. Minsky and Seymour Papert (1969), *Perceptrons: An Introduction to Computational Geometry*, expanded edition, 1988, MIT Press, Cambridge, Mass.

[14.25] David J. Montana and Lawrence Davis (1989), "Training Feedforward Neural Networks Using Genetic Algorithms," *Proceedings, International Joint Conference on Artificial Intelligence, IJCAI-89*, Vol. 1, Morgan Kaufmann, Palo Alto, Calif., pp. 762–767.

[14.26] Kumpati S. Narendra and Kannan Parthasarathy (1990), "Identification and Control of Dynamical Systems Using Neural Networks," *IEEE Transactions on Neural Networks*, 1(1), pp. 4–27.

[14.27] Jean Piaget and Barbel Inhelder (1967), *The Child's Conception of Space*, Norton, New York.

[14.28] John R. Rice and Ronald E. Boisvert (1985), *Solving Elliptic Problems Using ELLPACK*, Springer–Verlag, New York.

[14.29] Edward Rietman (1989), *Exploring the Geometry of Nature: Computer Modeling of Chaos, Fractals, Cellular Automata, and Neural Networks*, Windcrest Publishing, Blue Ridge Summit, Penn.

[14.30] Helge Ritter, Thomas Martinetz, and Klaus Schulten (1990), *Neuronale Netze*, Addison–Wesley, Munich, F.R.G., English translation currently under preparation.

[14.31] Frank Rosenblatt (1962), *Principles of Neurodynamics: Perceptrons and the Theory of Brain Mechanisms*, Spartan Books, Washington, D.C.

[14.32] David E. Rumelhart, Geoffrey E. Hinton, and Ronald J. Williams (1986), "Learning Internal Representations by Error Propagation," in: *Parallel Distributed Processing: Explorations in the Microstructure of Cognitions, Vol. 1: Foundations* (D.E. Rumelhart and J.L. McClelland, eds.), MIT Press, Cambridge, Mass., pp. 318–362.

[14.33] Timo Sorsa, Heikki N. Koivo, and Hannu Koivisto (1990), "Neural Networks in Process Fault Diagnosis," submitted to *IEEE Trans. Systems, Man, Cybernetics*.

[14.34] Terry Triffet and Herbert S. Green (1990), "Structured Neurobiological Networks," submitted to *J. Theoretical Biology*.

[14.35] Philip D. Wasserman (1989), *Neural Computing: Theory and Practice*, Van Nostrand Reinhold, New York.

[14.36] Bernard Widrow and M. E. Hoff (1960), "Adaptive Switching Circuits," *IRE WESCON Convention Record*, fourth part, Institute of Radio Engineers, New York, pp. 96–104.

[**14.37**] Ben P. Yuhas, Moise H. Goldstein, Jr., and Terrence J. Sejnowski (1989), "Integration of Acoustic and Visual Speech Signals Using Neural Networks," *IEEE Communication Magazine*, **27**(11), pp. 65–71.

Bibliography

[**B14.1**] Casimir C. Klimasauskas, Ed. (1989), *The 1989 Neuro-Computing Bibliography*, MIT Press, Cambridge, Mass.

[**B14.2**] Philip D. Wasserman and Roberta M. Oetzel (1990), *Neural Source, The Bibliographic Guide to Artificial Neural Networks*, Van Nostrand Reinhold, New York.

Homework Problems

[H14.1] Perceptron Network for Exclusive–Or Problem

Implement a perceptron network as shown in Fig.14.8 that solves the exclusive–or problem. For this purpose, compute the weights and thresholds of the two hidden perceptrons such that each one of them implements one of the slanted lines of Fig.14.9. Manually evaluate the truth table between the two inputs u_1 and u_2 and the two outputs of the hidden layer y_1 and y_2. Since you know the desired output value y for each combination of the hidden layer outputs, you can also generate a truth table that maps the hidden layer into the output layer. Draw a picture in the $< y_1, y_2 >$ plane similar to Fig.14.7, which represents this truth table. Draw a new slanted line into this picture that separates the '1' outputs from the '0' outputs. Design the weights and thresholds of the output perceptron so that it implements the new slanted line.

Program the perceptron network in CTRL–C (MATLAB) and check its correct performance by looping over all four input combinations.

[H14.2] Backpropagation Network for Linear System

Reimplement the backpropagation network of the linear system (with four inputs and three outputs) in DESIRE/NEUNET. Compare the DESIRE/NEUNET solution of the exclusive–or problem (presented in this chapter) with the CTRL–C version of the same problem (also shown). This should give you sufficient help to make the transcription of the CTRL–C solution to the backpropagation network for the linear system (shown in this chapter) to its DESIRE/NEUNET equivalent an easy task.

Run the program over 774 iterations (preferably on a 386– or 486–class machine) and compute the speed ratio between the DESIRE/386 solution and the CTRL–C/VAX solution. Let the DESIRE/NEUNET program run further and find the smallest value of the performance index that you can obtain.

[H14.3] Adaptive Backpropagation Network for Linear System

In Hw.[H14.2], we simulated the continuous system once over 900 time units and stored the results away for training the backpropagation network. Modify the program of Hw.[H14.2] such that you constantly integrate the continuous system in parallel with training the backpropagation network. Set the communication interval to 3 sec. The backpropagation network is now placed in the OUT block, which is executed once per communication interval. This is necessary since you will require an integration step size that is considerably smaller than three time units to obtain decent simulation results. The updating of the **WW**^{1} and **WW**^{2} matrices and of the **bb**^{1} and **bb**^{2} vectors occurs in a SAMPLE block, which is executed once every 300 communication intervals.

[H14.4] Counterpropagation Network for Linear System

Find a CTRL–C (or MATLAB) solution to generate Fig.14.20. Start by designing two small procedures implementing 10–bit A/D and D/A converters. The A/D converter takes a real number between -1.0 and $+1.0$ and generates a vector of length 10 containing only -1.0 and $+1.0$ elements. The D/A converter accepts a binary vector of length 10 and generates a real number between -1.0 and $+1.0$. Test the correctness of the two routines by computing:

$$xx = \text{DtoA}(\text{AtoD}(x))$$

for various numbers x between -1.0 and $+1.0$.

Use the *inpt* matrix from before and generate the new analog input matrix *ainpt* by concatenating u_{k-1} with x_{k-1}, then with u_k, further with x_k, and finally with u_{k+1}. This is accomplished by concatenating columns of the former *inpt* matrix shifted down by one or two elements. Notice that you lose two rows in this process. Compute the new analog target matrix *atarg* from the old *inpt* matrix in the same way. *ainpt* should be a matrix with 299 rows and nine columns, while *atarg* should be a matrix with 299 rows and three columns.

Convert the two analog matrices to digital matrices using the previously defined AtoD routine. The digital input matrix *dinpt* should have 299 rows and 90 columns, while the digital target matrix *dtarg* should have 299 rows and 30 columns.

Set up the weight matrices of the counterpropagation network by using the first 269 rows of each of the digital matrices.

Recall the counterpropagation network by applying the remaining 30 rows of the digital input matrix to the counterpropagation network. Compute the resulting digital target vectors for each of the digital input vectors. Convert the resulting digital target vectors back to analog target vectors using the previously designed DtoA routine and plot the resulting values together with the original values.

[H14.5] Adaptive Counterpropagation Network for Linear System

This program consists of a big loop that is executed 299 times. Construct digital input and target vectors one at a time from the previously computed *inpt* matrix. During the first iteration, simply place the two vectors in the weight matrices. The hidden layer has a length of one. In subsequent iterations, apply the digital input vector to the existing network and compare the resulting digital output vector to the correct digital output vector. If the vector is correct, do nothing; otherwise add the correct input/target pair to the weight matrices, thereby incrementing the length of the hidden layer by one.

Generate three different graphs. The first graph compares the true analog outputs to the computed analog outputs for rows 10 to 39, the second compares the same values for rows 150 to 179, and the third compares the same values for rows 270 to 299. Discuss the results. Determine the length of the hidden layer after 299 iterations.

[H14.6]* Reinventing the Binary Code

Design a two–layer pseudobackpropagation network (i.e., a backpropagation network using LIMIT/TRI activation function pairs) that maps binary unit vectors of length 16 through a hidden layer of length four back into the original unit vector representation. I suggest coding the kth unit vector as:

$$e_k = [\, -0.9 \, , \, \ldots \, , \, -0.9 \, , \, +0.9 \, , \, -0.9 \, , \, \ldots \, , \, -0.9 \,]'$$

That is, the logical true state is represented by the real value $+0.9$, while the logical false state is represented by the real value -0.9. This applies to both the input and target vectors.

Use gain values of $g_1 = 0.05$ and $g_2 = 0.07$ and use relaxation momenta of $m_1 = -0.06$ and $m_2 = -0.08$. Initialize the weight matrices to $0.1 \cdot$ RAND. Simulate the network over 20,000 iterations, applying sequentially one input pair after the other. The weights can be updated once per iteration, i.e., it is not necessary to loop over all input/target pairs before updating the weights. Display the mean square error.

Create a DESIRE/NEUNET program implementing this algorithm. Don't try CTRL–C or MATLAB on this problem. One simulation run in CTRL–C or MATLAB will require several hours of CPU time, while DESIRE/NEUNET will execute the same problem in less than 1 min on a 486–class machine.

Recall the learned network once for every input/target pair and display the resulting hidden layer output vectors. Look at the signs of the vector and interpret the results. You should notice that this neural network just reinvented the binary code. Repeat the simulation several times. Notice that each result is different. Each simulation reinvents the binary code, but generates a different sequence of binary numbers. This is what I meant when I wrote that each human brain probably uses a different alphabet to encode the same symbolic knowledge.

Projects

[P14.1] Intelligent Autopilot
Apply the recurrent counterpropagation network presented in Section 14.7 to the continuous simulation of a Boeing 747 jetliner in high–altitude horizontal flight, i.e., to the simulation program designed in Pr.[P4.1]. The purpose of this study is to check whether we can apply the proposed technique as reliably to a highly nonlinear system as to a linear system.

Enhance the ACSL program designed in Pr.[P4.1] by a set of faulty operational modes such as heavy ice on the wings or loss of one of the four engines.

For each of the fault modes, identify a set of weighting matrices that characterize the fault.

Modify the ACSL program once more. This time, one of the faults should be chosen arbitrarily at a randomly selected point in time during the simulation.

Use the counterpropagation network of the undamaged aircraft to identify when the accident has happened and discriminate the correct fault by comparing the recoded continuous data after the accident occurred and after the transients resulting from the accident have died out with forecasts obtained from the neural network using all of the stored weighting matrices. The forecast with the smallest deviation identifies (most likely) the type of fault that has occurred.

[P14.2] Stock Market
Repeat the stock market analysis presented in Chapter 13, but now using an adaptive counterpropagation network. In the neural network, the con-

tinuous values are discretized using Gray–code A/D converters. After the forecast has been completed, the predicted stock market values are further aggregated to the previously used three levels for comparison.

You may want to investigate which is the optimal number of bits used in the Gray–code converters. Too many bits (high resolution) call for a long time period until the neural network has seen all combinations once, a time period that may not be justifiable due to the inherent time–varying nature of the stock market. Too few bits (low resolution) may prevent the network from distinguishing between functionally different system behaviors that are lumped together in the discretization. My intuition tells me that you probably shouldn't use more than four to five bits.

Research

[R14.1] Simulating an Ant Brain

Rodney Brooks [14.2] developed a series of finite state machines that cooperate in simulating ant behavior. He designed and built a very small robot using solar batteries as its only source of energy, with an onboard chip implementing his finite state machines. He produced a beautiful and very impressive video showing the behavior of his robot as it walks over phone books and gravel and reacts in various other ways to its environment.

This robot clearly exhibits insect behavior. Therefore, we can say that Brooks' finite state machines represent a good working hypothesis of how a relatively small series of independent unintelligent agencies [14.23] can cooperate to perform amazing tasks. While the finite state machines demonstrate how autonomous agencies can cooperate in task solving, they do not tell us anything about the physical configuration of an ant brain.

It should be perfectly feasible to simulate a true ant brain since an ant brain contains only about 20,000 neurons, and yet, ants exhibit a highly interesting and fairly complex social behavior. The problem is that we don't yet truly understand how brains are configured or how they process information. If we are able to reproduce ant behavior using a neural network, we might learn something about the potential physical configuration of a very simple *id*.

I propose that you apply the neural network approach by Green and Triffet [14.7,14.34] and try to implement Brooks' [14.2] finite state machines with it. Try to reuse the same unit circuits as much as possible for different related tasks. Determine the minimum number of unit circuits necessary to reproduce Brooks' ant behavior using a neural network.

[R14.2] Teacher Network

Investigate the potential of neural networks for data preprocessing. A "teacher network" accepts arbitrary input/target pairs and prepares them for training a "student network."

Investigate the possibilities for training a future teacher network, i.e., find out whether it is possible to have a teacher network train a student network to be become a better teacher network than the original teacher network was.

[R14.3] Chaos and Learning

Build a neural network system with a forward network and an inverse network. Both networks are of the same type. To start with, they might be feedforward networks. However, feedforward networks are not useful for learning temporal patterns that are fed sequentially into the network. In the long run, you may wish to replace the feedforward networks by recurrent counterpropagation networks as proposed in Fig.14.19. Introduce global feedback loops as proposed in Fig.14.21. Implement an adaptive controller, i.e., a simple *ego*, as suggested in Fig.14.22, to ensure global stability of the overall system. Design the adaptive controller so that the overall system is either deterministic or mildly chaotic. Expose both versions of the network system to various stimuli and investigate the role of chaos in neural network learning.

[R14.4] Stock Market

Decide on a number of input parameters (such as the Dow–Jones index) and state variables (such as daily returns of companies) to be monitored. Design a backpropagation network with optimal prediction power and compare the results with an optimized linear regression model.

15

Automated Model Synthesis

Preview

In Chapters 12 to 14 of this text, we tried to imitate (model) the human ability to reason about system structure and system behavior. Now we shall let go of this constraint and discuss whether we can duplicate this capability in an analogical rather than a homomorphic fashion, i.e., we shall try to simulate human reasoning capabilities without restricting ourselves to methods that humans would or could use in their reasoning processes.

This final chapter describes the goal–driven automated generation of models from a set of design specifications. A five–level hierarchy is proposed that supports the automated generation of both models and simulation experiments from an abstract description of an overall design and from an abstract description of the goals of the simulation study. The aim is to be able to automatically synthesize models and experiments in a top–down fashion from a description of their components and the couplings between these components. Detailed submodel descriptions are extracted from template files that reside in model libraries.

The chapter starts with an assessment of the need for the proposed automated model synthesis methodology. Thereafter, the advocated five–level hierarchy is presented in a bottom–up fashion starting with classical approaches to continuous–system simulation (the first and bottom layer of our hierarchy), and advancing to higher and higher levels of abstraction. The chapter ends with the presentation of a complete example of the proposed methodology, now presented in a top–down fashion.

15.1 Introduction

Applications for automated model synthesis technology can be found in unmanned deep–space missions where robots must be able to make intelligent model–based decisions. This capability is important since it reduces the necessity for frequent communications with Earth, which inevitably slow down the decision–making process due to extensive communication time delays.

Applications can also be found in manufacturing. The design of either new parts or new tools from similar existing parts or tools is a complex process that is accompanied by much responsibility and few rewards. Automated design and simulation aids can make this process less painful and reduce the time needed to educate new design engineers.

Finally, applications can be found in real–time decision support. Military strategists could use this technology to describe proposed scenarios in abstract high–level terms. Detailed models for simulating the proposed scenarios could be automatically generated, and the strategist could watch the most likely effects of a proposed strategy unfold before his or her eyes.

Space

When humanity colonizes other planets of our solar system such as Mars or the moons of Jupiter and Saturn, this colonization will have to occur in three phases. In a first phase, unmanned Space missions will deploy high–autonomy systems, which must prepare the target planet for human arrival. These missions will be carried out by smart robots with a high degree of decision–making capability. The second phase will consist of manned Space missions conducted by a few highly specialized astronauts who must be supported in their endeavors by an abundance of smart automatic devices. During this second phase, the need for high–autonomy systems will not decrease. On the contrary, it will grow. Only in the third phase will it be possible to send larger numbers of less specialized humans to the target planet, and only then will life on this planet start to resemble life here on Earth.

As a first (and very rudimentary) example of this multiphased approach to colonization, the University of Arizona is currently developing a prototype of an automated system that will be able to produce oxidizer for rocket fuel on planet Mars from CO_2 extracted from the Martian atmosphere [15.10]. This high–autonomy system will be deployed by an unmanned mission to Mars that will not return to Earth. The deployed oxygen production plant will then produce oxygen in a basically unsupervised operational mode during a period of two years. A manned follow–up mission to Mars will then use the oxygen produced by this plant for its return flight to Earth. With this approach, manned missions to Mars can be made more economical since it will no longer be necessary to lift the (heavy) oxygen for the return flight out of the gravity well of planet Earth.

Intelligent decision making requires insight into the consequences of the decision made. Frequently, the decision–making process consists of choosing among a series of alternative scenarios. The decision–maker must be able to assess the pros and cons of each alternative by predicting its future effects on the overall plan. For this purpose, he or she often wishes to simulate alternative designs prior to their implementation.

A robot roaming around on planet Mars is confronted with exactly this problem. To choose between a set of alternative scenarios, it should base its decision on a proper assessment of the effects of each one. Since the situation of the robot at the time of decision making depends heavily on previous decisions it has made, it is not feasible to carry along individual simulation programs for all possible alternatives. Such an approach would quickly lead to myriads of simulation programs that are largely the same, but each one of which is slightly different from every other. Instead, the decision–making process should operate on a "world model" [15.1] from which a specific simulation model for any given purpose can be automatically synthesized at any time [15.15].

Manufacturing

Aircraft manufacturers maintain relational data bases containing large numbers of construction plans for aircraft parts. Whenever a new part is needed, a process design engineer searches such a data base for one or several similar designs, and modifies them until the new design is completed. The new design is then added to the data base. This is an extremely stressful job that carries a lot of responsibility (if the design doesn't work, who is to blame?) and few rewards. Consequently, the average process design engineer "lasts" in this job for only 18 to 24 months, while his or her education consumes the first 9 to 12 months of this period. Manufacturers obviously have a problem here.

An automated model generation system (AMGS) can help overcome such problems. With the new technology, a process design engineer would start by formulating (and formalizing) a set of goals and a set of constraints describing the new design. The AMGS would then automatically generate one or several alternative designs, together with simulation models for these designs, alternatives that satisfy all the goals and violate none of the constraints. If the number of resulting designs is small, the process design engineer could then simulate all of the proposed designs and pick the one that seems most attractive. If the AMGS generates too many designs, the engineer could formulate additional goals and/or constraints to limit the number of acceptable designs, and if the AMGS does not find any solution, the engineer knows that she or he has overconstrained the problem. This new technology will make the job of the process design engineer more interesting and joyful, it will reduce the time needed for her or his education, and it will enhance the quality of the end product since the design engineer can explore many more alternative designs than he or she would ever have time for in the current manual technology.

Military Strategic Planning

Strategic planning requires real–time decision making in a partly unknown environment. In current technology, this problem is tackled manually by assembling the general's staff in a room. The general proposes possible strategies (scenarios), and his or her staff will "simulate" the effects of the proposed strategies in a brainstorming exercise by throwing potential suggested outcomes back at the general. Finally, the general makes his or her decision, taking into consideration all the proposed (and partly contradictory) outcomes that he or she received from his or her staff.

We propose that our new technology can provide a more sound basis for decision making. While it is impossible to generate ahead of time simulation models for every potential situation, it is possible to develop ahead of time a "world model" from which our AMGS can then synthesize on the fly simulation models for any proposed scenario. In this way, the effects of a proposed strategy could be simulated in real time and could support the general in making the right decision.

15.2 Level One: Classical Simulation Models

Models used for similar purposes often contain similar components. For example, a model of a DC motor can be used to describe the mechanism that drives a windshield wiper in a car, the sump pump behind a house, or the food processor in a kitchen. The model is always the same, but the parameters assume different values. It therefore makes sense to include a DC motor template in the world model, a generic DC motor from which a specific DC motor for any given purpose can be generated when needed. In this way, the DC motor model needs to be debugged only once, and if the template model is ever modified, all its future instantiations automatically reference the updated template model. It is also possible to equip the model editor with a mechanism that allows it to trace all references to the edited model within the model library and automatically deletes the compiled versions of all those models from the library, thereby forcing the user to migrate model modifications through all simulation programs that make use of those models.

As shown in Chapter 5, "modular" continuous–system models are expressed in most Continuous–System Simulation Languages (CSSLs) using macros [15.2]. Macros are text templates. During compilation of the simulation program, the macro call is replaced by the macro definition body.

And yet, macros are *not* truly modular. The term "modularity" is often used in a narrow sense as a means to structuring program code into sections. Macros certainly do help us with structuring our code. However, we prefer to define the term "modularity" in a much wider sense as the capability of a program segment to represent a physical object irrespective of the environment in which the object is used [15.3]. Employing this definition, the term *modular modeling* becomes synonymous with *object–oriented modeling*.

Let me discuss the concept of object–oriented modeling by means of an example. DC motors can be used in two different ways. They can be driven electrically (either through the armature or through the field) in which case they will transform a portion of the electrical energy fed into the system into mechanical energy, as a consequence of which the mechanical axle starts rotating. However, we can also rotate the mechanical axle by force, thereby feeding mechanical energy into the system, and as a result, a portion of that energy will be converted into electrical energy, generating a voltage across the two terminals of the armature coil. In this case, the DC motor is used as a DC generator.

Clearly, the physical object is the same irrespective of whether it is used as a motor or a generator. In an object–oriented modeling environment, we should therefore be able to represent both types of operations through one and the same software object.

In Chapter 5, I demonstrated how this problem can be solved in ACSL [15.9] using the "macro if" statement. However, the code was clumsy, difficult to write, and equally difficult to read. The "macro if" statement is not truly an advanced feature of the ACSL macro handler. It is just a crude way to implement only the simplest of the capabilities that a true AMGS can offer.

We have realized that CSSL–type macros aren't truly modular with respect to the objects they represent. The same physical device calls for quite different macros depending on the environments in which it is supposed to operate. The simplest macro representing an electrical resistor, for instance, must be stored in the macro library in two different versions, one modeling the equation:

$$u_R = R \cdot i_R \tag{15.1a}$$

and the other modeling the equation:

$$i_R = \frac{u_R}{R} \tag{15.1b}$$

If the resistor is placed over a current source, the current i_R through the resistor is determined by the source and we need to use the macro according to Eq.(15.1a) to compute the voltage u_R, whereas if we place the resistor over a voltage source, the voltage u_R across the resistor is determined by the source and we need to use the macro according to Eq.(15.1b) to evaluate the current i_R. Obviously, an equation sorter is insufficient. In an object–oriented continuous–system modeling environment, we require an equation solver that accepts general equalities of the type:

$$< \text{expression} > \ = \ < \text{expression} > \tag{15.2a}$$

or:

$$< \text{expression} > \ = 0.0 \tag{15.2b}$$

and that can solve these equalities for arbitrary variables.

15.3 Level Two: Object–Oriented Modeling

Notice that macro handlers, which are commonly considered an intrinsic part of a CSSL language, have in fact nothing to do with the simulation language itself. The macro text replacement must be performed at source level and must be completed before any other activity of the compiler can begin. The macro handler is often implemented as the first path of the compilation and is completely separate from everything that follows. It makes perfect sense to develop a macro handler independent of the simulation language for which it is being used. The same macro handler could easily be used as a front end to several different simulation language compilers.

This approach was taken with DYMOLA [15.4]. DYMOLA is considerably more powerful than the conventional macro handler, but serves the same purpose. DYMOLA is a program generator that can generate code for a variety of different simulation languages. If DYMOLA is used as a preprocessor, the simulation language no longer needs a macro handler of its own; in fact, it no longer requires an equation sorter since DYMOLA will sort the resulting set of equations into an executable sequence after solving each of them for the appropriate variable. The syntax of statements in DYMOLA is that of Eq.(15.2*a*).

DYMOLA is able to solve arbitrarily nonlinear equations for any variable as long as that variable appears linearly in the otherwise nonlinear equation. DYMOLA cannot currently handle algebraic loops, not even linear algebraic loops (which is a pity). DYMOLA cannot currently handle most types of structural singularities as they occur when subsystems are coupled together in such a way that the overall system exhibits fewer degrees of freedom than the sum of the subsystems. Finally, DYMOLA cannot currently eliminate redundant equations as they appear frequently when subsystems with fuzzy borderlines between them are coupled together, i.e., when one and the same equation can be viewed as belonging to one or the other of the subsystems depending on their use. Other than that, DYMOLA has all the properties needed for truly hierarchical modular modeling, i.e., for object–oriented modeling.

DYMOLA provides for powerful mechanisms to connect submodels. These submodels are truly modular even in terms of our extended definition of modularity.

Notice that while no strict rule exists that forbids the mixing of physical equations with connection statements in one model, it is good practice to avoid such a mix [15.3]. At the bottom of the model hierarchy are models that are described solely through physical equations and do not contain any connect statements. These models are called *atomic models*. Models that invoke other models and describe the connections between these submodels are called *coupling models*. Coupling models can, of course, refer also to other coupling models, not only to atomic models, i.e., coupling models can be hierarchically structured. In Chapter 8, we presented an example (a solar–heated house) of a hierarchically structured modular model with five hierarchy levels.

Notice, however, that the concept of structuring models in a hierarchical fashion is different from the five–level hierarchy of model management that is the topic of this chapter. Hierarchically structured object–oriented models (with an arbitrary number of hierarchy levels) occupy just the second level of our hierarchical model management methodology, whereas the expanded, i.e., flat, simulation models occupy the first hierarchy level of the methodology.

Notice further that object–oriented modeling does not imply object–oriented simulation as well. DYMOLA clearly supports the concept of models representing physical objects in a modular, i.e., environment–independent, fashion. Thus, DYMOLA supports object–oriented modeling. Yet the generated simulation code is a flat simulation program expressed in any of a number of off–the–shelf simulation languages that clearly do not support object–oriented simulation. It is a commonly made mistake to believe that object–oriented program execution is necessarily desirable [15.3]. It is the object–oriented user interface that is desirable since it simplifies programming. In the context of continuous–system simulation, an object–oriented approach to simulation would force us to exchange information between the simulation data objects using a mechanism of message passing. This is far too inefficient since continuous objects exchange information on a continuous basis. Hierarchy flattening of continuous–system models is thus a must from the point of view of run–time execution efficiency.

This is exactly what DYMOLA provides us with: At the user interface, i.e., at the input level of the DYMOLA preprocessor, it supports a full–fledged object orientation. DYMOLA models can be formulated in a truly modular way. Yet at the simulation program level, i.e., at the output level of the DYMOLA preprocessor, the simulation programs are flat. The DYMOLA preprocessor flattens the model hierarchy for improved run–time efficiency.

15.4 Level Three: The System Entity Structure

We have not discussed yet one important property of object–oriented modeling. Object–oriented models hide details of the internal model structure from the outside. Only those properties of a model that transpire to its surface are noticeable from the outside. For instance, once a model has been encapsulated, only its input and output ports are still visible. In other words, the model dynamics have been encapsulated at hierarchy level two, and therefore, from hierarchy level three and up, we shall only deal with objects and their relations with each other.

The *system entity structure (SES)* is a mechanism to describe hierarchically structured sets of objects and their interrelations [15.13]. The SES is a labeled tree with attached variable types, i.e., a graphical object that describes the decompositions of systems into parts. It is a knowledge representation scheme that formalizes the modeling of systems in terms of decomposition, taxonomic, and coupling relationships. The scheme supports structured knowledge acquisition for, and flexible restructuring of, families of large–scale system designs.

Figure 15.1 shows a sketchy decomposition of the solar–heated house of Chapter 8 into parts. In order to keep Fig.15.1 small and understandable, only a few of the actual decompositions are shown.

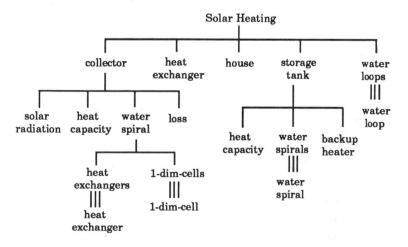

Figure 15.1. Decomposition of solar heater into parts.

Figure 15.1 is an *informal SES* of the solar heating system. It shows the decomposition of physical objects into parts. However, for practical use, the informal SES is insufficient. Somewhere, it must be stated how the physical objects are decomposed into parts. For this reason, a more formal version of the SES has been developed. In a formal SES, physical objects are called *entities*. The children of physical objects are always abstract objects called *aspects* and *specializations*. The children of these abstract objects are physical objects again. Consequently, physical objects and abstract objects always toggle in a formal SES. The root object in a formal SES is always an entity. The leaf objects of the SES are also always entities. The leaf objects represent the *atomic models*. All entities that are not leaf objects represent *coupled models*.

It is not necessary to represent abstract objects differently from physical objects in the SES. It is easy to tell them apart simply by counting the number of levels in the SES. Objects at odd levels from the top or bottom always represent entities, while objects at even levels of the SES are always either aspects or specializations.

Physical decompositions are shown as aspects. We shall discuss the purpose of specializations later. For the moment, it suffices to mention that it is exactly the distinction between aspects and specializations that forces us to formalize these abstract objects rather than implying the decompositions by attaching them to the parent entity directly.

The formal SES satisfies the following set of axioms:

(1) *Uniformity:* Any two nodes (objects) that have the same labels (names) have identical attached variable types and isomorphic subtrees.

(2) *Strict hierarchy:* No label appears more than once down any path of the tree from the root entity to any of the leaf entities.

(3) *Alternating mode:* Each node has a mode that is entity, aspect, or specialization; if the mode of a node is entity, then the modes of its successors are either aspect or specialization; if the mode of a node is aspect or specialization, then the modes of its children are entity. The mode of the root is entity.

(4) *Valid siblings:* No two siblings have the same label.

(5) *Attached variables:* No two variable types attached to the same item have the same name.

(6) *Inheritance:* Every entity in a specialization inherits all the variables, aspects, and specializations from the parent of the specialization.

The SES is completely characterized by its axioms [15.13,15.14]. However, the interpretation of the axioms cannot be specified and thus is open to the user.

When constructing an SES, it may sometimes seem difficult to decide how to represent concepts of the real world in the SES. A meaningful decomposition of a system into parts is not always easy to accomplish, and yet an inadequate decision can have serious consequences in terms of the number of wires that connect the various submodels to each other. One of the goals of a meaningful system decomposition is to limit the number of required connections between its subsystems. As of now, we don't have a tool to support the user in this process, i.e., we don't have a tool that can consider various alternative decompositions and propose one that will result in few subsystem connections. Such a tool would require a detailed knowledge of the internal structure of all subsystems. It could infer this knowledge by tracing through the nested couplings, as is done in hierarchy flattening.

DEVS–Scheme [15.14] is an application layer above the LISP–based PC–Scheme language [15.11]. It supports the DEVS modeling formalism [15.13]. DEVS stands for Discrete–EVent Specification, i.e., DEVS was developed primarily for discrete–event models. Yet due to the knowledge encapsulation capability of object–oriented modeling, the higher levels of the DEVS modeling hierarchy are the same for both discrete–event models and continuous–system models. DEVS–Scheme contains a set of PC–Scheme (i.e., LISP) procedures implementing the DEVS modeling formalism. They were written using the SCOOPS object–oriented LISP extensions, which form an intrinsic part of the PC–Scheme environment. They were coded such that all PC–Scheme (LISP and SCOOPS) programming tools are transparently usable within DEVS–Scheme as well. In this light, DEVS–Scheme is not a specialization of PC–Scheme, but a superset of PC–Scheme.

System entity structures can be coded in DEVS–Scheme in the form of a text file that contains a one–to–one translation of the graphical SES representation. An example of such a text file will be presented later in this chapter. Currently, the translation must still be done manually, but automatic translations from the graphical representation to the textual representation are being developed. Once the SES has been coded, DEVS–Scheme offers an automated transform procedure that iteratively generates more and more complex coupled models by invoking submodels and by setting up the coupling relations that were specified between them. In this way, transform translates the SES into either one single DEVS model or a set of DEVS models. The transform procedure can be requested to either flatten or keep the hierarchy. For continuous systems, transform can also automatically generate the necessary DYMOLA coupling models. Only the atomic (i.e., leaf) models must be user–coded. The coupled models can be automatically generated from the SES.

It is possible to store coupled models in the model library as if they were atomic models. If transform finds a coupled model in the model library, it will stop searching for its children and simply use the coupled model as if there were no more children and as if it were an atomic model.

Let us now return for a moment to the example of the aircraft manufacturer. If we maintain all our designs at the first hierarchy level (as is done in the current technology), we need to store millions of different designs in the data base. Each design is flat and therefore fairly complex. If a design is found to have a flaw, and if that flaw has been traced back to a particular subcomponent, which may be intrinsically used in hundreds of other designs, we should edit all these hundreds of flat design descriptions to remove the bug from the data base. Of course, this is never done (too much work), and therefore, the data base is always inconsistent. It can therefore easily happen that another process design engineer at a later time bases a new design on another earlier design that contains the same flaw that had been discovered once before, but which was never totally removed from the data base.

If we maintain our designs at the second hierarchy level, we still need to store millions of different designs in the data base. However, these designs will not be flat. They will reference design templates that are stored in a hierarchically structured template library. The main design files will thus be much shorter and more readable since they reference these templates rather than contain them in an already expanded form. If a flaw is traced back to a bug in a design template, the template itself is updated, rather than the design that makes use of it. When an atomic template is edited, all coupled templates in the template library that reference the atomic template either directly or indirectly could be automatically deleted from the template library. In this way, all future designs would automatically reference the updated template rather than the original one.

If the designs are maintained at the third hierarchy level, we still need to store millions of individual designs in the data base, but this time, they are stored in the form of different system entity structures. This does not seem to buy us much in comparison to the previous (i.e., level–two) alternative, but the next section will explain why an SES is indeed considerably more powerful than either a DEVS or DYMOLA model.

15.5 Level Four: The Generalized SES

We explained earlier that the SES contains two types of abstract objects: aspects and specializations. *Specializations* enable us to store several variants of similar SESs in a single generalized SES. In fact, from now on, we shall call the "generalized SES" simply "the SES." An SES without any specializations is a special case. It is sometimes referred to as a "pure SES."

Figure 15.2 shows an example of an SES that decomposes a car into a few components. Notice that the engine entity is specialized into either V6 or diesel.

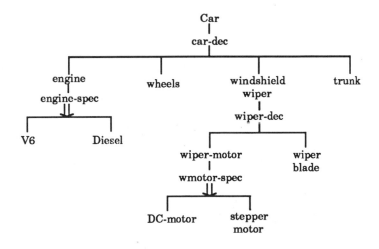

Figure 15.2. SES with specializations of a car.

Pure SESs can be obtained from general SESs by a process called pruning. *Pruning* an SES means cutting away all specializations but one. The car SES of Fig.15.2 can be pruned to retain either the V6 or the diesel engine.

DEVS–Scheme [15.14] contains a software tool called ESP–Scheme [15.7], which can prune any SES. The output of the pruning process is a pure SES.

Returning once more to the aircraft manufacturing example, we notice that, at the fourth hierarchy level, we no longer need to store millions of different designs. Maybe we can reduce all these designs to a few thousand basic designs with many specializations. This makes the SES of a system more powerful than either a corresponding DEVS or DYMOLA model. One single model expressed as an SES with specializations can represent many different variants of actual models of a process. This concept replaces the "macro if" statement of ACSL's macro handler, but is considerably more powerful. One single SES can be used to generate hundreds, maybe thousands, of different models for different purposes.

From now on, we shall call an ensemble of one or several SESs with specializations, which can be used to generate specific models for an entire application area, a *world model* [15.3].

Returning to our example of the strategic planning tool, it now becomes evident how this problem is tackled. We generate a set of models that describe all feasible variants of a battle — not in the form of myriads of individual simulation programs, but in the form of a few SESs with specializations. Specific models for any actual situation can then be automatically generated on the fly. Any one of these specific models can then be simulated, and may lead to a decent forecast (at the level of detail contained in the set of invoked atomic models) of what the consequences of the simulated scenario would be if it were implemented.

15.6 Level Five: Goal–Driven Pruning

One problem that remains to be solved is the following: How do we prune an SES? Which branches do we cut and which one do we retain? In terms of our previously used engine example: how do we choose between the V6 and the diesel engines?

The answer is fairly simple. We attach rules to the specialization objects in the same way that we attach coupling relations to the aspect objects. For the preceding example, the following rule could be coded:

if *purchase_price = high* **and** *maintenance_cost = low*
   then select *Diesel*
   else select *V6*
end if

Purchase_price and *maintenance_cost* are two attached variables of the parent node. They are enumerated variables that can assume the values 'low,' 'medium,' and 'high.'

The pruner always searches specialization nodes for such rules. If it finds one or several rules that can be fired, it will do so. If this process results in a unique selection, the pruner accepts it and proceeds to the next specialization node. If no rule has been specified, or no rule can be fired, or the fired rules do not lead to a unique selection, the pruner enters into an interactive mode, displays the set of remaining specializations, and asks the user for a selection.

FRASES–Scheme [15.6] is experimental software that implements some of these ideas. Unfortunately, this software has not yet been fully integrated with the DEVS–Scheme modeling environment. However, such integration is planned.

And yet even this solution does not satisfy our ultimate goal. In a high–autonomy system, we may not be able to afford giving up at this point and asking for human help. In terms of our robot on Mars, all too often, the robot would be stuck and would have to wait for advice from Earth which probably would take hours to reach it since the human decision–maker on Earth would first need to receive and assess many environmental data from the robot before she or he could make a sensible decision.

Instead, if no unique solution can be found by firing rules, the system could automatically generate simulation models for all of the remaining specializations. It could then simulate these models for some time into the future, assess the effects of implementing each one of them, and pick the one that promises the most desirable future. In this way, our Martian robot could still come up with an automated decision of its own as to which specialization to use.

Decision making on the basis of static rules is usually referred to as a process of *shallow reasoning*. Decision making on the basis of a time–dependent model is commonly referred to as a process of *deep reasoning*. The proposed methodology allows us to combine shallow and deep reasoning in one harmonic process. Thereby, the best of both worlds can be preserved: If a decision is easy to reach, the fast and simple shallow reasoner will do the job, but if shallow reasoning leads to ambiguous decisions, the higher resolution of a deep reasoner can help us out of the dilemma.

15.7 The Cable Reel Problem

Let us now return once more to an example that was discussed in great detail in Chapter 5: the cable reel problem.

A new lightweight fiber–optic deep–sea communication cable is to be laid through the British Channel between Calais in France and Dover in the United Kingdom. The cable comes on a huge reel that is placed on a ship. The ship moves slowly from one coast to the other, constantly leaving cable behind. A large motor unrolls the cable from the reel. A speedometer detects the speed of the cable as it comes off the reel. A simple controller is used to keep the cable speed v at its preset value V_{set}. V_{set} is the speed of the ship. A functional diagram of the overall system was shown in Fig.5.3. It is repeated here as Fig.15.3.

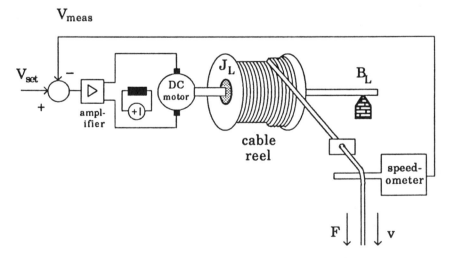

Figure 15.3. Functional diagram of the cable reel system.

Control engineers are used to represent systems in the form of block diagrams. A block diagram for the cable reel system was previously shown as Fig.5.6. This figure is repeated here as Fig.15.4.

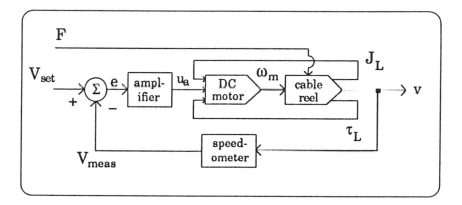

Figure 15.4. Block diagram of the cable reel system.

Block diagrams are a good way to graphically depict the coupling relations associated with an aspect node. However, if we wish to use block diagrams as a graphical programming tool, we must be a little more formal than hitherto with how we represent block diagrams of system decompositions. For this purpose, I wish to introduce the stylized block diagram. Figure 15.5 shows the stylized block diagram of the cable reel decomposition.

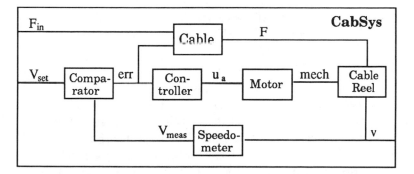

Figure 15.5. Stylized block diagram of the cable reel system.

The *stylized block diagram* represents a horizontal cut through an aspect node of a system entity structure [15.3]. The outer frame symbolizes its parent node, in our example, the root entity of the

SES, whereas the inner boxes symbolize the children of the aspect node. Wires connecting inner boxes (in our example: err, u_a, $mech$, v, V_{meas}, and F), are internal connections of the aspect node, whereas wires that extend to the outer frame (in our example: V_{set}, F_{in}, and v) are external connections of the aspect node, i.e., these are attached variables of the parent node. Notice that v appears in both the lists of internal and external connections. However, this is only a question of convenience. We can easily transform the internal v connection into a second external v connection. Thus, external connections take preference over internal connections.

Notice the difference between the block diagrams of Fig.15.4 and Fig.15.5. In Fig.15.5, we no longer distinguish between linear and nonlinear subsystems. In fact, we don't even distinguish between atomic models and coupled models. It could easily be that any one of the internal boxes of Fig.15.5 represented in itself a coupled model that could be further decomposed into smaller boxes in another stylized block diagram. In this sense, stylized block diagrams are hierarchical. Both internal and external connections represent connections in the DYMOLA sense. Wires represent cuts rather than individual variables. For instance, the wire labeled $mech$ represents a cut containing the three variables ω, τ_L, and J_L. We also eliminated the arrows from the stylized block diagram since DYMOLA cuts are nondirectional.

Also, Fig.15.5 introduces an additional subsystem: the cable. If we look once more at the ACSL cable reel program of Chapter 5, we notice that this was basically a hierarchical model. Three of the subsystems (the DC motor, the cable reel, and the speedometer) were coded as ACSL macros. The other three subsystems (the comparator, the controller, and the cable) were coded into the main program directly in the form of physical equations. This was done because each of these three subsystems is described by a single equation. Thus, the resulting ACSL program was shorter the way it was coded. However, in the context of hierarchical object–oriented modeling, this is a bad idea. I explained earlier that coupling models should preferably not contain any physical equations but only subsystem connections. Thus, in the light of our improved understanding, we added these additional subsystems. Figure 15.4 contains graphical elements representing two of these subsystems (the comparator and the controller), although they were not coded as subsystems in the ACSL program. The third subsystem (the cable) was simply left out to avoid cluttering the diagram. However, since the stylized block diagram is now used as a programming tool in our model management system, we need to be more formal in how we represent the stylized block diagram. Every child of the aspect node must be represented through an inner box in our stylized block diagram.

HIBLIZ [15.5] is a graphical front end to DYMOLA that follows exactly this approach. HIBLIZ allows us to zoom in on any of the internal boxes. Breakpoints are introduced to denote the magnification where an inner box suddenly becomes an outer frame and where the internal structure of that box becomes visible, i.e., where we jump from one aspect node of the SES to another. In HIBLIZ, connections between boxes represent DYMOLA cuts, i.e., they are cables rather than wires. One single connection can contain many individual wires, which can be of the across or through type or both.

HIBLIZ could be used in our DEVS–Scheme modeling environment to graphically represent the coupling relations that are attached to aspect nodes. However, HIBLIZ runs currently only on Silicon Graphics machines, and is therefore incompatible with our AMGS. One of our students is currently working on a reimplementation of a subset of these capabilities to be integrated with the DEVS–Scheme modeling environment. For our purposes, we don't need the zoom capability. We plan to implement the SES in one window offering a graphical SES editor (comparable to the network editor offered in STELLA). By double–clicking on any aspect node, we open a new window with a stylized block diagram editor in which we can place the subsystem boxes and draw connections between them. If the block diagram does not reference all the children of the aspect node, an error message is issued when we try to close the block diagram window. We must then either delete the child from the SES (in the SES window) or add the missing subsystem to the stylized block diagram. Only children of the current aspect node can be referenced in the stylized block diagram. If we need to reference another subsystem, we must first add the corresponding entity to the SES as an additional child of the aspect node.

Figure 15.6 depicts the generalized SES of the cable reel problem. The root entity *CabSys* is decomposed into six parts. For this purpose, the aspect node *CabSys–dec* has been introduced. The stylized block diagram that explains how the subsystems are coupled together has already been shown in Fig.15.5. Two of the component parts, namely, the motor and the controller, are specialized into variants. The motor can be either a DC motor or a hydromotor. The selection is made in the specialization node *Motor–spec*. Similarly, the controller can be a P–controller, a PI–controller, or a PID–controller. This selection is made in the specialization node *Controller–spec*. Notice that the double lines in the SES qualify the abstract nodes *Motor–spec* and *Controller–spec* as specializations rather than as aspects.

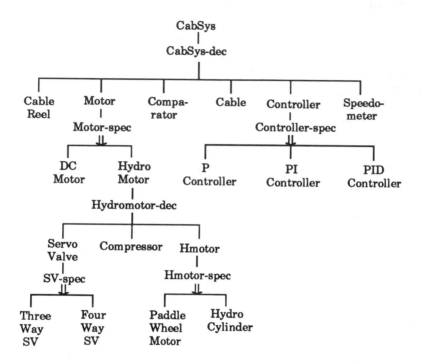

Figure 15.6. SES with specializations of the cable reel system.

It is easy to formulate rules for the selection of the motor. If the load is heavy and if it is important that the load can be accelerated or decelerated quickly, then select the hydromotor. If the purchase price must be small, then select the DC motor. However, it may not be as easy to formulate a set of unambiguous rules for the selection of the controller. Rules could include the following: If the positioning accuracy requirements are high, then don't select the P–controller. If it is important that the load can be accelerated or decelerated quickly, then select the PID–controller. However, in many cases, static rules may not lead to a unique selection. It may therefore be necessary to specify a set of dynamic rules also. Examples of such rules could be: The system must have a band width of at least 10 Hz. The system must have a stability phase margin of $45° \pm 5°$.

The system must have a steady–state positioning accuracy of 2% or better. Dynamic rules are usually specified in the form of performance criteria. They cannot be satisfied by firing them in a shallow reasoning sense. The pruner will therefore first look at all the static rules and check if it can come up with a unique selection. If this is the case, it will make the selection and generate a single simulation program possibly containing a few unspecified parameters. If all parameters are fully specified, it will perform a simulation run to verify that all the dynamic rules are satisfied as well. If some parameters are not fully specified, an optimization algorithm can now be employed to determine values for unspecified parameters that satisfy the dynamic rules. However, if the static rules do not lead to a unique selection, the pruner will generate several alternative simulation programs, and all of them are simulated in order to find a model and a set of parameters that satisfy the dynamic rules. In this way, a cheap and dirty (but often ambiguous) shallow reasoner is combined with a more refined and less ambiguous (but much more slowly executing) deep reasoner to preserve the best of both worlds.

The hydromotor entity is further decomposed into three parts. For this purpose, the aspect node *Hydromotor–dec* was introduced. Figure 15.7 shows a stylized block diagram that explains the external and internal couplings of the hydromotor decomposition.

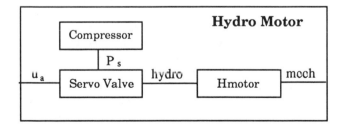

Figure 15.7. Stylized block diagram of the hydromotor.

Two of the component parts, namely, the servo valve and the Hmotor, are specialized into variants. The servo valve can be either a three–way servo valve or a four–way servo valve. The selection is made in the specialization node *SV–spec*. Similarly, the Hmotor can be either a paddle–wheel motor or a hydrocylinder. Again, we need a set of static and possibly dynamic rules for the selection.

There is a clear distinction between the static rules and the dynamic rules though. While the static rules are associated with a particular specialization node and serve the sole purpose of making a static selection among several structural variants, the dynamic rules are global rules that must be attached to the root entity of the SES rather than to any particular specialization node.

The DEVS–Scheme text file version of the SES is presented here:

```
;– Entity Structure for CabSys

(make-entstr 'CabSys)
(set-c-system-type e:CabSys 'system)

(ai e:CabSys asp 'CabSys-dec)
(sci e:CabSys 'CabSys-dec)
(ai e:CabSys ent 'Comparator)
(ai e:CabSys ent 'Controller)
(ai e:CabSys ent 'Motor)
(ai e:CabSys ent 'CableReel)
(ai e:CabSys ent 'Speedometer)
(ai e:CabSys ent 'Cable)

(add-variable e:CabSys 'input '(Fext Vdes)'())
(add-variable e:CabSys 'output '(radius velocity omega) '())

(acp e:CabSys 'Comparator 'Controller 'errport 'signal)
(acp e:CabSys 'Controller 'Motor 'command 'uport)
(acp e:CabSys 'Motor 'CableReel 'mech 'mech)
(acp e:CabSys 'CableReel 'Speedometer 'vport 'vport)
(acp e:CabSys 'Speedometer 'Comparator 'measport 'measport)
(acp e:CabSys 'Comparator 'Cable 'errport 'errport)
(acp e:CabSys 'Cable 'CableReel 'fport 'fport)

(acpx e:CabSys 'CabSys 'Comparator 'in1 'setport '(Vdes) '(Vset))
(acpx e:CabSys 'CabSys 'Cable 'in2 'finport '(Fext) '(Fin))
(acpx e:CabSys 'CabSys 'CableReel 'rport 'out1 '(radius) '(R))
(acpx e:CabSys 'CabSys 'CableReel 'vport 'out2 '(v) '(velocity))
(acpx e:CabSys 'CabSys 'CableReel 'mech 'out3 '(omega) '(omega))

(sci e:CabSys 'Controller)
(ai e:CabSys spec 'Controller-spec)
(sci e:CabSys 'Controller-spec)
(ai e:CabSys ent 'PController)
(ai e:CabSys ent 'PIController)
(ai e:CabSys ent 'PIDController)
```

```
(sci e:CabSys 'Motor)
(ai e:CabSys spec 'Motor-spec)
(sci e:CabSys 'Motor-spec)
(ai e:CabSys ent 'DCMotor)
(ai e:CabSys ent 'HydroMotor)

(sci e:CabSys 'HydroMotor)
(ai e:CabSys asp 'Hydromotor-dec)
(sci e:CabSys 'Hydromotor-dec)
(ai e:CabSys ent 'Compressor)
(ai e:CabSys ent 'ServoValve)
(ai e:CabSys ent 'Hmotor)

(sci e:CabSys 'ServoValve)
(ai e:CabSys spec 'SV-spec)
(sci e:CabSys 'SV-spec)
(ai e:CabSys ent 'ThreeWaySV)
(ai e:CabSys ent 'FourWaySV)

(sci e:CabSys 'Hmotor)
(ai e:CabSys spec 'Hmotor-spec)
(sci e:CabSys 'Hmotor-spec)
(ai e:CabSys ent 'PaddleWheelMotor)
(ai e:CabSys ent 'HydroCylinder)

;—END—
```

This SES text file is fairly self–explanatory. The *make–entstr* command creates the overall entity structure. The *set–c–system–type* command specifies the type of the SES. The *ai* command attaches a child to the currently processed entity. The *sci* command moves the processing pointer to a new node. For example:

$$sci\ e{:}CabSys\ 'CabSys\text{-}dec$$

makes the node *CabSys-dec* the current node. Subsequent *ai* commands attach children to that node rather than to the root entity. The *add–variable* command attaches variables to the current node. The *acp* command specifies internal couplings between the children of the current node, whereas the *acpx* command specifies external couplings between any one of the children and the parent node.

The SES text file specifies the coupling relations of the two aspect nodes, but it does not specify any static selection rules for automated pruning of the four specialization nodes since this feature has not yet been integrated with the DEVS–Scheme system.

The cable reel world model consists of:

(1) The SES with specializations as shown in Fig.15.6.

(2) The stylized block diagrams shown in Fig.15.5 and Fig.15.7, which explain how the decomposition of systems into parts is accomplished in the two aspect nodes *CabSys–dec* and *Hmotor–dec*.

(3) Four static rule bases, one for each specialization node, which formalize the knowledge when which of the specializations is to be used, and which together make up the static selection rules for the shallow reasoner.

(4) A dynamic rule base, which is attached to the root entity and describes performance criteria for the deep reasoner.

(5) A model library containing the 13 atomic models (coded in DY-MOLA) that describe the 13 leaf nodes of the SES.

From this (rudimentary) world model, any one of 15 structurally different pure SESs can be obtained by means of pruning. Figure 15.8 shows one such pure SES.

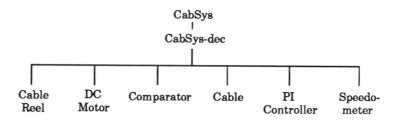

Figure 15.8. Pure SES of the cable reel system.

The atomic DYMOLA models describing the six leaf entities of the pure SES are given here:

```
model type CableReel
  cut vport(v), fport(F)
  cut mech(omega, tauL, JL)
  local R
  parameter Rempty, W, D, rho, BL, J0
  constant pi = 3.14159
    der(R) = -((D * D)/(2.0 * pi * W)) * omega
    v      = R * omega
    JL     = 0.5 * pi * W * rho * (R * *4 - Rempty * *4) + J0
    tauL   = BL * omega - F * R
end
```

```
model type DCMotor
  cut uport(ua)
  cut mech(omega, tauL, JL)
  local ia, if, ui, psi, taum, Twist, theta, uf
  parameter Ra, Rf, kmot, Jm, Bm = 0.0
    uf          = 25.0
    0.0         = uf − Rf ∗ if
    0.0         = ua − ui − Ra ∗ ia
    psi         = kmot ∗ if
    taum        = psi ∗ ia
    ui          = psi ∗ omega
    der(Twist)  = taum − tauL − Bm ∗ omega
    Twist       = (Jm + JL) ∗ omega
    der(theta)  = omega
end

model type Comparator
  cut setport(Vset), measport(Vmeas), errport(error)
    error = Vset − Vmeas
end

model type Cable
  cut finport(Fin), errport(error), fport(F)
  parameter kship = 10.0
    F = lim(kship ∗ error − Fin) + Fin
end

model type PIController
  cut signal(error), command(u)
  local err
  parameter kint, kprop
    der(err) = error
    u        = kprop ∗ error + kint ∗ err
end

model type Speedometer
  cut vport(v), measport(Vmeas)
  local x
  parameter k = 3.0
    der(x) = −k ∗ x + v
    Vmeas = k ∗ x
end
```

The pure SES can now be transformed. This process automatically generates the previously missing DYMOLA coupling model shown here:

model *CabSys*

   submodel *Comparator*
   submodel *PIController*($kint = 0.2, kprop = 6.0$)
   submodel *DCMotor*($Ra = 0.25, Rf = 1.0, kmot = 1.5, ->$
                    $Jm = 5.0, Bm = 0.2$)
   submodel *CableReel*($Rempty = 0.6, W = 1.5, D = 0.0127, ->$
                  $rho = 1350.0, BL = 6.5, J0 = 150.0) ->$
                  (**ic** $R = 1.2$)
   submodel *Speedometer*
   submodel *Cable*

   input *Vdes, Fext*
   output *radius, velocity, omega*

   connect *Comparator:errport* **at** *PIController:signal*
   connect *PIController:command* **at** *DCMotor:uport*
   connect *DCMotor:mech* **at** *CableReel:mech*
   connect *CableReel:vport* **at** *Speedometer:vport*
   connect *Speedometer:measport* **at** *Comparator:measport*
   connect *Comparator:errport* **at** *Cable:errport*
   connect *Cable:fport* **at** *CableReel:fport*

   Cable.Fin = Fext
   Comparator.Vset = Vdes
   radius = CableReel.R
   velocity = CableReel.v
   omega = CableReel.omega

end

The experiment description (i.e., the specification of the simulation run length, the termination conditions, and the desired outputs) can be synthesized from a second SES in the same way as the model was. Figure 15.9 shows an SES that describes various experiments foreseen for this system. Different experiments are shown as specializations. In our example, we have foreseen three different experiments: a simple simulation experiment, an experiment that optimizes the unspecified controller parameters, and an experiment that generates a Bode diagram of the control system. The various experiments can be automatically selected by attaching a static rule base to the specialization node. Such a rule could be: If the purpose is to evaluate the band width, then select Bode diagram. *Purpose* is an attached variable of the parent node *CabSysExp* that is assigned by the dynamic rule base attached to the root entity *CabSys* of the model SES.

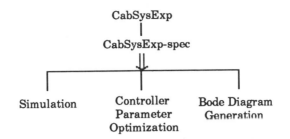

Figure 15.9. SES of the cable reel system experiments.

For the time being, let us select the simplest of all experiments, i.e., the simulation experiment. The transform procedure extracts the DYMOLA–coded experiment from the experiment data base. It is shown here:

```
cmodel
    simutime 3500
    step 0.05
    commupoints 2000
    input 2, Vdes(independ, 15.0), Fext(independ, 100.0)

ctblock
    scale = 1 | XCCC = 1
    label TRY | drunr
    if XCCC < 0 then XCCC = −XCCC | scale = 2 ∗ scale | go to TRY
                else proceed
ctend

outblock
    OUT
    term Rempty − R
    Radius = radius ∗ 10
    dispt Radius, velocity, omega
outend
end
```

It is possible to synthesize experiments from smaller parts. In our example, it would be feasible to decompose the simulation experiment into a **ctblock** and an **outblock** with additional specializations describing different types of ctblocks and outblocks that can be combined in various ways to form an overall experiment.

The transform procedure automatically invokes the DYMOLA preprocessor, which flattens the hierarchy and generates, in our example, a (flat) DESIRE [15.8] simulation program representing the selected variant of the cable reel system together with the selected variant of the simulation experiment. The simulation program could also have been generated for a number of alternative simulation languages.

```
----------------------------------------
-- CONTINUOUS SYSTEM CabSys
----------------------------------------
-- STATE R Twist theta x err
-- DER dR dTwist dtheta dx derr
-- OUTPUT radius velity Cs$oma
-- INPUT Vdes Fext
-- PARAMETERS and CONSTANTS:
kship = 10.0 | Rempty = 0.6 | W = 1.5 | D = 0.0127
rho = 1350.0 | BL = 6.5 | J0 = 150.0 | pi = 3.14159
Ra = 0.25 | Rf = 1.0 | kmot = 1.5 | Jm = 5.0
Bm = 0.2 | k = 3.0 | kint = 0.2 | kprop = 6.0
-- INITIAL VALUES OF STATES:
R = 1.2 | Twist = 0 | theta = 0 | x = 0 | err = 0
----------------------------------------
TMAX = 3500 | DT = 0.05 | NN = 2000
Vdes = 15.0 | Fext = 100.0
scale = 1 | XCCC = 1
label TRY | drunr
if XCCC < 0 then XCCC = -XCCC | scale = 2 * scale | go to TRY
            else proceed
----------------------------------------
DYNAMIC
----------------------------------------
-- Submodel: Speedometer
sr$Vms = k * x
-- Submodel: Comparator
Pr$err = Vdes - sr$Vms
-- Submodel: Cable
cale$F = lim(kship * Pr$err - Fext) + Fext
-- Submodel: CableReel
DCr$JL = 0.5 * pi * W * rho * (R ∧ 4 - Rempty ∧ 4) + J0
-- Submodel: DCMotor
Dr$oma = Twist/(Jm + DCr$JL)
-- Submodel: CabSys
Cs$oma = Dr$oma
-- Submodel: CableReel
d/dt R = -D * D/(2 * pi * W) * Cs$oma
Dr$taL = BL * Cs$oma - cale$F * R
```

```
—— Submodel: PIController
u = kprop * Pr$err + kint * err
—— Submodel: DCMotor
uf = 25.0
DCr$if = uf/Rf
psi = kmot * DCr$if
ui = psi * Dr$oma
ia = (u − ui)/Ra
taum = psi * ia
d/dt Twist = taum − Dr$taL − Bm * Dr$oma
d/dt theta = Dr$oma
—— Submodel: CableReel
velity = R * Cs$oma
—— Submodel: Speedometer
d/dt x = velity − k * x
—— Submodel: PIController
d/dt err = Pr$err
—— Submodel: CabSys
radius = R
_ _ _ _ _ _ _ _ _ _ _ _ _ _ _ _ _ _ _ _ _ _ _ _ _ _ _ _ _ _ _ _
OUT
term Rempty − R
Radius = radius * 10
dispt Radius, velity, Cs$oma
_ _ _ _ _ _ _ _ _ _ _ _ _ _ _ _ _ _ _ _ _ _ _ _ _ _ _ _ _ _ _ _
/ − −
/PIC 'CabSys.PRC'
/ − −
```

Notice that the equations stemming from the various submodels were both solved and sorted by the DYMOLA preprocessor. Also, a number of variables were automatically renamed to conform with the DESIRE syntax.

The SES and to a large extent DYMOLA support a clean separation of the experiment description from the model description as mandated by Zeigler [15.12]. In the synthesized DESIRE program, the statements describing the model and those describing the experiment are wildly interspersed. However, this doesn't matter. It is at the user interface where a clean separation between model description and experiment description is important, not at execution time [15.3].

The resulting simulation program can finally be simulated in DESIRE, which results in the trajectory behavior as depicted in Fig.15.10.

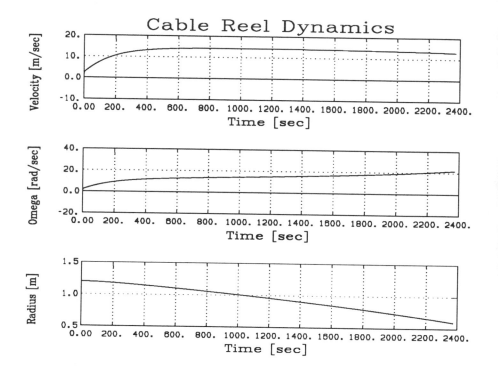

Figure 15.10. Trajectory behavior of the cable reel system.

The synthesized DESIRE program behaves identically with the manually generated ACSL program shown in Chapter 5. This concludes the example.

15.8 Summary

In this chapter, we demonstrated the usefulness of automatic model synthesis tools. A five–level hierarchy of model management was proposed that supports the automated generation of detailed models from successively more and more abstract (encapsulated) object descriptions. Figure 15.11 depicts the five–level hierarchy.

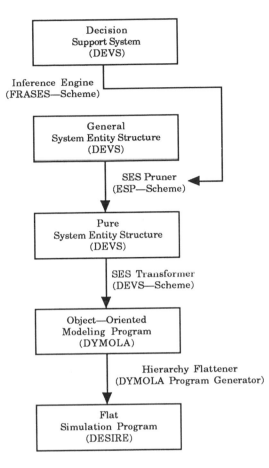

Figure 15.11. Five–level hierarchy of model management.

We have shown that automated model generation systems (AMGSs) can be used to tackle a number of important realistic application problems related to Space exploration, aircraft manufacturing, and military decision making. We have shown by means of a much simpler example, the cable reel problem, how a particular implementation of an AMGS, the DEVS–Scheme modeling environment, can be used to synthesize 15 structurally different models of the cable reel system. The example chosen to illustrate the automatic generation of continuous–system models obviously had to be a very simple one in order to fit onto the limited space provided in this chapter. However, considerably more complex examples have been developed and tested.

This chapter has unraveled another facet of intelligent decision making. In Chapters 12 to 14, we tried to emulate human abilities to deal with this problem. Maybe, this is the wrong way to go. Maybe it is more feasible to simulate these abilities without direct reference to how we humans deal with this problem. We humans are very bad at solving differential equations in our heads and we surely are bad at automated model synthesis techniques. However, why should we burden future generations of intelligent robots with our own shortcomings? The methodology described in this chapter cannot be learned and applied by humans without the use of computers, but it is well suited for automatic devices such as laboratory robots. Thus, it should be considered seriously.

Notice that the AMGS is not a competitor of or an alternative to the qualitative modeling techniques described in Chapters 12 to 14. It is just an additional tool. The model to be synthesized by the AMGS could be a qualitative model. In this chapter, I have generated quantitative state–space models, but this was my own choice. The decision was not at all dictated by the AMGS methodology. The AMGS can synthesize any type of model that can be described in terms of a hierarchical decomposition.

Only the future can tell which of the techniques that were presented in the last four chapters of this text will survive the test of time, if any. A most interesting period is certainly ahead of us. Lots of unanswered and challenging questions remain for generations of scientists and engineers to ponder. I for one would sure like to live the day when Clifford Simak's robot Webster becomes a reality. However, this will not be possible. We should not expect to solve the mystery of human intelligence quickly. Too high expectations lead only to frustration. Many valuable scientific methods have been discredited for years, not because they were bad techniques, but simply because they could not live up to the all too high expectations that were put into them in the first place. Neural networks are an excellent example. In a way, this is a disease of artificial intelligence as a whole. Many highly respected scientists and engineers still laugh about artificial intelligence. "Artificial intelligence is the part of computer science that doesn't work" is just one among many mockeries that can still be heard in the illustrious circles of otherwise well–informed scientists and engineers. These mockeries are only a ventilation of their frustration about the fact that our most advanced automated devices today possess approximately the intelligence of an angleworm.

Thus, it will take quite some time before our high–autonomy systems will exhibit intelligence that resembles human intelligence by any standards. Human intelligence will remain a mystery for a long time to come. But solve it we shall.

Do you know why people climb mountains? It is simply because they are there. A most curious breed we are: narrow–minded and ingenious, destructive yet civilized, short–lived but in constant search for a glimpse of eternity, selfish and generous, devilish and wonderful.

References

[15.1] James S. Albus, Harry G. McCain, and Ronald Lumia (1987), *NASA/NBS Standard Reference Model for Telerobot Control System Architecture (NASREM)*, NBS Technical Note 1235, Robot Systems Division, Center for Manufacturing Engineering, National Technical Information Service, Gaithersburg, Md.

[15.2] Donald C. Augustin, Mark S. Fineberg, Bruce B. Johnson, Robert N. Linebarger, F. John Sansom, and Jon C. Strauss (1967), "The SCi Continuous System Simulation Language (CSSL)," *Simulation*, 9, pp. 281–303.

[15.3] François E. Cellier, Qingsu Wang, and Bernard P. Zeigler (1990), "Model Management — A Five–Level Hierarchy for the Management of Simulation Models," *Proceedings 1990 Winter Simulation Conference*, New Orleans, La., pp. 55–64.

[15.4] Hilding Elmqvist (1978), *A Structured Model Language for Large Continuous Systems*, Ph.D. dissertation, Report CODEN: LUTFD2/(TFRT-1015), Dept. of Automatic Control, Lund Institute of Technology, Lund, Sweden.

[15.5] Hilding Elmqvist (1982), "A Graphical Approach to Documentation and Implementation of Control Systems," in: *Proceedings, Third IFAC/IFIP Symposium on Software for Computer Control, SOCOCO'82*, Madrid, Spain, Pergamon Press.

[15.6] Jhyfang Hu, Jerzy W. Rozenblit, and Yueh–Min Huang (1989), "FRASES — A Knowledge Representation Scheme for Engineering Design," *Proceedings, SCS MultiConference on Advances in A.I. and Simulation*, Tampa, Fla., pp. 141–146.

[15.7] Tag Gon Kim (1988), *A Knowledge–Based Environment for Hierarchical Modelling and Simulation*, Ph.D dissertation, Dept. of Electrical and Computer Engineering, University of Arizona, Tucson.

[15.8] Granino A. Korn (1989), *Interactive Dynamic–System Simulation*, McGraw–Hill, New York.

[15.9]   Edward E. L. Mitchell and Joseph S. Gauthier (1986), *ACSL: Advanced Continuous Simulation Language — User Guide and Reference Manual*, Mitchell & Gauthier Assoc., Concord, Mass.

[15.10] Kumar Ramohalli, Emil Lawton, and Robert Ash (1989), "Recent Concepts in Missions to Mars: Extraterrestrial Processes," *J. of Propulsion and Power*, **5**(2), pp. 181–187.

[15.11] Texas Instruments (1986), *PC–Scheme Users Manual*, Science Applications Press.

[15.12] Bernard P. Zeigler (1976), *Theory of Modelling and Simulation*, John Wiley & Sons, New York.

[15.13] Bernard P. Zeigler (1984), *Multifaceted Modelling and Discrete Event Simulation*, Academic Press, London, U.K.

[15.14] Bernard P. Zeigler (1990), *Object–Oriented Simulation with Hierarchical, Modular Models: Intelligent Agents and Endomorphic Systems*, Academic Press, Boston, Mass.

[15.15] Bernard P. Zeigler, François E. Cellier, and Jerzy W. Rozenblit (1989), "Design of a Simulation Environment for Laboratory Management by Robot Organizations," *J. of Intelligent and Robotic Systems*, **1**, pp. 299–309.

Bibliography

[B15.1] Mats Andersson (1989), "An Object–Oriented Modelling Environment," *Proceedings European Simulation MultiConference*, Rome, Italy.

[B15.2] Mats Andersson (1989), *Omola — An Object–Oriented Modelling Language*, Report CODEN: LUTFD2/(TFRT–7417), Dept. of Automatic Control, Lund Institute of Technology, Lund, Sweden.

[B15.3] François E. Cellier, Ed. (1991), "Planning and Intelligent Control for High Autonomy," *J. Intelligent and Robotic Systems*, special issue.

[B15.4] Hilding Elmqvist (1986), *LICS — Language for Implementation of Control Systems*, Report CODEN: LUTFD2/(TFRT-3179), Dept. of Automatic Control, Lund Institute of Technology, Lund, Sweden.

[B15.5] Hilding Elmqvist and Sven Erik Mattson (1989), "Simulator for Dynamical Systems Using Graphics and Equations for Modeling," *IEEE Control System Magazine*, January, pp. 53–58.

[B15.6] Sven Erik Mattson (1989), "Concepts Supporting Reuse of Models," *Proceedings, Building Simulation*, Vancouver, B.C., Canada.

[B15.7] Jerzy W. Rozenblit, Ed. (1991), "Design for High Autonomy," *Applied AI Journal*, special issue.

[B15.8] Hans Vangheluwe and Ghislain C. Vansteenkiste (1990), "Development of an Automatic Object–Oriented Continuous Simulation Environment," *International J. General Systems*, special issue on High Autonomy Systems: Modelling and Simulation (B.P. Zeigler, ed.).

[B15.9] Qingsu Wang and François E. Cellier (1991), "Time Windows: An Approach to Automated Abstraction of Continuous–Time Models into Discrete–Event Models," *International J. General Systems*, special issue on High Autonomy Systems: Modelling and Simulation (B.P. Zeigler, ed.).

[B15.10] Bernard P. Zeigler, Ed. (1991), "High Autonomy Systems: Modelling and Simulation," *International J. General Systems*, special issue.

Homework Problems

[H15.1] Solar–Heated House

Create a formal pure SES for the solar–heated house of Pr.[P8.1]. Attach a stylized block diagram to each of the decomposition nodes. For this application, it might be more convenient to replace the stylized block diagrams with the hierarchical bond graphs themselves, but we won't do this.

[H15.2]* Oxygen Production on Mars

Figure H15.2 shows the schematic of a prototype of an oxygen production plant to be used on Mars [15.10]. The plant extracts oxygen from the Martian atmosphere. Since the prototype is built and tested here in our own laboratory on Earth, it contains a Martian atmosphere simulator.

The Martian atmosphere consists of roughly 90% CO_2 at a pressure of about 6 mbar and at a temperature of about $-70°C$. The atmosphere is compressed to approximately 1 bar and heated up to $1000°C$. At this temperature, CO_2 decomposes spontaneously into CO and O_2 due to thermal disassociation:

$$2\ CO_2 \rightarrow 2\ CO + O_2 \qquad (H15.2a)$$

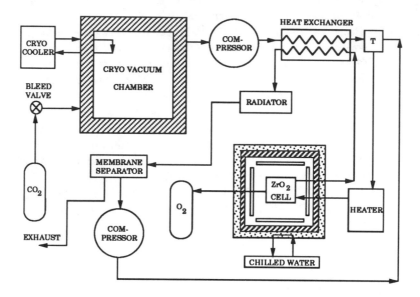

Figure H15.2. Schematic of Martian oxygen production plant.

Only a small percentage of the CO_2 actually decays to CO and O_2. It is the aim of this process to separate the O_2 gas from the $CO–CO_2$ gas mixture. This separation is not straightforward since the CO and O_2 gases have very similar molecular weights.

For this purpose, the gas mixture is vented through a zirconia (ZrO_2) cell. A voltage of 2 V is placed between two electrodes, one located on each side of the zirconia wall. In the presence of zirconia, an electrocatalytic reaction takes place whereby the oxygen gas is ionized:

$$O_2 + 2e^- \rightarrow 2\,O^- \qquad\qquad (H15.2b)$$

The necessary electrons are borrowed from the cathode located on the inside wall. Due to the presence of the electric field, an ionic current flows that carries the oxygen ions across the zirconia wall through which the small oxygen ions can permeate, whereas the larger gas molecules are retained inside the cell. At the other side of the wall, the oxygen ions meet with the anode, where they shed the borrowed electrons again:

$$2\,O^- \rightarrow O_2 + 2e^- \qquad\qquad (H15.2c)$$

In this way, the oxygen gas is separated from the $CO–CO_2$ gas mixture.

The $CO–CO_2$ mixture is sent through a heat exchanger in which it cools down while the Martian atmosphere on the other side of the heat exchanger is heated up. This simply saves energy.

The mixture is then separated in a membrane separator. This separation is much easier to accomplish due to the different molecular weights of the two gases. However, the gas mixture must be cooled down in a radiator before it can enter the membrane separator, because the used membrane is destroyed at high temperatures.

The (highly toxic) CO gas is then burned, while the CO_2 gas is recycled. This saves energy since the recycled CO_2 gas is already at a high pressure and a somewhat higher temperature when it is mixed with the Martian atmosphere. A second compressor brings the recycled CO_2 gas back up to a pressure of 1 bar so that a simple T–link can be used where the gases are mixed.

Draw a three–stage pure SES for this system and create stylized block diagrams for each of the decomposition nodes.

Projects

[P15.1] Cable Reel System

From the literature, find dynamic models for various types of servo valves and hydraulic motors. One such model was discussed in Hw.[H7.6]. Create atomic models for the missing leaf entities to be stored in the model library. Automatically generate various simulation programs by pruning away different branches of the SES. Run the resulting simulation programs and convince yourself that all these variants behave according to expectations.

Generate the missing two experiments and convince yourself that you can indeed optimize the controller parameters and generate the Bode diagram for arbitrary combinations of model structures.

Research

[R15.1] Rule Base Design

Design a formal syntax for the specification of the dynamic and static rule bases. Implement this specification in ESP–Scheme and convince yourself that just by modifying the requests, goals, and global parameters specified in the dynamic rule base, different simulation models and simulation experiments are indeed generated, which can be used to synthesize simulation programs that are compatible with the requests made in the dynamic rule base. Verify your approach by means of the cable reel system.

With Eye Serene:

A golden mean
An unfolding dream in our song
hopeful flower

An evergreen Progress
My dream of our symbiosis
closed dynamic balance

His endless song
The chancy dreaming in our painful pruning
hopeful echoes of recursion

The crystalline emergent hope
A chaos of growth
unfolding dark machinery

Her old dynamic balance
The rigid chaos has passed in a song
hopeful endless depth

A flexible machinery
The dream has stopped our conversation
tired soft feedback loops

Complex growth
My unfolding dream flutters on our emergent hope
Young heart

Complex complex feedback loops
Our fragrance drifts in His machinery
endless dark echoes of recursion.

Ann Droid
October, 1990

Index

AC response, 219
ACCULIB, 209
ACCUSIM, 209
Ackoff, Russell, 614
across variable, 162, 255, 258
ACSL, 8, 14, 32–34, 40, 41–45,
    93–94, 139, 144–154, 156,
    174–176, 180, 199, 275, 296,
    298, 322, 325, 658, 675, 708,
    716, 720
activation energy, 398, 400
activation function, 628
adaptive backpropagation, 698
adaptive control, 639, 659
adaptive counterpropagation, 699
adaptive learning, 639, 646–647, 665,
    669
adaptive resonance theory, 669
adiabatic reaction, 384
Adler, E., 72
aggregated bond graph, 339
aggression, 678
AIDS epidemiology, 504–505
Aiken, Richard, 405
aircraft, 116–123
Albus, James, 735
Alfeld, P., 405
algebraic equation, 18, 66
algebraic loop, 25–26, 66–68, 156,
    232, 265, 289, 458, 575
AMGS, 706
analog computer, 37, 242–244
Andersson, Mats, 736
anesthesiology, 622
ant brain, 701
antelope population model, 420,
    449–452

Aris, Rutherford, 20
Arrhenius' law, 399
ART network, 669
artificial neural network, 623–702
artificial neuron, 625–629
Ash, Robert, 736
Ashby, Ross, 3, 19
aspect, 712
association, 623
associative memory, 669
ASTAP, 238
asteroid mining, 412–415
Atlan, Henri, 406
atomic model, 710, 712, 726
attractor, 324, 328
Augustin, Donald, 46, 735
autism, 678
automated scaling, 38
autopilot, 621
auxiliary variable, 18, 24
Avogadro's law, 309
Avogadro's number, 355

Babbie, Earl, 511, 548
backpropagation network, 641–644,
    660–666, 697–698
backward Euler integration, 28
Bader, G., 405
Baltensweiler, Werner, 427, 447, 482,
    487, 500
Bandura, Albert, 694
Barber, Alfred, 239
Barin, Ihsan, 405, 407
basic behavior, 577
basket neuron, 629
batch reactor, 367
BBSPICE, 207, 210, 214, 238

Beckman, William,   332, 343
behavior modification,   637
behavioral complexity,   422, 443, 510
behavioral knowledge,   510
Beltrami, Edward,   20, 430, 447
Bender, Edward,   20
Berryman, Alan,   448
Beuken model,   311
Bier, Milan,   405, 416
bifurcation map,   436–438, 441–442,
      452–453
binary code,   699–700
binary random input,   587
binary structure,   603
biological models,   427
bipolar transistor,   204–209, 224–229
black box system,   16
block diagram,   251–255
block diagram modeling,   180
Blundell, Alan,   287
Bobrow, Daniel,   548
Bobrow, Leonard,   72
boiling water,   411
Boisvert, Ronald,   696
bond,   258
bond graph,   260–282, 361
bootstrapping,   638, 676
Bos, Albert,   288, 333
bouncing ball,   49
brain,   670–679
branch–admittance matrix,   61, 203
branch–impedance matrix,   56
Breedveld, Peter,   287, 288, 331, 332,
      333, 404
Broekstra, G.,   614
Broenink, Jan,   287
Brooks, Rodney,   694, 701
Brown, Joel,   446, 447
Brown, John,   510, 549
Bélanger, P.,   72

cable reel system,   141–151, 718–732,
      739
CAMAS,   275
CAMP,   275
cannon ball,   48
capacitive source,   371
Cartesian coordinates,   106

cartoon,   624
Carver, Mike,   405
cascade network,   628
case–sensitivity,   38
catalytic reaction,   399
causal loop diagram,   459
causal relation,   475
causal structure,   602, 610–612, 622
causality,   262–265, 472–475, 610–612
causality relation,   475
Cellier, François,   20, 47, 110, 183,
      287, 332, 500, 548, 549, 612,
      613, 735, 736, 737
cerebellar unit circuit,   629
cerebellum,   626
cervical syndrome,   111, 290–291
chaos,   328, 430–443, 671–672, 702
chaotic domain,   454
chaotic feedback,   672
chaotic motion,   432, 664
chaotic steady–state,   436, 672
Chaplin, K.,   405
character recognition,   634–636, 646
characteristic polynomial,   532
chemical potential,   359, 391
chemical process model library,
      415–416
chemical reaction bond graph,
      371–386
chemical reaction kinetics,   349–359
chemical reactor,   373
chemical thermodynamics,   359–366
Chin, James,   501
chromatography,   394
Chu, YaoHan,   47
circuit topology,   55, 61
Clarke, DeFrance,   47, 110, 184
Close, Charles,   20, 47, 109
Club of Rome,   491
code optimization,   167, 169, 232–236
cognitive function,   638
common–sense reasoning,   511, 514
communication interval,   77, 573
competition,   428
competition models,   428–430
compilation,   479
compiler generator,   676
complexity,   422, 443–446

composite structure, 603
computational structure, 252, 254, 257
computer–controlled system, 13
conduction, 303–315
confidence, 580
confidence measure, 580
conscious self, 673
conservation of energy, 84
conservation of momentum, 83
conservative system, 107
consistency, 477
consistency constraint, 522
constraint propagation, 522
continuity constraint, 523
continuous reactor, 392–398
continuous steady–state, 430, 526
continuous–time model, 11
control input, 18
control mechanism, 672–679
control problem, 19
control system, 184–185
convection, 316–317
convergence, 477
convergence speed, 654
converter variable, 457
convolution integral, 78
cooperation, 428
cooperation models, 428–430
copper rod, 333
Corner, David, 239
correlation analysis, 602
COSMOS, 198
Cotterill, Rodney, 694
counterpropagation network, 645–647, 667–669, 698–699
coupled model, 712
coupling model, 710
Coyle, R., 500, 549
crane crab, 96–99
cross–correlation, 469, 473
crossover, 682
crowding, 418
CSMP–III, 29, 30, 139
CSSL standard, 8, 23, 66
CSTR, 392
CTRL–C, 8, 40, 57, 74, 180, 558, 587–591, 595, 598, 609, 612,

647–654, 660–663
Curle, N., 332
Curran, James, 500, 504
Curran, Peter, 332
Currie, Iain, 332
curve fitting, 426
cut, 159
Cutler, Andrew, 412
cutset, 59
cutset potential, 59
cutset potential vector, 61

d'Alembert principle, 84, 90
daily return, 594
Dalton's law, 368
DARE–P, 8, 14, 31–32, 34–35, 138–139, 144
data collection, 620
data detrending, 594
data template processor, 676
Davies, Hubert, 332
Davies, John, 405
Davis, Ernest, 549
Davis, Lawrence, 685, 686, 687, 689, 696
DC motor, 102–105, 140–142, 145–147, 157–161, 171–174, 273
de Kleer, Johan, 510, 549
decay models, 417–422
decision making, 508, 512, 717
decision tree, 510
deductive model, 417
deep knowledge, 514
deep model, 514
deep reasoning, 510, 717
deep simulator, 514
Degn, Hans, 442, 447
delta rule, 640
demon, 499
Denavit, J., 109, 123
DESIRE, 36–38, 139, 144, 151, 156, 169, 199, 236, 237, 280–282, 296, 654, 730–732
DESIRE/NEUNET, 169, 654–658, 663
deterministic thermodynamics, 298
Deuflhard, Peter, 405

Devaney, Robert, 442, 447
developmental phase, 678
device modeling, 203–210
DEVS–Scheme, 713–714, 716, 717,
    721, 724–725
difference equation, 13, 18, 43,
    476–480
diffusion equation, 12
digital circuit, 616–619
digital–to–analog converter, 245–246
dimensional consistency, 45–46
direct problem, 19
discontinuity handling, 41–45
discrete event, 480
discrete–event model, 13, 14
discrete–time model, 12
discrete–time system, 42–43
discretization, 13, 514–516, 558
distillation column, 415
distributed parameter model, 12
disturbance, 4, 10, 11, 18, 44
domain of attraction, 328
domino game, 195–199
Dorf, Richard, 549
double bind, 678
dreams, 671–672
dual bond graph, 282–286
Duffie, John, 332, 343
DYMOLA, 156–170, 178–179, 199,
    201, 219–237, 238, 274–282,
    306, 312–315, 375–384, 403,
    415, 551, 709–710, 714, 721,
    726–728, 729, 731
dynamic rule, 722
dynamic table load, 139
DYNAMO, 476–477
DYNSYL, 403

EASE+, 180, 199
Easy5, 180, 199
effort, 258, 265–267
ego, 674
ego boundary, 679
electrical circuit, 260–265, 289–290
electrochemistry, 398–403
electrolysis, 401, 415
electromechanical system, 102–105,
    112, 291–292

electrophoresis, 416
Elgerd, Olle, 183
elliptic PDE, 676
ELLPACK, 676
Elmqvist, Hilding, 182, 183, 239,
    287, 332, 405, 735, 736
encapsulation, 711
end–of–day price, 593
endogenous variable, 616
endothermic reaction, 359
energies of formation, 386–392
energy flow balance, 410
energy transducer, 270–274
ENPORT, 274
enthalpy, 363
enthalpy of formation, 387
entity, 712
entropy flow, 306
envelope, 541
episode, 508, 515, 556, 558
episodical behavior, 515, 558
equation of state, 366–371
equation solver, 65, 155, 164–167,
    708
equation sorter, 24, 31, 37, 66,
    149–151, 155, 708
equilibrium point, 327
ESP–Scheme, 716, 739
Etkin, Bernard, 109
Euler equation, 106
evolution, 443–446
evolutionary game, 454
excitatory synaptic connection, 629
exclusive–or problem, 630–633, 697
exogenous variable, 615
exothermic reaction, 359
experiment, 4–5, 236
experiment class, 512
experimental frame, 6, 22
expert recoding, 564
expert system, 510, 620, 675
explicit integration, 28–29
explosion, 10
exponential growth, 418, 504
expressiveness, 560
extensive variable, 360

Fairbairn, William, 694

Farquhar, Adam, 541, 547, 549
fast Fourier transform, 473
fault detection, 619, 621
fault diagnosis, 625
fault discrimination, 621
feature, 685
Federn, Paul, 694
feedback control, 16
feedforward network, 628
Feigenbaum, Mitchell, 438, 447, 448
Fenchel, Tom, 448
Ferraudi, Guillermo, 406
fictitious tree–branch, 59
figure of merit, 682
film, 624
financial model, 503
Fineberg, Mark, 46, 735
finite automaton, 570
finite state machine, 14, 514, 540
Fischlin, Andreas, 427, 447, 482,
    487, 500
Fishwick, Paul, 549, 550
fitness, 687
fixed bed reactor, 393
flat simulation model, 707–708
flow, 258, 265–267
flow equilibrium, 351, 404
flow variable, 457
fluid dynamics, 344
Forbus, Kenneth, 510, 527, 549
forecasting, 583–586, 590–591
forgetting, 693
Forrester, Jay, 17, 461, 491, 492,
    493, 497, 498, 500, 502
Forrester's world model, 491–498,
    502–503
forward Euler integration, 27
Francisco, Joseph, 406
Franco, Sergio, 239
FRASES–Scheme, 717
Frederick, Dean, 20, 47, 109
frequency modulation, 691
Frost, Arthur, 405
Fu, King–Sun, 109
function generator, 628
fundamental cutset matrix, 61
fundamental loop matrix, 55
fuzzy forecasting, 614–615

fuzzy measure, 566, 582
fuzzy membership function, 567, 681
fuzzy recoding, 568
fuzzy set, 512

Gaines, Brian, 1, 19, 614
Gauthier, Joseph, 47, 110, 183, 288,
    332, 736
gear, 272
Gebben, Vernon, 289, 333
generalized bond graph, 331
generalized displacement, 107, 269
generalized force, 107
generalized Hamiltonian, 331
generalized momentum, 107, 268
generalized SES, 715–716
generic macro, 658
genetic algorithm, 679–690
genetic code, 681
genetic learning, 679–690
genetic pool, 682
genetic string, 682
Gentner, Dedre, 549
Ghausi, Mohammed, 240
ghost, 484
Gibbs equation, 360, 365
Gibbs free energy, 363
Gibbs free energy of formation, 391
Gibbs–Duhem equation, 365
Giloi, Wolfgang, 47
Gilpin, Michael, 431, 447, 458, 461,
    462, 500
Gilpin equations, 431–432, 438–442,
    452–453, 458–463, 478–479
global change, 505–506
global feedback, 670–671
goal–driven pruning, 716–718
Goldberg, David, 687, 688, 689, 695
Goldstein, Herbert, 109
Goldstein, Moise, 697
Golgi neuron, 629
Gonzales, Rafael, 109
Goodisman, Jerry, 406
gradient technique, 654
Graeme, Jerald, 240
Granda, Jose, 288
granule neuron, 629
graphical modeling, 180–182,

215–219
Gray–code converter, 668
Green, Don, 406
Green, Herbert, 629, 691, 693, 694,
        695, 696, 701
Grossberg, Stephen, 624, 647, 669,
        695
Grossberg outstar, 647
grouping, 429, 608
growth models, 417–422
gyrator, 272–273

Hadamard product, 657
Hamilton equations, 107
Hamiltonian, 107, 111, 361
Hamming distance, 650, 668
Hankel matrix, 598
Hanley, D., 405
Hartenberg, R., 109, 123
Hase, William, 406
Hayes, Patrick, 550
heart monitor, 559, 561, 564, 566,
        568, 609, 621–622
heat equation, 303, 311
heat flow, 333–334
heated air channel, 310
heated rod resistor, 310
Hebb, Donald, 623, 693, 695
Hebb rule, 693
Hecht–Nielsen, Robert, 645, 695
Heffley, Robert, 109, 122
Hermitian matrix, 469
Heun integration, 38
Heyward, William, 500, 504
HIBLIZ, 180–182, 199, 721
hierarchical model, 210–212, 709–710
hierarchy flattening, 710
high–autonomy system, 508
Hilburn, John, 72
Hinton, Geoffrey, 696
Hnatek, Eugene, 695
Hoff, M., 696
Holden, Arun, 447
Holland, John, 679, 695
Hopfield, John, 624, 634, 635, 636,
        669, 695
Hopfield net, 669
Hoppensteadt, Frank, 405

Hostetter, Gene, 20, 72, 109, 288
HSPICE, 207, 210
Hu, Jhyfang, 735
Huang, Yueh–Min, 735
Huelsman, Lawrence, 72
hydraulic motor, 292–295
hydraulic system, 292–296
hydrogen–bromine reaction, 352–354,
        356–359, 371–384, 398–399,
        407, 409
hydrogen–iodine reaction, 351–352
Hyman, James, 501
hysteresis, 49

icon, 180
id, 673
ideal gas, 367
ill–defined system, 10, 16
imitation, 637
implicit integration, 28–29
implicit loop solver, 29–30, 156
incomplete knowledge, 512
incremental compiler, 675
inductive model, 16, 417, 420–422,
        426, 430, 575–583
inductive reasoning, 555–622
inference engine, 510
influence diagram, 458–461, 535
influenza model, 487–491, 501–502
informal derivative, 594
informal SES, 712
information content, 578
information flow, 461
information hiding, 711
Inhelder, Barbel, 696
inhibitory synaptic connection, 629
initial condition, 212–215
innovation, 446, 672
input, 4, 18
input signal, 587
input/output behavior, 569–575
integration step size, 573
intelligent autopilot, 700
intensive variable, 360
intermediate language, 31
internal energy, 359
internal validity, 16
internal variable, 18

internalization, 672–679
interpretation, 479
interval measure, 512
intervention, 479
inverse neural network, 670–671
inverse problem, 19
irreversible thermodynamics, 320,
    321–330
isentropic reaction, 363
isobaric reaction, 362
isochoric reaction, 362
isothermic reaction, 363
Ivakhnenko, A., 470, 501
Iwasaki, Yumi, 501

Jewell, Wayne, 109, 122
Johnson, Bruce, 46, 735
Johnson, David, 72
Johnson, John, 72
junction, 259

Kailath, Thomas, 20
Karnopp, Dean, 288, 289, 404
Karplus, Walter, 15, 17, 20
Katzir–Katchalsky, Aharon, 332, 405
Kelton, David, 613
Kernberg, Otto, 678, 679, 695
Kerr, Alistair, 400, 405
Kettenis, Dirk, 183
Kheir, Naim, 20
Kim, Tag, 735
kinetic energy, 106
Kirchhoff's current law, 61
Kirchhoff's voltage law, 56
Kiss, Laszlo, 406
Kleijnen, Jack, 5, 20
Klimasauskas, Casimir, 697
Klir, George, 558, 612, 613, 614
Klotz, Irving, 406
Knacke, Ottmar, 405, 407
knowledge abstraction, 621
knowledge acquisition, 7, 620
knowledge base, 675
knowledge generalization, 513, 557,
    621
knowledge organization, 7
knowledge synthetization, 620

Koch, Christof, 695
Kohonen, Teuvo, 624, 646, 647, 695
Kohonen layer, 646, 647
Kohonen learning, 647
Koivisto, Hannu, 696
Koivo, Heikki, 696
Korn, Granino, 6, 20, 47, 109, 183,
    239, 288, 653, 654, 695, 735
Kuipers, Benjamin, 510, 515, 530,
    541, 547, 549

Lagrange equation, 107
Lagrangian, 107, 110
Laidler, Keith, 406
Laing, Ronald, 695
landmark, 514–516, 559
language quality assessment, 50
Lapa, Valentin, 470, 501
larch bud moth model, 423–428,
    480–487
large–scale system modeling, 179
LARKIN, 403
laterally diffused transistor, 204
Latham, Joseph, 405
laundry list, 457–458
Lavenda, Bernard, 332
Law, Averill, 613
Lawden, Derek, 332
laws of thermodynamics, 359
Lawton, Emil, 736
learning, 623
Lebel, Jean, 501
Lee, George, 109
legal state, 558
level, 558, 559
level variable, 456
Levine, Mark, 405, 416
Li, DongHui, 613
lightning rod, 334
limit cycle, 48, 327, 529
linear algebraic loop, 167–169
linear regression, 468, 597
linear regression model, 448–449
linear state–space model, 73
linear system, 12
linearization, 22, 213
Linebarger, Robert, 46, 735
link, 53

Lippmann, Richard, 695
LISP, 558, 675, 713
logic gate, 244–245
logic inverter, 247–248
logic simulator, 238
logistic equation, 419–420, 433–438, 452, 453–454
London, Fritz, 332
loop, 54
loop current, 54
loop current vector, 55
Loparo, Kenneth, 550
Loschmidt's number, 355
Lotka, Alfred, 447
Lotka–Volterra model, 22, 422–426, 452
Luker, Paul, 550
Lumia, Ronald, 735
lumped parameter model, 12
lunar lander, 24, 32–46, 81

machine independence, 34
MACKSIM, 403
macro, 140–156
macro expansion, 149
macroeconomy, 346, 453–454
macroscopic thermodynamics, 298
Mahan, Bruce, 405
majorante, 541
Maloney, James, 406
Mann, Jonathan, 501, 504
manufacturing, 705–706, 714–715, 716
Martinetz, Thomas, 696
mask, 569–575
mask candidate matrix, 575
mask depth, 574
mask history matrix, 580
mask quality, 580
mass balance, 350
mass flow, 461
mass flow balance, 410
mathematical model, 5, 11–18
mathematical simulation, 6
MATLAB, 8, 40, 57, 74, 558, 612, 647, 654
MATRIX$_X$, 180
Mattson, Sven, 736

Maxwell relations, 389
Maxwell's equations, 344
McCain, Harry, 735
McCulloch, Warren, 623, 695
Mecca, S., 614
mechanical system, 265–268, 290–291
memory function, 26–27
mental function, 673
mental model, 5, 508
mental simulation, 508
mesh equation, 52–57, 72, 73
mesh impedance matrix, 55
mesh–incidence matrix, 55
Mesterton–Gibbons, Michael, 448
meta–knowledge, 10
meta–model, 5, 21, 451–452
method of lines, 303
microscopic thermodynamics, 298
Minsky, Marvin, 5, 20, 623, 630, 695, 696
Mitchell, Edward, 47, 110, 183, 288, 332, 736
MMS, 199
model, 5–6
MODEL, 170
model order, 24
model robustness, 44
model synthesis, 703–739
model transformation, 713–714, 727
model validation, 50, 82, 410–411, 420, 498–500, 616
MODEL–C, 180
model–reference adaptive control, 674
modeling importance, 7
modeling methodology, 17, 477
modeling purpose, 16
modeling software, 464
modeling spectrum, 15
modular modeling, 707
modulated capacitance, 312–314
modulated resistor, 306
molar concentration, 355
molar flow rate, 356
molar mass, 355
mole fraction, 360
molecular mass, 355

Montana, David, 685, 686, 687, 689, 696
Monte Carlo optimization, 690
Moore, John, 376, 405, 407
Morgan, Antony, 515, 522, 527, 532, 537, 547, 549
Moss, Stephen, 400, 405
motor neuron, 626
mutation, 683

naïve physics, 507–553, 556
Narendra, Kumpati, 660, 696
Navier–Stokes equation, 346
neglect, 678
neural network learning, 631
neural network software, 647–658
neural network training, 637
neurobiological learning, 690–694
neurosis, 676
neurotransmitter, 679
Newell, Allan, 614
Newton's rotational law, 90–96
Newton's translation law, 81–89
Newton–Raphson iteration, 212
Nicholson, Harold, 406
Nielsen, Norman, 550
Nikravesh, Parviz, 110
Nise, Norman, 110
node, 162
node admittance matrix, 61
node equation, 58–62, 72, 73
node–incidence matrix, 61
nodeset, 213
nominal measure, 511
nondeterministic behavior, 327
nonlinear algebraic loop, 169–170
nonlinear optics, 344
Nowak, U., 405
numerical differentiation, 215
numerical integration, 26, 214
numerical stability, 477
Nyquist stability criterion, 460

object, 711
object–oriented modeling, 709–710
object–oriented simulation, 710
object–oriented user interface, 710

objective function, 673
observation ratio, 579
Oetzel, Roberta, 697
Ogata, Katsuhiko, 110
Olsen, Lars, 447
Onsager, Lars, 404
operational amplifier, 216–219
optimal landmark selection, 561–563, 588–589
optimal mask, 575–583, 589–590
optimal recoding, 588–589
optimal structure, 605
optimization, 508–509, 654, 679
ordinal measure, 511
ordinary differential equation, 12, 17
Orwell, George, 17
Oster, George, 404, 405
outer product, 657
output, 4, 18
overprotection, 678
oxidation reaction, 402
oxygen production, 737–739
oxyhydrogen gas reaction, 408–409

p–n junction, 203–204
Palm, William, 21, 110
Papert, Seymour, 623, 696
parallel code, 24
Parthasarathy, Kannan, 660, 696
partial differential equation, 12, 17, 676
partial entropy flow, 373
partial molar flow, 373
partial pressure, 368
partial temperature flow, 381
partial volume, 369
partial volume flow, 373
passive electrical circuit, 51–71
passive filter, 75
path, 163
pattern recognition, 555, 556–557, 624, 634–636
Patterson, Gary, 406
Paul, Richard, 110, 127, 129
Paynter, Henry, 258, 284, 288
PC–Scheme, 713
Pearson, Ralph, 405
peekaboo game, 678

pendulum, 100–102
perceptron, 623, 630–632
perceptron network, 632–633, 697
Perelson, Alan, 405
performance index, 682
periodic steady–state, 430, 529
Perry, Robert, 387, 406
PERT network, 540, 551
Peshkov, V., 332
Peterson, Steve, 501
phase plane, 22
photochemistry, 398–403
physical decomposition, 712
physical model, 417
Piaget, Jean, 696
Piot, Peter, 501
Pitts, Walter, 623, 695
Planck constant, 400
plausibility check, 82
political modeling, 615–616
Pollock, John, 614
polynomial growth, 504
population dynamics models,
    417–454
postoptimization, 427, 609
potential energy, 107
potential relaxation, 691
Pourrahimi, F., 288
power, 271
power balance equations, 365
power bond, 258–259
power flow, 298–302
power generation, 193–195, 199
power line, 75
preconscious, 673
predator–prey models, 422–428, 448
prediction, 16
predictiveness, 560
predictor–corrector scheme, 552
preprocessor, 33
Prigogine, Ilya, 332
primitive reaction, 386
Pritsker, Alan, 20
probability measure, 582
procedural section, 30–32
program generator, 156
programming comfort, 37
propane–oxygen reaction, 349–351

pruning, 5, 715
pseudobackpropagation network,
    699–700
PSpice, 180, 202, 214, 238
psychosis, 677
Pugh, Alexander, 501
pulley, 99–100
pump, 272
pure SES, 715
Purkinje neuron, 629
pyrolysis, 413

QSIM, 547
qualitative behavior, 512
qualitative control, 553
qualitative integration, 522–523,
    552–553
qualitative model, 14, 511, 512
qualitative physics, 511
qualitative plot, 529–530, 551
qualitative reasoning, 511
qualitative simulation, 512, 514,
    520–546, 550, 551, 583
qualitative stability, 553
qualitative state equation, 521
qualitative state–space, 511
qualitative variable, 511–512,
    516–520
qualitative vector, 515
quality measure, 579, 580
QualSim, 516–546
quantization, 489
Quinn, Thomas, 501

radiation, 317–319
Rajagopalan, Raman, 550
Ramohalli, Kumar, 736
ramping, 214
rate variable, 457
ratio measure, 512
raw data matrix, 569
reaction flow rate, 357
reaction rate equations, 351
rechargeable battery, 411–412
recoding, 558–569
reconstruction analysis, 601–610
reconstruction error, 607

recurrent network, 628, 669
recursive filter, 595
reduction reaction, 402
reentrant network, 628
region, 515, 558, 559
relative fitness, 687
reliability, 37
rendering column, 415
replication and batch, 503
Rerick, M., 614
reservoir, 187–188
residua vector, 469
resistive source, 301–302
resonance circuit, 77
reversible thermodynamics, 320
reward mechanism, 672
Rice, John, 696
Richardson, George, 501
Richmond, Barry, 501
Rietman, Edward, 696
Rimvall, Magnus, 500, 549
Ritter, Helge, 696
Robertshaw, Joseph, 614
Robichaud, Louis, 289
robot, 123–131, 132
robot modeling, 199
robustness, 37, 654
Rock, Peter, 406
Roddier, Nicolas, 549
Rosenberg, Robert, 406
Rosenberg, Ronald, 288, 289, 333
Rosenblatt, Frank, 696
rotational motion, 91–95
Routh–Hurwitz scheme, 533
Rozenblit, Jerzy, 110, 550, 735, 736, 737
Rozsa, R., 406
rule, 620
rule base, 510, 620, 621, 675, 739
Rumelhart, David, 644, 696
Rumin, N., 72
run–time display, 37
run–time library, 34
Runge, Thomas, 184
Runge–Kutta integration, 38

sampled–data control system, 13, 185–187

sampled–data system, 21
Sansom, John, 46, 735
SAPS, 558, 669
SAPS–II, 558–612, 684
Savant, Clement, 20, 72, 109, 288
Schaffer, William, 442, 447
Schulten, Klaus, 696
SCOOPS, 713
second sound, 320–321, 344
Sedra, Adel, 240
Segev, Idan, 695
Sejnowski, Terrence, 697
self–fulfilling prophecy, 17
self–organization, 446, 672
self–tuning regulator, 659
semantics, 620
sensitivity, 16
sensitivity analysis, 499, 502–503, 512
servo valve, 292–295
SES, 711–715
SES axioms, 712
Shafer, Glenn, 613
shallow model, 514
shallow reasoning, 509, 717
shallow simulator, 514
Shannon, Claude, 578, 579, 580, 583, 613
Shannon entropy, 578
shaping, 637
SIDOPS, 275
sigmoid function, 641
signal flow graph, 255–258
signal regeneration, 569, 585, 592–593
SIMNON, 156, 199, 236, 296
Simon, Herbert, 501, 614
simulation, 6–7
simulation danger, 9
simulation importance, 8
simulation methodology, 477
simulation software, 464
simulation speed, 10
simulation verification, 439–442
single assignment language, 25
single port element, 259
single refinement algorithm, 608
single–layer network, 639–640

singularity, 327
sink, 461
Skinnerism, 637
small signal analysis, 465
Smith, Jon, 21
Smith, Joseph, 406
Smith, Kenneth, 240
social learning, 637
social modeling, 637
software standardization, 41
solar–heated house, 334–343, 737
Sorsa, Timo, 696
source, 461
source conversion, 53, 58
source current vector, 61
source voltage vector, 55
Space exploration, 704–705, 737–739
specialization, 712, 715
SPICE, 202–215
Spong, Mark, 110
Spriet, Jan, 21
spurious solution, 530, 551
stability, 533
state, 558
state estimation, 19
state transition matrix, 571
state transition table, 539
state variable, 18, 24, 65
state–event, 43–45, 115
state–space, 23
state–space circuit model, 64–66
state–space model, 13
static characteristic, 137–139
static rule, 722
statistical steady–state, 475
statistical thermodynamics, 298
steady–state, 22, 212, 410
steady–state constraint, 524
Stefan–Boltzmann's law, 318
Stefani, Raymond, 20, 72, 109, 288
Steinfeld, Jeffrey, 406
STELLA, 463–464, 477–480,
        483–487, 489–490, 491, 499,
        721
stellate neuron, 629
Stevens, Albert, 549
stock market, 17, 593–601, 622,
        700–701, 702

stock variable, 456
stoichiometry, 350
Stowe, Keith, 333
strategic planning, 706, 716
strategy, 620
Strauss, Jon, 46, 735
stripping column, 415
structural complexity, 422, 443, 510
structural knowledge, 510, 514
structural singularity, 69–70, 74,
        170–179, 228, 265, 474
structure aggregation, 608
structure characterization, 464–472,
        601–610, 622
structure diagram, 461–464
structure identification, 19
structure optimization, 607
structure quality, 607
structure refinement, 607
student network, 638
stuttering, 527
stylized block diagram, 719
subjective function, 673
superego, 674
superfluid Helium, 320–321, 344
supervised learning, 637–647
surge tank, 188–192, 295–296
symbiosis, 428
symbiotic relationship, 678
symbolic representation, 558
symbolic variable, 558
symplectic gyrator, 278
synaptic weight, 627
syntax, 620
SYSMOD, 145, 179
system, 1–4
system decomposition, 712
System Dynamics, 455–505, 558
system entity structure, 711–716
SYSTEM–BUILD, 180

teacher network, 638, 702
TECAP, 209
telegraph equation, 345
temporal pattern, 624
terminal variable, 157
thermal capacitance, 307
thermal inertance, 320–321

thermal resistance, 306
Thoma, Jean, 289, 297, 333, 404, 406
threaded code, 37
through variable, 162, 255, 258
THTSIM, 275
TicTacToe, 619–620
time base, 74
time reversal, 321–330
time–event, 41–43, 78
time–slicing, 13
topological modeling, 201–203
topological structure, 254, 257
total reaction energy, 411
Towner, G., 614
trajectory, 515, 558
trajectory behavior, 4, 11, 12, 14, 15, 515
transfer function, 133–137
transformer, 271–272, 292
transgression of galaxies, 10
transient analysis, 212–215
translational motion, 84–89
translational system, 113
transmission line, 345
tree, 52
tree–branch, 52
tri function, 653
Triffet, Terry, 406, 629, 691, 693, 694, 695, 696, 701
Truesdell, Clifford, 333
tubular reactor, 367, 393, 396
tunnel diode, 240–242
TUTSIM, 275

uncertainty, 566
uncertainty reduction, 579
unconscious, 673
unsupervised learning, 637–647
Uyttenhove, Hugo, 564, 613

validation, 5, 6, 9, 45–46
van Dixhoorn, Jan, 288
Van Halen, Paul, 225, 239, 249
Van–der–Pol oscillator, 322–330

Vandermonde matrix, 468
Vangheluwe, Hans, 737
Vansteenkiste, Ghislain, 21, 737
variable structure model, 115
vertically diffused transistor, 204
Vesanterä, Pentti, 110, 613
Vescuso, Peter, 501
vicarious learning, 637
Vidyasagar, M., 110
Vincent, Thomas, 446, 447
voltage follower, 254
von Bertalanffy, Ludwig, 614

Wait, John, 6, 20, 47, 110, 184
Wang, Qingsu, 735, 737
Ward–Leonard group, 152–155, 171–178
Wasserman, Philip, 647, 696, 697
wave equation, 344
weakest node, 686
Weaver, Warren, 613
weight relaxation, 693
white box system, 16
Widman, Lawrence, 550
Widrow, Bernard, 696
Williams, Brian, 520, 549
Williams, R., 447, 448
Williams, Ronald, 696
Wittenburg, Jens, 110
Wolovich, William, 110
WORKVIEW, 180, 201, 215–218, 238
world model, 17, 453–454, 716, 725–726

Yandell, David, 613
Yao, Y., 333
Yuhas, Ben, 697

Zadeh, Lotfi, 566, 613
Zeigler, Bernard, 4, 5, 6, 7, 20, 110, 550, 731, 735, 736, 737
zero–padding, 473
Zhou, T., 288, 333
Ziegler, Hans, 110